109
Advances in Biochemical Engineering/Biotechnology

Series Editor: T. Scheper

Advances in Biochemical Engineering/Biotechnology

Series Editor: T. Scheper

Recently Published and Forthcoming Volumes

Biosensing for the 21st Century
Volume Editors: Renneberg, R., Lisdat, F.
Vol. 109, 2007

Biofuels
Volume Editor: Olsson, L.
Vol. 108, 2007

Green Gene Technology
Research in an Area of Social Conflict
Volume Editors: Fiechter, A., Sautter, C.
Vol. 107, 2007

White Biotechnology
Volume Editors: Ulber, R., Sell, D.
Vol. 105, 2007

Analytics of Protein-DNA Interactions
Volume Editor: Seitz, H.
Vol. 104, 2007

Tissue Engineering II
Basics of Tissue Engineering and Tissue Applications
Volume Editors: Lee, K., Kaplan, D.
Vol. 103, 2007

Tissue Engineering I
Scaffold Systems for Tissue Engineering
Volume Editors: Lee, K., Kaplan, D.
Vol. 102, 2006

Cell Culture Engineering
Volume Editor: Hu, W.-S.
Vol. 101, 2006

Biotechnology for the Future
Volume Editor: Nielsen, J.
Vol. 100, 2005

Gene Therapy and Gene Delivery Systems
Volume Editors: Schaffer, D. V., Zhou, W.
Vol. 99, 2005

Sterile Filtration
Volume Editor: Jornitz, M. W.
Vol. 98, 2006

Marine Biotechnology II
Volume Editors: Le Gal, Y., Ulber, R.
Vol. 97, 2005

Marine Biotechnology I
Volume Editors: Le Gal, Y., Ulber, R.
Vol. 96, 2005

Microscopy Techniques
Volume Editor: Rietdorf, J.
Vol. 95, 2005

Regenerative Medicine II
Clinical and Preclinical Applications
Volume Editor: Yannas, I. V.
Vol. 94, 2005

Regenerative Medicine I
Theories, Models and Methods
Volume Editor: Yannas, I. V.
Vol. 93, 2005

Technology Transfer in Biotechnology
Volume Editor: Kragl, U.
Vol. 92, 2005

Recent Progress of Biochemical and Biomedical Engineering in Japan II
Volume Editor: Kobayashi, T.
Vol. 91, 2004

Recent Progress of Biochemical and Biomedical Engineering in Japan I
Volume Editor: Kobayashi, T.
Vol. 90, 2004

Physiological Stress Responses in Bioprocesses
Volume Editor: Enfors, S.-O.
Vol. 89, 2004

Biosensing for the 21st Century

Volume Editors: Reinhard Renneberg · Fred Lisdat

With contributions by
D. Andresen · T. Balkenhohl · F. F. Bier · C. P. Chan · Y. Cheung
B. Danielson · S. Demin · E. Ehrentreich-Förster · G. Gauglitz
E. Hall · J. Henkel · J. Kafka · I. Karube · A. Lambrianou
S. Leimkühler · F. Ligler · F. Lisdat · H. Nakamura · S. Neugebauer
M. Nickisch-Rosenegk · O. Pänke · D. Pfeiffer · G. Proll · E. Reiß
R. Renneberg · R. R. Sathuluri · D. Schäfer · K. Schröder
W. Schuhmann · M. Seydack · M. Shimomura-Shimizu · R. Spricigo
L. Stoica · R. Strehlow · E. Tamiya · A. P. F. Turner · J. Wang
A. Warsinke · I. Willner · G. Wilson · U. Wollenberger
S. Yamamura · M. Zayats

Springer

Advances in Biochemical Engineering/Biotechnology reviews actual trends in modern biotechnology. Its aim is to cover all aspects of this interdisciplinary technology where knowledge, methods and expertise are required for chemistry, biochemistry, micro-biology, genetics, chemical engineering and computer science. Special volumes are dedicated to selected topics which focus on new biotechnological products and new processes for their synthesis and purification. They give the state-of-the-art of a topic in a comprehensive way thus being a valuable source for the next 3–5 years. It also discusses new discoveries and applications. Special volumes are edited by well known guest editors who invite reputed authors for the review articles in their volumes.
In references *Advances in Biochemical Engineering/Biotechnology* is abbeviated *Adv Biochem Engin/Biotechnol* and is cited as a journal.

Springer WWW home page: springer.com
Visit the ABE content at springerlink.com

Library of Congress Control Number: 2007937618

ISSN 0724-6145
ISBN 978-3-540-75200-4 Springer Berlin Heidelberg New York
DOI 10.1007/978-3-540-75201-1

Springer is a part of Springer Science+Business Media

springer.com

Cover design: WMXDesign GmbH, Heidelberg
Typesetting and Production: LE-TEX Jelonek, Schmidt & Vöckler GbR, Leipzig

Printed on acid-free paper 02/3180 YL – 5 4 3 2 1 0

Series Editor

Prof. Dr. T. Scheper

Institute of Technical Chemistry
University of Hannover
Callinstraße 3
30167 Hannover, Germany
scheper@iftc.uni-hannover.de

Volume Editors

Prof. Dr. Reinhard Renneberg

Department of Chemistry
The University of Science and Technology
Clear Water Bay, Kowloon, Hong Kong
chrenneb@ust.hk

Prof. Dr. Fred Lisdat

Biosystems Technology
Wildau University of Applied Sciences
15745 Wildau, Germany
flisdat@igw.tfh-wildau.de

Editorial Board

Advances in Biochemical Engineering/Biotechnology Also Available Electronically

For all customers who have a standing order to Advances in Biochemical Engineering/Biotechnology, we offer the electronic version via SpringerLink free of charge. Please contact your librarian who can receive a password or free access to the full articles by registering at:

springerlink.com

If you do not have a subscription, you can still view the tables of contents of the volumes and the abstract of each article by going to the SpringerLink Homepage, clicking on "Browse by Online Libraries", then "Chemical Sciences", and finally choose Advances in Biochemical Engineering/Biotechnology.

You will find information about the

- Editorial Board
- Aims and Scope
- Instructions for Authors
- Sample Contribution

at springer.com using the search function.

Attention all Users of the "Springer Handbook of Enzymes"

Information on this handbook can be found on the internet at springeronline.com

A complete list of all enzyme entries either as an alphabetical Name Index or as the EC-Number Index is available at the above mentioned URL. You can download and print them free of charge.

A complete list of all synonyms (more than 25,000 entries) used for the enzymes is available in print form (ISBN 3-540-41830-X).

Dedicated to the 65th birthday of
Frieder W. Scheller

Preface

To detect, quantify, and model biologically significant molecules is getting more and more important in our everyday life, in medicine, industry, and environment.

When a group of enthusiasts like Frieder Scheller started more than 40 years ago to develop biosensors, they would not foresee that biosensors are now available in every drug store, that the human genome sequence is available on the Internet, that DNA tests help in forensic cases, that we can track down the path of our ancestors from Africa...

This all is bioanalytics in practice!

Frieder Scheller is one of the pioneers of this field. So, we did not hesitate for long when Thomas Scheper, the Series Editor, asked us to compile a monograph written by the leading specialist to honor his 65th birthday with a fresh insight into the ever expanding bioanalytical field. Thanks to all contributors and thanks to the staff at Springer Verlag: Birgit Kollmar-Thoni, Ulrike Kreusel and Dr. Marion Hertel for their help!

Even in the 21st century biosensors will continue to play an important role in bioanalytics. By definition biosensors are characterized by a rather close contact of the biocomponent for recognition and the transducing element. Thus, the development of biosensors is a highly interdisziplanary field. Future developments can be particularly seen in sensitivity enhancement down to the molecular level, switchability of the sensing device, miniaturization and integration into microsystems, incorporation of new transduction and characterization methods and the use of artificial recognition elements.

This book tries to cover recent developments in order to illustrate the potential of this rather fascinating area of science.

At the start of the book, you will find a chapter about the history of biosensors and Frieder Scheller's contribution. It gives you an impression what this restless, unselfish and creative scientist has done. To underline the basic biosensor idea we start here with a cartoon about Frieder Scheller's work:

“Aller guten Dinge sind drei” (“‘all good things are 3”, as Germans believe). They all may partly characterize Frieder:

Chance favors only the prepared mind.
(Louis Pasteur)

The important thing is to create.
Nothing else matters; creation is all.
(Pablo Picasso)

The only truth that gets through will be
what we force through:
the victory of reason will be
the victory of people
who are prepared to reason, nothing else.
(Bertolt Brecht)

What a long way biosensors have come in such a short time and how fast the time flies.

Now, the prepared mind, the creative, smooth fighter for reason, Frieder Scheller, turns 65 years young...

On behalf of the worldwide biosensor community:
Happy birthday, Frieder!! Here is a special Chinese Firework for your birthday!

Hong Kong and Wildau, August 2007

Reinhard Renneberg
Fred Lisdat

Contents

Adv Biochem Engin/Biotechnol (2008) 109: 1–18
DOI 10.1007/10_2007_086

Published online: 9 November 2007

Frieder Scheller and the Short History of Biosensors

Reinhard Renneberg (✉) · Dorothea Pfeiffer · Fred Lisdat · George Wilson · Ulla Wollenberger · Frances Ligler · Anthony P. F. Turner

Department of Chemistry, The University of Science and Technology,
Clear Water Bay, Kowloon, Hong Kong, China
chrenneb@ust.hk

Abstract This is a first attempt at a brief sketch of the history of biosensors. It is far from complete and rather unsystematic. Many names are still missing, and we apologize for this. But the authors hope to have laid a humble cornerstone for a future "Complete History of Biosensors". We hope that many of our colleagues will contribute!

1 The Dawn of Biosensors

Without any doubt, Otto Warburg is the father of enzymatic analysis. His *optical test for the detection of NADH/NADPH* at 340 nm paved the way for the highly sensitive detection of dehydrogenases and their substrates in the mid-1930s. Hans-Ulrich Bergmeyer (Boehringer Mannheim) developed commercial optical enzyme tests based on Warburg's initial model.

The *father of biosensors* is Leland C. Clark, Jr. (Antioch College, Yellow Springs and the Children's Hospital in Cincinnati, Ohio, USA). He wanted to measure the reduction of oxygen with a platinum electrode to determine the oxygenation of blood. His first sensor failed because blood components were adsorbed on the electrode's surface, and this adsorption distorted the signal. Clark then had the ingenious idea of using the cellophane wrapper of a cigarette packet on his sensor. Only low molecular weight substances, mainly oxygen, could reach the electrode and be measured. The reduction current indicated the oxygen concentration, and so the Clark electrode was created. Today, Teflon is used as the membrane, and this sensor remains a key tool in medicine and environmental monitoring. To calibrate his sensor, Clark added an enzyme, glucose oxidase (GOD), to the solution. Clark then developed the sensor further by entrapping concentrated GOD with another

Fig. 1 Leland Clark Jr.

semi-permeable membrane in front of the electrode. The electrode could then be re-used for multiple glucose measurements. The enzyme layer became an integrated part of the sensor. Lee Clark went on to coin the term “enzyme electrode” at a meeting of the New York Academy of Sciences in 1962.

A few years later, Yellow Springs Instruments (YSI) developed their glucose analyzer, the YSI 23006, using Clark’s invention, although the company exploited electrochemical detection of hydrogen peroxide rather than monitoring oxygen. YSI launched their first instrument in 1973, but had to withdraw it due to interference problems. These were solved with the insertion of an additional membrane, and the company re-launched its (subsequently highly successful) series of laboratory analyzers in 1975.

Shortly after Clark published his enzyme electrode work, George Guilbault published a description of a *potentiometric urea electrode* using immobilized urease and a pH-sensitive sensor in JACS in 1969. In 1973, Mindt and Racine (Hoffmann la Roche) came up with the *first lactate sensor*. The natural redox partner of cytochrome b2 was replaced by the artificial redox mediator ferricyanide. Making the *mediated enzyme sensor*’s signal independent of the oxygen concentration in the sample was a decisive step in making future one-way biosensors possible. In 1976, Clemens and his co-workers incorporated an electrochemical glucose biosensor in a bedside artificial pancreas, and this was later marketed by Miles Laboratories as the Biostator.

The use of *thermal transducers* for biosensors was proposed in 1974. The new devices were christened by Klaus Mosbach and Bengt Daniellsson thermal enzyme probes or *enzyme thermistors*, respectively. The biosensor took a new evolutionary route in 1975, when C. Divies suggested that bacteria could be harnessed as the biological element to form *microbial sensors* for the measurement of alcohol.

Were these biosensors? What *is* a biosensor?

Karl Camman coined the term biosensor in 1977, but the IUPAC did not agree on the definition of a biosensor until 1997. The IUPAC committee, which included D.R. Thévenot, K. Toth, R.A. Durst, and G.S. Wilson, came up with the following definition:

A biosensor is a self-contained integrated device, which is capable of providing specific quantitative or semi-quantitative analytical information using a biological recognition element (biochemical receptor), which is retained in direct spatial contact with a transduction element. Because of their ability to be repeatedly calibrated, we recommend that a biosensor should be clearly distinguished from a bioanalytical system, which requires additional processing steps, such as reagent addition. A device that is both disposable after one measurement, i.e., single use, and unable to monitor the analyte concentration continuously or after rapid and reproducible regeneration should be designated as a single-use biosensor.

Biosensors may be classified according to their biological specificity-conferring mechanism or, alternatively, to their mode of physico-chemical signal transduction. The biological recognition element may be based on a chemical reaction catalyzed by, or on an equilibrium reaction with, macromolecules that have been isolated, engineered or are present in their original biological environment. In the latter cases, equilibrium is generally reached and there is no further, if any, net consumption of the analyte(s) by the immobilized biocomplexing agent incorporated into the sensor. Biosensors may be further classified according to the analytes or reactions that they monitor by directly monitoring the analyte concentration or by reactions producing or consuming such analytes; alternatively, an indirect monitoring of an inhibitor or the activator of the biological recognition element (biochemical receptor) may be achieved.

Groups all over the world had now joined biosensor research.

Frieder Scheller started in East-Germany (the German Democratic Republic, GDR) with his *amperometric glucose sensor* in 1975 and multi-functional biosensors and research on coupled enzyme reactions in combination with sensors in 1977.

In Lithuania (at that time part of the Soviet Union), Juozas Kulys studied *mediated enzyme sensors* employing dehydrogenases and organic metals almost at the same time. Ilya Berezin's group at Moscow State University and

Scheller's group in Berlin showed that even enzymes can communicate directly with an electrode. The electrochemical transformation of the active site can generate a current that is accelerated by the regeneration of the enzyme in the presence of a target substrate. This *mediator-free bioelectrocatalysis* laid the foundation for third generation biosensors.

Masuo Aizawa and Isao Karube in Japan (in 1983), and Christopher Lowe in the UK (in 1985) started to combine *various biological recognition elements* such as enzymes, antibodies, cells and nucleic acids with different transducers.

A major advance in the *in vivo application of glucose biosensors* was reported by Shichiri, who described the first needle-type enzyme electrode for subcutaneous implantation in 1982. In parallel, Uwe Fischer's group in Karlsburg began similar studies with the important goal of establishing the correspondence between blood and tissue glucose, the sensor being implanted in the subcutaneous tissue and not in a vascular bed.

Different transducers have been used. *Ion-selective field effect transistors* (ISFETs) are now used to integrate electronic signal conversion into the biosensor. The pioneer in this area was Piet Bergveld in Holland, starting as early as 1970.

The idea of building direct *immunosensors* by fixing antibodies to a piezoelectric or potentiometric transducer has been explored since the early 1970's, but it was Liedberg, Nylander, and Lundstrom who, in 1983, paved the way for commercial success. They described the use of *surface plasmon resonance* to monitor affinity reactions in real time. The BIAcore (Pharmacia, Sweden), launched in 1990, is based on this technology.

The first *optical biosensors* resulted from technical breakthroughs in optics and communications. The first sensors based on fiber optics were chemical sensors, but integration of biological recognition molecules offered far greater sensing specificity and sensitivity.

The first *fiber optic biosensors* were configured with an interrogation region at the end of the fiber. In 1982 Jerome Schultz and his colleagues developed a glucose sensor that measured the equilibrium exchange between fluorescent concanavalin A bound to glucose immobilized outside the light path and the same molecule bound to free glucose in a sample at the end of the fiber. This scheme was adapted for use with antibodies, but it was limited in that the reaction required a significant amount of time to reach equilibrium. The end-on, or optrode configuration was more successful initially for monitoring enzyme reactions. The enzyme could be captured in a membrane or gel at the end of the fiber, and low molecular weight targets and color-producing substrates could diffuse into the region occupied by the enzyme to generate a signal. Fluorescence in whole cells could be measured this way.

However, the optrode configuration really found its place in the biosensor world with the work of David Walt's group at Tufts. They combined highly multiplexed bead-based arrays with fiber bundles etched at the end to hold the beads. This technology, now marketed by Illumina, has rapidly captured a major share of the genetic testing market.

Shortly after the first optrode biosensor was reported in 1984, Thomas Hirschfeld combined the concept, originally demonstrated a decade earlier, with planar waveguides to demonstrate a biosensor based on evanescent wave sensing on the sides of a fiber. In 1975, M.N. Kronick and W.A. Little had immobilized antibodies or antigens on a planar waveguide and demonstrated that the fluorescence signal changed when a fluorescent-labelled antibody was bound to an antigen on the surface.

Hirschfeld, who had first demonstrated the concept of the evanescent wave as early as in 1965, demonstrated immunoassays on the surface of a small glass rod. He used the evanescent field to discriminate fluorescent complexes on the surface of the fiber from unbound fluorescent antibodies in solution.

Extensive work by Frances Ligler with groups at the US Naval Research Laboratory, the University of Utah, and Corning/Zeptosens in Switzerland, coupled with miniaturization of lasers and detectors, has led to the availability of evanescent wave-based biosensors employing both optical fibers and planar waveguides. They are now commercially available from several companies. In addition to being valuable research tools, these biosensors have found applications in diagnosing infectious disease, in food safety, in security scans, and in pollution monitoring.

The evolution of biosensor development since Clark's contributions has been heavily dominated by the importance of *measuring glucose* for detecting and managing diabetes. In 1967 Updike and Hicks described a glucose sensor, dubbed the enzyme electrode, which employed a gel-entrapped enzyme. Stuart Updike and his group at the University of Wisconsin continued to develop glucose sensors, especially implanted designs, and in 2000 described studies of sensors implanted in dogs for three months.

The Clark sensor evolved into the *Model 23 clinical analyzer* that was first marketed by Yellow Spring Instruments in 1975. It is still marketed to this day.

By the late 1960s it became obvious that a means for *self-monitoring of blood glucose* was needed. Tom Clemens at Ames Laboratories developed a colorimetric system based on glucose oxidase, where the patient compares the color intensity of a test strip with a printed color scale. This evolved into the Ames Reflectance Meter, used primarily in doctors' offices, though a version for patients appeared in 1985.

With the launching of the electrochemical *mediator-based test strip by MediSense* in 1987, the optical reflectance measurements of strips quickly shifted to electrochemical detection. There have subsequently been incremental improvements related to faster measurements, less blood required (now only 0.2 μL) and improved data management software.

One of the key pioneers in the biosensor area has long been Jerry Guilbault. In 1963 he demonstrated a *biosensor alarm system* using an immobilized enzyme, acetylcholine esterase. Some pesticides and nerve agents inhibit the activity of this enzyme, and this formed the basis for its function. Throughout the last 45 years, Guilbault, presently claimed by Ireland, has developed a wide variety of approaches to biosensor development, based most prominently on affinity techniques such as immunosensing combined with piezoelectric sensors.

Another long time contributor to biosensors is Garry Rechnitz, now Professor emeritus at the University of Hawaii. His contributions to biosensors are many, including the use of ion-selective electrodes for enzyme-catalyzed reactions and, most notably, the use of whole cells and tissues in sensors. His group has *interfaced crab antennae with a sensing system* and developed *tissue-slice biosensors* like the famous "bananatrode" (1985) for the analysis of catecholamines.

One of his former students, Mark Meyerhoff (University of Michigan), has made seminal contributions to ion-selective electrodes, and especially to the development of sensors that can measure ions reliably in a vascular bed. He pioneered the controlled release of NO to inhibit thrombosis on sensor surfaces.

In Switzerland, Horst Vogel (EFPL) has focused on *receptor-based biosensing* and made his group the most successful masters of this technology.

In 1984 A.E.G. Cass, D.G. Francis, H.A.O. Hill, W.J. Aston, I.J. Higgins, E.V. Plotkin, L.D.L. Scott, and A.P.F. Turner from Cranfield and Oxford Universities published a much cited joint paper on the use of ferrocene and its derivatives as an immobilized mediator for use with oxidoreductases. Their aim was an inexpensive enzyme electrode. This mediated electrochemical biosensor formed the basis for the screen-printed enzyme electrodes launched by MediSense (Cambridge, USA) in 1987 with a *pen-sized meter for home blood-glucose monitoring*. The electronics were redesigned into popular credit-card and computer-mouse formats, and MediSense's sales showed exponential growth, reaching US$175 million by 1996. The company was then purchased by Abbott for US$867 million.

The other market leaders in diabetes testing subsequently adopted similar mediated amperometric approaches, and Roche, Johnson and Johnson, LifeScan, and Bayer all now offer competing mediated biosensors. Three companies hold 85% of the world market for biosensors, which currently exceeds US$7 billion/yr. The technology developed by Cranfield and Oxford has largely displaced conventional reflectance photometry in home diagnostics.

Despite many attempts, successful implementation of *direct electron transfer* between an electrode and an enzyme has been very limited. In 1987, Adam Heller and his group at the University of Texas in Austin began to address this problem by creating a "wire": a string of mediator molecules embedded in the enzyme layer or coupled to the enzyme molecules. "Wiring" has evolved into a series of redox hydrogels that facilitate electron transfer between the enzyme and the electrode, but which are immobilized so that they do not migrate into the test medium.

TheraSense (USA) launched their FreeStyle coulometric glucose meter based on this technology in 1994. Because of improved patient acceptance, the FreeStyle was a significant step forward. Less than 300 nl of blood is enough for a test, which makes the procedure nearly painless and was an enormous help to diabetics. Abbott Diabetes Care purchased TheraSense to complement its earlier acquisition of MediSense.

Abbot combined sampling and measurement in their SoftSense unit as a further step to make blood glucose measurements reliable and simple. The Accu-Check Compact from Roche incorporates several strips in a single device. These two concepts have been combined into an innovative new product line by Pelikan Technologies Inc (Palo Alto) which integrates multiple sampling and measuring devices with a near painless electronic lancing system.

When, in the mid 1980s, Shichiri in Japan first reported on an indwelling needle-type glucose sensor for the *continuous monitoring of glucose*, George Wilson (University of Kansas) established a collaboration with Daniel Thévenot and Gerard Reach in Paris which led in 1993 to a sensor which was subcutaneously implanted for three days. This group then continued to address various sensor performance problems, including the relationship between blood and tissue glucose, sensor calibration, and biocompatibility. By the late 1990s, a number of continuous monitoring systems had been developed by firms including Medtronic/Minimed, DexCom, Abbott/TheraSense, and Cygnus/Animas. All were designed for 3–5 day implantation, and all require recalibration several times a day.

Although the majority of sensors have been implanted subcutaneously, there have been a few attempts to install a *sensor directly in the vascular bed*. David Gough (U.C. San Diego) in 1990 reported on a 3 month implant in a dog.

Joseph Wang (Arizona State University) has developed *screen printed and microfluidic biosensor systems* that offer promise for point-of-care (POC) applications.

2
The Development of Biosensors in the Divided Germany and During Re-unification

The authors of this chapter were lucky to obtain Frieder Scheller's manuscript for a German bioanalysis textbook (Renneberg, R. and Kayser, O. Bioanalytik für Einsteiger. SAV-Springer, Heidelberg 2008, in preparation). Some of the details that follow are therefore provided by Scheller himself, others have been added by the authors. We hope that Frieder will forgive us for intruding on his unpublished writings!

Frieder Scheller is another of the fathers of biosensors. However, in 1972, the *first German patent for a biosensor* was awarded to the private Research Institute Forschungsinstitut Manfred von Ardenne (Dresden, GDR). Ten years after the very first patent by Leland Clark, this East German patent (DE 101229) described a very similar combination of an oxygen electrode with an enzyme solution. The aim and potential application field described was the monitoring of hyperthermic therapy of cancer patients. Variation of oxygen saturation in venous blood, which may cause false glucose values, had been eliminated by a continuous air stream resulting in equalized oxygen saturation and a kind of stirring of the sample as well. However, the resulting foaming had to be reduced by adding an anti-foaming agent.

Frieder Wolfram Scheller was born in the middle of the terrible Second World War, in 1942. His name Frieder echoes the German word for peace, "Frieden". Frieder is living his name being a pacifist and seeking harmony in his team and with colleagues.

After his studies in Merseburg (1960) and his PhD work at Humboldt University, Scheller became a Group Leader of Electrochemistry in the Department of Enzymology of the Central Institute of Molecular Biology of the Academy of Sciences of the GDR in 1969. He was also appointed Professor of Biochemistry in 1984.

He focused on protein electrochemistry, at that time rather a niche research area at the Molecular Biology Institute in Berlin-Buch. His research in bioelectrochemistry laid a cornerstone to the following biosensor development. His book "Biological Electrochemistry" (with Karl Kadish, Glenn Dryhurst, and Reinhard Renneberg) is now a classical monograph.

In general, Frieder Scheller's case fits very well with Louis Pasteur's words: "*Chance favors the prepared mind!*" Scheller's mind was perfectly prepared for the biosensors!

Research on biosensors at the Research Center of Medicine and Biology in Berlin-Buch began "on planned schedule" (remember: East Germany had a "socialist planning economy"!) in September 1975 with the employment of a young PhD scientist, Dorothea Pfeiffer. The goal was not more and not less than an enzyme electrode for blood glucose. Another PhD student, Reinhard Renneberg joined Scheller in 1976 and came up with a simple scheme for the spatial integration of bioreceptors and transducers in 1978. This scheme is now in general use.

Ulla Wollenberger and Florian Schubert joined in 1978 and 1980 to complement the group of PhD students with their development of sensors with coupled enzyme systems, organelles and whole cells. By shuttling the substrate between two enzymes they achieved huge signal amplification. This was a hot subject in 1984, and within three weeks, similar patents had been filed, first by Scheller, followed by Kulys (Lithuania) and Mizutani (Japan).

The first electrochemical sensor was "hand-made" by the glassblower of the Institute: A platinum wire was placed into a glass tube, silver wire was wrapped around it as a reference/counter electrode and the enzyme "layer" was fixed in front of the electrode by flexible rubber. Material resources in the former GDR were very limited; thus, people had to be creative: Various ma-

Fig. 2 Scheller group in Berlin-Buch (around 1979)

terials for the immobilization procedures were tested to create a stable and reliable enzyme layer, e.g., nylon stockings (Dederon was the East German variety of it) and parachute silk (from the National People's Army of East Germany) for glutardialdehyde immobilization of glucose oxidase and acrylamide for photopolymerization. Finally, entrapment into photogelatin from the color film production of ORWO Wolfen (GDR) gave quite reliable results.

To realize the necessary arrangement to maintain the temperature, a measuring cell with the enzyme electrode and a stirrer was placed inside a water bath of a thermostat and the respective measurements were performed by injecting 50 μl of whole blood into a 2 ml stirred buffer solution. After finishing the measurement, the suck position of the thermostat was used for flushing the measuring cell and rinsing the enzyme electrode.

About 400 manual glucose analyzers of that Glukometer GKM, that used our glucose biosensors were produced by the Center for Scientific Equipment of the Academy of Sciences (ZWG) from 1982 to 1986 and sold to clinical laboratories in the former GDR. So, the Glukometer GKM was the first commercial enzyme electrode-based glucose analyzer developed and used commercially in Europe. As was joked at that time, the tiny GDR (17 million people) had at this time the highest biosensor density per capita in the whole world (see Fig. 3)!

Extensive clinical evaluations of the Glukometer GKM and the enzyme membrane electrodes had been conducted, e.g., at the Central Institute of Diabetes Karlsburg. The method and the instrument was then standardized and recommended for routine glucose monitoring by the Institute of Pharmaceutical Research Berlin.

After meeting all the sophisticated conditions required for analytical performance, e.g., a serial imprecision below 2.0% using blood samples, the method was included in 1987 in the GDR standard procedures "Arzneibuch".

Fig. 3 The glucose sensing system "Glukometer" developed by the group of Frieder Scheller

A red color was chosen for the enzyme layer to make it better visible. That was expected by colleagues from the West to be a necessity because it was a "socialistic (means red) product" ... However, the color choice was only because of availability. The group of Scheller learned later that their Japanese friends (and competitors) tried to find out without success, what red redox dye had been used in the highly stable East German enzyme membranes. The red glucose oxidase gelatin membrane was more than reliable: transport of a sample to Fidel Castro's Cuba and back by ordinary (!) mail at local temperatures and humidity did not cause any change in its analytical performance.

However, one of the "main antagonists of the socialist planning economy", the weather, did from time to time cause problems. During the very hot summer of 1985, the gelatin membranes immediately disaggregated on contacting the fluid. There was probably some microbial contamination exacerbated by the heat. A polyurethane shoe-glue was used for stabilization, and finally the gelatin was completely replaced by polyurethane. With the polyurethane version, in collaboration with B. Olsson, H. Lundbäck, and G. Johansson the world record was broken in 1986: 300 samples per hour using a FIA system was the most rapid procedure at that time anywhere in the world.

In parallel with the development of the Stat Glukometer GKM, the Scheller group conducted in 1981 a joint investigation on the adaptation of the glucose electrode to a flow system with the District Hospital of Neubrandenburg. This flow system idea remains trend-setting even by today's standards. The possible automatization of the flow system is a significant advantage as compared to a manual system based on a stirred measuring cell. An important side-result was the work with hemolysate. The glycolysis and the related glucose degradation of whole blood samples from patients were interrupted immediately after the sample was withdrawn and the resulting hemolysate samples remained stable for 24 hours at room temperature.

The instrumental development was done by PGW Medingen (Prüfgerätewerk). Beside the enzyme electrode containing the flow cell as the heart of the whole system, the blood glucose analyzer AM 300 consisted in general of a pump, a large automatic sampler, an amplifier, and a recorder. The whole apparatus required at least a table of one meter in length, which made it rather ungainly. However, its analytical performance was excellent and the people in the lab were enthusiastic.

Therefore, PGW Medingen decided to develop an automatic glucose measuring analyzer based on the flow-through principle. The Enzyme-Chemical Analyzer ECA 20 was based on an air-segmented continuous flow system resulting in a sample throughput of about 120 per hour. The analytical parameters demonstrated the superiority of the new instrument with respect to other enzyme electrode-based analyzers and PGW Medingen was awarded a Gold Medal for technology and design at the Leipzig International Trade Fair 1987.

Now, the West Germans finally "woke up": The company Eppendorf–Netheler–Hinz ENH (Hamburg) recognized the importance of new biosen-

Fig. 4 Frieder Scheller in his laboratory

Fig. 5 The core team of the glucose sensor development: Frieder W. Scheller, Dorothea Pfeiffer, Norbert Klimes, Bernd Fahrenbruch

sor technology from the central laboratories of East Germany and used the Leipzig Trade Fair as a point of contact. Via the official "State interface" Intermed, a cooperation agreement was created between PGW, ENH and the Academy of Sciences. PGW changed the instrument design for ENH "to improve the housing". (The difference between the East and West German devices was mainly the color of the housing: black for the East, white for the West...) The Academy of Sciences delivered the enzyme membranes (the convertible currency earned was used to buy enzymes for research and production from the West). ENH began to launch the ECA 20 in Western Europe under the brand name ESAT 6600 in 1988.

Another development between PGW and ENH resulted in a new generation of biosensor-based glucose analyzers. Just prior to the political changes in Germany in 1989, the first instrument from the EBIO family of instruments came on the market and the laboratory instruments based on the Academy of Sciences enzyme membrane biosensors became market leaders in Europe.

On the basis of research in *microbial biosensors* a different product line was developed in 1988 for the determination of Biological Oxygen Demand (BOD) by Klaus Riedel in Scheller's group. Sensors were based on *Bacillus subtilis* or *Trichosporon cutaneum* immobilized in polyvinyl alcohol (PVA) and combined with a Clark oxygen electrode. As compared to the former Five-day procedure of the BOD5 test, the very short response time of the new BOD biosensor (15 to 30 seconds) reduced the measuring time enormously. Later, these sensors were used for an industrial instrument of the German company Dr. Lange GmbH.

In parallel to extensive attempts to bring biosensors into application, Scheller' s group started with fundamental research in bioanalytics. Coupled enzyme systems have been combined with different transducers to create enzyme sequence based biosensors for detection of, for example, ATP and NAD+ and to reduce interference in biosensing elements by forming anti-interference layers (e.g., to eliminate glucose in sucrose solutions or reducing substances in glucose solutions).

Enzyme sequences have now been used to amplify the sensitivity of enzyme sensors as much as several thousand-fold! One technique is based on recycling substrate molecules between pairs of enzymes. The substrate is shuttled between the two enzymes several times, thus forming much more active product in the enzyme layer.

This principle originates back to Otto Warburg, who used this cycling principle as early as 1935 to measure NADP in solution. It was later shown by O.H. Lowry to be "trouble free and give surprisingly reproducible results". Amplification factors up to 4100 were demonstrated in 1987 measuring lactate with an electrode bearing lactate oxidase and lactate dehydrogenase. Enzyme cascades have also been explored to achieve high sensitivity. This principle is widely implemented in nature to trigger response to very tiny chemical signals.

Another development also had its origin within this period—the miniaturization of sensor electrodes. Mainly Rainer Hintsche (now at the Fraunhofer Institute of Silicon Technology) has been connected to this development and the use of microelectrodes for bioanalysis.

After the political changes in 1989, the great potential of the biosensor knowledge in the former East was not exploited to create a real biosensor center in Germany. A great opportunity was lost!

Various attempts by serious companies along with former employees to acquire PGW Medingen from the German Trust (Treuhand, responsible to "sell" East German industries to interested parties from the West) to continue the biosensor instrument process failed because of unrealistic conditions. Finally, PGW was sold to a (mainly) real-estate and environmental technology group

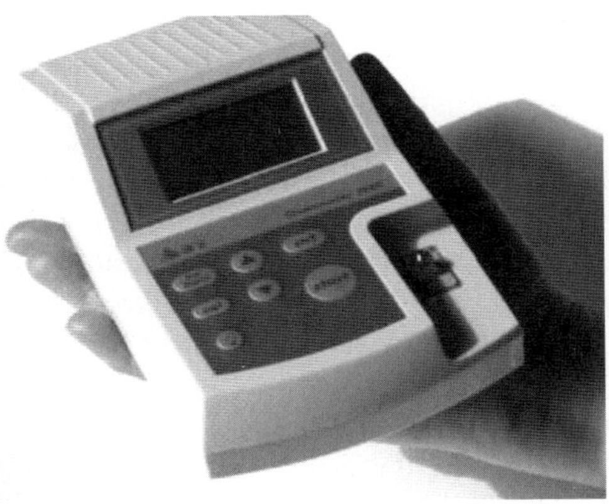

Fig. 6 One of Scheller's latest brainchildren: the Glukometer 3000

(Preiss–Daimler) and that was the signal to start to destroy the old structures. A significantly diminished group continued the production of analyzers for ENH Hamburg and PGW itself and in parallel Dr. Müller GmbH was founded by two former PGW employees.

In the early 1990s two more companies started with the same business in Germany: EKF GmbH Magdeburg was outsourced out of Elektroapparatewerk EAW Berlin/Germany and Care diagnostica GmbH was founded near Düsseldorf. These various companies brought various families of instruments to the market. Preiss–Daimler and ENH produced the EBIO and ESAT families. Dr. Müller GmbH produced the SuperG family and EKF produced the Biosen family, Care diagnostica followed with the Eco family.

Many families, but one common ancestor: Frieder Scheller!

All instruments were based on air segmented flow-through systems and the Academy of Sciences biosensors. The Academy's enzyme membranes were the basis for the formation of the company BST Bio Sensor Technologie GmbH in 1991. All the membranes for the different instruments were produced by BST.

In the 1990s, more than 25 000 membranes per year were sold. BST became the market leader in Germany and Europe and is still one of the very few companies dealing with multi-use biosensors worldwide today mainly based on thick film technology. More than 40 000 multi-way biosensors are now sold per year. One glucose sensor chip is applicable for about 1000 samples.

Besides BST GmbH in Berlin, similar companies were founded from the Institute of Biotechnology Leipzig by Bernd Gründig (SensLab GmbH with a lactate biosensor) and the Institute of Diabetes in Karlsburg.

The general situation in the early 1990s in Germany was characterized by permanent changes. In the field of biosensors the opening of the former East indeed caused a brain drain into the former West.

Just after the official political reunification of Germany, the Institute of Molecular Biology of the Academy of Sciences Berlin-Buch, the Society of Biotechnological Research (GBF) Braunschweig and the National Ministry of Research and Technology organized the First German Biosensor Meeting at the International Education Center in Bogensee near Berlin. More than 150 scientists from various German research institutions and companies participated. These scientists represented enormous developments in the field. The program and the monograph indicated the leading role of the former East in biosensor research before reunification.

In the Western part of Germany, biosensor research started about 1983.

Schindler and Schindler published a booklet "Biochemische Membran-Elektroden" (Biochemical membrane electrodes) and Hanns-Ludwig Schmidt started in 1976 dealing with enzyme electrodes based on NAD-dependent dehydrogenases and their coupling to electrodes via mediators at TU Munich/Weihenstephan.

This work was continued later by Wolfgang Schuhmann (now at Bochum University).

Rolf D. Schmid started to deal with biosensors by initiating an excellent meeting in 1987 in Braunschweig (GBF Braunschweig). The main topic of the meeting was protein technology to create optimal enzymes and their combination with planar thick film technology. GBF Braunschweig was also the place that sequenced glucose oxidase. Dietmar Schomburg's team at GBF came up with the 3-D structure of GOD. Rolf Schmid is now using biosensors for biotechnology research at Stuttgart University.

When Karl Cammann visited Garry Rechnitz in 1977 in Buffalo, New York he used the term *biosensor*. He actually started to work in biosensors himself much later. His Institute of Chemo- and Biosensors (ICB) started its work in the middle of the 1990s and it became the largest German institution for biosensors. The brain drain from the East brought him Reinhard Renneberg, Christa Dumschat and Bernd Gründig. Renneberg is now doing biosensor research in Hong Kong, Christa Dumschat with EKF diagnostics in Madgeburg and Bernd Gründig is back to Leipzig with his own company.

The biosensor group in Tübingen concentrated on optical transducers. The work under the direction of Günter Gauglitz with optical-affinity sensors has been very successful and has resulted in a commercial product (RIFS analyzer). After Otto Wolfbeis moved to Regensburg and took his professorship on biosensors, his group developed a powerful center for research on fluorescent bioanalytics.

Commercialized biosensors are doing well in Germany now: Compared to the 1960s, the instruments that are now available to diabetes patients for self monitoring using single-use strips or sensors are a great step forward in reducing secondary diseases and improving the quality of life. However, not only in centralized laboratories at hospitals but also for the point-of-care market the analysis should be of lab quality.

Based on Frieder Scheller's work ABT Advanced Bioanalytical Technology GmbH and Care diagnostica GmbH brought the very first pocket instrument based on multi-use biosensors to the market. The Glukometer 3000 and LactatProfi 3000 (ABT GmbH) have been on the worldwide market since 2005 and the Glukometer 3000 under the brand EcoSoloII has been available in Germany since 2004 (Care GmbH).

These pocket instruments are reliable and give high-quality results. One multi-use glucose sensor can be used 1000 times within 30 days. The multi-use concept is an alternative way for POCT to improve quality and to reduce costs and it seems to have high potential for improving patient care, too.

3
Biosensor Research at Potsdam University

After the political changes in the former GDR and the unification of Germany, the Academy of Science was dissolved and reorganized drastically. Frieder

Scheller had to reorientate. Since 1993 he has been holding a Chair professorship of Analytical Biochemistry at the newly founded University of Potsdam. Here, he was again able to build up a strong research group with some of his former coworkers, several new postdocs and very talented PhD-students. Frieder has extended his fundamental and applied research on enzyme-based electrodes to other bioanalytical systems.

The early successful research on protein electrochemistry (at that time predominantly mercury electrodes were used) is now being continued with novel surface chemistry and modern biochemistry. His research has been supported by a large research grant from the Deutsche Forschungsgemeinschaft. Ulla Wollenberger (from the former Berlin-Buch group) and Fred Lisdat (now at Wildau University) have been his core people. This work has led toward a deeper understanding of the electrical behavior of redox enzymes and proteins but always with an eye on potential analytical applications. So, the group was the first who could show the time course of superoxide radical formation in vivo during reperfusion with a cytochrome c based sensor.

With the availability of enzymes of high purity, activity and desired stability and by the help of recombinant enzyme techniques it was possible to further develop coupled enzyme sensors (by Ulla Wollenberger, Alexander Makower, and Fred Lisdat) and to create also more recently very sensitive detectors for enzyme inhibitors (Alexander Makower, Walter Stöcklein). By using enzymatic or bioelectrocatalytic recycling it has been possible to develop ultrasensitive enzyme detectors for neurotransmitters and immunoassays. For example, by combining lactase and glucose dehydrogenase on an oxygen electrode, a zeptomole detection limit for the pesticide 2,4-dichlorophenoxyacetic acid (2,4-D) was achieved.

Electrochemical immunosensors have been developed by Axel Warsinke for metabolites (like creatinine) and proteins of medical importance. Affinity-based techniques such as SPR and QCM have been used in enzyme assays for the sensitive determination of inhibitors of choline esterases and metalloproteinases.

Optical detectors and DNA sensing were established when Frank Bier (now at the Fraunhofer Institute of Biomedical Engineering) joined the group. At about this time, the first aptamer-based biosensor was reported by Scheller's group. It detected adenosine using total internal reflection fluorescence.

Biosensing using (artificial) biomimetic systems has been investigated by Axel Warsinke, and more recently Martin Katterle. Warsinke has also developed catalytic antibodies. One of Frieder's recent research interests are MIPs—molecular imprinted polymers. Here, he has developed catalytic active MIPs and has shown their activity as artificial recognition elements by a thermistor-based approach. Frieder will actively pursue this research even after his retirement as a guest researcher at the Fraunhofer Society.

In addition to pure science, Frieder Scheller was also very active in initiating cooperations between research and industry. In 1997 he founded the

Fig. 7 Scheller group 1998 Berlin-Buch

Interdisciplinary Research Association Bioanalytics on a regional level in Berlin-Brandenburg. He won a competition initiated by the German government in 1999 in order to support regional development in science and industry in 2000. "Biohybrid Technologies" (BioHyTec) became a brand name for this development and was mentioned in 2003 by the World Technology Evaluation Council of the US Government as one of the 20 leading centers in the world.

Because of the continuously growing international interest in biosensors in 1990, the "biosensor community" initiated a World Congress on Biosensors. Under the chair of A.P.F. Turner the debut was in Singapore in 1990. The enormous resonance even now underlines this was the right idea at the right time. Since Singapore, the "World Congress on Biosensors" has been held every other year with increasing success. Since New Orleans 1994 Frieder Scheller has been a member of the organizing committee and is the Program Chairman for Europe.

At the end of the 1990s, the idea was born to install a new series of national conferences focusing on new developments in the field of biosensors and bioanalytics. Frieder Scheller was from the beginning one of the initiators and organized in 2003 the 3rd German Biosensor Symposium in Potsdam.

Facing retirement from his professorship this year, he has been creative in initiating new groups of young scientists at the University of Potsdam (InnoProfile group "Integrated protein chips for point-of-care diagnostics") and the Fraunhofer Society ("Biohybrid systems on a supramolecular basis") and thus enlarged the basis for a fruitful development of bioanalytics and biosensorics.

4 Conclusion

What a long way biosensors have come in such a short time and how fast time flies.

Now the prepared mind, the creative, smooth fighter for reason, Frieder Scheller, turns 65 years young ...

On behalf of the worldwide biosensor community:

Happy birthday, Frieder!!

Adv Biochem Engin/Biotechnol (2008) 109: 19–64
DOI 10.1007/10_2007_083

Published online: 11 October 2007

Protein Electrodes with Direct Electrochemical Communication

Ulla Wollenberger (✉) · Roberto Spricigo · Silke Leimkühler · Katrin Schröder

Institute of Biochemistry and Biologie, University of Potsdam,
Karl-Liebknecht-Strasse 24–25, 14415 Golm, Germany
uwollen@uni-potsdam.de

Abstract Electrochemistry using direct electron transfer between an electrode and a protein or an enzyme has developed into a means for studying biological redox reactions and for bioanalytics, biosynthesis and bioenergetics. This review summarizes recent work on direct protein electrochemistry with special emphasis on our results in bioelectrocatalysis using isolated enzymes and enzyme-protein couples.

Keywords Bioelectrocatalysis · Bioelectrochemistry · Biosensor · Direct electron transfer · Modified electrodes · Protein electrochemistry

Abbreviations

CCP Cyt c peroxidase
CNT Carbon nanotube

CP	Carbon paste electrode
CYP	Cytochromes P450
Cyt	Cytochrome
Cyt c	Cytochrome c
DDAB	Didodecyldimethylammonium bromide
DET	Direct electron transfer
DMSOR	Dimethyl sulfoxide reductase
DTSSP	3,3′-Dithiobissulfosuccinimidyl-propionate
EDC	*N*-(3-dimethylaminopropyl)-N-ethylcarbodiimide
EDTA	Ethylene diaminetetraacetic acid
EPG	Edge-plane oriented pyrolytic graphite
FAD	Flavin adenine dinucleotide
FMN	Flavin mononucleotide
G	Graphite electrode
GC	Glassy carbon electrode
H_4B	(6*R*)-5,6,7,8-Tetrahydrobiopterin
Hb	Hemoglobin
HOPG	Highly oriented pyrolytic graphite
ITO	Indium/tin oxide
Mb	Myoglobin
Moco	Molybdenum cofactor
MP	Mercaptopropanol
MPA	Mercaptopropionic acid
MPS	Mercaptopropane sulfonate
MPT	Molybdopterin
MU	Mercaptoundecanol
MUA	Mercaptoundecanoic acid
NHA	*N*-hydroxy-L-arginine
NHE	Normal hydrogen electrode
nNOS	Neuronal nitric oxide synthase
NOS	Nitric oxide synthase
PAA	Poly-allylamine hydrochloride
PDDA	Poly-dimethyldiallylammonium chloride
PEI	Poly-ethylene imine
PEO	Poly-ethylene oxide
PG	Pyrolytic graphite electrode
PS I	Photosystem I
PSS	Poly-styrene sulfonate
SAM	Self-assembled monolayer
SERRS	Surface enhanced Raman resonance spectroscopy
SO	Sulfite oxidase
SOD	Superoxide dismutase
SP	Spectroscopic graphite
XDH	Xanthine dehydrogenase
XOD	Xanthine oxidase
XOR	Xanthine oxidoreductase

1
Introduction

Electrochemical methods provide rapid measurements of redox properties of proteins and the kinetics of electron transfer and coupled reactions and may thus improve our understanding of biological electron transfer processes. Electrochemistry is also powerful for the exploitation of these systems in biosensors, effective biofuel cells and selective biosynthesis routes [1–6]. Traditionally, soluble small mediators were used to facilitate the communication between the redox-active center and the electrode. They may, however, be involved in unspecific side reactions that cause erroneous results, and therefore direct electrochemical methods were developed. Methods based on the direct electron transfer (DET) do not involve electroactive mediators for the exchange of electrons between a biomolecule and an electrode.

The first reports on a reversible DET between redox proteins and electrodes were published 30 years ago, and showed that cytochrome c (Cyt c) is reversibly oxidized and reduced at tin doped indium oxide [7] and gold in the presence of 4,4′-bipyridyl [8]. At the same time the DET between bare electrodes and enzymes was reported for laccase, peroxidase, and P450-enzymes (CYP-enzymes) [9–12]. Since then a key issue has been, and still is, to interface redox-active biomolecules to electrode surfaces in a way that the protein does not become denatured and characteristic properties such as redox properties and catalytic activity are not impaired, while the interfacial electron transfer is fast, which requires small reorganisation energy and effective electronic coupling between active site and electrode surface [13–15]. The electronic coupling between the protein and the electrode can be influ-

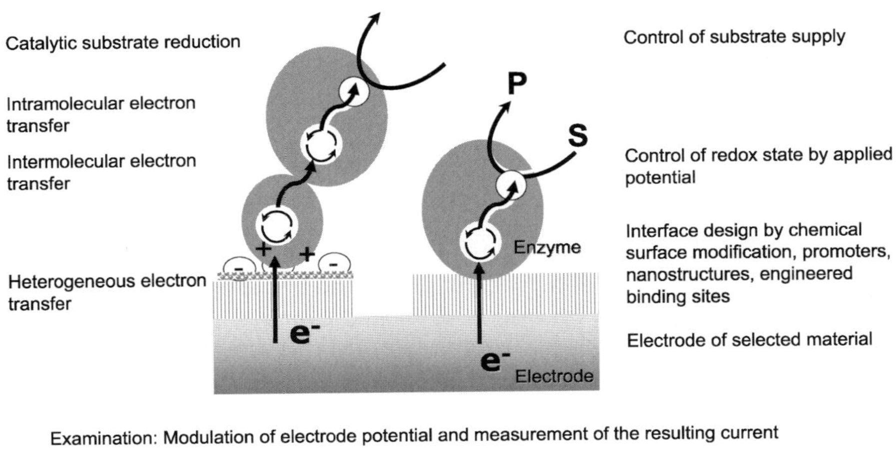

Fig. 1 Illustration of direct electron transfer contact between an enzyme and an electrode and bioelectrocatalysis

enced by using different electrode modifiers, thus controlling the orientation of the protein and the distance between the redox active site of the protein and the electrode. The applied electrode potential is related to the driving force and therefore the rate of electron transfer. In the presence of enzyme substrate the electrochemical reaction is coupled to the enzyme catalysis by regenerating the redox active site of the enzyme directly or with additional proteins in DET contact involved (Fig. 1).

During the past decade, direct electrochemistry of proteins has also developed into a means for studying more complex biological redox reactions including DNA-repair enzymes [16], and membrane protein complexes [17, 18]. Furthermore, the heterogeneous electron transfer between proteins and electrodes was coupled with other reactions where proteins act as vectorial mediators [19–21]. An example will be given where Cyt c mediates the biocatalysis of sulfite oxidase and also its molybdopterin-containing catalytic domain.

This review is a survey of the progress made in the research on the DET between proteins and electrodes during the last 5–10 years. It summarizes some recent work on protein electrochemistry of particularily Cyt c, CYP-enzymes, NO-synthase, peroxidase, and molybdopterin-containing enzymes. Furthermore, some analytical applications are discussed.

2 Protein Electrochemistry

2.1 The Role of the Interface

For an effective bioelectric contact, the surface of the electrode, the material and pretreatment are of critical importance. The most successful electrodes are noble metal modified with various adsorbates and carbon materials and metal oxides that have natural surface functionalities. Self-assembling monolayers (SAM) of functionalized thiols on gold, polyelectrolyte and surfactant films on carbon, electron-transfer promoters in connection with various electrode material and metal oxides were shown to be suitable tools [2, 19, 22–27]. In particular, SAMs spontaneously formed on clean gold surfaces from solutions of thiols with functional terminal groups facilitate the anisotropic immobilization of proteins and enzymes [22, 26, 27]. For example Cyt c, which is positively charged at pH 7 and contains a number of lysine residues surrounding the heme edge, adsorbs on negative charged surfaces that have been created by self-assembling carboxy terminated alkylthiols [28]. The conversion of the charge by, for example, a polycation enables the DET to negatively charged redox proteins such as adrenodoxin [29]. The distance between the functional head group and the electrode surface, and thus the distance over which the electron has to travel, can be controlled by the chain length of the alkyl spacer between thiol and terminal group [28, 30]. The non-covalent adsorption to the modified surface at low ionic strength keeps the protein mobile; the mobility of the protein is an important factor for a further coupling to a protein/enzyme partner [20, 31]. Carbon is the most suitable electrode material for investigation of peroxidases and oxygenases. Simple mechanical and electrochemical surface pretreatment generates hydrophilic surface oxide groups that facilitate electron transfer to peroxidases and other enzymes. Colloidal layers and surfactant films provide more hydrophobic surfaces, which may enhance DET to a number of enzymes, such as CYP2B4, nNOS and XDH [32–34]. Recent works also show the potential of nanoscaled material such as carbon nanotubes and nanoparticles because of their small size and remarkable electrocatalytic activity [35–38].

Progress has been made not only in surface chemistry, but also in engineering of proteins to achieve rapid and efficient DET reactions [39–45]. For example, the direct binding of cysteine(s) exposed from the protein shell or histidine tags to metall chelating agents on the electrode were described to create surface layers of proteins with distinguished orientation.These attachment sites can specifically be engineered in the protein. Novel methods for anisotropic immobilizing redox active membrane proteins, such as cytochrome c oxidase, combine the attachment of the protein to the electrode surface by engineered anchor groups on selected positions in the protein and reconstitution in lipid bilayers [17].

2.2 Proteins at Electrodes

In biological redox processes a vectorial electron transfer with a relatively small driving force is guaranteed by spatial arrangements and structural complementarity of the interacting protein partners. The charge is transferred either by redox mediators, or by direct intermolecular interaction of enzymes. Small redox proteins are often the carriers of the redox equivalents in electron transfer chains, and in biological systems are membrane associated, mobile (soluble) or associated with other proteins. Their molecular structure ensures specific interactions with other proteins and enzymes. The major classes of soluble redox active proteins are heme proteins, ferredoxins, flavoproteins and copper proteins. In most cases, they function as natural electron carriers to or between enzymes catalyzing specific transformations. Those enzyme proteins typically have redox active centers accessible to the outer surface on the protein and are therefore able to communicate with electrodes (Table 1). In addition, some proteins that are naturally not involved in redox processes but carry redox active sites (e.g. hemoglobin and myoglobin) show reversible electron exchange at functionalized electrodes.

A few of these proteins specifically interact with small signal molecules. This interaction has been used for biosensing in particular using c-type

Table 1 Classes of redox proteins, examples, some reaction partners and sensor application

Redox active group	Protein	DET	Sensor	Enzyme/protein partner	DET
Heme	Cyt c	+	H_2O_2, O_2^-, antioxidants	CCO, CCP, CDH, NR,	+, +, +, +
				SO, Cyt b2, Theophylline oxidase	+, +, +
	Cyt c_{550}	+		QH Amine dehydrogenase	+
	Cyt b_{562}	+		QH Glucose dehydrogenase	+
	Cyt c3	+		Fumarate dehydrogenase	+
	Cyt b_5	+		(Moco)SO, myoglobin	+, +
	Hemoglobin	+	NO, NO^{2-}, CN, CO, H_2O_2	Methemoglobin reductase	n.d.
	Myoglobin	+	NO, styrene, H_2O_2, halooganics		
Iron-sulfur complex (Ferredoxins)	Putidaredoxin	+		CYP101, Putidaredoxin reductase	+, n.d.
	Adrenodoxin	+		CYP11A1, 15-β-Hydroxylase,	+, +
				Adrenodoxin reductase	n.d.
	Ferredoxin	+		PS I	+
	Rubredoxin	+		Rubredoxin reductase	n.d.
Copper	Azurin	+		Cyt c3, *p*-cresolmethyl hydroxylase	+, +
	Plastocyanin	+		Cyt *bf*-complex, PSI	+, +
	Amicyanin	n.d.		Methylamine dehydrogenase	+
Flavoproteins	Flavodoxin	+		Reductase	n.d.
Thiol/disulfide	Thioredoxin	+		Thioredoxin reductase	+
				Ribonucleotide Reductase	n.d.

cytochromes, hemoglobin and myoglobin for the analysis of superoxide anion [46–51], NO [52–55] and antioxidants [49, 56]. The reaction is a feature of the central iron atom. A number of studies reported the coupling of the heterogeneous electron transfer of a redox protein to complex biological electron-transfer reactions, in which the redox protein acts as vectorial redox mediator [3, 20, 21, 31, 57, 58] and electrochemistry provides the tool for examining both kinetics and mechanisms. This combination results often in a more efficient bioelectrocatalysis of complex enzymes, as will be illustrated below (Sect. 5.2).

An efficient transfer of electrons from the redox electrode directly to the enzyme substitutes electron donors proteins. Among the roughly 3000 (wild type) enzymes currently known, of which about 1060 are oxidoreductases, only a small number of enzymes are capable of communicating directly with an electrode [19, 59–62]. The reason is found in the structure of the enzyme. To ensure high selectivity towards its substrate the catalytic center is deeply buried in the polypeptide, and thus it is almost impossible to transfer electrons directly to or from an electrode without any conformational change. The enzymes capable of being in direct contact (without additional redox relays) with electrodes are those proteins that have protein interaction partners. Examples of such enzymes interacting with proteins are Cyt c peroxidase (CCP) and Cyt c, Cyt c3 and azurin, and CYP101 and putidaredoxin (Table 1). If the electrode is designed to resemble the surface characteristics of the protein (put simply, the net charge or hydrophobicity) the enzyme may bind to such a surface without dramatic structural changes and if the distance between the redox center and the electrode is low enough direct electrochemistry may be observed [63]. Many of the redox enzymes for which efficient DET with electrodes has been observed are metalloenzymes (Table 2) and have macromolecular redox partners, or react on large substrates. The evidence for DET has not always been presented by direct electrochemical measurements, but also indirectly by measurement of a substrate dependent catalytic current, i.e. bioelectrocatalysis.

Various metabolites ranging from sugars such as fructose, cellobiose and gluconate [62], lactate [64], amines like methylamine and histamine [65], and drugs such as benzphetamine [66] can be measured with enzymes in direct contact with an electrode. The bioelectrocatalytic reduction of peroxide is one of the most important reactions for the determination of peroxide(s) in various media and also of substrates of coupled oxidases [60, 61].

3 Heme Proteins

Heme proteins are widely distributed in nature. Some of these proteins have oxygen carrier (hemoglobin) and storage (myoglobin) function, while many others are part of redox reaction chains, either as electron transfer mediators.

Table 2 Redox enzymes for which direct mediator-free reactions with electrodes have been shown

Enzyme	Prosthetic Group	Substrates	Electrode
Ascorbate oxidase	4 Cu,	O_2	Modified Au
Bilirubin oxidase	4 Cu	O_2	Modified Au, ITO, CNT
Laccases	4 Cu	O_2	CP, PG, G, SP, modified Au
Theophylline oxidase	Cu, unknown	O_2, Cyt c	PG
Superoxide dismutase	Cu–Zn Fe Mn	O_2^-	Modified Au
Glucose oxidase	FAD	Glucose	Hg, CNT
Putidaredoxin reductase	FAD, FMN	Putidaredoxin	HOPG, promotor
Pentachlorophenol hydroxylase	FAD	O_2, PCP	PG promoter
Photolyase	FAD	Oxidized nucleotides	G-DNA
Methylamine dehydrogenase	TTQ	Methylamine	Hg, Au, HOPG, PG, GC with various modifiers and promoters
Phospholipidhydro-peroxide glutathione peroxidase	Selenocysteine	Glutathione, H_2O_2	Modified Au
Catalase	Heme *b*	H_2O_2/O_2	GC
Cytochromes P450	Heme *b*	O_2, cholesterol, styrene, aminopyr-ine, benzphetamine, verampil, ...	Hg, HOPG, GC, Au with various modifiers, also nanoparticles
NOSynthase (oxygenase domain)	Heme *b*	O_2, arginine	PG, GC with surface modifier
Methane monooxygenase	Binuclear heme	acetonitrile, methane	Modified Au
Peroxidases	Heme	H_2O_2	Various carbon material – CP, PG, SP, CP, SP, CNT Hg, Au,
Multi center enzymes			
Amine oxidase	Cu, topa quinone	Amines	SP
Cytochrome c oxidase	Cu_A, heme a, heme a_3, Cu_B	O_2, Cyt c	Modified Au
Nitrite reductase	Cu Multi heme	Nitrite	PG, modified Au
Cellobiose dehydrogenase	FAD, heme c	Cellobiose, lactose	SP, PG, modified Au
p-Cresolmethyl-hydrolase	FAD, heme c	*p*-Cresol	PG promoter, modified Au

Table 2 (continued)

Enzyme	Prosthetic Group	Substrates	Electrode
L-Lactate dehydrogenase (flavocytochrome b_2)	FMN, heme b_2	Lactate	Carbon material
Flavocytochrome *c*3	FAD, heme c	Fumarate	HOPG
Fumarate reductase	FAD, Fe–S FAD, heme c	Fumarate/succinate	PG PG, promoter
D-Gluconate dehydrogenase	FAD, heme c, Fe–S	D-Gluconate	CP, PG, Au, Pt
Alcohol dehydrogenase	PQQ, heme c	Ethanol	CP, SP, G, Au/SAM
Aldose dehydrogenase	PQQ, heme	Aldose	G
D-Fructose dehydrogenase	PQQ,heme c	D-fructose	CP, Pt, GC
Amine oxidase	Cu, topa quinone	Amines	SG
Cytochrome c oxidase	Cu_A, Cu_B, heme a3,	O_2, cyt c	Modified Au
Hydrogenases	Fe–S and Ni, (Se), Fe–S	H_2, H^+, NAD	PG, HOPG
EcoIII	Fe–S	Oxidized oligonucleotides	G/DNA
Nitrogenase	Fe_4S_3, $-MoFe_3S_3$ (FeMoCo), 4Fe–4S, 8Fe–7S	N_2	PG, promoter
DMSO-reductase	Mo-pterin, Fe–S	DMSO	G
Sulfite oxidase	Mo-pterin Heme b_5	Sulfite	PG, Au/SAM
Sulfite dehydrogenase	Mo-pterin, Heme c	Sulfite	PG/DDAB
Arsenite reductase	Mo-pterin, 3Fe–4S 2Fe–2S	Arsenite	PG with promoter, PG/DDAB
Aldehyde oxidoreductase	Mo-pterin, 2Fe–2S	Aldehydes	PG, GC, Au with promoter
Xanthine oxidase	Mo-pterin, FAD, 2Fe–2S	Xanthine	PG
Xanthine dehydrogenase	Mo-pterin, FAD, 2Fe–2S	Xanthine	PG/DDAB

Abbreviations: PG – pyrolytic graphite, HOPG – highly oriented pyrolytic graphite, G – graphite, GC – glassy carbon, CP – carbon paste, SP – spectrographic graphite, CNT – carbon nano tubes, Au – gold, Hg – mercury, SAM – self assembled monolayer

e.g. Cyt b_5, Cyt c, or as catalysts, e.g. peroxidase, catalase, cytochrome P450-enzymes (CYP). Several enzymes contain heme(s) along with other cofactors such as flavin, copper, iron-sulfur cluster(s), molybdopterin or PQQ, where the heme domain has typically charge-transfer function. Examples of these multi-cofactor enzymes which have been coupled with electrodes are the

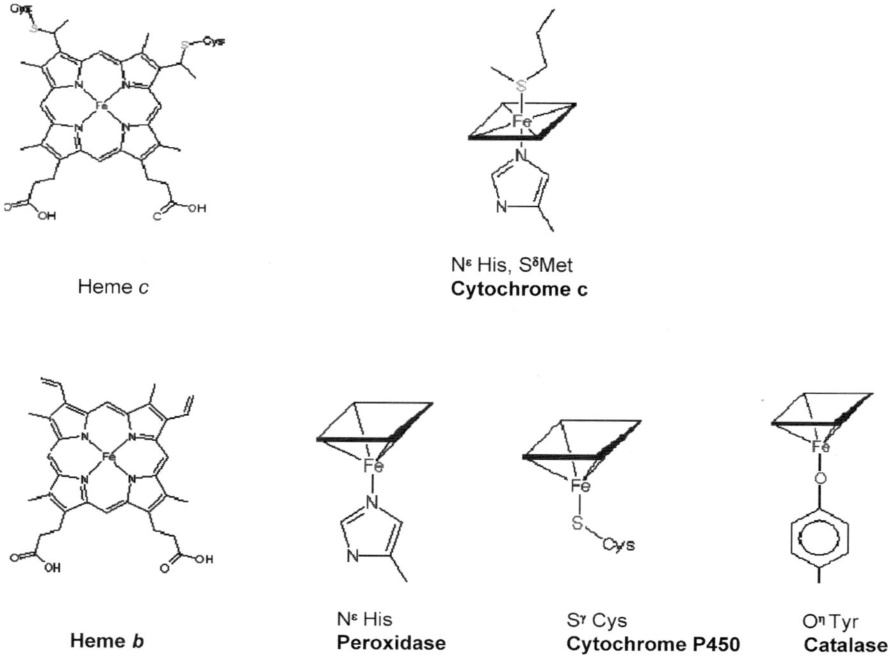

Fig. 2 Illustration of heme, heme coordination in cytochrome c, P450, catalase and peroxidase

flavohemoproteins flavocytochrome b_2, flavocytochrome *c*3, *p*-cresolmethyl hydroxylase, and cellobiose dehydrogenase, the molybdenum containing enzyme sulfite oxidase, and the quinohemoproteins fructose dehydrogenase, gluconate dehydrogenase and alcohol dehydrogenase (Table 2).

A prominent example of the copper-containing heme enzymes is cytochrome c oxidase.

Common to all heme proteins is the iron porphyrin prosthetic group (Fig. 2). Four of the six coordination positions of the heme iron are occupied by nitrogen from the porphyrin ring. The other two coordination positions on either side of the heme plane are available for further ligands which strongly influence the redox potential and reactivity of the heme protein [6].

3.1 Cytochrome c

Cytochrome c (Cyt c) is involved in electron transfer pathways such as the respiratory chain in the mitochondria, where it transfers electrons between membrane bound cytochrome reductase complex III and Cyt c oxidase. The active site is an iron porphyrin (heme) covalently linked to the protein at Cys14 and Cys17 through thioether bonds (heme c). The iron itself lies in the plane of the porphyrin ring, the two axial positions of the iron are occupied by sulfur of Met80 and nitrogen of His18 [67].

Cyt c from horse heart is a small globular protein of 12.4 kDa with 108 amino acids in one polypetide chain. At pH 7 Cyt c bears an overall positive charge of +7, due to an excess of basic lysine residues. There is also a large dipole moment because of the charge distribution on the protein. In the vicinity of the solvent exposed heme edge is a cluster of lysine residues whose charge is largely uncompensated. This domain is thought to be the interaction site for physiological redox partners. The rest of the heme is located in a hydrophobic environment that is responsible for the considerable positive redox potential (+260 vs. NHE) of Cyt c compared to isolated heme in aqueous milieu.

Various electrode modifiers allow a (quasi) reversible redox conversion of Cyt c (for review see [68, 69]). Among the many modifiers of gold electrodes, carboxyalkylthiols, in particular mercaptoundecanoic acid (MUA), have found most suitable to promote a fast electrode reaction of Cyt c. This promoter layer is negatively charged at neutral pH and thus able to interact with the heme environment of Cyt c. The interaction with the modified surface did not dramatically influence the redox potential of Cyt c, indicating that the protein kept its native configuration. For Cyt c in solution the heterogeneous electron transfer rate constants decreased with the layer thickness from 4×10^{-3} cm s^{-1} (cysteine) to 4×10^{-6} cm s^{-1} (MUA). The interactions with the surface and reaction kinetics have been studied in detail using various techniques, such as voltammetry, electroreflectance measurements and surface enhanced Raman resonance spectroscopy (SERRS) [70, 71]. For monolayers on gold assembled from long chain thiols of the structure HS-$(CH_2)_n$-COOH (with $n > 9$), the interfacial electron transfer rate decreased exponentially with chain length and the tunnelling parameter β was in the order of $\beta = 1.09$ per CH_2-unit [28, 30]. This shows that here the rate of electron tunnelling through the alkanethiol layer is limiting the overall electron transfer process between Cyt c and the gold electrode. For modifiers of shorter chains the electron transfer approaches saturation and experimentally determined rates were up to 3000 s^{-1} for adsorbed Cyt c. The strength of interaction limits the rate of rotation of the protein at the monolayer and therefore influences the reaction kinetics. Dilution of the carboxylic acid terminated thiols with methyl or hydroxy terminated alkanethiols (similar chain length) enhanced electron transfer rates by factors of 5–6 [50, 72, 73]. At 1 : 3 mixed MUA and mercaptoundecanol (MU) the ks increased to 70 s^{-1} compared to 23 s^{-1} for only MUA [74]. The adsorbed Cyt c was still capable for interprotein electron transfer to coupled enzymes. [19, 20, 31, 58]. Covalent coupling of Cyt c to the surface reduced the rate by a factor of 2–6 and shifted the formal potential slightly toward negative potentials. The electrode reaction was faster when Cyt c was coupled to mixed MUA/MU-layers with $k_{MUA/MU} = 3.1 \times 10^4$ M^{-1} s^{-1} [74].

Much higher loading of electroactive protein has been reached with multilayers of Cyt c and the weak anionic polyelectrolyte polyaniline sulfonate (PASA) formed on gold electrodes using the layer-by-layer (LBL)-approach [75]. In this case, the electron transfer has been proposed to involve protein-protein self-exchange. From a practical point of view, the highest loading of electroactive Cyt c with fast electrode reaction will enhance the sensitivity of Cyt c based sensors. Analytical application of this bioelectrochemical process has been reported for the determination of superoxide anion and antioxidants [19, 56, 75, 76].

Free radical biosensors may help to investigate the role of reactive oxygen species as cellular messengers by the direct, real-time measuring of free radical production directly as cell signal and in relation to stimuli to which the cell is exposed. Many assays for O_2^- exploit Cyt c, because the radical is quickly reducing the protein [3, 74] according to:

$$O_2^{-\cdot} + \text{Cyt c(FeIII)} \rightarrow \text{Cyt c(FeII)} + O_2 \,. \qquad (1)$$

In a sensor the immobilized Cyt c is subsequently oxidized at an appropriate electrode potential. Most stable sensors contain Cyt c covalently coupled directly to activated surface modifiers such as 3,3′-dithiobissulfosuccinimidyl-propionate (DTSSP) and DSP or by couplig to mercapto propionic acid/mercapto propanol (MPA/MP) or MUA/MU by activation with *N*-(3-dimethyl-aminopropyl)-*N*-ethylcarbodiimide (EDC) [31, 77].

The mixed monolayers of MUA/MU have the advantage of forming rather dense layers, and thus provide an efficient barrier for potentially interfering substances.

At a constant potential of +150 mV, the Cyt c electrode is sensitive to superoxide in the nanomolar concentration range. The sensitivity of the sensor is critically limited by the amount of electroactive protein. One way to avoid this limitation is the multilayer assembly of Cyt c on gold electrodes using sulfonated polyaniline as electrostatic glue between the Cyt c layers [75]. Changes of $O_2^{-\cdot}$ are indicated within seconds. The specificity of the sensor signal can be easily tested by additions of superoxide dismutase (SOD), which is the most effective natural scavenger of superoxide radicals. Well prepared sensors showed an instantaneous decline of the sensor current to the baseline.

Superoxide sensors were applied, for example, in various cell cultures and brain [78] and in vivo to follow the superoxide level during ischemia and reperfusion [48]. Another important field of application of superoxide sensors is the determination of antioxidative efficiency due to the increasing use of antioxidants in food products, drugs and cosmetics. This has been done simply by measuring the depletion rate of superoxide with the Cyt c sensor in the presence of antioxidants such as additives to cosmetic formulars [79], herbs and SOD [49]. For these assays it was necessary to generate the radical in situ. The integration of a superoxide generating enzyme, e.g. xanthine oxidase (XOD) and Cyt c, simplified the assay for antioxidant capacity [76].

3.2 Cytochrome c′

Cytochrome c′ (Cyt c′) is a dimeric heme protein found in phototropic, sulfur oxidizing and denitrifying bacteria [80]. The heme moiety is bound to two cysteins, similar to the binding found in mitochondrial Cyt c. However, the iron atom of Cyt c′ is mainly high spin pentacoordinated. The sixth ligand site of ferrous Cyt c′ is deeply buried within the protein and closely surrounded by aromatic and hydrophobic amino acid residues, restricting access to exogenous compounds except to small molecules such as carbon monoxide and nitric oxide [81].

The quasi-reversible redox responses of Cyt c′ from *Rhodospirillum rubrum* in solution was observed on a surface-modified gold electrode [82]. At a mercaptosuccinic acid modified gold electrode Cyt c′ from *Chromatium pinosum* and *Rh. Gelatinosus* could be adsorbed [52]. The adsorption was found to be dominated by non-ionic interactions, since adsorption could be observed at higher ionic strength (100 mM) and was facilitated at low pH (pH 5–6) where both the promoter and the protein became less charged. This electrode showed a quasi-reversible, diffusionless electrochemical redox behaviour of Cyt c′ with a formal potential of about –132 mV vs. Ag/AgCl and a heterogeneous electron transfer rate constant in the range of 30–50 s^{-1} [77]. The charge transfered was calculated from the CV to be 4.5 $\mu C\,cm^{-2}$ for the total amount of bound protein of 16 $pmol\,cm^{-2}$. These values support that both parts of the Cyt c′ molecule dimer are electroactive.

When the cyt c′ functionalised electrode was exposed to a NO, an increase in the reduction current was observed due to coordination and subsequent oxidation followed by an electrocatalytic reduction of the biomolecule. This can also be used in an amperometric mode for NO-biosensing [52, 77].

3.3 Myoglobin

Myoglobin (Mb) is an oxygen-carrying protein located in the muscles of vertebrates. It is a monomeric heme protein of about 16 kDa [83, 84]. The heme group is located in a crevice of the molecule and polar propionate side chains of the heme are on the surface. The heme plane is surrounded almost entirely by non-polar residues, except for two histidines. The native state of myoglobin features a proximal histidine bound to the fifth heme iron coordination site and the other axial heme iron position remains essentially free for oxygen-coordination. A distal histidine stabilizes a water-ligand to ferric iron, and in the ferrous state this conformation stabilizes bound oxygen as the sixth ligand and suppresses autoxidation of the heme [85]. The outside of the protein has both polar and apolar residues with an isoelectric point of 7.0.

Direct electrochemistry of Mb has been of interest for a long time, and numerous papers have been published and reviewed [2]. Among the electrode modification are various films on carbon electrodes, particularly surfactants, clay and polyelectrolyte films in which Mb exhibited direct electrochemistry for the redox transition of the heme group [86–90]. The role of surfactants in this process is suggested to be partly orientation of the protein and inhibition of oligomeric adsorbates. Higher protein loading can be created by alternating adsorption of protein and surfactant or polyelectrolyte in multilayers. The formal potential of myoglobin is sensitive to the nature of the film material used and also to the presence of heme ligands. For example for Mb films on carbon electrodes values were determined between –0.1 V and –0.21 V vs. SCE for the anionic surfactants didodecyldimethylammonium bromide (DDAB), sodium dodecyl sufate or cetyltrimethylammonium bromide [86, 90, 91], and –0.380 vs. SCE for montmorillonite [55, 87] and –335 mV for polyacrylamide [89] (values in pH 7.0 buffers). Intermediate values are found for combinations of surfactant and clay. For Mb in solution, a formal potential of –0.297 V vs. Ag/AgCl was reported [6]. A negative shift has also been found for Mb in bentonite and surfactant films [92]. The rates of electron transfer of Mb adsorbed to clay are fast at high ionic strength. In 200 mM Tris buffer k_s values of 450 s^{-1} were estimated [55]. The electron transfer rate is unusually fast for a protein that is naturally not involved in electron transfer. Together with the deviation of the formal potential compared with the value in solution, this strongly suggests conformational changes due to the interaction with the modifier. In addition, studies on His64 mutants of sperm whale Mb allowed the conclusion that a partly hydrophobic environment of a surfactant can disrupt the hydrogen bonding network of His64 and thus lower the reorganisation energy for the electron transfer [85].

Mb on electrodes possesses pseudo-enzymatic activity towards O_2, H_2O_2, nitrite and organohalides. Under aerobic conditions, oxymyoglobin (MbO_2) is also active in oxygenation reactions. Thus peroxide reduction, styrene oxygenation and dehalogenation of pollutants were followed electrocatalytically on Mb-modified electrodes. Furthermore, NO can bind to ferric Mb in the absence of oxygen and MbO_2 reacted rapidly and stoichometrically with nitric oxides [55, 89, 90].

3.4 Hemoglobin

Hemoglobin (Hb) functions physiologically in the storage and transport of molecular oxygen in mammalian blood. Human Hb is a 64 456 kDa protein and comprises four subunits (2 alpha and 2 beta chains), each of which has a heme within molecularly accessible crevices [93]. The heme is five-coordinated (four nitrogen ligands and one histidine), while the sixth site is available for binding oxygen. The electrochemistry of Hb has recently been comprehensively reviewed [94].

In general it is difficult to obtain DET between Hb and electrodes because of the large three-dimensional structure of Hb with the burried heme center and electrode passivation due to protein adsorption. Numerous efforts have been made to improve the electron transfer characteristics by using mediators or promoters [95]. The most efficient way was to modify preferentially carboneous electrodes with polymeric and membrane forming films of, for example, surfactants [25], clay [53], and composites of both. The redox potential of immobilized Hb was determined between −100 to −380 mV vs. SCE.

We could demonstrate that Hb adsorbes to a layer of colloidal clay nanoparticles prepared from sodium montmorillonite mixed with a hexachloroplatinate-polyvinyl alcohol colloid onto glassy carbon electrodes [53]. Hb displayed a quasi-reversible one-electron transfer process with an $Eo' = -370$ mV (vs. Ag/AgCl), $\Delta E_P = 130$ mV and $k_s = 70\,s^{-1}$. The linear dependence of the peak currents from the scan rate is indicative of a surface process. After introduction of CO, a positive potential shift was observed, which was used as indication that the DET of Hb was indeed from its heme iron. Because the affinity of ferrous heme for CO is about 200 times as high as for oxygen the addition of CO to the high-spin ferrous Hb yielded the heme ferrous-CO complex. This has a slightly higher redox potential. Other authors exploited the same reaction as an indicator for CO. Hb does also posess enzyme-like catalytic activity [96, 97], and therefore, common to all Hb-modified electrodes is their ability for electrocatalytic oxygen and peroxide reduction. Furthermore ferrous oxygenated Hb (HbO_2) reacts with NO to form nitrate and MetHb [98]. This reaction has been explored for peroxide and NO sensors.

The binding of NO to heme centers of Cyt c′ [52], Mb [55], and Hb [53] is the basis for NO biosensors.

Ge et al. developed an amperometric NO detector with Cyt c′ from *Chromatium vinosum* immobilized to mercaptosuccinic acid modified gold electrodes [52]. When the Cyt c′ electrode was polarized at −220 mV the current increases proportionately up to 500 nM NO. The electrode reaction was found to be rather fast with response times of a few seconds. The interaction of NO with Cyt c′ in solution was also confirmed spectroscopically. In another NO sensor, the reaction of Hb has been exploited [99]. The direct reaction between NO and HbO_2 with a rate of $k = 3.7 \times 10^7\,M^{-1}\,s^{-1}$ yielded nitrate and MetHb, which was electrochemically reduced and then again bound either oxygen or NO. In a sensor with clay immobilized Hb the NO-dependent MetHb-reduction current was accompanied by a cathodic peak shift when the NO concentration exceeded 50 nM. The shift of the cathodic peak potential is due to formation of nitrosylhemoglobin (HbNO) and was also observed and confirmed at a Hb/SnO_2-electrode [54]. The rate of NO binding to ferrous deoxyhemoglobin to form HbNO is $k = 2.5 \times 10^7\,M^{-1}\,s^{-1}$ [99].

The NO heme reactions predominate in the deoxygenated blood to form either Hb(III) or Hb(II)NO [100]. However, the presence of O_2 causes formation of *S*-nitrosohemoglobin [101] leading to a shift of the Hb reduction potential. Although this reaction (rate constant: $6 \pm 2 \times 10^6\,M^{-2}\,s^{-1}$) is not as fast as the oxidation of HbO_2 to MetHb, it may interfere in the determination of NO by causing a shift in reduction potential [100]. To avoid this problem, we used iodoacetate to block the reactive SH groups on the Hb. The modification did not change the formal potential but resulted in a slight increase in k_s to a value of $185\,s^{-1}$. A catalytic response on NO was observed with a detection limit of 0.5 μM NO [102].

Mb-modified electrodes respond in a similar way as Hb electrodes [55, 87]. In the presence of small amounts of NO a reversible redox couple appeared around −870 mV that is most likely the Fe^{II}/Fe^{I}-couple. At higher NO concentrations a catalytic reduction current is generated around −700 mV. It is assumed that this is the result of N_2O formation.

In oxygenated solutions, nitrate is formed and Mb(Fe^{II}) liberated can either bind NO or oxygen, which have reaction orders of the same order of magnitude. Thus in oxygenated solutions, sensitivity can be provided for micromolar NO concentrations [55]. In oxygen-free solutions, Mb/clay-modified electrode can be applied for the NO detection in the nanomolar concentration range by evaluating of the shift of the formal potential of the Mb(Fe^{II})/Mb(Fe^{III}) redox couple. Upon exposure of the Mb electrode to higher NO concentrations (> 2.5 μM), the reduction peak at −750 mV increased.

Nitrite is also catalytically reduced with Mb or Hb modified electrodes. A catalytic reduction peak current was found at −750 mV (pH 5.5). It increased with the concentration of NO^{2-} in the linear range of 0.17–3.2 mM [89].

3.5 Heme Oxygenases

Cytochromes P450 (CYP) and nitric oxide synthases (NOS) belong to the class of monooxygenases that catalyze the insertion of an oxygen atom into various compounds. Both are thiolate heme proteins, which are characterized by a very similar heme iron coordination sphere, where the proximal protoporphyrin IX axial coordination sphere is occupied by a cysteine (thiolate proximal ligand, fifth ligand). The sixth coordination position is occupied by different small molecules or atoms depending on the state of the protein within the reaction cycle [103, 104]. Despite some similarities, the protein structure, domain composition and the substrate specificities of NOS and CYP are quite different. But both are involved in very important physiological processes.

3.5.1 P450-Monooxygenase

CYP form a large family of heme proteins predominantly involved in drug metabolism and in the biosynthesis of steroids, lipids, vitamins and various endogeneous compounds. Many of these compounds are inducers for CYPs expression in different organs [103–105]. A variety of chemical reactions including epoxidation, hydroxylation and heteroatom oxidation are catalyzed by the monooxygenases.

The overall hydroxylation reaction of the monooxygenases is insertion of one atom of an oxygen molecule into a substrate RH, the second atom of oxygen being reduced to water while consuming two reducing equivalents under formation of ROH:

$$RH + O_2 + 2e^- + 2H^+ \rightarrow ROH + H_2O\,. \tag{2}$$

The single reaction steps have been investigated in detail [105, 106]. The resting enzyme is in the hexa-coordinated low-spin ferric enzyme with water beeing the sixth ligand. Substrate binding excludes water from the active site, which causes a change to the 5-coordinated high-spin state. The decrease of polarity is accompanied by a positive shift of the redox potential (about 130 mV), that makes the first electron transfer step thermodynamically favourable. The transfer of one electron from a redox partner reduces the ferric iron to the ferrous form. This can now bind molecular oxygen, forming a ferrous-dioxy (Fe^{II}-O_2) complex. The second electron is transferred, gaining an iron-hydroperoxo (Fe^{III}-OOH) intermediate. The O–O bound is cleaved to release a water molecule and a highly active iron-oxo ferryl intermediate. This intermediate abstracts one hydrogen atom from the substrate to yield the one-electron reduced ferryl species (Fe^{IV}-OH) and a substrate radical. Then follows immediately enzyme-product complex formation and release of product ROH to regenerate the initial low-spin state. The iron-oxo intermediates may also induce the formation of protein radicals [107]. The electrons for the reactions are naturally delivered from NAD(P)H by associated flavoproteins or ferredoxin-type proteins.

CYPs are highly relevant to the bioanalytical area. The main problem, however, is the complexity of the monooxygenase systems with the required electron supplying cofactors. Furthermore, oxygen and NAD(P)H are not only consumed in the P450-catalyzed substrate conversion, but also in a parasitic peroxide release without product formation (uncoupling). This makes it difficult to relate NAD(P)H consumption to substrate conversion. Direct cathodic reduction of the terminal oxidase in the presence of substrate can overcome the NAD(P)H dependent uncoupling. The generation of catalytic currents, i.e., increasing reduction currents upon addition of substrate, is a direct indicator of CYP dependent electrocatalysis.

Attempts to substitute the biological electron donators by artificial ones to reduce the terminal oxidase include electrochemical [32, 66, 108–110] and photochemical systems [111, 112] (Fig. 3).

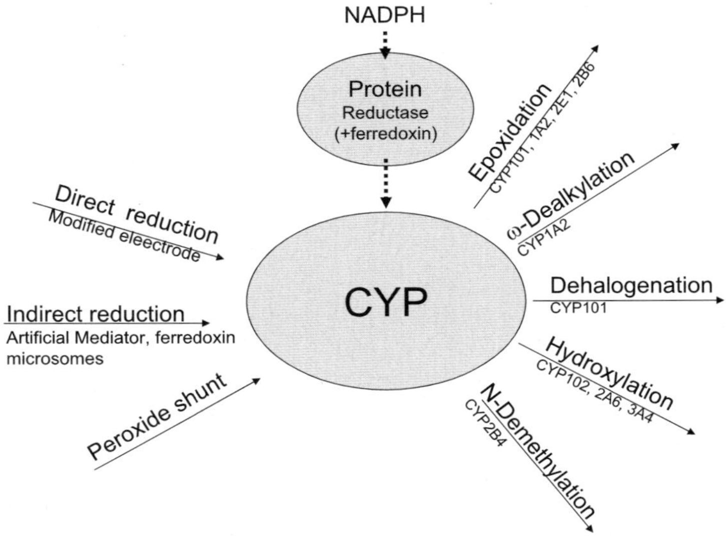

Fig. 3 Reaction types of CYP-bioelectrocatalysis and ways of introduction of reducing equivalents compared with the natural complete monooxygenase system with NADPH and proteins

Mediated spectroelectrochemistry based on the use of antimony-doped tin oxide electrodes has been used to determine the influence of mutations on the redox potentials [109]. The ultimate approach is the direct unmediated electron supply from a redox electrode to the heme iron of CYP. The first paper on DET to CYP reported a reduction of solubilized CYP from rabbit liver at a mercury electrode at –580 mV (versus SCE) [12] and catalytic currents were obtained upon addition of substrates such as benzphetamine [113] and deoxycorticosterone [114]. However proteins adsorb irreversibly at mercury and therefore the signals are impeded by partially denatured protein.

Only a very few more observations of direct redox-reaction of CYP on unmodified electrodes were published. CYP2E3 could be directly reduced at GC electrodes at a heterogeneous electron transfer rate of 5 s^{-1}, indicating that the enzyme was adsorbed on the electrode surface but the electron transfer was restricted [115]. Reversible direct one-electron transfer has been reported for CYP101 [116] at edge-plane graphite (EPG) by using CYP101 in solution. The authors proposed an interaction of the graphite surface with positively charged Arg-72, Arg-112, Arg-364 and Lys-344 residues on the protein surface. The formal potential of camphor free CYP101 was –526 mV (versus SCE). Addition of camphor caused a positive shift of the potential. Later the same group showed for mutant enzymes where cysteins were replaced by inert alanins that electron transfer is not affected by these cystein residues [117]. The introduction of surface modifications enabled DET for various different CYP isoenzymes (Table 3).

Most CYPs are membrane associated, and therefore the incorporation into surfactant and lipid-films on electrodes facilitate a heterogeneous electron transfer [110, 118, 119], while for the soluble bacterial enzymes hydrophilic charged layers of colloids and polyelectrolytes are also effective [32, 120–122]. Importantly, fast electron transfer is, with a few exeptions, only possible when carboneous electrode material was used. Modification of CYP with polyethyleneoxide (PEO) improved the thermostability as was reported for CYP119A2 from thermostable microorganism [123, 124].

On gold electrodes with self-assembled modifier layers, the reduction of P450 is very slow [125]. In a spectroelectrochemical experiment at modified gold for substrate-bound CYP101 $E^{0\prime}$ of –373 mV was determined. Although the electron transfer was very slow, the surface interaction and direct electrochemical transformation did not effect the enzyme structure as was confirmed spectroscopically. Both, upon direct electrochemical reduction and upon ligand binding the spectral changes clearly indicated the native state of CYP101 during reversible redox reaction at modified electrodes [125]. Reduction of CYP has also been reported for modified gold electrodes with alternating CYP101/polycation multilayers [126].

For the non-mediated first electron-redox transition in a number of CYPs and at different electrodes, the redox potential is in the range of –450 mV to –100 mV vs. Ag/AgCl. For example, at montmorillonite modified glassy carbon electrodes, reversible and very fast heterogeneous reduction of immobilized substrate-free CYP101 with a formal potential of –361 mV (vs. Ag/AgCl) was obtained [120]. The heterogeneous electron transfer rate constants reached values as high as 152 s^{-1} compared with rates between 27–84 s^{-1} reported for the transfer of the first electron from putidaredoxin to CYP101 [127, 128]. This similarity suggests that the intercalation into the negatively charged clay structure [129] may hold the CYP in a productive orientation without deterioration of the enzyme. The active site of the immobilized CYP101 is still accessible for small iron ligands like CO and dioxygen and also the larger metyrapone, as is indicated by a positive shift of the formal potential after its addition. The apparent surface coverage of the electroactive CYP101 was less than a monolayer.

The formal potential of substrate-free CYP101 is approximately 160 mV more positive than the solution value, but close to the value of the camphor-bound species in solution ($E^{0\prime}$ =– 407 mV vs. Ag/AgCl 1M KCl). Thus the interaction with the matrix may force displacement of solvent in the local environment of the heme or conformational changes. Changes of the secondary structures, however, were not identified with IR-spectroscopy [120].

Table 3 Survey of CYP-enzymes in DET contact

CYP species	Electrode/ modification	Formal potential [mV], SCE	Comments	Refs.
CYP1A2	Carbon cloth/ Multilayers with PSS	−310	Styrene epoxidation, electrolysis at −600 mV. Peroxide shunt, catalase inhibited catalysis	[118]
CYP1A2 **CYP2B4**	Screen printed graphite/DDAB		Au nanoparticles enhance DET, catalysis, inhibition of catalysis shown	[142]
CYP2B	Hg	−580	Demethylation of benzphetamine, *p*-nitroanisole, aminopyrine Hydroxylation of deoxycorticosterone	[113, 114]
CYP2B4	GC/clay/detergent	−292– −305	Demethylation of aminopyrine, benzphetamine, product analysis $\Gamma = 40.5\ \mathrm{pmol\,cm^{-2}}$, $k_{cat} = 1.54\ \mathrm{min^{-1}}$	[66]
CYP2C9	Au/chelate		His-tag protein, reversible 1 electron transfer, binding of torsemide, CO, O_2	[143]
CYP2D6	Polyaniline doped GCE	−120	Anodic shift of $E^{0\prime}$ in the presence of fluoxetine, catalysis, $K'_m = 3,7\ \mu\mathrm{mol/l}$	[144]
CYP2E1	GC, Au	−334	Catalysis on Au, *p*-nitrophenol oxidation	[115]
CYP2E1	Bare, GC/DDAB Au/HS-x	−334	*p*-nitrophenol oxidation	
CYP3A4	Au–MPS–PDDA multilayers	−342 (CV) −335 (SWV)	Dealkylation of verapamil and medazolam Amperometry at −500 mV vs. Ag/AgCl. $K'_m = 271-1082\ \mu\mathrm{M}$	[119]
CYP11A1	Langmuir Blodgett films (mono- and multilayers), ITO, glass	−295 to −318	$k_s = 0.45\ \mathrm{s^{-1}}$, $E'_0 = -470$ mV vs. Ag/AgCl. Binding of cholesterol seen. Binding also characterized with X-Ray, QCM, CD, Ellipsometry, Brewster angle microscopy	[147]

Table 3 (continued)

CYP species	Electrode/ modification	Formal potential [mV], SCE	Comments	Refs.
CYP101	EPG	–526 (ls) –390 (hs)	Binding of camphor seen as shift of potential upon camphor addition	[116]
CYP101	Clay – GC	–368	$\Gamma = 3.54$ pmol cm^{-2}, $k_s = 5–152$ s^{-1}	[120]
CYP101	PG/DMPC PG/DDAB	–357 (DMPC) –250 (DDAB)	$\Gamma = 7.2$ (DMPC), 4.9 mol cm^{-2} (DDAB), $k_s = 25$ (DMPC), 26 (DDAB) s^{-1}. Catalytic reduction of O_2 and TCA	[122]
CYP101	GC/DDAB GC/BSA-glutardialdehyde	–260	Amperometry, reduction currents after addition of camphor, adamantanone, or fenchone K'_m 1.41–3.9 mM	[121]
CYP101	Au-MPS/Multilayers with PEI, PDDA, PSS	–250	Catalysis with styrene, Au-MPS-(PEI/PSS-CYP101) pH 5.2, PDDA instead of PEI	[126]
CYP101	PG/Multilayers with PEI	–250	Styrene epoxidation, $\Gamma = 0.1$ nmol cm^{-2}, turnover rate = 6.3 h^{-1}	[138]
CYP101	PG/PEI multilayers	–270	Catalysis with O_2 and H_2O_2 $\Gamma = 0.15$ nmol cm^{-2}, $k_s = 5–152$ s^{-1}	[221]
CYP101	Au-MPS/PDDA, DDAB, multilayers carbon cloth	–250	CYP101 immobilised and in solution Styrene and *cis*-β-methylstyrene oxygenation, turnover rates higher in multilayers that in solution	[137]
CYP101	Au/dithionite and aldrithiol	–380	Native state of enzyme during electrolysis	[125]
CYP101 wt and mutants	EPG Au	–428 to –449	Surface Cys replaced with Ala; no effect on electron transfer	[117]
CYP101 wt CYP101K334C	Polycrystalline Au		CYP101K334C binds onto Au, DET is faster than for wild type, STM, QCM	[218]

Table 3 (continued)

CYP species	Electrode/ modification	Formal potential [mV], SCE	Comments	Refs.
CYP102	PG/DDAB/PSS	–330	O_2 -reduction	[134]
CYP102	EPG/DDAB	–188	$k_s = 221\ s^{-1}$, O_2- and H_2O_2-reduction	[145]
CYP102	Graphite/DDAB	–260	Spectroscopy reveals P420 in DDAB Pyrene wired BM3 domain	[217]
CYP102	Graphite/DDAB/ PSS	–330 (wt) –193 (mutant)	$k_s = 30\ s^{-1}$, (wt) $k_s = 250\ s^{-1}$ (mutant BM3) O_2-reduction	[218]
CYP119	Carbon/DDAB	–250 vs. SSE (20 °C) –50 vs. SSE (80 °C)	Electrochemistry also at 80 °C. ΔE (20 °C) = 90 mV, ΔE (80 °C) = 30 mV	[201]
CYP119A2	Carbon electrode/PEO	–120 vs. Ferri/ Ferrocyanide	Thermostable up to 120 °C, $k_s = 35.1\ s^{-1}$	[124]
CYP119A2	Carbon electrode	–103 (wt); –114 (PEO-mod) vs. Ferri/ Ferrocyanide	PEO2000 modified enzyme, thermo stabilized	[123]
CYP176A1	EPG/DDAB	–360	1-e^-- transfer to $Fe^{II/III}$	[146]

The liver microsomal phenobarbital induced CYP2B4 has also been incorporated in montmorillonite on GC-electrodes [66]. In contrast to CYP101 this enzyme has a flavoenzyme as redox partner and does not need a ferredoxin for electron supply. Using CV at low scan rates a reduction peak was observed at around −430 mV (vs.Ag/AgCl), which disappeared at higher scan rates. The electron transfer reaction is obviously very slow. However this process has been enhanced by the non-ionic detergent Tween 80 because CYP2B4 is a membrane bound enzyme and is monomerized by the detergent [130] as was confirmed also by AFM-studies. The amount of 40.5 pmol cm^{-2} electroactive protein was calculated from the CVs. The voltammetry also revealed a reversible one-electron surface redox reaction with a formal potential of about −295 mV vs. Ag/AgCl and heterogeneous electron transfer rate constant of 80 s^{-1}. As in many other published cases (e.g. [122]) where surfactants have been employed, the formal potential of a CYP at the electrode is more positive than the potential in solution and at the electrode without the detergent.

A positive shift of the redox potential may be an indication of low to high spin-state conversion ascribed to strong interaction of CYP with surfaces [70]. Positive shifts of the redox potential are generally observed when water is excluded from the heme pocket, as in the case of camphor binding [131, 132], and therefore we suggested that the surfactant leads to a dehydration of the heme environment. This is supported by combined spectroscopic and electrochemical characterization of CYP102 in DDAB, where it was found that BM3 adopts a compact conformation within DDAB that in turn strengthens hydrogen bonding interactions with the heme axial cysteine producing a P420-like species with decreased electron density around the metal center [133, 134]. Recent SERRS measurements showed also substantial alteration of the protein structure of CYP101 at SAM-modified Ag-electrodes induced by the electrostatic interaction [135].

The studies of direct heterogeneous electron transfer have been carried out in most cases by CV and SWV, where almost exclusively first of the two electrons required for the catalytic reaction has been transferred.

CYP101 predominantly catalyzes the regio- and stereo-specific hydroxylation of (1*R*)-camphor to exclusively 5-exo-hydroxycamphor. Other compounds than camphor, such as compounds of environmental and industrial interest, have also been identified as substrates for CYP101. On indium-tin oxide electrodes, CYP101 conducts camphor hydroxylation mediated by putidaredoxin [57] and dehalogenation of haloalkanes with spinach ferredoxin in the presence of polylysine as promoter [136].

Using only the monooxygenase, electrochemically driven styrene epoxidation was observed at polyion-multilayers containing CYP101 [137] or CYP1A2 [118]. The mechanism involves a peroxide activated reaction step. In an electrocatalytic oxygen reduction, peroxide is formed, which activates the enzyme for styrene epoxidation. The role of peroxide has been proved by lack of styrene oxide formation in the presence of catalase [118, 138]. Acceleration of styrene epoxidation and dehalogenation of hexachloroethane, carbon tetrachloride and other polyhalomethanes was successful with mutated CYP101 [109, 139, 140].

Human CYP3A4 was electrocatalytically active at a polycationic film-loaded gold electrode prepared with poly-dimethyldiallylammonium chloride (PDDA) adsorbed to mercaptopropane sulfonate activated gold [119]. This immobilization caused a drastic anonic potential shift of enzyme to about +98 mV (vs. NHE), indicating conformational changes. This electrode was used for drug sensing. Addition of millimolar amounts of verapamil or midazolam to oxygenated solution increased the reduction current. The product analysis after electrolysis at –500 mV under aerobic conditions confirmed the demethylation and dealkylation of verapamil at a rate of about 4 to 5 min^{-1}.

We succeeded in CYP2B4 bioelectrocatalytic demethylation by immobilizing surfactant monomerized CYP2B4 in montmorillonite [66]. When aminopyrine was added to air saturated buffer solution, there was an increase in the reduction current. In chronoamperometry the detection limit was 1 mM of aminopyrine and 1.2 mM of benzphetamine. Evidence for demethylation was delivered by product (formaldehyde) analysis after 1 h of controlled potential electrolysis at –500 mV vs. Ag/AgCl. The apparent catalytic rate related to the amount of electroactive protein was $k'_{cat} = 1.54\ min^{-1}$ comparable to the value $k_{cat} = 3.5\ min^{-1}$ of the microsomal system [110]. Furthermore, the reaction was inhibited by metyrapone, which was also indicative of the electrocatalytic activity of CYP2B4. Another recent work demonstrated a promoting effect of gold nanoparticles on the bioelectrocatalysis of steroid hydroxylating P450 on rhodium graphite electrodes [141]. In summary, preferentially when CYP enzymes were immobilized with polyelectrolytes and surfactants electrocatalytic enhancement of direct reduction current was measured upon addition of substrates and oxygen.

Figure 3 summarizes the observed reaction types and illustrates how the reaction is related to electrode processes. Only two examples describe hydroxylation of substrates. Other catalyzed reactions are demethylation, dealkylation, dehalogenation and epoxidation. Many of the reactions may use peroxide, which is always formed in side reactions.

3.5.2
Nitric Oxide Synthase

Nitric oxide (NO) is involved in neurotransmission, cardiovascular regulation, immune host-defense and many other (patho)physiological processes. In vivo NO is produced by nitric oxide synthase (NOS). The family of NOS enzymes is divided into three isoforms characterized by different localization, regulation and catalytic properties with 51–57% homology between the human ones. The isoforms are known as: nNOS (type I) in neuronal tissue, iNOS (type II) being the isoform that is inducible in a wide rage of cells and tissue and the eNOS (type III), first found in vascular endothelial cells [148, 149].

All NOS isoenzymes catalyse the formation of NO and L-citrulline from L-arginine via *N*-hydroxy-L-arginine (NHA) in two NADPH dependent oxygenation reaction.

$$\text{L-Arg} \xrightarrow{\text{NADPH}+\text{O}_2-\text{H}_2\text{O}} \text{NHA} \xrightarrow{0.5\text{NADPH}+\text{O}_2-\text{H}_2\text{O}} \text{L-Citrulline} + \text{NO}\,. \quad (3)$$

The known NOS enzymes are flavocytochromes, which exhibit a bi-domain structure, in which an oxygenase domain is linked by a calmodulin recognition site to a reductase domain. The C-terminal reductase domain contains flavin adenine dinucleotide (FAD), flavin mononucleotide (FMN) and binding sites for NADPH. The oxygenase is a thiolate heme protein with a cysteine axial heme-iron ligand and binding sites for (6*R*)-5,6,7,8-tetrahydrobiopterin (H_4B) and L-arginine.

The mechanism of the first turnover from L-arginine to NHA is a P450-type monooxygenase reaction. It starts with substrate binding to the resting ferric heme of the oxygenase and introduction of the first electron followed by oxygen binding and a second reduction to heme peroxo and heme ferryl species [150] consecutively followed by the formation of a highly reactive iron-oxo intermediate (similar to compound I) [151] that directly attacks L-arginine to produce NHA. In the second reaction cycle, NHA binding is followed by a single electron reduction of the ferric heme. Subsequent oxygen binding generates the oxy-complex. The reaction with NHA generates a ferric hydroperoxy complex and a NHA radical. The NHA radical is rebound and reacts with another NHA radical to NO, L-citrulline, water and the ferric heme protein [148].

Unique to the NOS is that it requires H_4B for its activity. Among the several important functions of H_4B are allosteric effects for substrate binding, promotion of dimer formation, and stimulation of low to high-spin heme transition. H_4B supports also coupling between electron transfer and NO-synthesis and inhibition of superoxide and peroxide formation and is probably also involved in the transfer of single reducing equivalents [150, 152–155]. It has been proposed that NOS uses H_4B as an electron donor because it provides the second electron to the heme ($Fe^{II}O_2$-intermediate) at a speed that keeps oxygen activation coupled to substrate oxidation [150, 154, 156].

Interestingly, the separate domains are catalytically active and the NOS can be reconstituted by combining reductase and oxygenase domains of different origin [148].

Using spectroelectrochemical titration, the midpoint potentials of the ferric heme iron in combination with different binding partners have been determined [152, 157]. For iNOS upon binding of H_4B and/or arginine, the reduction potential shifted (from −347 mV (vs. NHE)) 50 to 110 mV positive and thus made subsequent electron transfer from the flavoprotein thermodynamically more favoured. For nNOS, however, only a minor effect of arginine binding on the redox potential was observed. The resting enzyme has already a 100 mV higher redox potential than the iNOS. Almost similar values for the midpoint potentials of −180–210 mV (NHE) have been determined for iNOS, nNOS and eNOS using potentiometric titration [158].

Efforts have been made to substitute the reductase domain by an electrode to study the redox behaviour of the oxygenase and to generate NO without the need for NADH. Up to now only immobilization in a bilayer of the surfactant DDAB [159, 160] and in colloidal layers [33] on graphite and glassy carbon have been successful for direct electrochemical investigation of NOS.

Bayachou et al. [159] embedded nNOS oxygenase domain (nNOSoxy) in DDAB onto a pyrolytic graphite (PG) electrode. The forward and backward square wave voltammogram showed reversible redox couples at −197 mV (vs. Ag/AgCl) for heme $Fe^{II/III}$ and −1092 mV (vs. Ag/AgCl) for heme $Fe^{I/II}$. Accompanied UV-VIS studies did not indicate changes in the structure of the enzyme. The presence of H_4B shifted the reduction potential by 100 mV in a positive direction, indicating the higher affinity of H_4B for the reduced (Fe^{II}) form of nNOSoxy. They also determined for the Fe^{II}/Fe^{III}-couple a pH-dependency of −54 mV $E_{1/2}$/pH, which is typical for a proton coupled electron transfer reaction.

Also for iNOS in DDAB on basal plane graphite electrode two redox couples were observed at −191 mV and −1049 mV (vs. Ag/AgCl), potentials similar to the values of nNOSoxy in DDAB [160]. With redox titrations of iNOS in solution, they assigned a potential of −544 mV for heme $Fe^{II/III}$. The immobilization in the surfactant layer resulted in anodic shift of several hundred millivolts, as was also observed for P450-enzymes in DDAB. Binding of CO and imidazole produced shifts of +62 mV and +20 mV, respectively, for the $Fe^{II/III}$-couple. Furthermore, the authors observed an increase of a second reduction peak with increasing scan rates while the original reduction peak decreases with increasing scan rate. The two reduction peaks differ by approximately 133 mV, the typical potential difference between five and six coordinated heme iron in iNOS. From the pH-dependence and the scan rate dependence of these reduction peaks it has been deduced that the two peaks result from water-free Fe^{III} and water-bound Fe^{III} [160]. At high scan rates only one peak is observed, because the reduction of water-free Fe^{III} occurs faster than water ligation. Conversely, scanning slowly allows water enough time to ligate Fe^{III} before reduction. $E^{0\prime}$ and k_s for the 5-coordinated heme were experimentally determined to be −150 mV and 370 s^{-1} and for the 6-coordinated heme −250 mV and 10 s^{-1}. This potential shift is often observed for heme proteins immobilized in surfactant films [66, 122] and is most likely due to dehydration [132]. It was shown for P450-enzymes [120] and for nNOS [33] that the effect is much smaller when clay-film, such as Na-montmorillonite, was used to modify the electrode surface.

Rapid direct heterogeneous electron transfer was achieved in the absence of mediators by immobilizing nNOSoxy in a colloidal layer onto a glassy carbon (GC) electrode [33]. For this study, the wild type oxygenase domain (wt nNOSoxy) and a deletion mutant (D290nNOSoxy) were investigated. The oxygenase domain nNOSoxy, containing at the N-terminus an additional protein stretch domain (PDZ) that is linked via a small peptide stretch (PIN) and a deletion mutant (D290nNOSoxy), where the PDZ plus PIN (residues 1–290) are missing, have been expressed and purified [151]. The clay colloid has been composed of sodium montmorillonite colloid, poly(vinyl alcohol) (PVA) and colloidal platinum (Pt) as in [120].

From CV and SWV of wt nNOSoxy in a clay colloid [33], the reversible oxidation and reduction peaks with a midpoint potential of –346 mV (vs. Ag/AgCl) at 12 V s^{-1} were estimated and assigned to the heme Fe^{II}/Fe^{III}-couple. A slightly higher formal potential of –297 mV was found for the truncated protein (Fig. 4a). One possible reason is that the reduction of the protein shell by 30 kDa may account for a decrease of the electron transfer barrier and thus in an increased reduction potential. The peak separation for wt nNOSoxy of 53 mV and $|E_p - E_{p/2}|$ of 62 mV is typical for a reversible one electron transfer. Integration of the reduction and oxidation peaks revealed amounts of electroactive protein of 4.7 ± 3.0 pmol cm^{-2} (surface area of 0.07 cm^2), which is in the order expected for a monolayer. The peak currents were proportional to the scan rate, which is an indication for a surface process. From the scan rate dependence of the CV the heterogeneous electron transfer rate constant of the first reduction was calculated to be $k_s = 316\ s^{-1}$ using the theory of Laviron. The subsequent electron transfers were slower. The kinetics of ferrous-dioxy complex-formation rate of nNOS has been found to be 81 s^{-1}, the transition to the ferric enzyme 14.5 s^{-1} and the rate of $Fe^{II}O_2^-$ disappearance is 5 s^{-1} [156]. This means that the introduction of the first electron to reduce the ferric heme iron seems not to limit the reaction. In the presence of oxygen a pronounced catalytic reduction current is observed. The reduction peak increased and the oxidation peak disappeared completely. This can serve as evidence of the catalytic activity of the oxygenase, as it is known that reduced ferrous nNOSoxy can bind dioxygen with high efficiency [156].

In the presence of H_4B and NHA in the measuring solution, an additional reduction wave evolved at around –720 mV in the SWV [33]. This reduction could be due to the formation of NO. When a small amount of NO gas that was chemically generated from $NaNO_2$, $FeCl_2$ and H_2SO_4 was added to the electrochemical cell, a sharp reduction peak at –725 mV appeared in the SWV. The position of the peak may allow the conclusion that NO has been formed and a direct reduction of the heme by the electrode and/or a reduction mediated by H_4B is possible. Initial experiments with a NO-sensor and at constant potential electrolysis at –550 mV further support this assumption [to be published]. In addition no NO-formation was observed when H_4B was

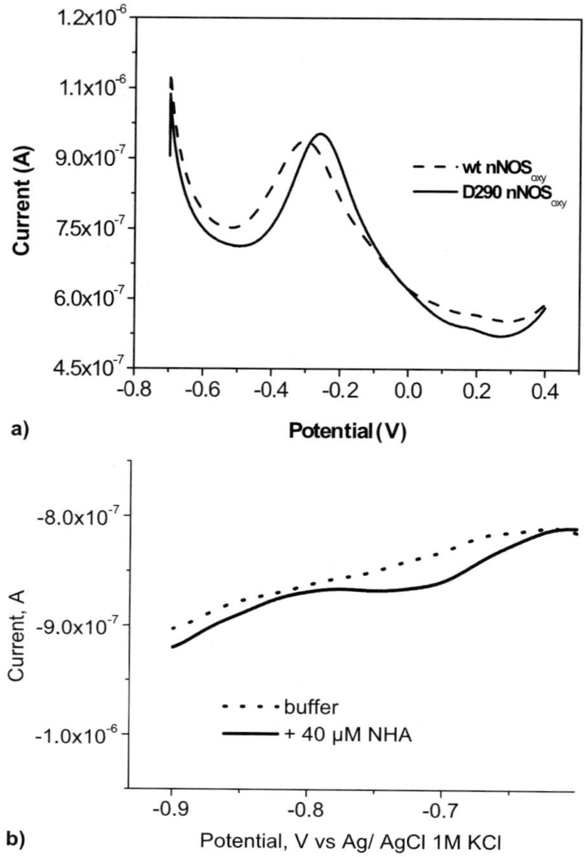

Fig. 4 **a** SWV (1 mV increment, 50 mV amplitude, 10 Hz frequency) of nNOSoxy and D290 nNOSoxy immobilized in montmorillonite on a GC electrode. Measurements in argon-purged 0.1 M K-phosphate buffer, 0.3 M NaCl, pH 7.4 and **b** section of the SWV of D290 nNOSoxy clay GG electrode with 24 μM O_2 in the presence of 16 μM H_4B (**a**) and **b** after addition of 40 μM *N*-hydroxy arginine (after [33])

missing. Thus, it was shown that H_4B may act as a mediator and activator of the electrode-driven catalysis of nNOS.

4 Peroxidases

The majority of peroxidases are glycoproteins of 20–70 kDa molecular weight, which contain ferric protoporphyrin IX as a prosthetic group [161]. Peroxidases (EC 1.11.1.7) catalyze the removal of hydrogen peroxide or alkyl hydroperoxides, while a wide range of substrates act as electron donors. The mechanism of peroxidase catalyzed reactions has been intensively studied and reviewed [161–165]. The reaction kinetics reveal a ping-pong mechanism. In the first step, the peroxide binds to a free coordination site of the resting heme iron (Fe^{III}), where it is reduced to water (or an alcohol ROH) in a rapid two-electron process, whereby compound I is formed as the reactive primary intermediate. Compound I is the oxy-ferryl species (($Fe^{IV} = O$) $P^{\cdot +}$) formed with one oxygen atom from the peroxide, one electron from iron and one electron withdrawn from the heme group to form a porphyrin π cation radical (or a protein radical in the case of chloroperoxidase and cytochrome c peroxidase). In the next step, the cation radical is reduced by a one-electron donor to give compound II (($Fe^{IV} = O$) P), which is subsequently reduced by a second one-electron donor molecule back to the resting ferric enzyme. The last two steps are accompanied by the uptake of two protons and release of a water molecule. Formation of compound I is fast with apparent second order rate constants in the order of 10^6–10^7 M^{-1} s^{-1} [162], the reduction of compound I is 10–100 times faster than of compound II. Depending on the nature of the peroxidase, a number of compounds reduce the higher oxidation state intermediates of the enzyme back to its native form. Hydroquinone and *o*-phenylene diamine are some of the best electron donor substrates for HRP with almost diffusion limited reaction rates of 3×10^6 M^{-1} s^{-1} and 5×10^7 M^{-1} s^{-1} [166, 167]. HRP is also active in a number of organic solvents.

For an appropriately immobilized peroxidase on an electrode, compound I may be reduced by the electrode. Thus the electrode can be considered as substrate of the enzyme, and peroxide is electrocatalytically reduced. This has been demonstrated for peroxidase from horseradish and from many other sources [60, 168, 186].

Observation of non-turnover direct electrochemistry of compound I and II have not been reported. Most studies on non-turnover DET of peroxidases involve the redox transformation of the $Fe^{II/III}$-couple, which is observed in a rather negative potential region of < -0.1 V vs. Ag/AgCl. Formal potentials of HRP Compound I/II and Compound II/POD(Fe^{III}) redox couples are much more positive, i.e. close to 0.7 V (vs. Ag/AgCl). Therefore catalytic peroxide reduction currents are measured at very small overvoltages.

In most publications carboneous material (e.g. [10, 170–172]) and gold (e.g. [40, 173–175]) was used and the oxygen functionalities of the carbon material were suggested to be the reason for the efficiency of the DET process on such surfaces [171, 176]. The material ranges from carbon black, graphite powder, spectroscopic and pyrolytic graphite, epoxygraphite to carbon fibers and fullerens to which peroxidases are physically adsorbed, electropolymerized [169] or covalently coupled. Simple thermal pre-treatment is often sufficient to activate the carbon material for adsorption and reaction with peroxidases. When incorporated into a carbon paste, it is important to adsorb the enzyme before binder material, usually oil or grease, is added [170, 172]. Various additives have proved to enhance the sensitivity of the composite electrodes. Among the most effective are polyelectrolytes and polyalcohols such as polyethyleneimine and lactitol. HRP incorporated into electrodes of pyrolyzed organic material catalyze also electroreduction of H_2O_2 [177]. Gold has been applied as gold colloid [174] or as disk and wire electrodes [173, 175, 178]. The rate constant for the heterogeneous electron transfer of native HRP on graphite and modified gold electrodes was 1–2 s^{-1} and was enhanced when the non-glycosylated HRP is applied [40]. The glycosylation obviously restricts the DET. This has been also supported by the observation that native HRP bound via the sugar moiety to boronic acid activated gold surface catalyzed peroxide reduction only in the presence of soluble mediators [179]. Non-glycosylated (recombinant wild-type) HRP adsorbed on graphite reached values of 7.6 s^{-1}. The rate of the recombinant wild-type HRP $k_s = 18\ s^{-1}$ [173] was higher at freshly cleaned gold electrodes, and could be further increased when terminal His-tags were introduced to the protein. In addition, the catalytic rate was higher for the recombinant enzymes adsorbed on gold compared to graphite electrodes and native HRP. Systematic studies on how the DET properties of HRP can be influenced involved electrode material and pH [180], single point mutations and glycosylation [181], cystein mutants and introduction of a (His)6-tag [40, 41, 173]. Further investigations also included other plant and fungal peroxidases adsorbed onto graphite and modified gold electrodes [98, 178, 183, 184]. The transfer of electrons from a SAM–modified gold electrode to tobacco peroxidase is much faster than to HRP. The most effective DET of the native enzyme was reported for sweet potatoe peroxidase where 91% efficiency of DET was reached [40]. Higher efficiency was demonstrated only for the HRP active site mutants Asn70Val and Asn70Asp, which showed an almost 100% effective DET electron transfer [186]. This behavior could lead to more selective peroxide biosensors. A model of the DET kinetics in peroxidase electrocatalysis can be found in [185].

The reaction of peroxide with ferrous heme iron is the basis of electrocatalytic peroxide sensors. Most papers on DET-based biosensors are related to peroxide detection in a variety of environments with peroxidases, but also hemoglobin, Cyt c, microperoxidase and hemin were explored for peroxide measurement, whenever less efficient. Not only hydrogen peroxide but also organic hydroperoxides [170] are indicated, although the highest reaction rate is obtained for hydrogen peroxide.

Many of the developed sensors based on peroxidases have lower detection limits in the range 10–100 nM. For Cyt c, which has a considerably lower catalytic rate with peroxide, the detection limits are around 10 μM. Only for a multilayer approach with a conducting underlayer could peroxide concentrations as low as 3 nM be indicated [187].

For the measurement, a moderate reduction potential between −100 mV and +100 mV vs. Ag/AgCl is sufficient. In this region the potential for electrochemical interferences is very low. However, the biggest problems arise from the high reactivity of compound I and II with reducing substrates (electron donators), which compete with the electrode for the reduction of peroxidase. Ascorbic acid, naturally occurring phenolics and aromatic amines are among those compounds. The competitive reaction of reductants should be minimized when all enzyme molecules are in DET contact to the electrode. From the comparison of peroxidases from different sources, sweet potato peroxidase showed the highest percentage of DET on graphite electrodes. Still less interference was found for Asn70Asp-mutated HRP on graphite [181, 188]. Unfortunately, the stability of the sensors with recombinant enzymes was only a few hours. Longer shelf life was obtained when peroxidase is in a composite or covalently attached. The most stable peroxide detector had HRP embedded with activated carbon in a screen printed paste [171]. The material is extremely rich in oxygen. Real application was reported for peroxide measurement in various samples including on-line monitoring of cerebral peroxide [189].

The use of peroxide sensors has been extended to the determination of glucose, lactate, alcohol, oxalate, glutamate and other amino acids by coimmobilization of the respective oxidase on top of the peroxidase [59, 60, 190, 191] and furthermore to affinity-based assays [171, 192–194]. An octane sensor was created by layering a porphyrin type P450-mimics on a screen printed HRP-modified carbon electrode [195]. The biomimetic catalyst (iron(III)-meso-tetrakis-(pentafluorophenyl)-β-tetrasulfonatoporphyrin chloride) was linked to the electrode with polyallylamine on the basis of the strong electrostatic interaction between the anionic polyelectrolyte and the polyhalogenated iron porphyrin. The basis of sensing is a competition between peroxidase and the biomimetic for cumene hydroperoxide. A decrease of about 5 μM peroxide/mM octane is obtained at up to 20 mM octane [195].

5
Molybdenum-containing Enzymes

Molybdenum (Mo) is the only second-row transition metal that is required for cofactors of redox enzymes that catalyze reactions at carbon, sulfur and nitrogen atoms. There are two distinct types of molybdoenzymes: molybdenum nitrogenase has a unique molybdenum-iron-sulfur center called FeMoco, and catalyzes the reduction of atmospheric dinitrogen to ammonia. All other molybdoenzymes are oxidoreductases that transfer an oxo group or two electrons to or from the substrate. They bind the molybdenum cofactor (Moco) in which Mo is coordinated to a dithiolene group on the 6-alkyl side chain of molybdopterin (MPT). Moco containing enzymes catalyze the oxidation and reduction of various small molecules including the oxidation of xanthine, aldehydes, sulfite and arsenite and the reduction of dimethyl sulfoxide

(DMSO) and nitrate [196–198]. Several molybdoenzymes such as xanthine dehydrogenase (XDH), transfer oxygen derived from water to different small substrate molecules (R).

$$R + H_2O + A \leftrightarrow RO + AH_2 . \tag{4}$$

This reaction is accompanied by reduction of an electron acceptor A in a two or one electron/proton transfer step. These enzymes have a complex overall architecture, and generally contain multiple redox-active centers. The Mo center (oxidation states between VI and IV) acts as transducer between the two-electron redox reaction on the substrate and two single electron oxidation-reductions at the other cofactors. On the basis of their sequence and the structures of their oxidized molybdenum centers, these mononuclear molybdenum enzymes are categorized into the xanthine oxidoreductase (XO) family, the sulfite oxidase (SO) family and the dimethyl sulfoxide reductase (DMSOR) family [199, 200] (Fig. 5).

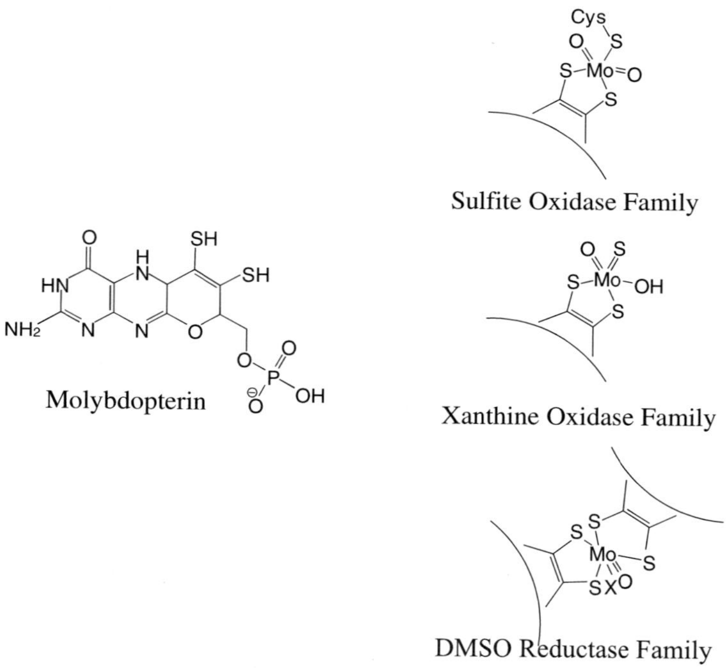

Fig. 5 Families of Mo-pterin enzymes and illustration of their oxidized Mo-sites according to [199]

To date, more than 50 enzymes containing Moco were identified in bacteria, fungi, plant and animal, but investigations on DET are limited (Table 4) [34].

Table 4 Survey of molybdoenzymes in DET contact

Protein	Enzyme Family	Cofactors	Electrode	Substrate	Note	Refs.
Xanthine dehydrogenase	XO	Moco FAD 2 × [2Fe–2S]	PG	Xanthine	MPT might have an active role in catalysis Prokaryotic	[203]
Xanthine oxidase	XO	Moco FAD 2 × [2Fe–2S]	PG/DDAB	Xanthine	FAD electroactive (– 292 mV), catalytic wave at + 300 mV (NHE) Eukaryotic	[221]
Aldehyde oxidoreductase	XO	Moco 2 × [2Fe–2S]	Au+ neomycin PG, GC	Benzaldehyde and other aldehydes	Active cofactors: Moco at Au [2Fe–2S] at C Prokaryotic	[204]
Sulfite dehydrogenase	SO	Moco Heme c	PG/DDAB	Sulfite	Prokaryotic	[212]
Sulfite oxidase	SO	Moco Heme *b*5	PG Au/HS-C_6OH Au/MU Au/mercapto-undecanamine Au/MUA/MU/PAA	Sulfite	Chicken at – 120 mV direct redox couple $Fe^{II/III}$ Human Single domains	[211] [213] [214]
DMSO reductase	DMSOR	Moco	PGE + DDAB	Sulfoxides, nitrate, tertiary amines oxides, carboxylates	Single peak at 632 nm in the UV spectrum upon electrochemical reduction	[208] [207]
Arsenite reductase	DMSOR	Moco High potential [3Fe–4S] Rieske [2Fe–2S]	PG+ polymyxin PG/DDAB PG/adsorbed enzyme	Arsenite	$Mo^{VI/IV}$ is highly cooperative	[209]

5.1 Enzymes of the Xanthine Oxidase and Dimethyl Sulfoxide Reductase Families

Enzymes of the XO family catalyze the oxidative hydroxylation of a vast group of heterocyclic compounds including pyridines, purines, pteridines, pyrazolopyrimidines and various aromatic and aliphatic aldehydes [202]. The best studied enzyme of this family is xanthine oxidoreductase (XOR). XOR comprise four redox active cofactors: a five-coordinated mononuclear Mo ion, two non-identical [2Fe-2S] clusters and FAD. A voltammetric study of a member of the XO family was performed with the *Rhodobacter capsulatus* XDH at a DDAB-modified PG-electrode [203]. Three distinct couples were observed in a CV at potentials comparable with those found by potentiometry. The highest reversible redox peak was observed at –300 mV (vs. NHE) which resulted probably from an overlap of the FAD/$FADH_2$ and the first iron-sulfur center $[2Fe–2S]^{2+/+}$. The weak signal at around –400 mV was attributed to the second $[2Fe–2S]^{2+/+}$. The peaks appearing between –510 mV and –550 mV were assigned to $Mo^{VI/V}$ and $Mo^{V/IV}$, because they were absent when the Moco-free enzyme was studied. EPR-studies supported this interpretation of the CV results.

Upon the addition of xanthine, a catalytic current evolved starting at +400 mV (vs. NHE) indicative of electrochemically driven enzyme turnover. This potential was about 600 mV more positive than the potential where the XDH redox-active centers were fully oxidized. At lower potential, no catalysis was obtained. From these findings, it has been postulated that the enzyme requires oxidative activation to restore catalytic activity and the site of this oxidation is the pterin ring [203]. Reversible redox transformation of the Mo active site at the electrode leads to anodic activation and cathodic deactivation of XDH. In this process, the tricyclic pyranopterin backbone of Moco may be seen as a bridge between the Mo active center and the other cofactors, permitting the intramolecular electron exchange.

Recently, aldehyde oxidoreductase (AOR) was studied by direct electrochemistry at carbon and gold electrodes, using neomycin as promoter [204]. Only one reduction reaction was obtained at carbon electrodes for soluble and adsorbed AOR with $E^{0\prime}$ between –257 mV and –280 mV (vs. SHE, pH 7.6). This was assigned to the more exposed iron-sulfur cluster. Reduction of the Moco occured at a much more negative potential at gold electrode modified with neomycin.

Enzymes of the DMSOR family catalyze two-electron oxo- or hydride-transfer reactions of sulfoxides, nitrate, tertiary amine oxides and carboxylates. In these enzymes, the Moco consists in an oxomolybdenum group coordinated to two bidentate pterin-dithiolene ligands. Additional ligands at the Mo site in this family are serine, cysteine or selenocysteine [198, 205, 206].

The voltammetric response from a DMSO reductase was obtained using the protein isolated from *R. capsulatus* [207]. The communication between the Moco and the electrode was possible by the usage of the surfactant DDAB on the PGE. The $Mo^{VI/V}$ wave exhibited a pH dependence of ca. –59 mV/pH unit in the range 5–9, which consisted in a single electron/proton transfer reaction in this pH domain. The calculated pK_a (9.0) was assigned to the protonation of $Mo^V = O$ to yield Mo^V–OH. Interestingly, the potential of the $Mo^{V/IV}$ couple was constant, in contrast to the behavior of the DMSOR from

E. coli, where each single transition was coupled with a sequential single proton transfer reaction. Upon electrochemical reduction, a single peak at 632 nm was visible in the UV spectrum. After adding the substrate, it was possible to detect an increase of the reduction current. Also voltammetric studies of the catalytic activity of DMSOR adsorbed at a graphite electrode indicated that the electron pathway is directed through Mo^{V} [208].

Different results were obtained for arsenite oxidase from *Alcaligenes faecalis*, belonging also to the DMSOR family [209]. In this study of the adsorbed enzyme, only a single couple of peaks was identified assigned to $Mo^{VI/IV}$. No EPR signals due to Mo^{V} were detected, supporting the conclusion that the $Mo^{VI/IV}$ is highly cooperative. In addition, the couple position was pH sensitive with a one to one proton/electron ratio. The main difference of arsenite oxidase and DMSO reductase was that in DMSO reductase the Mo atom is coupled to a serine residue, while in arsenite oxidase this position is taken by an alanine, which is not coordinated to the Mo atom.

5.2 Sulfite Oxidase

Enzymes of the SO family comprise the plant assimilatory nitrate reductases and sulfite-oxidizing enzymes.The latter can be separated into the sulfite oxidases (SO) mainly found in eukaryotes, and the sulfite dehydrogenases (SDH) found in bacteria, distinguished by their ability to transfer electrons to molecular oxygen. However, both types can use cytochromes as electron acceptors for the oxidation of inorganic sulfite to sulfate:

$$SO_3{}^{2-} + H_2O \rightarrow SO_4{}^{2-} + 2H^+ + 2e^- . \qquad (5)$$

In mammals, SO comprises a Moco-containing subunit connected by a flexible polypeptide chain region to a second subunit with a b-type heme [210]. The two-electron oxidation of sulfite to sulfate occurs at the Moco with concomitant reduction of Mo^{VI} to Mo^{IV}. The intramolecular electron transfer between the Moco and the heme cofactor ($Fe^{II/III}$) is fundamental for the Mo reoxidation. For this reaction to occur, a backbone rearrangement was postulated that moves the heme and Moco domains close together to achieve the electron exchange, which otherwise is unlikely due to the large distance between the two redox centers. The natural electron acceptor for SO is Cyt c.

The potential employment of SO in amperometric biosensors has long been recognized. Sulfite is an important preservative in foods and drinks and its analytical determination in solution and in gas phase has been the subject of many investigations. Many strategies have been studied in order to develop a SO biosensor with small redox mediators or Cyt c. Recently, the proof of principle for sulfite biosensors exploiting direct interaction between electrode and chicken liver SO was delivered [211, 212].

Spectroelectrochemical results with thiol-modified Au electrodes and CV studies at pyrolytic graphite and thiol-modified Au electrodes demonstrated that SO communicates directly with the electrode through its heme domain [211, 213]. In the absence of substrate at 11-mercaptoundecanol-modified Au electrode the oxidation and reduction peaks corresponded to the heme $Fe^{II/III}$ redox couple with $E^{0\prime} = -120$ mV [213] and a heterogeneous electron transfer rate constant in the order of $k_s = 15\ s^{-1}$. The oxidation peak current increased proportional with sulfite concentrations between 10 and 100 μM and approached saturation at 3 mM. Catalytic sulfite oxidation was demonstrated also at pyrolytic graphite, at mercapto-6-hexanol modified polycrystalline gold [211] and 11-mercaptoundecanol/mercaptoundecanamine monolayers on gold [213]. At PG-electrodes the rate is about 5 times slower than at modified gold, but still slower than in the homogeneous solution. The pH-dependence of the bioelectrocatalysis indicated also a backbone rearrangement for the electron transfer at the electrode.

We were able to show DET of human SO at gold electrodes modified with a monolayer of MUA/MU and covered with the cationic polyelectrolyte polyallylamine to obtain an overall positive charged surface [214]. A single reversible redox couple with a midpoint potential of –166 mV (SCE) was visible that changes to a catalytic oxidation current upon addition of sulfite. Using the same electrode modification for the single SO-heme domain an oxidation occurs at –154 mV very close to the value for the holo SO. However, no catalytic current was observed in the presence of sulfite. This clearly showed that the heme domain is the electron transducing unit, but has no catalytic function. A further elucidation of the function of the single domains was delivered by investigating the electrochemical behaviour of both domains and the holo-enzyme involving Cyt c as mediator [214]. In this investigation, we exploited the fast direct electrochemistry of Cyt c ($Fe^{II/III}$) on gold electrodes modified with MUA/MU (Sect. 3.1) and its ability to act as electron acceptor of SO. For this study it was also important that Cyt c could freely rotate between electrode and protein reaction partners as has been earlier reported for its interaction with other enzymes at electrodes [19, 20, 31].

In the presence of Cyt c and SO, the CV showed clearly a single redox couple assigned to the oxidation and reduction of Cyt c. At the negatively charged electrode, SO did not show any electrochemical activity. A pronounced catalytic current was observed after addition of sulfite, that is both dependent on SO activity and sulfite concentration. The second order rate constant for the SO catalyzed Cyt c reduction was calculated to be $k = 4.47 \pm 0.13 \times 10^6\ M^{-1}\ s^{-1}$ (Fig. 6).

The same experiment was performed using the SO-Moco domain, and interestingly the interaction was still possible (Fig. 7). The rate constant of the reaction between the SO-Moco domain and Cyt c is much lower than for holo-SO ($k = 0.19 \pm 0.05 \times 10^6\ M^{-1}\ s^{-1}$). The electron transfer between Moco and Cyt c appears to be the limiting step, since the rate of reduction of the

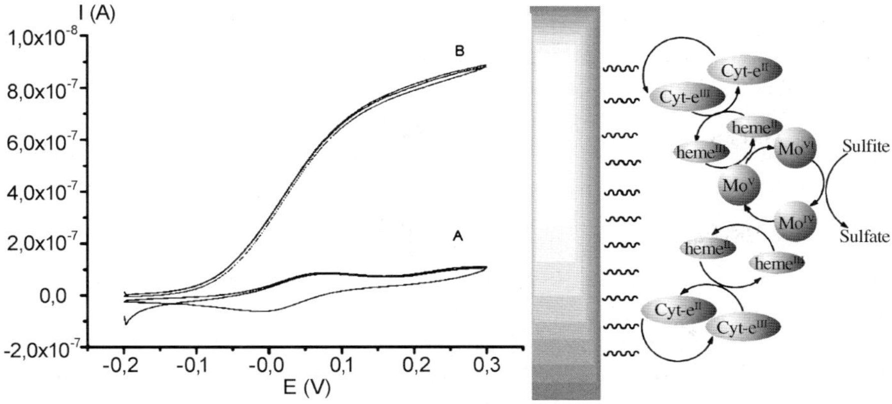

Fig. 6 Scheme and CV of the bioelectrocatalytic sulfite oxidation in the **A** absence and **B** presence of SO mediated by Cyct c measured with a MU/MuA (3 : 1)-modified Au wire in 5 mM K-phosphate buffer, pH 8.5, 10 mM sulfite, 30 μM Cyt-c, scan rate 5 mV/s, Reference electrode + 326 mV vs. NHE

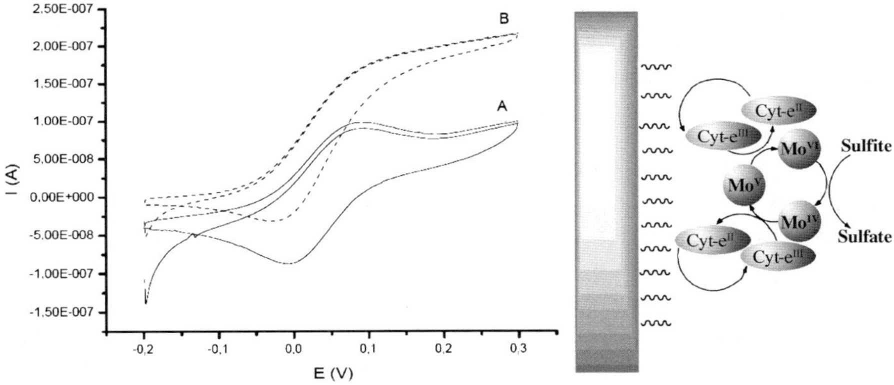

Fig. 7 Scheme and CV of the electrocatalytic sulfite oxidation by 4.84 μM Moco domain **B** mediated by Cyct c other conditions as Fig. 6. **A** CV without Moco domain in the presence of Cyt c and sulfite

SO and Moco by the substrate is expected to be the same. Furthermore, for this reaction Mo is essential, because no catalysis was recorded for the MPT-containing protein lacking the Mo atom. Also, in this experiment the heme domain was catalytically inactive supporting the above discussed conclusions. Details of this interaction and its application for biosensors will be published elsewhere.

The voltammetric response of the Mo center of the bacterial SDH from *Starkeya novella* in the presence and absence of substrate has also been reported [212]. Bacterial SDH is a heterodimer of a Moco containg domain and a Cyt c containing domain and shows some characteristics of eukaryotic SOs, like the inhibition by small anions. However, some differences were observed like the absence of the eukaryotic characteristic low-pH EPR spectrum [215]. For this investigation a PGE surface was modified by DDAB and signals from the single redox sites were recorded. The higher potential $Mo^{VI/V}$ couple exhibited pH dependence in agreement with the uptake of a single proton upon one-electron reduction. While the potential of $Mo^{V/IV}$ couple was pH independent in the range 5–10 and the $Fe^{III/II}$ showed a potential in between these two Mo transitions. Introduction of sulfite to the electrochemical cell resulted in a pronounced increase of oxidation current. From the pH dependence of the catalytic current, it was possible to find the pK_a values of 7.2 and 10. The higher pK_a value was explained by deprotonation of the tyrosine residue close to the active site that is responsible for stabilizing the substrate at the active site by hydrogen bonds. The origin of the lower pK_a could be due to the substrate pK_a (HSO_3^-/SO_3^{2-}), or a shift of the $Mo^{VI/V}$ potential, which below pH 7 becomes significantly more positive than the heme couple. The involvement of an amino acid residue in the switching of the electron transfer between the two cofactors remains also possible. These studies on SDH gave an overview on this complex system; however, they still did not clarify the role of the pterin ring and showed differences to the model for eukaryotic SO.

6
Conclusion

In recent years, a toolbox of methods has been developed that enables to investigate and exploite the direct electronic communication with redox proteins and complex enzymes. Self-assembled monolayers of desired charges, colloids and surfactants are among the well established approaches for the design of the sensor surfaces, where the enzymes are studied in different level of integration.

Hemeoxygenases catalyze various important reactions are therefore among the most attractive targets for bioelectrochemists, who could demonstrate steroid hydroxylation, demethylation, dealkylation, dehalogenation and epoxidation. However, many of the reactions may also proceed with peroxide, which is always formed in side reactions. Possibly the assembly of microsomes with enriched CYP-enzymes on an electrode [223] will overcome limitations due to uncoupling for hydroxylation of otherwise inert carbons in substances of pharmaceutical interest.

With most enzymes, bioelectrocatalytic substrate conversion is only seen in μmolar to millimolar range. This restriction is due to a kinetic limitation of the heterogeneous electron transfer between the enzyme active site and the electrode, which can be overcome by engineering specific attachment regions and electron transfer bridges. Sulfite oxidase is an example where the electron transduction is much faster when Cyt c is introduced.

The DET to small redox proteins and to peroxidases is of particular interest for bioanalytics as they show interaction with reactive (oxygen) species. We already now witness research application for in vivo studies of pathophysiological processes, evaluation of the quality of food, cosmetics, and pharmaceutical formulas.

In recent years, we have profited from molecular biology, nanotechnology and surface chemistry because they provide tools for the molecular design of proteins and nanostructuring of surfaces. Important impact is also coming from novel methods of spectroscopy and in situ imaging techniques for mapping and control of single-molecule activity or visualization of molecules during their redox processes. The discovery of microorganisms loaded with cytochromes in their outer membranes [216] will broaden the scope of DET sensors to whole cell-based assays.

The understanding of the molecular mechanism may lead to tailor made catalytic systems. These do not necessarily have to be of pure native origin. Recent work indicates the potential of engineering sites for surface binding and coupling of redox active dyes, de-novo designed redox proteins and genetic chimeras as well as of fully synthetic structures and biohybrid synthetic structures with catalytic redox activity.

Acknowledgements We are very grateful to Frieder Scheller for his encouragement and many stimulating ideas. The work was in part supported by International Max Planck Research School on Biomimetic Systems (IMPRS), and German Federal Ministry of Economics and Technology (BASUMA 01MT308).

References

1. Wilson G (ed) (2002) Bioelectrochemistry In: Bard AJ, Stratmann M (eds) Encyclopedia of Electrochemistry, vol 9. Wiley-VCH, Weinheim
2. Armstrong FA, Wilson GS (2000) Electrochim Acta 45:2623
3. Scheller F, Wollenberger U, Lisdat F (2005) Application of electrically connected enzymes for biosensors. In: Willner I, Katz E (eds) Bioelectronics. Wiley-VCH, Weinheim, p 99
4. Armstrong FA (1990) Struct Bond 72:137
5. Willner I, Katz E (2000) Angew Chem Int Ed 39:1180
6. Dryhurst G, Kadish KM, Scheller FW, Renneberg R (1982) Biological Electrochemistry, vol 1. Academic Press, Inc., New York
7. Yeh P, Kuwana T (1977) Chem Lett 1145
8. Eddowes MJ, Hill HAO (1977) J Chem Soc Chem Comm 771
9. Varfolomeev SD, Berezin IV (1978) J Mol Catal 4:387
10. Yaropolov AI, Malovik V, Varfolomeev SD, Berezin IV (1979) Dokl Akad Nauk SSSR 249:1399
11. Tarasevich MR (1979) Bioelectrochem Bioenerg 6:587
12. Scheller FW, Renneberg R, Strnad G, Pommerening K, Mohr P (1977) Bioelectrochem Bioenerg 4:500

13. Jeukens LJC (2003) Biochim Biophys Acta 1604:67
14. Marcus R, Sutin N (1985) Biochim Biophys Acta 811:265
15. Moser CC, Keske JM, Warncke K, Farid RS, Dutton PL (1992) Nature 355:796
16. Gorodetsky AA, Boal AK, Barton JK (2006) J Am Chem Soc 128:12082
17. Ataka K, Giess F, Knoll W, Naumann R, Haber-Pohlmeier S, Richter B, Heberle J (2004) J Am Chem Soc 126:16199
18. Jeuken LJC, Connell SD, Henderson PJF, Gennis RB, Evans SD, Bushby RJ (2006) J Am Chem Soc 128:1711
19. Scheller FW, Wollenberger U, Lei C, Jin W, Ge B, Lehmann C, Lisdat F, Fridman V (2002) Rev Molec Biotechnol 82:425
20. Fridman V, Wollenberger U, Bogdanovskaya V, Lisdat F, Ruzgas T, Lindgren A, Gorton L, Scheller FW (2000) Biochem Soc Trans 28:63
21. Lebedev N, Trammell SA, Spano A, Lukashev E, Griva I, Schnur J (2006) J Am Chem Soc 128:12044
22. Schuhmann W (2002) Rev Mol Biotechnol 82:425
23. Bain CD, Whitesides GM (1989) Angew Chem Int Ed 28:506
24. Ulman A (1996) Chem Rev 96:1533
25. Rusling JF (1998) Accounts Chem Res 31:363
26. Gooding JJ, Hibbert DB (1999) TRAC 18:525
27. Finklea HO (1996) In: Bard AJ (ed) Electroanalytical Chemistry, vol 19. Marcel Dekker, New York, p 109
28. Song S, Clark RA, Bowden EF, Tarlov MJ (1993) J Phys Chem 97:6564
29. Jin W, Wollenberger U, Bernhardt R, Stöcklein W, Scheller F (1998) Bioelectrochem Bioenerg 47:75
30. Feng ZQ, Imabayashi S, Kakiuchi T, Niki KJ (1997) Chem Soc Faraday Trans 93:1367
31. Jin W, Wollenberger U, Kärgel E, Schunck WH, Scheller FW (1997) J Electroanal Chem 433:135
32. Bistolas N, Wollenberger U, Jung C, Scheller FW (2005) Biosens Bioelectron 20:2408
33. Wollenberger U, Bistolas N, Schroeder K, Richter M, Jung C (2007) In: Kyriakopoulos A, Michalke B, Graebert A, Behne D (eds) Metallobiolomics. Herbert Utz Verlag, Munich
34. Bernhardt PV (2006) Aust J Chem 59:233
35. Gooding JJ, Wibowo R, Liu J, Yang W, Losic D, Orbons S, Mearns FJ, Shaper JG, Hibbert DB (2003) J Am Chem Soc 125:9006
36. Hartmann M (2005) Chem Mater 17:4577
37. Murphy L (2006) Curr Opin Chem Biol 10:177
38. Weigel MC, Tritscher E, Lisdat F (2007) Electrochem Comm 9:689
39. Gilardi G, Fantuzzi A, Sadeghi SJ (2001) Curr Opin Struct Biol 11:491
40. Presnova G, Egorov A, Ruzgas T, Lindgren A, Gorton L, Börchers T (2000) Faraday Discuss 116:281
41. Ferapontova E, Schmengler K, Börchers T, Ruzgas T, Gorton L (2002) Biosens Bioelectron 17:953
42. Gilardi G, Meharenna YT, Tsotsou GE, Sadeghi SJ, Fairhead M, Giannini S (2002) Biosens Bioelectron 133
43. Andolfi L, Bruce D, Cannistraro S, Canters GW, Davis JJ, Hill HAO, Crozier J, Verbeet MP, Wrathmell CL, Astier YJ (2004) Electroanal Chem 565:21
44. Wong TS, Schwaneberg U (2003) Curr Opin Biotech 14:590
45. Fantuzzi A, Meharenna YT, Briscoe PB, Sassone C, Borgia B, Gilardi G (2006) Chem Comm (12):1289
46. Cooper JM, Greenough KR, McNeil CJ (1993) J Electroanal Chem 347:267

47. McNeil CJ, Smith KA (1989) Free Rad Res Comm 7:89
48. Scheller W, Jin W, Ehrentreich-Förster E, Ge B, Lisdat F, Büttemeier R, Wollenberger U, Scheller FW (1999) Electroanal 11:703
49. Lisdat F, Ge B, Ehrentreich-Förster E, Reszka R, Scheller FW (1999) Anal Chem 71:1359
50. Gobi V, Mizutani F (2000) J Electroanal Chem 484:172
51. Lisdat F (2004) In: Mirsky VM (ed) Ultrathin electrochemical chemo- and biosensors: Technology and performance. Springer Verlag, Berlin
52. Ge B, Meyer T, Schöning MJ, Wollenberger U, Lisdat F (2000) Electrochem Commun 2:557
53. Lei C, Wollenberger U, Bistolas N, Guiseppi-Elie A, Scheller FW (2002) Anal Bioanal Chem 372:235
54. Topoglidis E, Astuti Y, Duriaux F, Grätzel M, Durrant JR (2003) Langmuir 19:6894
55. Kröning S, Scheller FW, Wollenberger U, Lisdat F (2004) Electroanal 16:253
56. Krylov AV, Sczech R, Lisdat F (2007) Analyst 132:135
57. Reipa V, Mayhew MP, Vilker VL (1997) Proc Natl Acad Sci USA 94:13554
58. Jin W, Wollenberger U, Bier F, Makower A, Scheller F (1996) Bioelectrochem Bioenerg 39:221
59. Ghindilis AL, Atanasov P, Wilkins E (1997) Electroanal 9:661
60. Gorton L, Lindgren A, Larsson T, Munteanu FD, Ruzgas T, Gazaryan I (1999) Anal Chim Acta 400:91
61. Wollenberger U (2005) In: Gorton L (ed) Biosensors and modern biospecific analytical techniques. Elsevier, Amsterdam, pp 65–130
62. Ikeda T (1992) Bull Electrochem 8:145
63. Guo LH, Hill HAO (1991) Adv Inorg Chem 36:341
64. Kulys JJ, Svirmitskas GY (1979) Dok Phys Chem 245:208
65. Niculescu M, Ruzgas T, Nistor C, Frebort I, Sebela M, Pec P, Csöregi E (2000) Anal Chem 72:5988
66. Shumyantseva VV, Ivanov YD, Bistolas N, Scheller FW, Archakov AI, Wollenberger U (2004) Anal Chem 76:1646
67. Scott RA, Mauk AG (eds) (1996) Cytochrome c: A multidisciplinary approach. University Science Books, Sausalito
68. Fedurco M (2000) Coord Chem Rev 209:263
69. Allen PM, Hill HAO, Walton NJ (1984) J Electroanal Chem 178:69
70. Niki K (2002) Bioelectrochemistry. In: Wilson G, Bard AJ, Stratmann M (eds) Encyclopedia of Electrochemistry, vol 9. Wiley-VCH, Weinheim, p 341
71. Murgida DH, Hildebrandt P (2001) Angew Chem Int Ed 40:728
72. Kasmi AE, Wallace JM, Bowden EF, Binet SM, Linderman RJ (1998) J Am Chem Soc 120:225
73. Arnold S, Feng ZQ, Kakiuchi K, Knoll W, Niki K (1997) J Electroanal Chem 438:91
74. Ge B, Lisdat F (2002) Anal Chim Acta 454:53
75. Beissenhirtz MK, Scheller FW, Stöcklein WFM, Kurth D, Möhwald H, Lisdat F (2004) Angew Chem Int Ed 43:4357
76. Dronov R, Kurth DG, Möhwald H, Scheller FW, Lisdat F (2007) Electrochim Acta; doi: 10.1016/j.electacta2007.02.044
77. Lisdat F, Ge B, Meyerhoff ME, Scheller FW (2001) Probe Microscopy 2:113
78. McNeil CJ, Manning P (2002) Rev Molec Biotech 82:443
79. Beissenhirtz M, Scheller FW, Lisdat F (2003) Electroanal 15:1425
80. Barker SLR, Kopelman R, Meyer TE, Cusanovich MA (1998) Anal Chem 70:971
81. Ren Z, Meyer T, McRee DE (1993) J Mol Biol 234:433

82. Erabi T, Ozawa S, Hayase S, Wada M (1992) Chem Lett 11:2115
83. Stryer L (1995) Biochemistry, 4th ed. Freeman, New York
84. RCSB Protein Data Bank http://www.rcsb.org/pdb/explore/biologyAndChemistry.do?structureId=1MBN, last visited: Oct. 1 2007
85. Van Dyke BR, Saltman P, Armstrong FA (1996) J Am Chem Soc 118:3490
86. Nassar AEF, Willis WS, Rusling JF (1995) Anal Chem 67:2386
87. Lei C, Wollenberger U, Scheller F (2000) Quim Anal 19:28
88. Bayachou M, Lin R, Cho W, Farmer PJ (1998) J Am Chem Soc 120:9888
89. Shen L, Huang R, Hu N (2002) Talanta 56:1131
90. Lin R, Bayachou M, Greaves J, Farmer PJ (1997) J Am Chem Soc 119:12689
91. Kamau GN, Guto MP, Munge B, Panchagnula V, Rusling JF (2003) Langmuir 19:6976
92. Zhou Y, Hu N, Zeng Y, Rusling JF (2002) Langmuir 18:211
93. Weissbluth M (1974) Molecular Biology: Biochemistry and Biophysics, vol. 15. Springer, New York
94. Scheller FW, Bistolas N, Katterle M, Jänchen M, Wollenberger U (2005) Adv Colloid Interface Sci 116:111
95. Dong S, Chen T (1997) In: Scheller WF, Schubert F, Fedrowitz J (eds) Frontiers in Biosensorics I. Birkhäuser, Basel, p 209
96. de Montellano PRO, Catalano CE (1985) J Biol Chem 260:9265
97. Wade RS, Castro CE (1973) J Am Chem Soc 95:231
98. Murphy ME, Noack E (1994) Method Enzymol 233:240
99. Gow AJ, Luchsinger BP, Pawlowski JR, Singel DJ, Stamler JS (1999) Proc Natl Acad Sci USA 96:9027
100. Wink DA, Nims RW, Darbyshire JF, Christodoulou D, Hanbauer I, Cox GW, Laval F, Laval J, Cook JA, Krishna MC, DeGraff WG, Mitchell JB (1994) Chem Res Toxicol 7:519
101. Gow AJ, Stamler JS (1998) Nature 391:169
102. Bistolas N (2005) PhD-Thesis, Univ Potsdam
103. Lewis DFV (1996) Cytochromes P450 – structure, function and mechanism. Taylor & Francis Ltd., London
104. Jung C, Ghosh DK (2004) Cell Mol Biol 50:335
105. de Montellano PRO (ed) (2005) Cytochrome P450: structure, mechanism, and biochemistry, 3rd edn. Plenum Press, New York
106. Sligar SG, Makris TM, Denisov IG (2005) Biochem Biophys Res Com 338:346
107. Jung C, Schünemann V, Lendzian F, Trautwein AX, Contzen J, Galander M, Böttger LH, Richter M, Barra AL (2005) J Biol Chem 386:1043
108. Estabrook RW, Faulkner KM, Shet MS, Fisher CW (1996) Adv Enzymol 272:44
109. Reipa V, Mayhew MP, Holden MJ, Vilker VL (2002) Chem Comm 318
110. Shumyantseva VV, Bulko TV, Usanov SA, Schmid RD, Nicolini C, Archakov AI (2001) J Inorg Biochem 87:185
111. Hintz MJ, Peterson JA (1980) J Biol Chem 255:7317
112. Contzen J, Jung C (1999) Biochemistry 38:16253
113. Scheller F, Renneberg R, Schwarze W, Strnad G, Pommerening K, Prümke HJ, Mohr P (1979) Acta Biol Med Germ 38:503
114. Renneberg R, Scheller F, Ruckpaul K, Pirrwitz J, Mohr P (1978) FEBS Lett 96:349
115. Fantuzzi A, Fairhead M, Gilardi G (2004) J Am Chem Soc 126:5040
116. Kazlauskaite J, Westlake ACG, Wong LL, Hill HAO (1996) Chem Commun 2189
117. Lo KK, Wong LL, Hill HAO (1999) FEBS Lett 451:342
118. Estavillo C, Lu Z, Jansson I, Schenkman JB, Rusling JF (2003) Biophys Chem 104:291

119. Joseph S, Rusling JF, Lvov YM, Friedberg T, Fuhr U (2003) Biochem Pharmacol 65:1817
120. Lei C, Wollenberger U, Jung C, Scheller FW (2000) Biochem Biophys Res Commun 268:740
121. Iwuoha EI, Joseph S, Zhang Z, Smyth MR, Fuhr U, Ortiz de Montellano PR (1998) J Pharm Biomed Anal 17:1101
122. Zhang Z, Nassar AEF, Lu Z, Schenkman JB, Rusling JF (1997) J Chem Soc Faraday Trans 93:1769
123. Matsumura H, Wiwatchaiwong S, Nakamura N, Yohda M, Ohno H (2006) Electrochem Comm 8:1245
124. Matsumura H, Nakamura N, Yohda M, Ohno H (2007) Electrochem Comm 9:361
125. Bistolas N, Christenson A, Ruzgas T, Jung C, Scheller FW, Wollenberger U (2004) Biochem Biophys Res Comm 314:810
126. Lvov YM, Lu Z, Schenkman JB, Zu X, Rusling JF (1998) J Am Chem Soc 120:4073
127. Fisher MT, Sligar SG (1985) J Am Chem Soc 107:5018
128. Mouro C, Bondon A, Jung C, Hui Bon Hoa G, De Certaines JD, Spencer RGS, Simonneaux G (1999) FEBS Lett 455:302
129. Ege D, Ghosh PK, White JR, Equey JF, Bard AJ (1985) J Am Chem Soc 107:5644
130. Kiselyova OI, Yaminsky IV, Ivanov YD, Kanaeva IP, Kuznetsov VY, Archakov AI (1999) Arch Biochem Biophys 371
131. Poulos TL, Finzel BC, Howard A (1986) J Biochem 25:5314
132. Jung C, Kozin A, Canny B, Chervin JC, Hui Bon Hoa G (2003) Biochem Biophys Res Comm 312:197
133. Wiwatchaiwong S, Matsumura H, Nakamura N, Yohda M, Ohno H (2007) Electroanal 19:561
134. Udit AK, Hagen KD, Goldman PJ, Star A, Gillan JM, Gray HB, Hill MG (2006) J Am Chem Soc 128:10320
135. Todorovic S, Jung C, Hildebrandt P (2006) J Biol Inorg Chem 11:119
136. Wirtz M, Klucik J, Rivera M (2000) J Am Chem Soc 122:1047
137. Zu X, Lu Z, Zhang Z, Schenkman JB, Rusling JF (1999) Langmuir 15:7372
138. Munge B, Estavillo C, Schenkman JB, Rusling JF (2003) Chem Biochem 4:101
139. Mayhew MP, Reipa V, Holden MJ, Vilker VL (2000) Biotechnol Prog 16:610
140. Walsh ME, Kyritsis P, Eady NAJ, Hill HAO, Wong LL (2000) Eur J Biochem 267:5815
141. Shumyantseva VV, Carrara S, Bavastrello V, Riley DJ, Bulko TV, Skryabin KG, Archakov AI, Nicolini C (2004) Biosens Bioelectr 19:971
142. Shumyantseva VV, Bulko TV, Rudakov YO, Kuznetsova GP, Samenkova NF, Lisitsa AV, Karuzina II, Archakov AI (2007) J Inorg Biochem 101:859
143. Johnson DL, Lewis BC, Elliot DJ, Miners JO, Martin LL (2005) Biochem Pharm 69:1533
144. Iwuoha EI, Smyth MR (2003) Biosens Bioelectron 18:237
145. Flemming BD, Tian Y, Bell SG, Wong LL, Urlacher V, Hill HAO (2003) Eur J Biochem 270:4082
146. Aguey-Zinsou KF, Bernhardt PV, De Voss JJ, Slessor KE (2003) Chem Comm 3:418
147. Nicolini C, Erokhin V, Ghisellini P, Paternolli C, Ram MK, Sivozhelezov V (2001) Langmuir 17:3719
148. Alderton WK, Cooper CE, Knowles RG (2001) Biochem J 357:593
149. de Montellano PRO, Nishida C, Rodriguez-Crespo I, Gerber N (1998) Drug Metabol Dispos 26:1185
150. Werner ER, Gorren ACF, Heller R, Werner-Felmayer G, Mayer B (2003) Exp Biol Med 228:1291

151. Jung C, Lendzian F, Schünemann V, Richter M, Böttger LH, Trautwein AX, Contzen J, Galander M, Ghosh DK, Barra A-L (2005) Magn Reson Chem 43:S84
152. Stuehr DJ, Wei CC, Wang Z, Hille R (2005) Dalton Trans (21):3427
153. Stuehr DJ (2004) J Nutrit 134:2748
154. Stuehr DJ (1997) Ann Rev Pharmacol Toxicol 37:339
155. Feng CJ, Tollin G, Holliday MA, Thomas C, Salerno JC, Enemark JH, Ghosh DK (2006) Biochemistry 45:6354
156. Wang ZQ, Wei CC, Santolini J, Panda K, Wang Q, Stuehr DJ (2005) Biochemistry 44:4676
157. Presta A, Siddhanta U, Wu CQ, Sennequier N, Huang LX, Abu-Soud HM, Erzurum S, Stuehr DJ (1998) Biochemistry 37:298
158. Ghosh DK, Abusoud HM, Stuehr DJ (1995) Biochemistry 34:11316
159. Bayachou M, Boutros JA (2004) J Am Chem Soc 126:12722
160. Udit AK, Belliston-Bittner W, Glazer EC, Le Nguyen YH, Gillan JM, Hill MG, Marletta MA, Goodin DB, Gray HB (2005) J Am Chem Soc 127:11212
161. Dunford HB (1999) Heme peroxidases. Wiley-VCH, New York
162. Anni H, Yonetani T (1992) Mechanism of Action of Peroxidases. In: Siegel H, Siegel A (eds) Metal Ions in Biological Systems. Marcel Dekker, New York, p 219
163. Everse J, Everse KE, Grisham MB (eds) (1992) Peroxidases in Chemistry and Biology, vol 1 and 2. CRC Press, Boca Raton
164. Poulos TL (1993) Curr Opin Biotech 4:484
165. Ryan O, Smyth MR, O'Fagain C (1994) Essays Biochem 28:129
166. Ator MA, de Montellano PRO (1987) J Biol Chem 262:1542
167. Barman TE (ed) (1992) Enzyme Handbook, vol 1. Springer-Verlag, Berlin
168. Christenson A, Dimcheva N, Ferapontova EE, Gorton L, Ruzgas T, Stoicha L, Shleev S, Yaropolov AI, Haltrich D, Thorneley RNF, Aust SD (2004) Electroanal 16:1074
169. Wollenberger U, Bogdanovskaya VA, Bobrin S, Scheller F, Tarasevich MR (1990) Anal Lett 23:1795
170. Wollenberger U, Wang J, Ozsoz M, Gonzalez Romero E, Scheller F (1991) Bioelectrochem Bioenerg 26:287
171. Ho WO, Athey D, McNeil CJ, Hager HJ, Evans GP, Mullen WH (1993) J Electroanal Chem 351:185
172. Ruzgas T, Csöregi E, Emnéus J, Gorton L, Marko-Varga G (1996) Anal Chim Acta 330:123
173. Ferapontova EE, Grigorenko VG, Egorov AM, Börchers T, Gorton L (2001) J Electroanal Chem 509:19
174. Crumbliss AL, Perine SC, Stonehuerner J, Tubergen KR, Zhao JG, Henkens RW (1992) Biotechnol Bioeng 40:483
175. Zimmermann H, Lindgren A, Schuhmann W, Gorton L (2000) Chem Eur J 6:592
176. Gorton L, Jönsson-Pettersson G, Csöregi E, Johansson K, Dominguez E, Marko-Varga G (1992) Analyst 117:1235
177. Loew N, Bogdanov Scheller F, Wollenberger U, Katterle M (2006) Electroanal 18:2324
178. Gaspar S, Zimmermann H, Gazaryan I, Csöregi E, Schuhmann W (2001) Electroanal 13:285
179. Liu S, Wollenberger U, Halamek J, Leupold E, Stöcklein W, Warsinke A, Scheller FW (2005) Chem Eur J 11:4239
180. Ferapontova EE, Gorton L (2002) Bioelectrochemistry 55:83
181. Lindgren A, Tanaka M, Ruzgas T, Gorton L, Gazaryan I, Ishimori K, Morishima I (1999) Electrochem Commun 1:171

182. Ferapontova EE, Reading NS, Aust SD, Ruzgas T, Gorton L (2002) Electroanal 14:1411
183. Munteanu FD, Lindgren A, Emnéus J, Gorton L, Ruzgas T, Csöregi E, Ciucu A, van Huystee RB, Gazaryan IG, Lagrimini LM (1998) Anal Chem 70:2596
184. Lindgren A, Ruzgas T, Gorton L, Csöregi E, Ardila GB, Sakharov IY, Gazaryan IG (2000) Biosens Bioelectron 15:491
185. Andreu R, Ferapontov EE, Gorton L, Calvente JJ (2007) J Phys Chem B 111:469
186. Ferapontova EE (2004) Electroanal 16:1106
187. Yu X, Sotzing GA, Papadimitrakopoulos F, Rusling JF (2003) Anal Chem 75:4565
188. Dock E, Lindgren A, Ruzgas T, Gorton L (2001) Analyst 126:1929
189. Mao LQ, Osborne PG, Yamamoto K, Kato T (2002) Anal Chem 74:3684
190. Kacaniklic V, Johansson K, Marko-Varga G, Gorton L, Jönsson-Petterson G, Csöregi E (1994) Electroanal 6:381
191. Kulys J, Bilitewski U, Schmid RD (1991) Sens Actuator 3:227
192. Ghindilis AL, Morzunova TG, Barman AV, Kurochkin IN (1996) Biosens Bioelectron 11:873
193. McNeil CJ, Athey D, Renneberg R (1997) Immunosensors for Clinical Diagnosis. In: Scheller FW, Schubert F, Fedrowitz J (eds) Frontiers in Biosensorics II, Fundamental Aspects. Birkhäuser Verlag, Basel, p 17
194. Zeravik J, Ruzgas T, Franek M (2003) Biosens Bioelectron 18:1321
195. Wollenberger U, Neumann B, Scheller FW (1998) Electrochim Acta 43:3581
196. Rajagopalan KV, Johnson JL, Wuebbens MM, Pitterle DM, Hilton JC, Zurick TR, Garrett RM (1993) Adv Exp Med Biol 338:355
197. Hilton JC, Rajagopalan KV (1996) Arch Biochem Biophys 325:139
198. Hilton JC, Temple CA, Rajagopalan KV (1999) J Biol Chem 274:8428
199. Hille R (1996) Chem Rev 96:275
200. Hille R (2002) Met Ions Biol Syst 39:187
201. Oku Y, Ohtaki A, Kamitori S, Nakamura N, Yohda M, Ohno H, Kawarabayashi Y (2004) J Inorg Biochem 98:1194
202. Garattini E, Mendel R, Romão MJ, Wright R, Terao M (2003) Biochem J 372:15
203. Aguey-Zinsou KF, Bernhardt PV, Leimkühler S (2003) J Am Chem Soc 125:15352
204. Correia dos Santos MM (2004) Eur J Biochem 271:1329
205. Garton SD, Temple CA, Dhawan IK, Barber MJ, Rajagopalan KV, Johnson MK (2000) J Biol Chem 275:6798
206. Bell AF, He X, Ridge JP, Hanson GR, McEwan AG, Tonge PJ (2001) Biochemistry 40:440
207. Aguey-Zinsou K-F, Bernhardt PV, McEwan AG, Ridge JP (2002) J Biol Inorg Chem 7:879
208. Heffron K, Leger C, Rothery RA, Weiner JH, Armstrong FA (2001) Biochemistry 40:3117
209. Hoke KR, Cobb N, Armstrong FA, Hille R (2004) Biochemistry 43:1667
210. Kisker C, Schindeline H, Pacheco A, Wehbi WA, Garrett RM, Rajagopalan KV, Enemark JH, Rees DC (1997) Cell 91:973
211. Elliot SJ, McElhaney AE, Feng C, Enemark JH, Armstrong FA (2002) J Am Chem Soc 124:11612
212. Aguey-Zinsou KF, Bernhardt PV, Kappler U, McEwan AG (2003) J Am Chem Soc 125:530
213. Ferapontova EE, Ruzgas T, Gorton L (2003) Anal Chem 75:4841
214. Spricigo R, Leimkühler S, Wollenberger U, Scheller F (2007) Proc XIX Int Symp Bioelectrochem Bioenerg 67p

215. Kappler U, Bennett B, Rethmeier J, Schwarz G, Deutzmann R, McEwan AG, Dahl C (2000) J Biol Chem 275:13202
216. Tayhas G, Palmore R (2004) Trends Biotech 22:99
217. Udit AK, Hill MG, Bittner VG, Arnold FH, Gray HB (2004) J Am Chem Soc 126:10218
218. Udit AK, Hindoyan N, Hill MG, Arnold FH, Gray HB (2005) Inorg Chem 44:4109
219. Sultana N, Schenkman JB, Rusling JF (2005) J Am Chem Soc 127:13460
220. Davis JJ, Djuricic D, Lo KKW, Wallace ENK, Wong LL, Hill HAO (2000) Faraday Diss 116:1
221. Rusling JF, Zhou L, Munge B, Yang J, Estavillo C, Schenkmann JB (2000) Faraday Disc 116:1
222. Bernhardt PV, Honeychurch MJ, McEwan AG (2006) Electrochem Comm 8:257
223. Sultana N, Schenkman JB, Rusling JF (2005) J Am Chem Soc 127:13460

Adv Biochem Engin/Biotechnol (2008) 109: 65–96
DOI 10.1007/10_2007_080

Published online: 25 October 2007

Protein Engineering and Electrochemical Biosensors

Andreas Lambrianou · Soren Demin · Elizabeth A. H. Hall (✉)

Institute of Biotechnology, University of Cambridge, Tennis Court Road,
Cambridge CB2 1QT, UK
lisa.hall@biotech.cam.ac.uk

Abstract Protein engineered biosensors provide the next best step in the advancement of protein-based sensors that can specifically identify chemical substrates. The use of native proteins for this purpose cannot adequately embrace the limits of detection and level of stability required for a usable sensor, due to globular structure restraints. This review chapter attempts to give an accurate representation of the three main strategies employed in the engineering of more suitable biological components for use in biosensor construction.

The three main strategies in protein engineering for electrochemical biosensor implementation are: rational protein design, directed evolution and de novo protein design. Each design strategy has limitations to its use in a biosensor format and has advantages and disadvantages with respect to each. The three design techniques are used to modify aspects of stability, sensitivity, selectivity, surface tethering, and signal transduction within the biological environment.

Furthermore with the advent of new nanomaterials the implementation of these design strategies, with the attomolar promise of nanostructures, imparts important generational leaps in research for biosensor construction, based on highly specific, very robust, and electrically wired protein engineered biosensors.

Abbreviations

αHL	α-Hemolysin
α-TNTscFV	Anti-TNT antibody fragment
AChE	Acetylcholinesterase
β-CD	β-cyclodextrin
BP	Binding protein
BR	Bacteriorhodopsin
Cyt *c*	Cytochrome *c*
DDI	DNA directed immobilization
dsDNA	Double-stranded DNA
ELISA	Enzyme-linked immunosorbent assay
epPCR	Error-prone PCR
ET	Electron transfer
FAD	Flavin adenine di-nucleotide
FRET	Fluorescence resonance energy transfer
FMN	Flavin mononucleotide cofactor
GBP	Glucose binding protein
GDH	Glucose dehydrogenase
GOx	Glucose oxidase
HRP	Horseradish peroxidase
MBP	Maltose binding protein
NA	Neutravidin
NTA	Nitrilotriacetic acid
PCR	Polymerase chain reaction
PQQ	Pyrroloquinoline quinone
QBP	Quinone binding protein
RC	Reaction centre
SAM	Self assembling monolayer
SDM	Site directed mutagenesis
SiNW	Silicon nanowires
ssDNA	Single-stranded DNA
SWNT	Single-walled nanotubes
TMADH	Trimethylamine dehydrogenase
TNT	2,4,6-Trinitrotoluene
WT	Wild type

1 Introduction

Biosensors offer the potential to detect widespread diverse molecular species in a complex mixture. Even though protein-based biosensors have been successfully developed, the task of identifying a protein with the desired analyte specificity and transducer mechanism can be arduous [1], with the detector instrumentation also needing to be changed to suit the unique signal transduction properties of the natural protein. A powerful tool currently being used to expand the biosensor family is protein engineering; this could eliminate the limitation of creating biosensors only with naturally occurring biological macromolecules. The ultimate vision may be a generic biosensor, with a modular molecular-engineering system in which the integrated signal transduction function and the specific binding site could be modified separately without hindrance to the interaction between them.

In the first instance, it is necessary to identify the characteristics of different approaches that may achieve this. Three prevailing methods in protein engineering exist: rational protein design, directed evolution and de novo design. The first two methods have become some of the most powerful and widely used tools in enzyme biotechnology. Rational protein engineering involves the creation of defined mutations (substitution, deletion or insertion) in the sequence of a gene coding for a protein that results in specific changes in amino acid sequence of the protein, leading to desirable changes in function. It uses methods such as site directed mutagenesis and megaprimer PCR mutagenesis (variation of site directed mutagenesis), while directed evolution uses DNA shuffling, error prone PCR, chemical and radiation mutagenesis, which produce a random generation of mutations in the gene sequence that ultimately, through screening of thousands of mutants, presents a modified version of the protein that may have enhanced features for any desired phenotype. Figure 1 illustrates the methods for rational protein engineering. De novo protein design is characterized as the complete construction of a novel protein structure encompassing all protein folding concepts, including the physical forces that stabilize native protein structures (van der Waals interactions, electrostatics and environment free energy) and the understanding of all amino acid interactions.

To use rational redesign of a target protein for biosensor development requires certain pieces of information; the most obvious being the gene coding sequence. However, information on the structure and chemistry (either crystal or NMR structure), in sufficient detail, needs to be available so that specifically desired amino acid changes in the sequence can begin to be designed. Changes in the amino acid sequence, although made at single points may not be inconsequential for the whole protein. The basic forces (van der Waals and electrostatic interactions, hydrogen bonds, the hydrophobic forces, and the favourable packing interactions associated with the condensed state of protein interiors and exteriors) that determine the non-covalent interactions within the polypeptide chain, with the surrounding solvent, and with ligands will be affected by any such changes and need to be well understood.

The ultimate goal in protein engineering for electrochemical biosensor implementation is the creation of stable, highly selective protein macromolecules with an up to attomolar-sensitive analyte binding entity, complemented by a high throughput direct electron transfer link to an electrode surface. Each of the protein engineering techniques could be used in the development of functional biosensors, either by engineering an appropriate signal transduction function or appropriate analyte specificity. The protein design process can thus be divided into different steps: the design of the

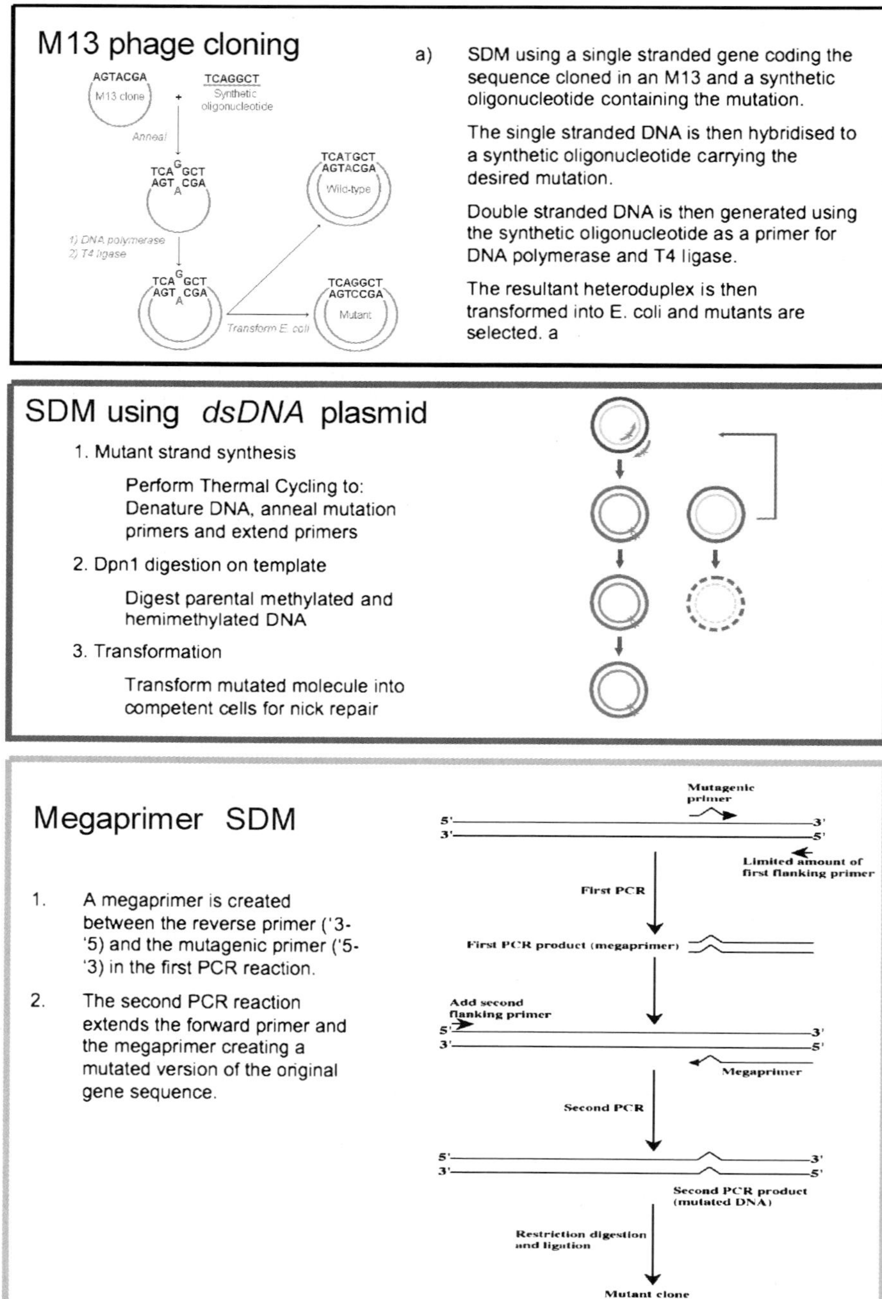

Fig. 1 The techniques used for rational protein engineering. These include M13 phage cloning, SDM via dsDNA plasmid and megaprimers

Table 1 Summary of protein engineering techniques and uses in biosensor applications

Protein engineering technique	Engineering purposes
Rational protein design	Stability, sensitivity, selectivity, signal transduction, surface tethering
Directed evolution	Stability, sensitivity, selectivity
De novo protein design	Stability, sensitivity, selectivity, signal transduction, surface tethering

overall scaffold and its transduction capability and the design of active sites. Table 1 summarizes the uses of the three protein engineering approaches on biosensor applications.

2 Rational Protein Design

To understand these approaches in the context of protein engineered biosensors, used in conjunction with electrochemical transduction methods, the results of research using these techniques must be considered. A great deal of work has focused on several aspects of this vision, but mostly without biosensor applications in mind. It is clear that while protein engineering is an attractive way in which to create the designer protein biosensor, the design process is not trivial. This can be seen for example if we consider how rational protein design might be used to improve biosensor performance. As always, rational protein engineering of suitable macromolecules begins with the structural analysis of the protein, found in the Protein Data Bank or Swiss-Prot, using visual software (SwissPDB, Insight, Rasmol, etc.).

2.1 Engineering Protein Stability

One of the main problems of many biosensors is their lack of intrinsic stability. In some notable cases this has not been a drawback, allowing for example, the successful commercialization of glucose biosensors world-wide. However, the majority of enzymes, with existing functionality suitable for biosensors, are labile and require stabilization to produce viable devices [2, 3].

The rational design of a mutation that will increase the ΔG of folding requires a deep understanding of the forces underlying the energy balance of protein and solvent in the folded as well as the unfolded state. Trivially, $\Delta G = \Delta H - T\Delta S$ indicates that there are principally two ways to stabilize a protein, via ΔH or ΔS. In practice, this mission is still difficult to design and it is nearly impossible to decompose the effect of a mutation by reasoning. For example, a Ser → Pro mutation in a surface loop is not likely to remove or add any special atomic interactions. This mutation is more likely to affect the entropy of the unfolded protein and is therefore, often referred to as an example of "entropic stabilization" [4]. Entropic stabilization is a term used to describe the relationship of the entropy between the folded and unfolded states, where decreasing the latter results in the folded state becoming more energetically favourable. If the ΔH and ΔS of such a mutation are measured, however, it will be seen that enthalpy does play a role.

It is obvious however, that all terms that contribute to ΔH and/or ΔS contribute to ΔG and thus, to the stability of a protein. Some important terms are van der Waals interactions, hydrogen bonds, salt bridges, torsion potentials, bond stretching, planarity of conjugated systems, π–π stacking, the entropy of water (this is by far the biggest term of all, and probably the least well understood), interactions with ions, loop tension, helix dipole interactions, and disulfide bridges.

An interesting example of a rational approach to engineering an enzyme for a biosensor application is described by Yoshida and Sode [5]. Using the glucose dehydrogenase (GDH) enzyme, containing pyrroloquinoline quinone (PQQ) as its prosthetic group, a series of substitution mutations were made to investigate the thermal stability of the protein. The mutant GDH, Ser231Cys, showed higher thermal stability than the wild type. The Ser231 residue was consequently replaced with a series of amino acids, and their thermal stability and kinetic activity analyzed. The amino acid variants included Ser231Met, Ser231His, Ser231Asp, Ser231Asn, Ser231Leu, and Ser231Lys. All the variants showed similar stability to the wild type except the Ser231Lys mutant which showed a higher thermal stability. The specific activity of the mutant increased slightly (10% increase) compared to the wild type. The authors furthermore, suggest this mutation constitutes a better enzyme system for use as a glucose sensor as it retains an almost 10-fold higher activity than any glucose oxidase. However, the accessibility of the active site and advantages in interference of the PQQ prosthetic group in an electrochemical environment have not been established. Thermal stability of proteins is a major research area for use in sensor applications as thermal stability often results in a longer shelf life for viable biosensors while maintaining result integrity [2, 6–9].

2.2
Engineering Protein Sensitivity

There is an enormous potential in the genetic engineering of enzymes to create more selective and sensitive biomolecules that will eventually improve the detection limit of current biosensors and aid in the development of superior analytical characteristics. There are many examples of rationally designed proteins for increased sensitivity and selectivity in the literature [9, 10], usually focused on the design of residues around the binding site, but also associated with the prosthetic group. The most relevant to biosensors are the advances in acetylcholinesterase-based biosensors. For example, the design of a biosensor for the detection of dichlorvos at attomolar levels was described in Sotiropoulou et al. [11], where a highly sensitive double mutant (E69Y Y71D) of the *Drosophila melanogaster* acetylcholinesterase (AChE) was created with a decrease in the detection limit for dichlorvos to 10^{-17} M. This work uses the rational design approach (Fig. 1) employing computer modeling and site directed mutagenesis to select and mutate individually selected residues close to or in the active site. In this instance residues 69 and 71 are targeted (Fig. 2).

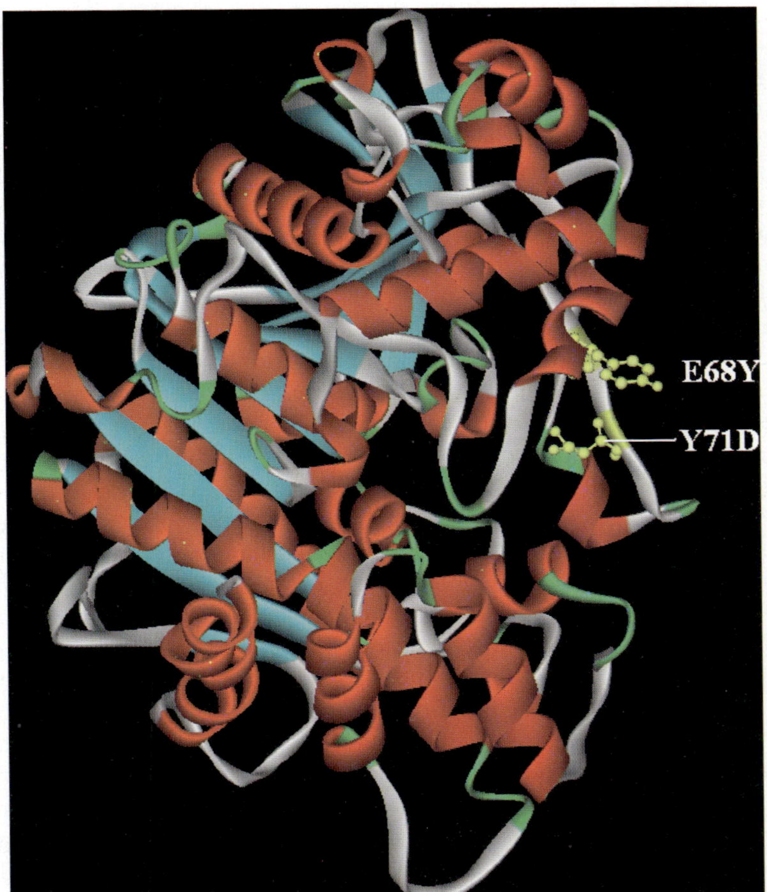

Fig. 2 Mutations E69Y and Y71D completed by Sotiropoulou et al. [11] on *Drosophila melanogaster* acetylcholinesterase

2.3 Engineering Tethering Capability in Proteins

The interface between the protein and the transducer in a biosensor is possibly the most critical junction that needs to be designed. For an electrochemical amperometric biosensor, the properties of this interface may also be relevant to the charge transfer process between the protein and the electrode [12, 13]. Classically enzyme–electrode attachment in biosensors has involved chemical or physical methods that have not involved engineering the protein [14–17]. However, facilitating tethering as an inherent property of the protein can potentially offer certain advantageous features such as site directed orientation. Several strategies have been widely used to capture or anchor biomolecules at surfaces depending on the type of protein [18]. The common approach, however, is the engineering of a component onto the C or N-terminal sequence of the protein, providing a "tail" generally protruding from the protein for interaction. These include histidine tags, polylysine residues, and DNA hybridization [1, 9, 12, 19–24].

The addition of a histidine tag on the end terminal sequence of a protein enables coupling with a chemically modified electrode surface, for example, with a self-assembled monolayer of hydroxyl- and Ni(II)-nitrilotriacetate-terminated headgroups to which the protein is site-specifically coordinated via attachment of the histidine tag to the Ni(II) ion [25, 26]. This type of attachment has been used for sensors based on both enzyme and binding protein biomolecules [12, 13, 20, 24]. Much of the enzyme work using this strategy remains relatively unspectacular in terms of the enzyme properties itself, as the active redox site remains buried within the tertiary protein structure and a mediator is still required. However, it is an effective and well-tried reversible immobilization method that has been exhaustively optimized for protein purification (metal affinity chromatography).

An excellent example of a directly tethered protein substrate onto an electrode surface is illustrated by Trammel [20]. Here the immobilization described is of a photosynthetic reaction centre (RCs) from *Rhodobacter sphaeroides* onto a gold surface with the RC primary donor looking towards the substrate by using a genetically engineered polyhistidine tag (His_7) at the C-terminal end of the M-subunit and a Ni^{2+}-NTA terminated self-assembled monolayer (SAM). With the addition of an electron acceptor, ubiquinone-10, illumination of this RC electrode generated a cathodic photocurrent. The action spectrum of the photocurrent coincides with the absorption spectrum of RC and the photocurrent decreases in response to the herbicide, atrazine.

Cyclic voltammetry of gold modified electrodes containing immobilized RC between 0.2 and 0.9 V versus a control at 50 mV/s revealed an irreversible oxidation with a peak potential at 0.7 V. This peak was assigned to the oxidation of the primary donor. This approach was successful in proving that Ni^{2+} NTA terminated SAM on gold provides a useful platform for immobilization of proteins; however, it does not achieve any charge "communication" between the protein and transducer surface to generate an amperometric signal.

Clearly, the polyhistidine tag for immobilization is not the critical element for this; any amino acid tag could be added to the end terminal of the protein, by protein engineering for immobilization, if it creates a useful chemistry for immobilization. For example, a polylysine chain can also be useful to attach to a carbon-based electrode as the lysine residues adhere easily by hydrophobic interaction, or can be chemically coupled onto say, a carbon electrode surface.

Beissenhirtz et al. [27] demonstrates a tethering approach, for electrochemical sensing of superoxide, using cysteine mutations on the surface of a superoxide dismutase monomer. These mutations allowed for attachment to a gold surface via the new thiol group present on the surface of the monomer. The mutant monomers when attached to a gold surface showed quasi-reversible direct electron transfer. In particular mutant Ser145Cys showed a noticeable shift in the potential towards more positive values compared to the WT. This method for immobilization onto an electrode surface provides an alternative to using his-tag sequences and can specifically orientate the engineered protein depending on where the mutations are designed.

2.4 Towards Charge Transfer Communication

From these simple his-tag immobilization systems, methods still need to be designed to create and integrate the transduction component. If the protein is to be designed for electrochemical amperometric measurements, then understanding the principle of electron transfer (ET) in proteins is important in finding strategies for facilitating electrical communication between proteins and their surrounding environment. One important factor that influences the electron transfer rate is the distance between the redox site and the electrode.

Direct electron transfer between protein and an electrode is controlled mainly by three factors [28]:

- Reorganization energies
- Potential differences and orientations of the redox-active sites involved in each oxidation state
- Distances between redox-active sites and charge transfer agent/mediator

In proteins, the electron transfer rates drop by 10^4 when the distance between an electron donor and acceptor is increased from 8 to 17 Å [29]. Analogies can be gained from examination of an optical Förster resonance energy transfer system where distance between acceptor and donor determine the fluorescence coupling efficiency [30]. For example, Medintz et al. [21] reported a prototype biosensor, which uses a maltose binding protein (MBP) biorecognition element as the basic sensing unit. NTA Ni^{2+} binding provides a generic site for tethering his-tag proteins, so that MBP with the C-terminal engineered with a histidine tag for attachment onto the surface could be anchored at a biotin-NTA coated surface, self-assembled on NeutrAvidin (NA).

In this instance, MBP binds the cyclic sugar β-cyclodextrin (β-CD), and a competitive binding assay is set up whereby a dye labelled β-CD is coimmobilized on the same surface as the MBP on a modular "tether arm" (Fig. 3). For the FRET system to work, rational selection of a residue close to the active site is required in order to attach a donor molecule. Thus, residue 95 was mutated to Cys and CY3 dye attached.

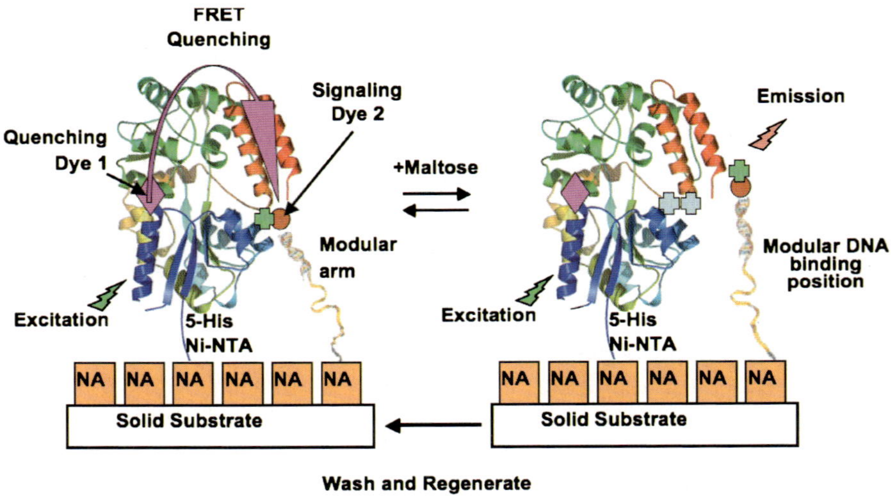

Fig. 3 The design of a maltose sensor using a histidine tagged mutant MBP with residue 95 substituted for a cysteine residue for attachment to a CY3 dye. The MBP is tethered to the gold surface via Ni^{2+}-NTA attachment while the hybridized DNA arm is fluorescently labelled to institute a FRET signal when maltose is not present. With the addition of maltose the β-CD on the DNA arm is replaced reducing the overall FRET

By using DNA as part of the tether arm, the target analogue and signal transduction functionalities (β-CD and acceptor dye) could be anchored to the sensor surface via DNA directed immobilization (DDI), using hybridized DNA interactions with modified oligonucleotides. The transduction principle then follows competitive displacement of β-CD with maltose, at the MBP binding site. Movement of the β-CD analogue into or out of the protein-binding site is coupled with obligatory movement of the integrated signaling dye, and this spatial displacement forms the basis for quantitative reporting of analyte sugar concentration by changes by FRET.

By using protein engineering techniques in designing the binding site Medintz et al. [22] implemented a general design strategy continuing from this work. This was demonstrated for example, in the development of a TNT sensor. The binding site is created using an antibody fragment, α-TNTscFv (derivative of an antiTNT antibody) with a polyhistidine tag introduced via a cloning vector onto the C-terminal of the protein. This also demonstrates a suitable strategy for use in the absence of direct structural information on the α-TNTscFv and its polyhistidine tail. A homology model was formulated based on the degree of sequence homology between the α-TNTscFv and scFv crystal structure. Location of point mutations were selected based on:

- Surface-exposed residues to facilitate labeling
- Proximity to peptide turns
- Locations distal from the binding site
- Locations distal from already present cysteines
- Selection of residues which would have minimal effect on overall structure

Four mutations at residues G13C, L145C, A210C, and A241C (Fig. 4) were undertaken. However, mutations are not necessarily inconsequential: 241 was found to interfere with the histidine tail and mutation at residue 13 resulted in inclusion bodies. Finally after ELISA screening mutation 145 was found to have suitable activity.

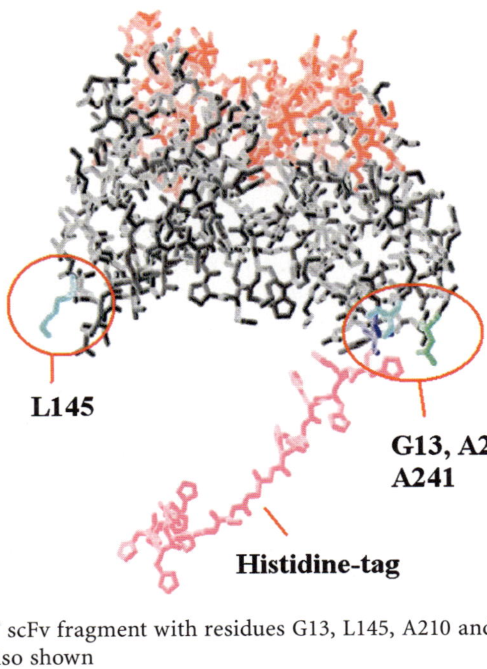

Fig. 4 The a-TNT scFv fragment with residues G13, L145, A210 and A241 shown. The 12x histidine tag is also shown

Thus, this mutant could be dye-labelled with AFF 532 and the acceptor and donor positions suitably engineered to achieve the same FRET scheme described above: competitive replacement of TNB (attached to the DNA modular arm) with TNT reduces FRET (Fig. 5).

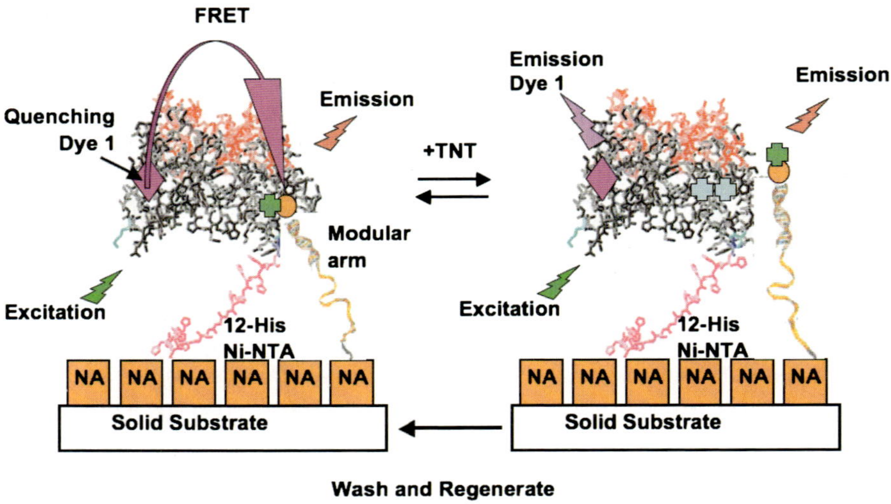

Fig. 5 TNT sensor design with a-TNT scFv antibody fragment attached to the Ni^{2+}-NTA via a 12x histidine tag. AFF 523 dye is covalently attached to mutated residue 145 that creates a FRET response when the DNA modular arm is in close proximity (TNT analogue molecule bound to DNA). On addition of TNT the TNT analogue is replaced removing the DNA arm from the antibody binding site and reducing the overall FRET

2.5 Artificial Cofactors and Reporter Groups

Interestingly, the analogous electrochemical amperometric system does not seem to have been reported as yet, and similar models that bring redox "donor" and "acceptor" within close proximity through protein engineering are less common. However, Benson and Trammell [19, 31] also used his-tag coupling to anchor their binding proteins (BPs), and needed to employ a covalently bound redox system to act as a reporter of the binding process. They describe the use of the ligand-mediated hinge-bending conformational change in the construction of a chemical sensor, by modulating electronic coupling between the surface of a self-assembled nanostructure and a covalently linked redox-active reporter. MBP, GBP, and QBP have all been used in the construction of such amperometric biosensors using this self-assembled nanostructure (Fig. 6). The MBP was tethered to the gold surface by the introduction of a classical polyhistidine tag onto the C-terminal sequence of the protein using a cloning vector with C-terminal his-tag present, but the most interesting feature of this engineered enzyme was that SDM was used to substitute residue Gly174 for cysteine, so that the synthetic Ru(II) "cofactor" could be covalently linked to the Cys174 residue.

Thus, a multicomponent nanostructured layer was self-assembled onto a gold electrode consisting of a thiol-reactive, hydroxyl-terminated mono-

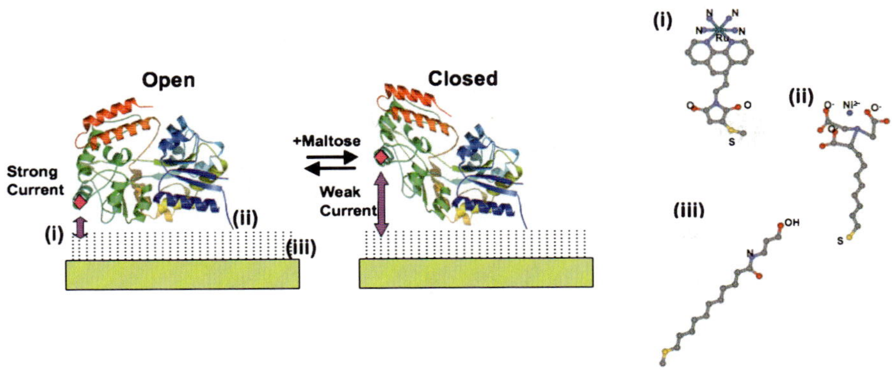

Fig. 6 The design for a reagentless electrochemical maltose sensor. Comprised of a self-assembled layer on a gold electrode consisting of a thiol-reactive, hydroxyl-terminated monolayer (**iii**), doped with Ni-NTA groups (**ii**), to which an oriented MBP monolayer is attached via a C-terminal oligohistidine tag. A redox reporter group (**i**) is covalently attached to a cysteine, such that it is positioned between the electrode and MBP surfaces. In the absence of maltose, the open conformation permits strong electronic coupling between the electrode surface and the reporter group, whereas in the closed form, this interaction is weakened

layer, doped with Ni-NTA groups, to which the his-tag oriented binding protein (e.g. maltose binding protein, MBP) monolayer was attached via the C-terminal oligohistidine sequence. The synthetic redox reporter group, introduced by covalent attachment to the cysteine, was designed such that it was positioned between the electrode and MBP surfaces. In the absence of maltose, the open conformation permitted strong electronic coupling between the electrode surface and the reporter group, whereas in the closed form with bound maltose, this interaction was weakened and binding followed amperometrically, with the strongest current being observed in the absence of maltose at the redox potential of the reporter group. As with the comparable FRET systems, an important feature of this model is the distance between the redox reporter and the electrode, depending on binding. Since the binding protein itself does not contain redox groups, this approach is then not feasible by protein engineering alone.

On the other hand, this is to some extent reminiscent of the fusion protein approach using fluorescent conjugates based on FRET between green fluorescent protein (GFP) variants fused to the N and C-termini of the maltose binding protein (MBP) [32], glucose binding protein (GBP) [33], and ribose binding protein (RBP) [34]. These fusion proteins show ligand-dependent changes in the fluorescence intensity emission of the FRET acceptor. Although the absolute changes are not large, the fusion proteins can be used as sensors by monitoring changes in the ratio of donor and acceptor emission intensities. The most important advantage of these sensor constructs is that the entire sensor is genetically encoded and hence can be used in in vivo studies. Similar fusion of electron transfer proteins could conceivably produce the "electrochemical" equivalent of this system.

2.6 Communication with Existing Redox Groups

Lysine has been a particularly versatile mutation because it offers a way to introduce electron transfer mediators (ferrocene) to anchor onto the polylysine chain. Achieving direct electron transfer between the active centre of the enzyme and electrode is crucial for the further development of amperometric enzyme biosensors [35, 36]. Direct electron transfer with wild type enzymes has been observed on small redox proteins such as cytochrome *c* [14], horseradish peroxidase (HRP) [37], and lactase [15]. However for most oxidoreductase enzymes, their active site redox centres are embedded deep inside the glycoprotein shell, such that charge transport communication is weak. For example, the FAD group of GOx is deeply embedded within a protective protein matrix (Fig. 7), so that the matrix or glycoprotein shell surrounding the redox site creates an effective kinetic barrier for electron transfer.

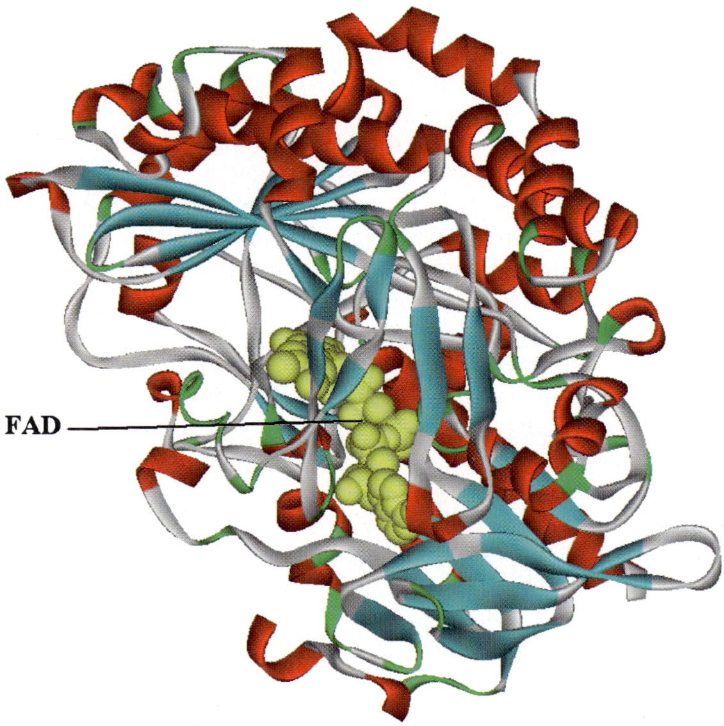

Fig. 7 Glucose oxidase (GOx) molecule showing the FAD cofactor embedded within the protein matrix

Chen et al. [38] described a polylysine chain cloned into an expression vector containing a glucose oxidase (GOx) sequence, using 10 lysine residue primers with HindIII and Not1 restriction sites. These restriction sites were used to insert the sequence of lysines into an open reading frame between the HindIII and Not1 restriction sites, yielding a GOx-lys fusion. Chemical modification of this GOx-lys fusion protein by an EDC coupled reaction then linked to ferrocenecarboxylic acid. This modification improved the signal level from 5 μA for the wild type GOx to 12 μA for the mutant at a 10 mM glucose solution. The linear range of the sensor was also doubled from 0–20 mM for wild type to 0–45 mM for the GOx-lys mutant sensor. Clearly, the ferrocene is acting here as a mediator or electron shuttling agent and the marked improvement in signal with ferrocene suggests that even when the enzyme is covalently anchored to the electrode, it is not correctly "formatted" to achieve direct electron transfer between the protein redox group and the electrode without these extra surface redox groups.

X-ray structure studies by Hecht and coworkers [39] suggested that the minimum distance from the redox-active centre, flavin adenine di-nucleotide (FAD), of GOx to the surface of the enzyme is 8 Å and the minimum calculated distance between the nitrogen (N7 of the isolloxazine) part of the FAD where the redox reaction occurs is 13–18 Å. Thus, electrochemical communication between enzyme and electrode is usually inhibited due to the buried electroactive sites [40].

This problem was initially resolved with the introduction of ferrocene and other mediators to mediate the electron transfer between the redox site and an electrode surface [36]. However, understanding of electron transfer in proteins has advanced considerably in recent years for both unimolecular and biomolecular processes. In addition to novel methods to control and manipulate surface chemistry of electrodes facilitating heterogeneous electron transfer with biomolecules, the possibility of engineering the enzyme must also be considered.

Loechel et al. [41, 42] described a rational design approach for the site-specific redox coupling of the protein trimethylamine dehydrogenase (TMADH) to facilitate charge transfer between enzyme and an electrode. Protein engineering and site-specific chemical modification were used to extend the electron pathway from the protein surface to a redox polymer. Critical to this example was the knowledge of the electron transport pathway through the enzyme, allowing the correct residues to be selected for mutation. After creating a cysteine mutation on residue Y442, the mutant was successfully attached to the redox polymer, poly-$[Fe(5\text{-}NH2\text{-}phen)3]^{2+}$, which linked with the electrode surface. This design enabled direct electrical communication between the enzyme and electrode (Fig. 8).

The key to this success was that the Y422C mutant was designed for wiring from the protein close to the 4Fe–4S centre, but the initial electron transfer is activated on the opposite side of the protein, close to the FMN prosthetic group,

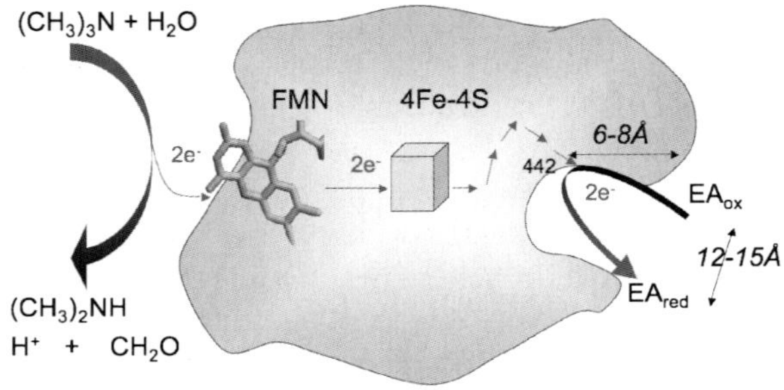

Fig. 8 Schematic of electron transfer pathways through TMADH, from the substrate active site to external acceptor docking site, showing positions of residues 442 and 648

by substrate (trimethylamine) binding at the active site. It is a two-electron process resulting in the reduced anion. The iron-sulfur centre is normally shifted to more positive potentials, giving a facilitated electron transfer pathway through the enzyme and sequential one-electron transfers to the 4Fe–4S cluster ($[4Fe-4S]^{2+}$ to $[4Fe-4S]^{+}$) achieve the reoxidation of the FMN.

The tethering of redox relay units on proteins was reported as a general means to shorten intra-protein electron transfer distances, and to "electrically wire" the redox centres of enzymes with bulk electrode supports [40]. It was found that implanting of redox relay units in inner-protein sites, close to the active site, is important to attain electron transfer paths between the enzyme's redox site and the macroscopic environment. Furthermore, the chain length that tethers the relay unit to the protein was found to be central to the effectiveness of electrical contacting, where long, flexible, bridging chains were found to facilitate the intra-protein electron transfer [43, 44].

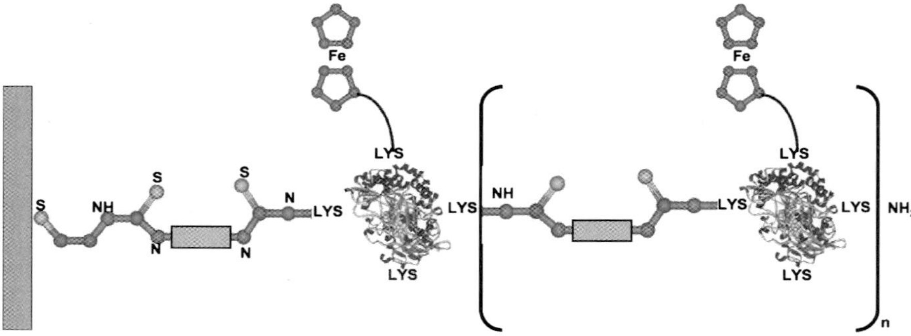

Fig. 9 The completed design of a multilayer assembly of GOx, on a gold electrode. Covalent linkage of GOx, on the gold electrode, followed by crosslinking of the enzyme layers with the bifunctional disulfonato-diisothiocyanate-*trans*-stilbene. Further tethering of the ferrocene electroactive units to the protein, electrically contacts its active site with the electrode

Figure 9 shows the completed design of a multilayer assembly of GOx, on a gold electrode. Covalent linkage of GOx, on the gold electrode, followed by crosslinking of the enzyme layers with the bifunctional disulfonato-diisothiocyanate-*trans*-stilbene, yields enzyme assemblies with a controllable number of layers. Further tethering of the ferrocene electroactive units to the protein electrically contacts its active site with the electrode. The attachment of an electron-relay unit to the redox-protein shortens the electron-transfer distances, and facilitates the electrical communication between the redox-centre and the electrode. [45]. Although in this example no protein engineering was conducted, a good strategy for future biosensor development through introduction of a tethered redox molecule can be made much easier and site directed using a rational approach to engineer specific mutations [46–50].

3
Future Protein Evolutionary Approaches

While rational design or redesign is an effective tool for optimizing and altering protein function, one drawback of rational design is that the structure and mechanism of the proteins to be designed must be understood. Unfortunately, the number of structures in the protein databases is increasing dramatically while knowledge about the protein structure and function is still quite limited. Furthermore, it is extremely difficult to predict long-range effects of residues far from the active site on the structure and function. More studies show that those long-range effects are important in protein design.

4
Directed Evolution

In contrast to the rational design, directed evolutionary design or design through combinatorial methods requires little prior knowledge of the protein structure. Random mutations can be generated in the whole gene by error-prone PCR (epPCR) or at specific residues by introducing oligonucleotides with degenerated sequences. The use of degenerate oligonucleotides is also the method of choice if an entire sequence is subject to randomization. In contrast, if several randomly distributed mutations are to be introduced into a gene, random point mutagenesis by epPCR is commonly applied [51]. However, the inherent bias of nucleotide transitions over transversions and the restrictions imposed by the genetic code do not allow for the introduction of all 20 amino acids at each position of the protein by epPCR. In fact, for any given amino acid residue, an average of 5.7 amino acid substitutions are accessible by epPCR [52]. The effect of these mutations can be enhanced with DNA shuffling.

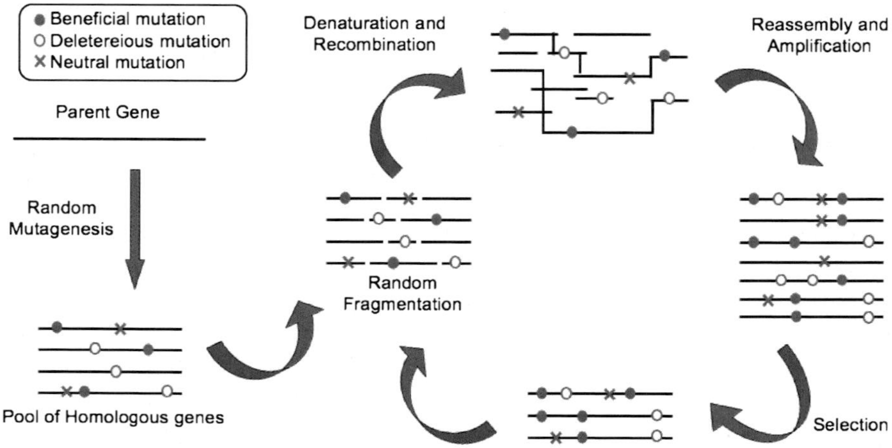

Fig. 10 The DNA shuffling method. A parent gene can be either directly treated with DNaseI or mutagenized, e.g. by epPCR, before fragmentation, to create a pool of homologous variants that contains beneficial, deleterious or neutral point mutations. Novel recombinations accompany reassembly of the random fragments into full-length genes. Improved variants are selected and provide the starting point for another round of shuffling

Table 2 Different recombination techniques

Group	Technique	Refs.
Shuffling	Recombination of small fragments based on homology in the sequence between mutations that stem from all kinds of mutagenesis strategies or different family members	Shuffling [53] Family shuffling [60] RPR RETT [61]
Full-length parent shuffling	Recombination of small fragments from different origin using one or more full-length parent strands	RACHITT [63]
Single cross-over	Recombination of non-homologous genes by ligating front and back of two different genes, selection of new genes on size	ITCHI [64] SCRATCHY [65]
Domain swapping	Recombination of structural, functional or less homologous parts of different family members	Exon shuffling [66]
In vivo recombination	Recombination using gap repair system of yeast ot recE/recT system of *E. coli*	CLERY [67]
Synthetic shuffling	Recombination of (un)known mutations in synthetic oligonucleotides	[68]

The invention of DNA shuffling by Stemmer et al. [53, 54] has stimulated much research in the area of directed evolution. DNA shuffling entails molecular mixing of similar genes in vitro, and allows mutations to be exchanged among sequences by recombination (Fig. 10). Initially, DNA from a single mutant or a pool of mutants is cleaved with DNaseI, resulting in a pool of overlapping DNA fragments of different sizes. These random fragments are purified and reassembled into full-length genes in a self-priming PCR-like reaction. During this self-priming chain extension process, single-stranded fragments anneal to each other based on their complementarity; recombination occurs when two fragments from different parent genes prime each other. The results are combinations of mutations from different copies. The enormous combinatorial potential that is realized by enabling pool-wise recombination from multiple parental sequences allows DNA shuffling to rapidly access vast areas of sequence space. Table 2 summarizes several DNA shuffling techniques that are used to enhance the mutagenesis rate in directed evolution.

4.1 Directed Evolution for Stability

In other fields, directed evolution has been used on many occasions for the development of novel mutants with increased stability [2, 52, 55, 56], sensitivity and selectivity [57, 58]. More specific to biosensor applications is the work done by Minagawa et al. [59] with lactate oxidase. Increasing the thermostability of lactate oxidase significantly prolongs the lifetime of these biosensors. Originally having obtained the more thermostable mutant of lactate oxidase via a rational design approach with two mutations (E160G/V198I), further improvement was made to the thermostability of E160G/V198I lactate oxidase using directed evolution. A library of mutant lactate oxidase genes were constructed via error-prone PCR and DNA shuffling, and screened for thermostable mutant lactate oxidase using a plate-based assay. After three rounds of screening a thermostable mutant lactate oxidase, which had six mutations (E160G/V198I/G36S/T103S/A232S/F277Y) was found. The half-life of this lactate oxidase at 70 °C was about 2 times that of E160G/V198I and about 36 times that of the wild type enzyme. The amino acid mutation process suggests that the combined neutral mutations are important in protein evolution [60–68].

4.2 Directed Evolution for Transduction

Evolution of in-built redox transduction properties has not been well applied to biosensors, although manipulation of mid-point potentials, through changes in the amino acids close to the redox prosthetic groups, has been explored, especially in the context of the study of enzyme kinetics. In terms of biosensors, we can obtain further insight into the possibilities for transduction manipulation by looking again at examples where an optical system has been employed. Bacteriorhodopsin (BR) for example, has a number of intrinsic characteristics that have made it a useful component for consideration in transduction systems. It illustrates therefore the potential capability of protein engineering that may also be applied for redox transduction. The native BR protein is already remarkably stable across a wide range of conditions [69]. The protein possesses unique photophysical properties which are attractive for manipulation and demonstrate how this might be achieved more widely. Wise et al. [7] used directed evolution on BR for improvement of light transduction. In this work five different engineering approaches were

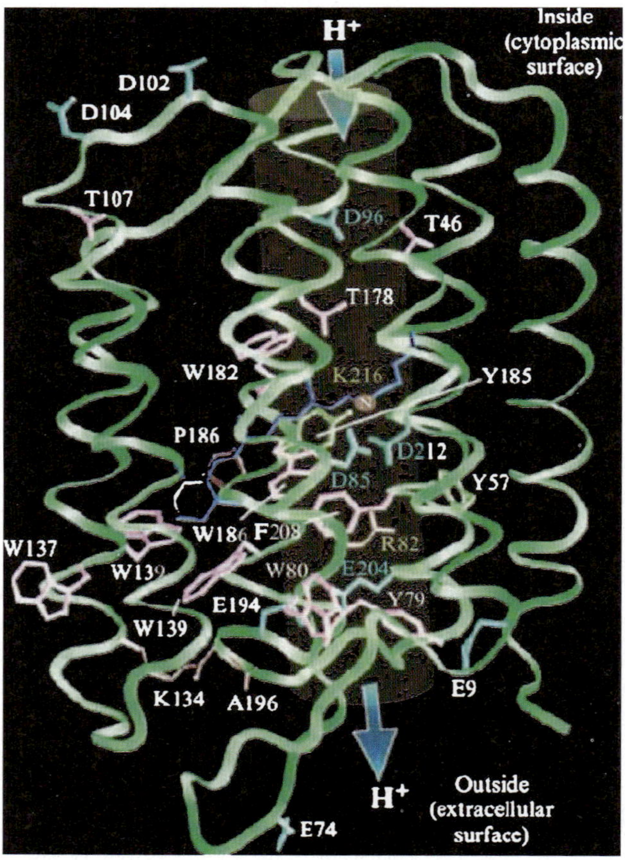

Fig. 11 Br is an ion pump that results in the net transfer of a proton from the intracellular to the extracellular surface. Key residues are highlighted, and the proton transfer channel is shown

examined, including single residue replacement through to randomized replacement and finally directed evolution. Using the first two alone a 700-fold improvement was achieved in the performance of the protein. Such "tuning" can also be combined with other properties. Helebrecht et al. [70] for example described the direct evolution of thermostability and photophysical characteristics of BR (Fig. 11).

Directed evolution is best envisioned in terms of a hypothetical "mutational landscape" [7]. An uneven "photochemical" landscape is a result of the localized nature of the characteristics being optimized; thermal stability involves a much larger portion of the protein and can be accomplished via a variety of complementary mutations.

5 De Novo Design

The assembly of smaller peptide motifs (maquettes) into de novo designed proteins is another way to impact the protein design. This design strategy has several advantages in that the maquette stabilizes the desired folding of the designed protein and also acts as a probe. The maquette construct has a defined specificity and so allows the use of different maquettes for varying metal ion sensors. Several successes have been published concerning the design of α-helical bundle proteins that incorporate hemes or iron-sulfur clusters [71–73]

A different and quite successful tactic in designing peptide models for metalloprotein active sites is to mount the metalloprotein ligands on a more rigid scaffold of known secondary and tertiary structure. Self-associating helices are the scaffold of choice, upon which metal ligands have been arranged with the aid of computer design algorithms [74]. This appears attractive for future electrochemical amperometric biosensors, where the metal redox group in the protein is engineered to be directly accessible to an electrode scaffold material [75]. Recent developments in nanostructure materials, including electrode materials (e.g. carbon nanotubes), make the feasibility of achieving the molecular scale integration of these scaffold more reasonable.

Nanostructure electrode materials are increasingly becoming more popular as novel materials for biosensors. Nanotechnology refers to research and technology development at the atomic, and molecular scale, leading to the controlled manipulation and study of structures and devices with length scales in the 1 to 100 nanometer range. Objects at this scale, such as "nanoparticles", take on novel properties and functions that differ markedly from those seen in the bulk scale. The small size, surface tailorability, improved solubility, and multifunctionality of nanoparticles open many new research avenues for biologists [76]. Some of the most recent biosensor developments using nanoscale materials seem very promising but rather quite primitive in terms of protein engineered biological substrates and site-specific immobilization.

Cui et al. [77] was one of the first to describe a highly sensitive, real-time electrically based sensor for biological and chemical species based on boron-doped silicon nanowires (SiNWs). Biotin-modified SiNWs were used to detect streptavidin in a picomolar concentration range, where biotin was covalently attached to a monomeric layer covering the SiNWs. In addition, antigen-functionalized SiNWs showed reversible antibody binding and concentration-dependent detection in real-time.

Amperometric sensor attempts using CNT have focused mostly on protein adsorption onto the surface [78]. To date these results are promising, showing the possibility of a better protein–electrode interface for electron transfer than using more classical electrode materials, but they have not yet fully joined the possibilities of protein engineering with nanostructured materials.

Cytochrome *c* for example, which has previously been adsorbed onto macroelectrodes [14], has been adsorbed onto a single-walled carbon nanotube (SWNT)-modified electrode. The direct electrochemistry of cytochrome *c* (Cyt *c*) was studied by cyclic voltammetry and the results indicated that Cyt *c* remained in its original structure and did not undergo structural change after its immobilization on the SWNT. Further results demonstrated that the SWNT had promotional effects on the direct electron transfer of Cyt *c* and also indicated that the immobilized Cyt *c* retained its electrocatalytic activity to the reduction of H_2O_2.

Alternatively, taken together with the most recent of the peptide designed systems these nanostructured materials offer enormous potential for future novel versatile biosensor capability. This has the ability to include not just primary metal ligands, but also have secondary ligand shells that may stabilize metal-ligating residues through hydrogen bonds or hydrophobic interactions. Lombardi et al. [79] targeted the class of carboxylate-bridged di-iron proteins that includes methane monooxygenase and the R2 subunit of ribonucleotide reductase amongst other well-known examples. Upon comparing active site sequences and geometries in a "retrostructural analysis" of the native proteins, a Glu–X–X–His metalbinding motif was found to reside within a four-helix bundle, with defined and well-conserved geometries, despite very low overall sequence homologies. Based on this, a "primordial" four-helix bundle was designed with histidine and glutamic acid metal ligands placed in positions *a* and *d* of the classic coiled-coil heptad repeat sequence.

This motif therefore offers a "primordial" redox centre for electrochemical communication. Furthermore, included in the design was a second shell providing hydrogen-bonding interactions to the ligating residues, and access to the metal "active site". These peptide scaffolds provide the perfect opportunity to mount electron donor and acceptors within a biomimetic framework. This framework can be used essentially for the creation of a vast number of biosensors with a wide range of detectable substrates, not only metal ions but also the α helical bundle creates an excellent electron transfer pathway from substrate-specific protein to the signaling species.

The use of de novo protein design in the construction of four-helix bundle maquettes for metal ion sensitivity and electron transfer, from active protein to electrode surface, herald the most promising work in this field of protein engineering centered on electrochemical biosensors. With a de novo designed scaffold almost perfected the next stage includes the addition of a de novo designed biorecognition element to complement a completed de novo designed biosensor (Fig. 12). De novo designed biorecognition substrates have not been used in electrochemical sensor design as yet but multiple substrates have already been generated using computational design [80] utilizing information obtained from known enzyme active site structures and interactions.

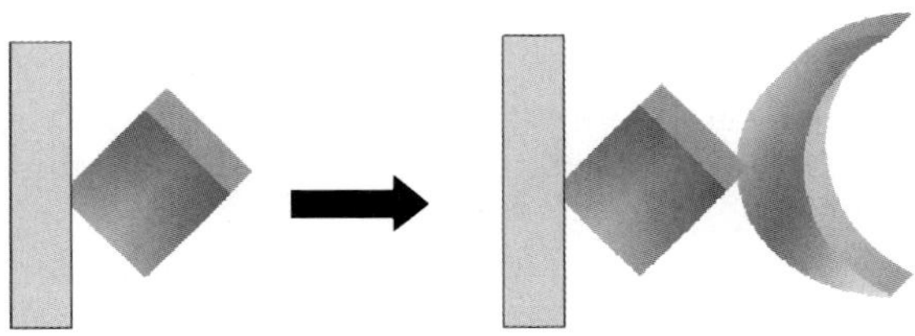

Fig. 12 De novo designed macromolecules for biosensor design. Where currently only electrode + scaffold have been utilized and electrode + scaffold + biorecognition design present the next step in de novo designed biosensors

Case et al. [81] describes the incorporation of an amino-terminal bipyridyl ligand into a designed three-helix bundle. An electron transfer system with a variable intra-helical distance between redox partners was further accomplished, with the donor and acceptor distance estimated at 6 Å.

A multilayered functional model for electron transfer is illustrated by Willner et al. [82]. The structure incorporates designed heme–peptides on a gold electrode, generating vectorial electron transport that can be coupled to redox-active proteins. A four-helix bundle, oriented by attachment to a cyclic peptide at one end and with cysteine tags at the other end, was attached to the surface of a gold electrode via a covalently attached disulfide bond formation and a linker (Fig. 13b). Two hemes were incorporated into the four-helix bundle, and stepwise, vectorial electron transport from the electrode to the proximal and then distal heme could be observed. Proteins such as cytochrome *c*, nitrate reductase, and Co(II)-substituted myoglobin were associated with the solvent-exposed end of the designed electron transfer peptide. When protein layers were covalently crosslinked using glutaraldehyde, a stable, integrated electron-transfer multilayer could be formed with biocatalytic

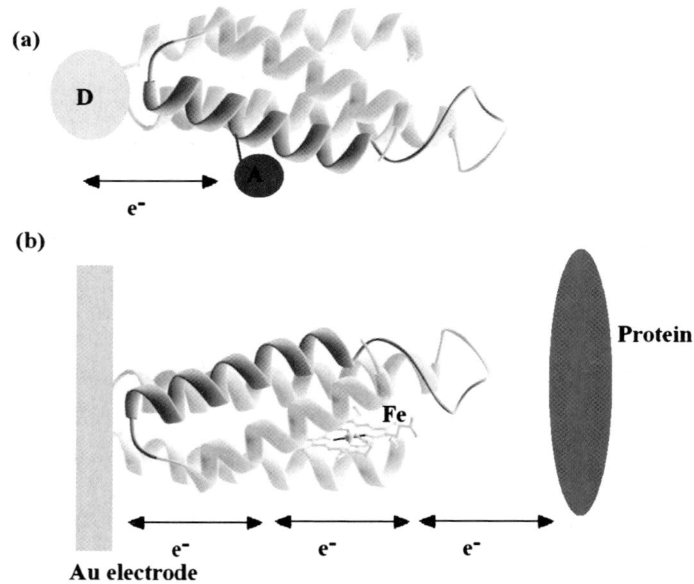

Fig. 13 **a** Fluorescent de novo design of a four helix bundle with variable distance between donor and acceptor species. **b** A layered electrode-supported biocatalytic assembly, in which the protein is a natural redox-active catalyst such as nitrate reductase

properties. Restricting the metal ion concentration to reduce electron transfer subsequently controlled this multilayer. Figure 13a illustrates the design of a four-helix bundle, utilizing existing fusion protein concepts to provide a de novo designed FRET sensor [82].

6 Engineering Protein Structure for Electrochemical Actuation and Transduction

Although this review has focused on potential opportunities to communicate with redox systems in proteins, as transducers of a biosensor analyte activated signal, protein engineering opens up the possibility of designing other electrochemically measurable signals. For example, an intrinsic signal-transduction mechanism is provided by the measurement of electrical currents resulting from the opening and closing of protein pores in membranes. The construction of such biosensors requires the formation of a multicomponent macromolecular assembly and usually involves a membrane protein, and so requires the fabrication of a robust lipid bilayer. The attention to lipid bilayers in biosensors is an extensive area, with the opportunity for peptide design and synthesis recognized quite early in its evolution. For example, α-Hemolysin is a 293-residue monomer that is water-soluble, binds spontaneously to a variety of lipid membranes, and self-assembles to form heptameric pores from identical monomers [83, 84]. These pores have further been engineered to respond to different chemical or physical switches. A β-barrel structure forms the basis of the pore and this is ideal as histidine residues can be introduced at the bottom of the pore to form the primary coordination of a metal binding site [85].

In an effort to rationally design and construct sensors for specific chemicals in solution, various types of analyte binding sites were engineered into the channel formed by *Staphylococcus aureus* α-hemolysin (αHL) [85–87]. Binding sites for the divalent cations were formed by altering the number and location of coordinating side chains, e.g. histidines and aspartic acids, between positions 126 and 134 inclusive, using SDM.

Since the experimentally determined narrowest segment of the channel's pore diameter is 2 nm [84] the pore is large enough to accommodate engineered analyte binding sites. However, the pore is sufficiently small that the binding of an analyte to a receptor site inside the pore will cause marked changes in the channel's current. In addition, unlike many voltage-gated channels, the αHL channel remains in the fully open state for long periods of time.

Kasianowicz and Bezrukov [88] have shown that the wild type αHL (WT-αHL) can sense different ions in solution and can be used to determine their concentration. For example, the reversible binding of hydrogen or deuterium ions to the channel causes rapid current fluctuations. The rate constants and equilibrium constants for these reactions were determined using the pH dependence of the current spectral density, and provided the analytical basis to use αHL in sensing applications.

Detection of transition metal divalent cations was demonstrated by Kasianowicz et al. [89], where five consecutive amino acids in a glycine rich region of the protein (residues 130–134, inclusive) were replaced with histidines using SDM and a rational design approach. The mutated channel forms a pore regardless of the mutations and can be blocked via the addition of 100 mM Zn(II) added to either side of the membrane [86]. The channel can be blocked from either side of the pore as the Zn(II) ions, in affect, clog the lower end of the barrel where the histidine residues have been introduced. Further work by Kasianowicz et al. [87] identified several residues that were near one mouth of the pore, and suggested that the pore's secondary structural motif is β-sheet. This was completed via a series of histidine scanning mutations, where residues 129, 130, 131, 132, 133, 134 (Fig. 14) were substituted to histidine and screened for pore forming activity as a function of Zn(II) concentration.

The application of engineered "DNA-nanopores" to sense individual DNA strands with single-base resolution was demonstrated in Howorka et al. [90]. Each nanopore was built by covalently attaching an individual ssDNA oligomer within the lumen of an engineered version of the αHL pore

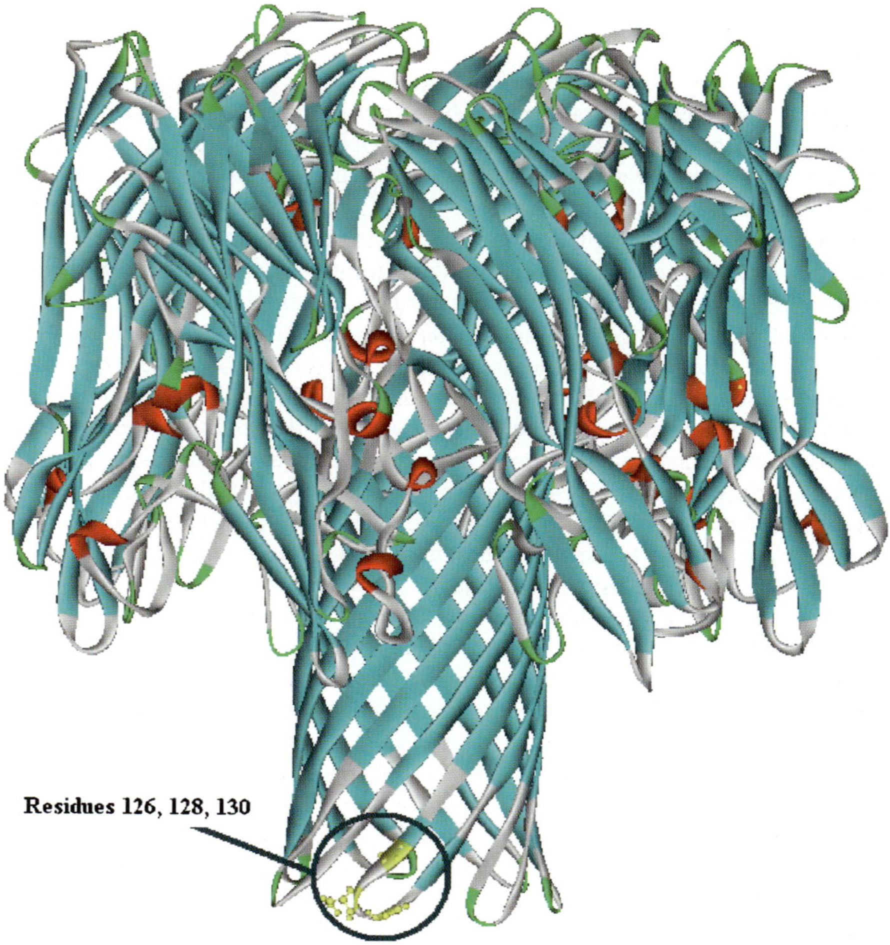

Fig. 14 Molecular structure of *Staphylococcus aureus* α-hemolysin (aHL) with residue mutations 126, 128 and 130 shown at the bottom of the β-barrel

from *Staphylococcus aureus*. This biosensor element was capable of identifying individual DNA strands with single-base resolution. The binding of single-stranded DNA (ssDNA) molecules to the tethered DNA strand caused changes in the ionic current flowing through a nanopore. On the basis of DNA duplex lifetimes, the DNA-nanopores were able to discriminate between individual DNA strands up to 30 nucleotides in length differing by a single base substitution. This was exemplified by the detection of a drug resistance-conferring mutation in the reverse transcriptase gene of HIV. Furthermore, the approach was used to sequence a complete codon in an individual DNA strand tethered to a nanopore.

7 Conclusion

Progress in biosensor development has had a major overhaul over the past 5 years. Proteins used in the development of sensing apparatus can no longer remain in their native wild type state as growing demands for increased sensitivity, stability, selectivity and ease of diversification cannot be met within the current collection of native peptides nature has to offer. The implementation of protein engineering techniques into biosensor substrate optimization has provided the solution.

With the increasing protein data bank collection the use of rationally designed proteins can only increase exponentially. Principally with our enhancement in the understanding of amino acid interactions, protein folding and overall protein structural integrity commitments, generic alterations can be prepared to hundreds of protein constructs within a single family, creating a wide variety of host peptides with engineered properties, useful for any given biosensor requirement. Furthermore, attachment of redox relays for charge transfer [40, 42], histidine tags for protein tethering [12, 24], polylysine residues for mediator adsorption [38] and in-built coupling of electron acceptors [19, 91] provides the next best step for biosensor development. SDM has successfully been used in the construction of a direct electron transfer pathway from the electrode to enzyme active site [42] either via molecular wiring or nanoparticle electron shuttling [35], constituting the dawn of the next generation biosensor [15].

Directed evolution provides perhaps the finest method of obtaining augmented protein characteristics. The ease of obtaining mutations has not been kept up by the screening techniques, leaving daunting numbers of mutations uncharacterized [57]. More recently however, high throughput methods have been introduced, drastically reducing the time and effort involved in acquiring a desired outcome [92]. Even with the desired mutations a rational approach is always necessary for the final attachment and optimization of the protein to the transduction material.

The ultimate goal of protein engineered biosensors is to specifically design the entirety of the protein construct [72, 74]. In doing this the minimal amount of protein matrix is used in obtaining a specific binding or catalytic activity. Still in its initial stages de novo protein engineering shows incredible potential as a tool for the synthesis of all future biosensor applications eliminating the need for wild type enzyme optimization.

With the introduction of nanoscale materials a new field of research has commenced to incorporate our understanding of proteins with the advantages offered from working in the nanoscale range [76, 93]. Bionanosensors or Nanobiosensors are fast becoming the preferred mode of research when dealing with sensor detection ranges below the attomolar range [77]. From an electrochemical point of view, a nanoscale electrode does not suffer from diffusion related problems and ideally individual nanoelectrodes can be tethered to single peptide molecules. In the instance of CNT arrays hundreds of different species of proteins, each with differing substrate selectivity, can be wired up in the construction of a multisubstrate biosensor.

This review has attempted to give the reader an insight into the basics concerning protein engineering for biosensor implementation in a fast developing arena. The vast amount of research currently being conducted throughout the world emphasizes the potential and opportunities in this field.

References

1. Marvin JS, Hellinga HW (1998) Engineering Biosensors by Introducing Fluorescent Allosteric Signal Transducers: Construction of a Novel Glucose Sensor. J Am Chem Soc 120(1):7–11
2. O'Fagain C (2003) Enzyme stabilization – recent experimental progress. Enzyme Microb Technol 33(2–3):137–149
3. Mezzasalma TM et al (2007) Enhancing Recombinant Protein Quality and Yield by Protein Stability Profiling. J Biomol Screen 12(3):418–428
4. Matthews BW, Nicholson H, Becktel WJ (1987) Enhanced Protein Thermostability from Site-Directed Mutations that Decrease the Entropy of Unfolding. PNAS 84(19):6663–6667
5. Sode K et al (2000) Increasing the thermal stability of the water-soluble pyrroloquinoline quinone glucose dehydrogenase by single amino acid replacement. Enzyme Microb Technol 26(7):491–496
6. Minagawa H, Shimada J, Kaneko H (2003) Effect of mutations at Glu160 and Val198 on the thermostability of lactate oxidase. Eur J Biochem 270(17):3628–3633
7. Wise KJ et al (2002) Optimization of bacteriorhodopsin for bioelectronic devices. Trends Biotechnol 20(9):387–394
8. Bryan PN (2000) Protein engineering of subtilisin. BBA Protein Struct M 1543(2):203–222
9. Gilardi G, Fantuzzi A, Sadeghi SJ (2001) Engineering and design in the bioelectrochemistry of metalloproteins. Curr Opin Struct Biol 11(4):491–499
10. Hock B, Seifert M, Kramer K (2002) Engineering receptors and antibodies for biosensors. Biosensor Bioelectron 17(3):239–249
11. Sotiropoulou S, Fournier D, Chaniotakis NA (2005) Genetically engineered acetylcholinesterase-based biosensor for attomolar detection of dichlorvos. Biosensor Bioelectron 20(11):2347–2352
12. Maly J et al (2002) Immobilisation of engineered molecules on electrodes and optical surfaces. Mater Sci Eng C 22(2):257–261
13. Kroger D et al (1999) Immobilization of histidine-tagged proteins on gold surfaces using chelator thioalkanes. Biosensor Bioelectron 14(2):155–161
14. Wei J et al. (2002) Direct Wiring of Cytochrome c's Heme Unit to an Electrode: Electrochemical Studies. J Am Chem Soc 124(32):9591–9599
15. Gorton L et al (1999) Direct electron transfer between heme-containing enzymes and electrodes as basis for third generation biosensors. Anal Chim Acta 400:91–108
16. Davis JJ, Halliwell CM, Hill HAO, Canters GW (1998) Protein adsorption at a gold electrode studied by insitu scanning tunnelling microscopy. New J Chem 22(1119–1123)

17. Klis M et al (2006) Electroreduction of laccase covalently bound to organothiol monolayers on gold electrodes. Surface Imaging/Spectroscopy at Solid/Liquid Interface (ISSIS). In: Selection of papers from the International Symposium on Surface Imaging/Spectroscopy at Solid/Liquid Interface (ISSIS), 28 May – 1 June 2006, Krakow, Poland. Electrochim Acta 52(18):5591–5598
18. Torrance L et al (2006) Oriented immobilisation of engineered single-chain antibodies to develop biosensors for virus detection. J Virol Meth 134(1–2):164–170
19. Trammell SA et al (2001) Synthesis and Characterization of a Ruthenium(II)-Based Redox Conjugate for Reagentless Biosensing. Bioconjugate Chem 12(4):643–647
20. Trammell SA et al (2004) Orientated binding of photosynthetic reaction centers on gold using Ni–NTA self-assembled monolayers. Biosensor Bioelectron 19(12):1649–1655
21. Medintz IL et al (2004) A fluorescence resonance energy transfer-derived structure of a quantum dot-protein bioconjugate nanoassembly. PNAS 101(26):9612–9617
22. Medintz IL et al (2005) Self-Assembled TNT Biosensor Based on Modular Multifunctional Surface-Tethered Components. Anal Chem 77(2):365–372
23. Marvin JS, Hellinga HW (2001) Manipulation of ligand binding affinity by exploitation of conformational coupling. Nat Struct Mol Biol 8(9):795–798
24. Maly J et al (2004) Reversible immobilization of engineered molecules by Ni-NTA chelators. Proceedings of the XVIIth International Symposium on Bioelectrochemistry and Bioenergetics. Bioelectrochemistry 63(1–2):271–275
25. Brewster JD, Lightfield AR, Bermel PL (1995) Storage and Immobilization of Photosystem II. Reaction Centers Used in an Assay for Herbicides. Anal Chem 67(7):1296–1299
26. Nakamura C et al (2003) Rapid and specific detection of herbicides using a self-assembled photosynthetic reaction center from purple bacterium on an SPR chip. Selected papers from the 7th World Congress on Biosensors Kyoto, Japan 15–17 May 2002. Biosensor Bioelectron 18(5–6):599–603
27. Beissenhirtz MK, Scheller FW, Viezzoli MS, Lisdat F (2006) Engineered Superoxide Dismutase Monomers for Superoxide Biosensor Applications. Anal Chem 78(3):928–935
28. Marcus RA, Sutin N (1985) Electron transfers in chemistry and biology. BBA Rev Bioenerget 811(3):265–322
29. Mayo S et al (1986) Long-range electron transfer in heme proteins. Science 233(4767):948–952
30. Van de Meere GC, Chen SYS (1994) Resonance Energy Transfer: Theory and Data. Wiley, New York
31. Benson DE et al (2001) Design of Bioelectronic Interfaces by Exploiting Hinge-Bending Motions in Proteins. Science 293(5535):1641–1644
32. Fehr M, Frommer WB, Lalonde S (2002) From the Cover: visualization of maltose uptake in living yeast cells by fluorescent nanosensors. PNAS 99(15):9846–9851
33. Fehr M et al (2003) In Vivo Imaging of the Dynamics of Glucose Uptake in the Cytosol of COS-7 Cells by Fluorescent Nanosensors. J Biol Chem 278(21):19127–19133
34. Lager I et al (2003) Development of a fluorescent nanosensor for ribose. FEBS Lett 553(1–2):85–89
35. Willner I, Willner B, Katz E (2002) Functional biosensor systems via surface-nanoengineering of electronic elements. Rev Mol Biotechnol 82(4):325–355
36. Cass AE et al (1984) Ferrocene-Mediated Enzyme Electrode for Amperometric Determination of Glucose. Anal Chem 56:667–671
37. Hui Yao NL, Wei Y-L, Zhu J-J (2005) A H_2O_2 Biosensor Based on Immobilization of Horseradish Peroxidase in a Gelatine Network Matrix. Sensors 5:277–283

38. Chen L-Q et al (2002) Genetic modification of glucose oxidase for improving performance of an amperometric glucose biosensor. Biosensor Bioelectron 17(10):851–857
39. Hecht HJ, et al (1993) Crystal Structure of Glucose Oxidase from Aspergillus niger Refined at 2·3 A Reslution. J Mol Biol 229(1):153–172
40. Heller A (1990) Electrical wiring of redox enzymes. Accounts Chem Res 23:128
41. Loechel C, Jaswir Basran AB, Scrutton NS, Hall EAH (2003) Using trimethylamine dehydrogenase in an enzyme linked amperometric electrode, Part 1. Wild-type enzyme redox mediation. Analyst 128:166–172
42. Loechel C, Jaswir Basran AB, Scrutton NS, Hall EAH (2003) Using trimethylamine dehydrogenase in an enzyme linked amperometric electrode, Part 2. Rational design engineering of a wired mutant. Analyst 128:889–898
43. Willner I, Riklin A (1994) Electrical Communication between Electrodes and NAD(P)+-Dependent Enzymes Using Pyrroloquinolinequinone-Enzyme Electrodes in a Self-Assembled Monolayer Configuration: Design of a New Class of Amperometric Biosensors. Anal Chem 66:1535–1539
44. Habermüller K, Mosbach M, Schumann W (2000) Electron-transfer mechanisms in amperometric biosensors. Fresen J Anal Chem 366:560–568
45. Badia A et al (1993) Intramolecular electron-transfer rates in ferrocene-derivatized glucose oxidase. J Am Chem Soc 115:7053–7060
46. Padeste C et al (2003) Redox labelled avidin for enzyme sensor architectures. Biosensor Bioelectron 19(3):239–247
47. Anicet N et al (1998) Electron Transfer in Organized Assemblies of Biomolecules. Construction and Dynamics of Avidin/Biotin Co-immobilized Glucose Oxidase/Ferrocene Monolayer Carbon Electrodes. J Am Chem Soc 120:7115–7116
48. Trojanowicz M, Miernik A (2001) Bilayer lipid membrane glucose biosensors with improved stability and sensitivity. Electrochim Acta 46(7):1053–1061
49. Wilson JR, Caruana DJ, Gilardi G (2003) Engineering redox functions in a nucleic acid binding protein. Chem Commun 356–357
50. Astuti Y et al (2004) Proton-Coupled Electron Transfer of Flavodoxin Immobilized on Nanostructured Tin Dioxide Electrodes: Thermodynamics versus Kinetics Control of Protein Redox Function. J Am Chem Soc 126:8001–8009
51. Fromant M, Blanquet S, Plateau P (1995) Direct Random Mutagenesis of Gene-Sized DNA Fragments Using Polymerase Chain Reaction. Anal Biochem 224(1):347–353
52. Armold FH et al (1999) Directed Evolution of Mesophilic Enzymes into Their Thermophilic Counterparts. Ann NY Acad Sci 870(1):400–403
53. Stemmer W (1994) DNA Shuffling by Random Fragmentation and Reassembly: In vitro Recombination for Molecular Evolution. PNAS 91(22):10747–10751
54. Stemmer WP (1994) Rapid evolution of a protein in vitro by DNA shuffling. Nature 370:389–391
55. Vincent GH, Eijsink SG, Borchert TV, van den Burg B (2005) Directed evolution of enzyme stability. Biomol Eng 22:21–30
56. Tamakoshi M et al (1997) A new Thermus – *Escherichia coli* shuttle integration vector system. J Bacteriol 179:4811–4814
57. Kaur J, Sharma R (2006) Directed Evolution: An Approach to Engineer Enzymes. Crit Rev Biotechnol 26(3):165–199
58. Hamamatsu N et al (2006) Directed evolution by accumulating tailored mutations: Thermostabilization of lactate oxidase with less trade-off with catalytic activity. Prot Eng Design Sel 19(11):483–489
59. Minagawa H et al (2007) Improving the thermal stability of lactate oxidase by directed evolution. Cell Mol Life Sci 64(1):77–81

60. Crameri A et al (1998) DNA shuffling of a family of genes from diverse species accelerates directed evolution. Nature 391(6664):288–291
61. Lorimer IAJ, Pastan I (1995) Nucleic Acids Res 23:3067–3068
62. Miyazaki K (2002) Random DNA fragmentation with endonuclease V: application to DNA shuffling. Nucl Acids Res 30(24):e139
63. Coco WM et al (2001) DNA shuffling method for generating highly recombined genes and evolved enzymes. Nat Biotech 19(4):354–359
64. Ostermeier M, Shim JH, Benkovic SJ (1999) A combinatorial approach to hybrid enzymes independent of DNA homology. Nat Biotech 17(12):1205–1209
65. Lutz S et al (2001) Creating multiple-crossover DNA libraries independent of sequence identity. PNAS 98(20):11248–11253
66. Kolkman JA, Stemmer WPC (2001) Directed evolution of proteins by exon shuffling. Nat Biotech 19(5):423–428
67. Abecassis V, Pompon D, Truan G (2000) High efficiency family shuffling based on multi-step PCR and in vivo DNA recombination in yeast: statistical and functional analysis of a combinatorial library between human cytochrome P450 1A1 and 1A2. Nucl Acids Res 28(20):e88
68. Ness JE et al (2002) Synthetic shuffling expands functional protein diversity by allowing amino acids to recombine independently. Nat Biotech 20(12):1251–1255
69. Shen Y et al (1993) Stabilization of the membrane protein bacteriorhodopsin to 140 °C in two-dimensional films. Nature 366(6450):48–50
70. Hillebrecht JR et al (2004) Directed Evolution of Bacteriorhodopsin for Device Applications. Method Enzymol 388:333–347
71. Gibney BR et al (1996) Ferredoxin and ferredoxin-heme maquettes. PNAS 93(26): 15041–15046
72. Gibney BR, Rabanal F, Dutton PL (1997) Synthesis of novel proteins. Curr Opin Chem Biol 1(4):537–542
73. Rabanal F, DeGrado WF, Dutton PL (1996) Toward the Synthesis of a Photosynthetic Reaction Center Maquette: A Cofacial Porphyrin Pair Assembled between Two Subunits of a Synthetic Four-Helix Bundle Multiheme Protein. J Am Chem Soc 118(2):473–474
74. DeGrado WF et al (1999) De novo design and structural characterization of proteins and metalloproteins. Ann Rev Biochem 68(1):779–819
75. Kornilova AY et al (2000) Design and Characterization of A Synthetic Electron-Transfer Protein. J Am Chem Soc 122:7999–8006
76. McNeil SE (2005) Nanotechnology for the biologist. J Leukocyte Biol 78:585–594
77. Cui Y et al (2001) Nanowire Nanosensors for Highly Sensitive and Selective Detection of Biological and Chemical Species. Science 293:1289–1292
78. Yin YPW, Lü Y, Du P, Shi Y, Cai C (2007) Immobilization and direct electrochemistry of cytochrome c at a single-walled carbon nanotube-modified electrode. J Solid State Electrochem 11(3):390–397
79. Lombardi A et al (2000) Inaugural Article: Retrostructural analysis of metalloproteins: Application to the design of a minimal model for diiron proteins. PNAS 97(12):6298–6305
80. Dwyer MA, Looger LL, Hellinga HW (2004) Computational Design of a Biologically Active Enzyme. Science 304:1967–1971
81. Case MA, Ghadiri MR, Mutz MW, Mc Lendon GL (1998) Stereoselection in designed three-helix bundle metalloproteins. Chirality 10(1–2):35–40
82. Willner I et al (1999) Integration of a Reconstituted de Novo Synthesized Hemoprotein and Native Metalloproteins with Electrode Supports for Bioelectronic and Bioelectrocatalytic Applications. J Am Chem Soc 121:6455–6468

83. Gouaux J et al (1994) Subunit Stoichiometry of Staphylococcal α-Hemolysin in Crystals and on Membranes: A Heptameric Transmembrane Pore. PNAS 91(26):12828–12831
84. Song L et al (1996) Structure of Staphylococcal α-Hemolysin, a Heptameric Transmembrane Pore. Science 274(5294):1859–1865
85. Braha O et al (1997) Designed protein pores as components for biosensors. Chem Biol 4(7):497–505
86. Walker B et al (1995) An intermediate in the assembly of a pore-forming protein trapped with a genetically-engineered switch. Chem Biol 2(2):99–105
87. Kasianowicz JJ et al (1999) Genetically Engineered Metal Ion Binding Sites on the Outside of a Channel's Transmembrane β-Barrel. Biophys J 76(2):837–845
88. Kasianowicz JJ, Bezrukov SM (1995) Protonation dynamics of the α-toxin ion channel from spectral analysis of pH-dependent current fluctuations. Biophys J 69(1):94–105
89. Kasianowicz JJ et al (1994) Genetically engineered pores as metal ion biosensors. MRS Symp 330:217–223
90. Howorka S, Cheley S, Bayley H (2001) Sequence-specific detection of individual DNA strands using engineered nanopores. Nat Biotech 19(7):636–639
91. Ferapontova E et al (2002) Effect of cysteine mutations on direct electron transfer of horseradish peroxidase on gold. Biosensor Bioelectron 17(11–12):953–963
92. Aharoni A et al (2006) High-throughput screening methodology for the directed evolution of glycosyltransferases. Nat Meth 3(8):609–614
93. Valentini F et al (2007) Coupling of single-wall carbon nanotube and l-histidine on Ag microwires: New architectures for the assembling of NAD^+ sensors. Sens Actuator B Chem 123(1):5–9

Adv Biochem Engin/Biotechnol (2008) 109: 97–122
DOI 10.1007/10_2007_088

Published online: 6 November 2007

Artificial Receptors

Bengt Danielsson

Lund University Box 124, S-22100 Lund, Sweden
bengt.danielsson@tbiokem.lth.se

Abstract Herein I will provide a brief overview of artificial receptors with emphasis on molecularly imprinted polymers (MIPs) and their applications. Alternative techniques to produce artificial receptors such as in silico designed and modelled polymers as well as different receptors designed using libraries of more or less natural composition will also be mentioned. Examples of these include aptamers and bio-nanocomposites. The physical presentation of the receptors is important and may depend on the application. Block polymerization of MIPs and grinding to particles of suitable size used to be the preferred technique, but today beaded materials can be produced in sizes down to nanobeads and also nanofibers can be used to increase available surface area and thereby capacity. For sensor applications it may be attractive to include the artificial receptors in surface coatings or in membrane structures. Different composite designs can be used to provide additional desirable properties. MIPs and other artificial receptors are gaining rapidly increasing attention in very shifting application areas and an attempt to provide a systematic account for current applications has been made with examples from separation, solid-phase extraction, analysis, carbohydrate specific experiments, and MIPs-directed synthesis.

Keywords Aptamers · Combinatorial libraries · Ion channel sensors · MIP catalysis · MIPs · Molecularly imprinted polymers · Solid-phase extraction

1 Introduction

Nature has provided us with a large number of highly efficient natural molecules that can perform very efficient recognition and/or catalytic reactions. These are typically not isolated reactions but in some context as one of a sequence of reactions or for triggering or confirming some event. Therefore these molecules, usually more or less complicated protein structures, such as enzymes and membrane receptors, can be virtual molecular machines containing several functionalities in sophisticated dynamic operation, often concerted with other molecular systems. Only part of the molecule is involved in the actual molecular recognition or catalytic reaction. Because of flexible conformation and mechanisms such as allosteric and transducing functions there is more of a risk for lower stability and even specificity than with smaller molecules with a singular function. The most common reason for opting for artificial receptors is operational stability and robustness. Other reasons could be limited availability and difficult isolation of the natural systems. It can also be problematic to find or develop natural receptors for all kinds of targets, especially for smaller molecules.

Herein I will provide a brief overview of artificial receptors with emphasis on molecularly imprinted polymers (MIPs) and their applications. Template-driven synthesis of MIPs has some drawbacks, such as availability of template or necessity to reuse the template for economical reasons, toxicity, which make designer approaches attractive. Alternative techniques to produce artificial receptors such as in silico designed and modelled polymers as well as different receptors designed using libraries of more or less natural composition will therefore be mentioned. Examples of these include aptamers and bio-nanocomposites. The physical presentation of the receptors is important and may depend on the application. Block polymerization and grinding to particles of suitable size used to be the preferred technique, but today beaded materials can be produced in sizes down to nanobeads and also nanofibers can be used to increase available surface area and thereby capacity. For sensor applications it may be attractive to include the artificial receptors in surface coatings or in membrane structures. Different composite designs can be used to provide additional desirable properties.

MIPs and other artificial receptors are gaining rapidly increasing attention in very shifting application areas and an attempt to provide a systematic account for current applications has been made. The early applications were mostly in separation and today solid phase extraction is becoming an important application field with a number of commercial products. Other common applications are in analysis and in sensors. Carbohydrate specific receptors are an interesting emerging area and so is MIP-directed synthesis.

2 Types of Artificial Receptors

2.1 MIPs

The usual way to produce molecularly imprinted polymers is to employ the target molecule (or a similar derivative of it) as a template around which interacting and cross-linking monomers are arranged and co-polymerized to form a block that contains a number of cavities each with a template molecule inside (Fig. 1). Since the monomers initially have arranged spontaneously through noncovalent or forced by covalent interactions to form a complex with the template the resulting binding sites will be complementary to the template in size, shape, and position of the functional groups [1]. After polymerization, the polymer block is ground into small particles of a size suitable for the application and the template removed. The binding sites that have been imprinted on the polymer are now capable of selectively rebinding to the template or the target molecule.

In comparison with biological receptors the MIPs, although they can recognize and specifically bind their target molecules, are large, rigid, and insoluble, whereas their natural counterparts are smaller, flexible, and, in most

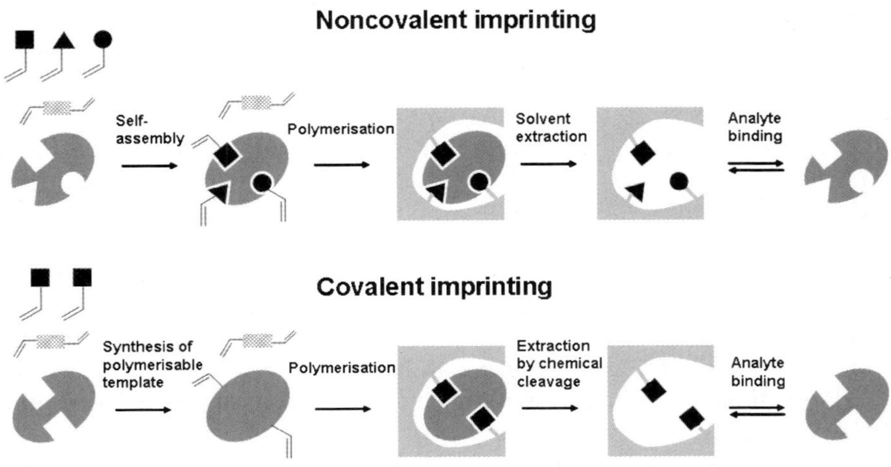

Fig. 1 Schematic representations of two different molecular imprinting principles: In *noncovalent imprinting* the preassembly of the functional monomers around the template is driving their molecular interactions. In covalent imprinting a polymerizable derivative of the imprint molecule is synthesized by covalent coupling of the monomers to the template. Removal of the template by solvent extraction or chemical cleavage results in binding sites complementary in size and chemical functionality to the template. Reproduced from [1]

instances, soluble. MIPs particles can have thousands or millions of binding sites, whereas biological receptors may have just one. Because of the way the binding sites are created in MIPs, especially in those imprinted using noncovalent monomer-template interactions, their distribution, accessibility and binding properties are heterogeneous. This may not always be problematic, but it is obvious that it can be difficult for this type of artificial receptor to substitute for natural receptors in certain applications. Nevertheless, to date MIPs synthesized by radical polymerization of functional and cross linking monomers with vinyl or acrylic groups using noncovalent interactions with the template are the most common form of artificial receptors. Their synthesis is straightforward and there is a vast choice of available monomers: carrying basic (e.g., vinylpyridine) or acidic (e.g., methacrylic acid) groups. They could be permanently charged (e.g., 3-acrylamidopropyltrimethylammonium chloride) or hydrophobic (e.g., styrene) as well as such leading to hydrogen bonding (e.g., acrylamide). Addition of a solvent to the polymerization mixture that induces a porous structure in the polymer facilitates the access of the analyte molecules to the imprinted sites. The use of such porogens will be further discussed in the next section. There are many, more detailed overviews in literature describing different ways to prepare MIPs. One recent review has compiled a large number of applications with relation to the composition of the MIPs used [2].

Newer imprinting approaches include other materials, such as polyphenols and polyurethanes that may be easier to synthesize in the desired form or be better suited to a given application. On the other hand these materials have the drawback of a more limited choice of functional monomers. Sol–gels such as silica and titanium dioxide are also finding wider use as imprinting matrices.

2.2 Combinatorial Libraries

In the MIPs technology a template is needed for the receptor synthesis. Different approaches have been attempted to avoid this drawback. One possibility is to use combinatorial chemistry to produce libraries of potential receptors from which the most suitable one is selected by some screening procedure [3]. Combinatorial chemistry became very popular in the 1990s in the pharmaceutical industry as a tool for producing drug candidates and has since matured in various directions. The general concept of combinatorial libraries was first developed for peptides and involved the generation of all possible permutations of peptide sequences of a given length in connection with subsequent screening and selection to identify the peptide with the best binding activity to a certain target in the presence of less active peptides. The field of applications for combinatorial chemistry has now expanded to include proteins, synthetic oligomers, small molecules, and oligosaccharides.

Preparation of libraries with the split and mix method. This is the most common method to prepare immobilized libraries generating one-bead–one-compound libraries. For a peptide library this means one sequence of amino acids on a single 0.1 mm diameter bead. The split synthesis results in a collection of beads, each of which contains one specific peptide sequence consisting of every possible combination of every amino acid used. The synthesis that is fast and easy to carry out is, however, only applicable to the synthesis of peptides that can be sequenced. One bead carries only ca. 100 pmol of peptide which is sequenceable with modern Edman sequencing of small peptides, but with other methods it is difficult to establish the identity of the active library members. This is especially a problem for libraries of nonpeptide components for which unique chemical tags to encode the structure of the library compounds on each resin bead have to be developed, which can then be decoded by sequencing or some other analysis, such as GC or HPLC. This is, however, problematic because tags need to be chemically inert and reliably analyzed on femtomolar scales from a single bead which makes screening of such libraries and the subsequent decoding tedious and challenging and in most cases only indirect.

Dynamic combinatorial libraries. Dynamic combinatorial chemistry relies on the reversible connection of building blocks to give access to libraries whose composition is not yet fixed but can change in response to its surroundings [4, 5]. If bond formation is reversible, the library can rearrange, controlled by the thermodynamics of the whole molecular ensemble. This should lead to an amplification of host molecules with high affinity to the added target molecule but this is not always the case. It has been shown that statistical reasons can lead to the amplification of hosts with lower affinity. The need for sufficient solubility of every single library member, the lack of suitable reversible chemistry and the low level of reaction control are other limitations. Most dynamic libraries are based on reversible chemistry such as imine or disulfide formation. Research in the field of dynamic combinatorial chemistry is, however, still at an early stage and interesting developments can be expected, for example in glycobiology [5].

The step from combinatorial libraries to supramolecular chemistries is not long. With supramolecular functionality the artificial systems become even more similar to natural counterparts in that they can have more functions built in, such as signal transduction mechanisms [6]. Even if the synthesis of supramolecular complexes to a great part relies on self-assembly and template effects, it may become too complex and the resulting molecules too fragile to serve as robust artificial receptors.

2.3
Aptamers

Functional nucleic acids are on the borderline between artificial and natural molecules. DNA and RNA can have many different functions besides their well-known common roles. Nucleic acids can act as enzymes (deoxyribozymes and ribozymes) and as receptors (aptamers). These functional nucleic acids can either be found in nature or isolated from libraries of random nucleic acids. Nucleic acid aptamers have been found to be highly useful molecular tools and the study of functional nucleic acids has become a very important part of chemical biology. Aptamers can offer many of the properties that are desirable for artificial receptors such as flexibility, stability, ease of immobilization and susceptibility to various chemical modifications and labelling. They can be generated via "in vitro selection" or by SELEX (systemic evolution of ligands by exponential enrichment) to bind diverse targets also those for which antibodies are difficult to obtain (such as toxins).

Nucleic acid libraries are easily obtained via combinatorial chemistry synthesis. Each sequence synthesized represents a linear oligomer of unique sequence and the molecular diversity is dependent on the number of randomized nucleotide positions. Typically libraries of ca. 10^{15} independent sequences are produced, with a variable region of 30 bases flanked by primers of 20–25 bases. A random sequence oligonucleotide library is incubated with the target of interest using techniques such as affinity chromatography or filter binding for separation. The isolated population of sequences is then amplified for use in the following selection/amplification cycle. After affinity saturation has been achieved, typically after 8–15 cycles, the enriched library is cloned and sequenced. When the desired sequence has been identified the aptamer can be produced in any quantities by chemical synthesis, which is a great advantage.

Another advantage is that the selection mechanism can be used to produce aptamers under any pre-defined conditions, making it possible to design receptors for tasks that cannot be met with proteins. Moreover, aptamers can also be easily immobilized to provide a custom-made surface for specific applications. They can refold to their native conformation following one round of sensing thereby being simple to implement in reusable devices. The change in shape upon binding to their target can be conveniently coupled with various signaling mechanisms, such as molecular beacons and quantum dots, for easy monitoring of molecular recognition events [7].

2.4
Recognition by Designed Molecules

It would be very attractive to be able to design artificial receptors directly based on computer modelling and design instead of screening large libraries of potential binders. Especially peptide and protein recognition by designed molecules has been intensively studied because of the central role protein–protein interactions have in biological processes. Understanding of such interactions is of central importance to our knowledge about all normal and abnormal cellular events and how diseases develop and how they can be treated.

To understand the nature of protein–protein interfaces particular emphasis has been placed on structure, energetics, electrostatic complementarity, and kinetics of protein–protein interactions. An important breakthrough was the identification of "hot spots" on protein surfaces [8]. A hot spot is a defined locale of ca. 600 Å^2 on the surface of a protein at or near the geometric center of the protein–protein interface. The residues comprising the hot spot contribute significantly to the stability of the protein–protein complex. Surrounding the hot spot is an area of residues that contribute slightly less to the stability of the complex. This outer area has been compared to an O-ring that excludes solvent from the protein–protein interface, stabilizing the complex (Fig. 2).

There are many approaches to protein recognition by synthetic molecules [8, 9]. The most promising routes involve specific metal coordination, epitope-docking on miniature proteins, aptamer selection, non-natural pep-

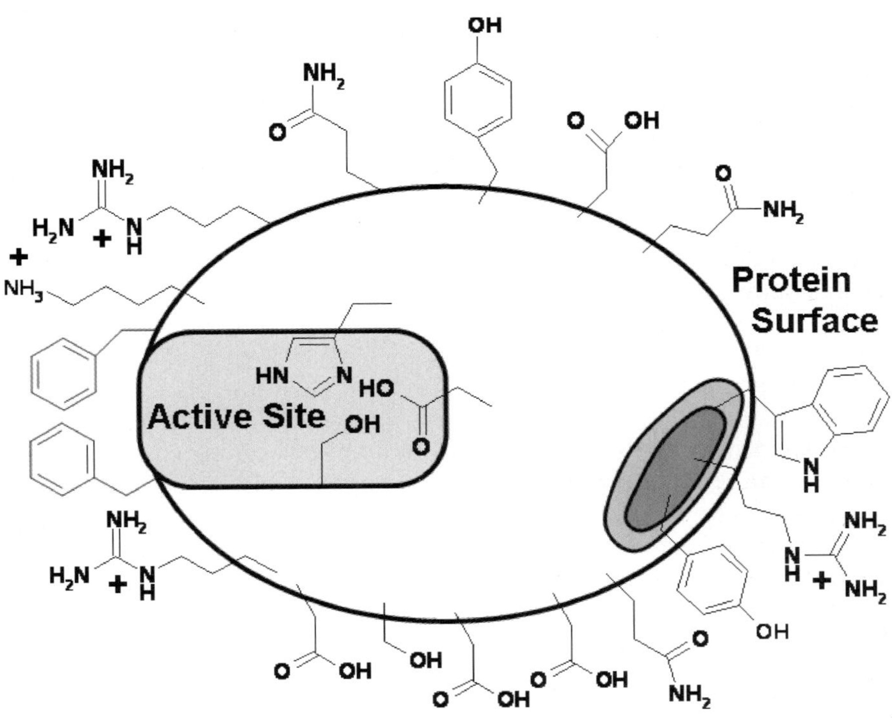

Fig. 2 Schematic representation of a protein (enzyme): interior, exterior, and the hot spot. Reproduced from [8]

tide isosteres, functionalized platforms, secondary structure mimetics, molecular imprinting and receptors embedded in lipid layers. Here will be mentioned only a few techniques while (monoclonal) antibodies that are surprisingly effective at protein surface recognition and the powerful phage display techniques for producing peptides for the same purpose will be left out.

Functionalized platforms are a powerful concept for the development of artificial protein receptors. A complementary image of the desired hot spot on the protein surface can be built up by functionalizing a large, preferably flat platform like tetraphenylporphyrin with peptides [8]. Other useful platforms are calix[4]arene, ruthenium bipyridine complexes, and beta-cyclodextrin dimers.

Secondary structure mimetics. Strategies for alfa-helix and beta-sheet recognition and stabilization have received much attention. Monomeric alfa-helices may be stabilized through covalent attachment of N-terminal templates which spatially orient H-bond acceptors for the amides at the first turn of the helix. Rational recognition of the side chain functionality found on protein surfaces has been studied using model alfa-helical peptides and molecules designed for binding in competitive solvents. One approach is based on the hydrogen-bonding complementarity between peptide carboxylate groups and synthetic receptors containing guanidinium sites. The spacing of the functionalities on the complementary structure is critical in order to keep in phase with the functional groups on the alfa-helix. Monosaccharides have been employed as templates for the imitation of beta-turns and polypyrrolinones as beta-sheet mimetics. Thus, the cyclic hexapeptide somatostatin was imitated based on a beta-D-glucose skeleton and while the backbone of 3,5-connected pyrrolinones imitates a beta-strand /beta-sheet conformation that allows formation of intermolecular hydrogen bonds, beta-turns can be imitated with alternating D,L-polypyrrolinones. More information on these topics can be found in [8] and [9].

2.5 Natural Structures

There are many ways by which natural, more or less complex molecular assemblies can serve as a platform for or be combined with artificial receptors. There are no clear borderlines between natural structures, structures used by nature or artificial mimics. An example is lipid membranes. Natural receptors are usually embedded in lipid bilayers and can for instance serve as specific channels for effector molecules through cell membranes. It is, however, for several reasons difficult to use the natural systems as they are, poor stability being one main reason and difficult signal transduction another. Therefore, methods have been developed to prepare stable biological membranes that couple the biomimetic properties of bilayer lipid membranes (BLMs) with the high stability of hybrid bilayer membranes (HBMs) resulting in a mixed

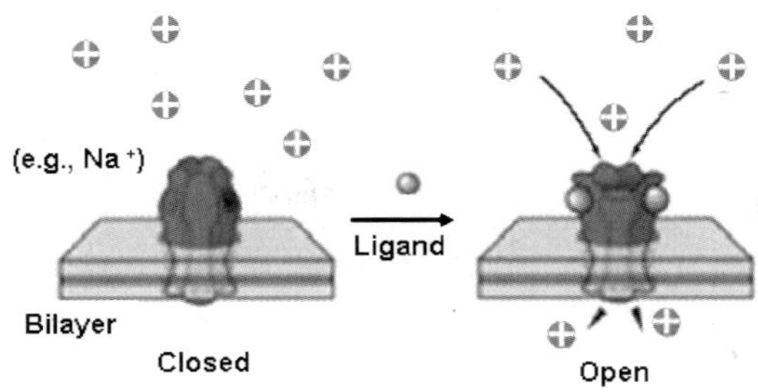

Fig. 3 A ligand-gated ion channel protein. Complexation of the ion channel protein with ligands allows the transport of transmembrane ions across the lipid bilayer. Reproduced from [13]

hybrid bilayer lipid membrane (MHBLM). One example in which a natural glutamate receptor was built into such an MHBLM was reported to yield a linear transmembrane current in response to 1–100 nM glutamate under flow conditions [10]. *Different ways to modify BLMs* with artificial and natural receptors are summarized in short overview [11]. BLMs and single layer lipid membranes can also be stabilized by a supporting porous filter. As an example repetitive detection of adrenaline was reported using stabilized BLMs with calix[4]resorcinarene as the receptor [12].

Ion channel sensors are particularly suited to detect large, hydrophilic ionic species with high multiple charges. Typical analytes are bioactive substances such as peptides, proteins, polysaccharides, and oligonucleotides as well as multivalent inorganic ions. The working principle of this biomimetic sensor type is similar to the transmembrane signaling used by ligand-gated ion channels in biological membranes (Fig. 3). The transport of transmembrane ions can be affected in many different ways by chemical modification of the ion channel [13]. There are two different types of response mechanism for ion channel sensors: (i) the analyte molecules form inclusion-type complexes with the receptors and block physically the intramolecular channels, preventing redox markers from accessing the electrode surface. This approach closely mimics ion channel proteins and is the basis for detecting electrically neutral, redox-inactive analytes. (ii) The second mechanism is based on electrostatic attraction or repulsion. When ionic analytes bind to the synthetic receptor layer, they alter the net charge of the layer which regulates the oxidation or reduction of easily charged electroactive ions (redox markers) at a receptor monolayer that coats the electrode.

Ion channel sensors based on synthetic receptors were developed for hydrogen and metal ions, simple organic analytes such as phthalate and glucose, and more complex analytes such as antibodies and concanavalin derivatives. Phosphate esters, antibiotics, oligopeptides, DNA dendrimers, cyclodextrins, calixarenes, and antigenic groups are typical examples from the wide range of synthetic receptors that have been used in ICSs [13].

Metalloporphyrines are another type of natural molecule that offers almost unique opportunities to design artificial receptors. They can be tailored at the synthetic level by changing the sensor selectivity in a controlled way so that it can be oriented towards a desired group of analytes. They have therefore found use in sensor arrays [14] and can be combined with mass-sensitive detection, such as QCM (quartz crystal microbalance), as well as potentiometric and optical transducers.

3 Physical Arrangements

The physical arrangement and presentation of artificial receptors are very diverse and becoming more adapted to the application and the variation is largest for MIPs. From the beginning MIPs were prepared as block polymer monoliths and mechanically ground to smaller sized particles (25–50 μm) and packed into columns. Although simple this method has its drawbacks in being time consuming and labor intensive, wasteful in material and producing particles of irregular size. Micro- to nanospheres can be produced using polymerization by precipitation under careful control of the parameters governing the process (porogen amount, temperature, cross-linker etc). Typically 2–10 times more porogen is used than in bulk polymerization. The beads precipitate as the polymerization proceeds and the growing polymer chains become insoluble in the liquid phase [15].

A heterogeneous suspension polymerization method was developed in Lund that could produce spherical beads of μm to mm size. In this method the organic-based polymerization mixture is suspended as droplets in an excess of continuous dispersion phase (water or perfluorocarbon) by stirring in the presence of a stabilizer – a suspending agent. These techniques were recently reviewed [16] and one development of the suspension polymerization allowed the polymers to be prepared under UV-light directly in solid-phase extraction columns resulting in a time-saving automatable process that is very convenient for screening purposes [17]. Other in situ polymerization methods involved MIPs monoliths produced in capillary electrophoresis columns [18]. By anchoring the monoliths to the inner wall of fused silica capillaries using silanizing agents, frits at the column ends can be omitted [19]. Very small particles with a molecular weight in the order of 10^6 (towards the size of a protein molecule) can be produced by adjusting the precipitation polymerization conditions properly [20].

An alternative to nanobeads for achieving high surface area-volume ratio may be to use nanofibers. Molecular recognition sites could be imprinted on electrospun polymer nanofibers. The fibers were prepared from a mixture of PET [poly(ethylene terephthalate)] and polyallylamine in the presence of 2,4-dichlorophenoxyacetic acid (2,4-D) as template and analyte [21].

Imprinting at surfaces is especially useful in sensor applications since most sensors available have a planar design. In most cases to date lipid membranes have been used. Imprinted materials with binding sites situated at or close to the surface of the imprinting matrix have many advantages – the sites are more accessible, mass transfer is faster, the binding kinetics may be faster, and target molecules conjugated with bulky labels can still bind. Unfortunately, their preparation is less straightforward than for bulk polymers, and they require specially adapted protocols. A polymer can be applied as a thin film on a surface with a number of standard techniques, such as spin or spray coating. Soft lithography [22] can be used to create patterned surfaces for multianalyte sensors and high throughput screening systems, but may not be compatible with all current methods for MIPs preparation.

MIPs have been synthesized at an electrode surface by electropolymerization [23] or at a nonconducting surface by chemical grafting [24]. Radio-frequency glow-discharge plasma deposition was employed to form polymeric thin films around proteins coated with disaccharide molecules. The disaccharides were covalently attached to the polymer film, creating polysaccharide-like cavities that exhibited highly selective recognition for a variety of template proteins, including albumin, immunoglobulin G, lysozyme, ribonuclease, and streptavidin. Protein molecules were adsorbed on a mica surface and a sugar layer of 10–50 Å was spin-casted over the adsorbed protein molecules. Finally, plasma deposition of C_3F_6 was used to form a 10- to 30-nm-thick fluoropolymer thin film. The resulting plasma was fixed to a glass cover slip with epoxy resin and oven-cured. With this method it could be possible to develop highly selective MIPs for the proteins and high molecular weight compounds [25].

It is also possible to generate imprints of cells and even larger objects. Unless a specific cell surface molecule is targeted, this cannot be called mo-

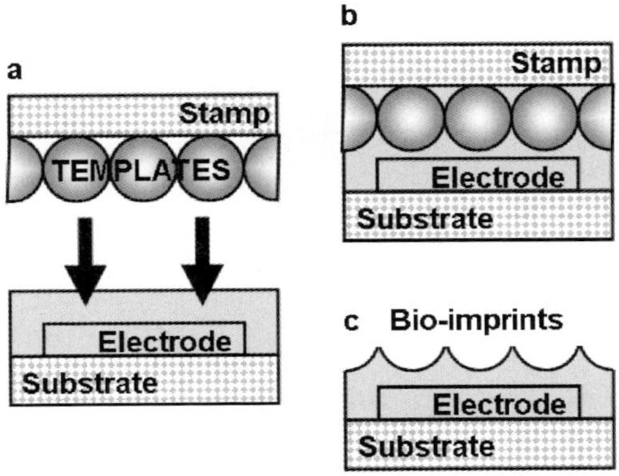

Fig. 4 Schematic illustration of surface imprinting using biotemplates applied directly on a pre-coated transducer. Reproduced from [26]

lecular imprinting because of the size of the template. Anyway, template-directed molding of polymer thin films can be performed to generate artificial receptors directly on pre-coated piezoacoustic devices and interdigitated capacitors (IDCs). This approach is generally applicable and useful for the fabrication of robust nano- and microstructured sensor coatings. Label-free detection of yeast and mammalian cells has been shown with these biomimetic receptors even down to single cell detection [26]. The principle of such surface imprinting is shown in Fig. 4. A pre-polymer film is applied on the transducer covering the electrode structure. A stamp covered with a layer of the template components (cells) is pressed onto the polymerizing thin film during curing and then removed leaving a moulded sensor layer that reflects the geometry of the bio-templates.

4 Applications

4.1 Separation, Enantiomer Separation

Applications of artificial receptors in separation have mostly been based on MIPs and it is indeed one of the most widely used areas of MIPs. Especially for the chiral separation of racemic mixtures, which is a common need in the pharmaceutical industry, there was an early and great interest and also good success. Many enantiomeric separations with baseline resolution were real-

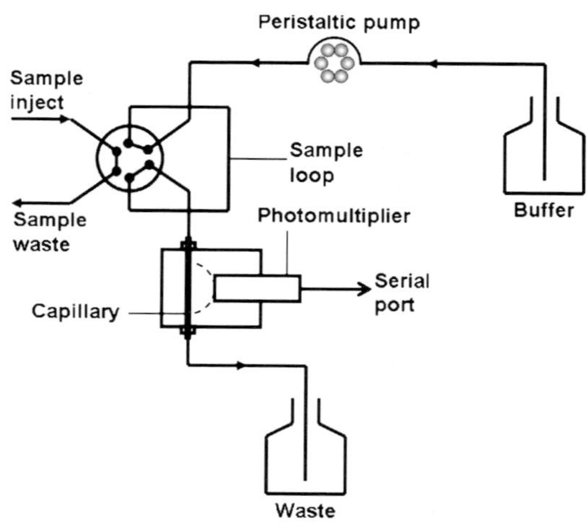

Fig. 5 Experimental setup for flow injection analysis of 2,4-D using glass capillaries coated with 2,4-D MIPs and chemiluminescent detection. From [53]

ized. A separation factor of near 18 was obtained for the racemic resolution of a dipeptide using only nonionic, noncovalent interactions [27].

Highly efficient MIPs for chiral discrimination were prepared under favorable conditions by use of trifunctional cross-linkers such as trimethylolpropane or pentaerythritol triacrylate instead of the difunctional cross-linker normally used. This resulted in hard and stable MIPs with increased binding site density and thereby increased load capacity as well as better resolution with a significantly lower percentage of cross-linker needed [28]. One gram of such material could resolve between 0.1 and 1 mg of racemic mixtures of peptides with baseline separation.

There is a great interest in using artificial receptors and especially MIPs in environmental control and food analysis. More about that will follow in the

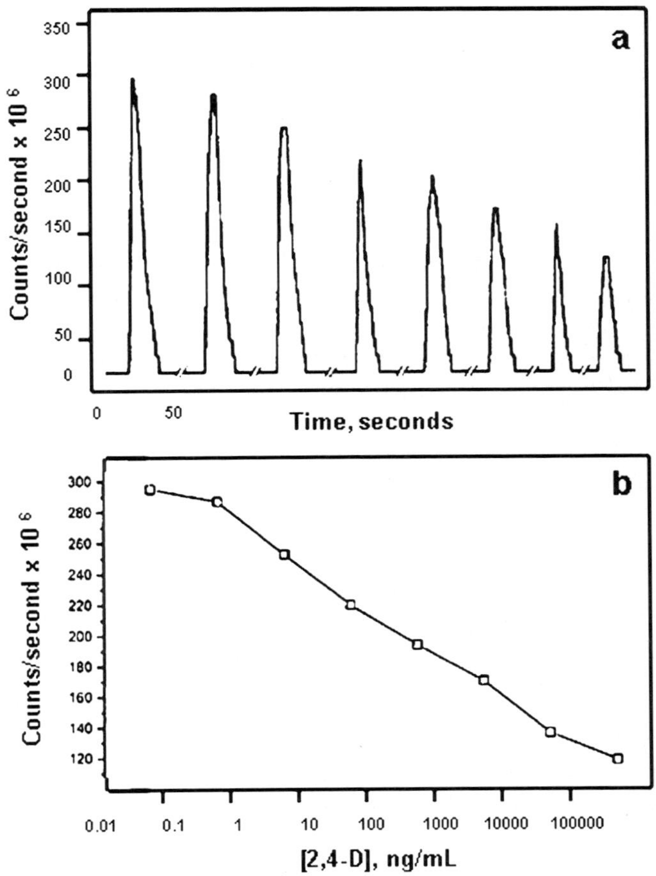

Fig. 6 Results from flow injection analysis of 2,4-D using 2,4-D MIPs in competitive mode with peroxidase as enzyme label. From [53]

sensing section but one example of separation could be given here. Ochratoxin is a well-known natural contaminant in cereals, rice, peanuts, coffee beans, cottonseed, and decaying vegetation, which eventually enters the food chain. It is highly carcinogenic and there is a great need to separate and quantify it in food materials. In one study ochratoxin-A MIP particles were synthesized from *N*-phenylacrylamide and packed into a microcolumn for selective extraction of ochratoxin-A from wheat [29]. Pulsed elution using methanol/triethylamine allowed good quantitative desorption of ochratoxin-A with a recovery of ochratoxin-A from wheat extracts of $103 \pm 3\%$. Detection with fluorescence at 385/445 nm afforded a detection limit of 5.0 ng/mL and the analysis took less than 5 min.

A health-related, successful separation of nicotine from tobacco smoke is noteworthy. In this case the MIP was prepared using methacrylic acid (MAA) and ethylene glycol dimethacrylate (EDMA) and chloroform as a porogen. The binding capacity for nicotine was very high (90 mg/g) and it was pointed out that it is much more capable of removing nicotine from tobacco smoke than commercial filter tips [30].

Although micro beaded materials are gaining popularity for good reasons most of the works on separation using MIPs have been done with cross-linked macroporous monolithic polymers which have been ground to small particles and sieved into suitable fractions. Drawbacks are time-consuming preparation with moderate yield of useful imprinted particles. A further disadvantage with these particles is their irregular shape, which is less suitable for efficient packing and use in chromatographic separations. A good alternative are as mentioned above MIPs in the form of micro spheres. Micro beaded MIPs with good selectivity and affinity against theophylline, caffeine, and 17-estradiol were prepared using the mentioned three molecules as templates with dilute solutions of MAA and trimethylolpropane trimethacrylate [31]. This type of MIPs is highly desirable and will also find application in competitive ligand binding assays, solid-phase microextraction (SPME), sensor development, and capillary electrophoresis.

Still a novel technique for preparing MIPs in bead form down to nano-size involves use of supercritical carbon dioxide as the medium. A recent study used propranolol as a model template with MAA as the functional monomer and DVB (divinylbenzene) as the cross-linker [32]. The imprinted binding sites were chiral-selective with a cross-reactivity against ®-propranolol of $<5\%$. The overall binding performance of the imprinted nanoparticles was comparable to imprinted polymers prepared in organic solvents. The use of the nontoxic supercritical carbon dioxide as medium can greatly reduce the consumption of volatile organic compounds and minimize environmental pollution.

Although it is a clear advantage of the MIP technology that effective imprints can be made of small molecules for which antibodies may be more difficult to obtain, there is a considerable interest in developing MIPs capable of selectively binding proteins. Well functioning protein imprinting would have great biological and technological importance. In contrast to small molecules with few specific features to give specific imprints proteins have too many and to large a variability and complexity of structure and composition to make it an easy task to prepare protein-specific MIPs. Because of their size protein templates risk becoming trapped inside the polymer structure when traditional imprinting techniques are used. Among alternative strategies the surface imprinting approach that was developed in Lund has been one of the most effective strategies for the imprinting of proteins. The first protein imprinting was performed on the surfaces of porous silica particles using a silane-boronate monomer and glycoprotein transferrin as the print molecule [33]. Later, this same strategy was applied to another protein, bovine pancreas ribonuclease A (RNase A). A metal-binding monomer, *N*-(4-vinyl)benzyl iminodiacetic acid, was used to coordinate metal ions with the imidazole groups on the histidines on the surface of RNase A [34]. In the presence of metal ions, the MIP of RNase A showed a stronger affinity for RNase A than the control silica-polymer hybrid particles prepared under the same condition as the MIP but using BSA as the template did.

A monolayer surface-imprinting approach was also developed for imprinting the glycoprotein RNase B on the surface of silica beads [35]. The polysaccharide group of RNase B was first connected covalently with 3-aminophenylboronic acid, the compound thus formed being coupled directly to the surface of the epoxy-activated silica particles. A recent review article describes the state of the art of MIPs for the recognition of proteins [36] by structuring the field into six areas:

1. Recognition through the placement of few constraints on the polymer;
2. Shape recognition with polyacrylamide gels;
3. Epitope approach: exploiting a small structural element of the protein for its whole recognition;
4. Molecularly imprinted polymers grafted on surfaces;
5. Immobilizing the template on a surface;
6. Silica-based materials imprinted for protein recognition.

Examples from the different areas were given, some of them also cited above. A promising approach based on the use of the silane monomers, conjugating the silica material to the concept of template immobilization (area 6) was recently proposed [37]. The protein template (hemoglobin) is at first covalently immobilized on the derivatized silica surface and then the polymerization takes place. A comparison with imprinted silica beads obtained with a conventional nonimmobilized hemoglobin template showed only a slightly higher binding capacity for the proposed technique, but a far better performance in competitive binding experiments: MIPs were able to bind the template specifically and it seemed that the binding cavities were more homogeneous. A further advantage is that the immobilized template could be used for repetitive re-creation of imprinted layers, which is of value for a mass production of recognition layers, for instance for sensors.

4.2 Solid-Phase Extraction

There is no clear distinction between separation and sample clean-up and sample preparation for biomedical, environmental, and food analysis as well as product recovery for industrial processes have become the primary target for commercialization of artificial receptors, in particular MIPs. There are many small companies now producing solid-phase extraction materials, MIP Technologies AB in Lund, Sweden probably being the largest with ca. 30 employees (www.miptechnologies.com). They have a capacity to produce MIP phases in a scale of up to 500 kg. The SPE sorbents developed by the company are marketed under the trade mark SupelMIP™ by Supelco (Sigma–Aldrich) who is the worldwide distributor.

MIPs seem to be a very suitable material for SPE. In comparison with biological sorbents for immuno- or affinity-extraction MIPs are relatively stable, have a higher adsorption capacity, can more easily be obtained for small molecules, and are likely to be cheaper to produce. There is, however, an inherent purity problem connected with the template-driven synthesis of MIPs in that some of the template trapped in the material may gradually leak out during use leading to erroneous results. That potential problem may be avoided if the imprint is created not with the target compound itself, but with a structurally related molecule. The polymer can still bind the target analyte, but it may be possible to eliminate any leaking template in the following separation steps. A recent review on selective sample treatment using molecularly imprinted polymers [38] compiles a large number of MIPs for SPE of compounds from real matrices coupled off-line or on-line with analytical methods, mainly LC. Most of the studies involved only one target, but MIPs can also be developed for a group of structural analogues. The importance of the selection of template in such cases was pointed out. For the extraction of triazines (methoxy-, thiomethyl- and chloro-triazines) MIPs were made with chlorotriazine (terbutylazine) or thiomethylazine (ametryn) templates. The ametryn-MIP strongly retained the ten studied triazines with 60–100% recovery while low recoveries were observed with the terbutylazine-MIP for triazines that did not belong to the chlorotriazine group [39]. Similar results had actually been obtained when using immunosorbents with antibodies against the same targets. The situation was similar with MIPs produced for phenylurea herbicides for which the recoveries were good for most phenylureas with isoproturon as the template, while use of linuron as the template led to low recoveries [40]. Other examples of class-selective MIPs for SPE include phenolic acids and degradation products from organophosphorous nerve agents [41] and anti-inflammatory drugs [42].

4.3 Binding Assays, Antibody Mimics, Receptor Mimics, Sensors

A systematic approach to this section is difficult because different authors use different definitions for sensors, assays and for what artificial sensors could be. Again, most activities are found with MIPs and in particular MIPs used as replacement for antibodies in ELISA-like assays and in biosensor-like applications. The practical applications of artificial receptors in membrane-bound designs are fewer, maybe because of larger technical constraints, but there are examples and possibly this is a field where nanotechnology fits in well. Attractive features of MIPs in analysis are their potential robustness and the possibility to relatively easily make MIPs against small molecules and also lower cost expectations. Thus, in most cases where you traditionally have used antibodies, and sometimes enzymes, you could try an artificial receptor, like a MIP. There are many reviews, such as [2, 8, 13, 43] that can give some ideas, but otherwise it is your own fantasy that places limits on what could be tried. Some examples will be given in the rest of this section.

Label-free detection in assays and sensor applications could be a problem but mass-sensitive detection of binding using piezoelectric devices such as QCM is a common approach as is surface plasmon resonance (SPR). Conductometric, capacitive and even calorimetric techniques have also been used. A QCM device was used for highly sensitive detection of 2-methylisoborneol, an off-flavor compound produced by a variety of microorganisms causing odor problems in drinking water and in fish. By placing an intercalating nylon layer between the QCM electrode and the MIP detection limits down to 10 ng/L could be obtained [44].

A *piezoelectric sensor* was also used for the analysis of microcystin LR, which is a highly toxic compound produced by freshwater cyanobacteria that could be present in aqueous samples. Microcystin MIPs were used for up to 1000-fold preconcentration by SPE from the water samples and as recognition receptors in the piezoelectric sensor. The minimum detectable concentration of toxin for this sensor was 0.35 nM which is below the required detection limit (1 nM) for microcystin LR in drinking water [45].

QCM- and *SAW* (*surface acoustic wave*)-*devices* have been the transducers of choice in MIP-based detection of larger components like cells and viruses. Yeast imprints were made on QCM electrode surfaces using the titanium oxide sol–gel stamping method. The sensor surface showed a regular honeycomb-like yeast imprint coating 1 μm deep imprints. The yeast cells could be measured up to 21 g/L in growth media [46]. This detection system was proposed for the fermentation and biotechnological industries for cell counting and contamination detection. Viruses can also be detected using this general surface imprinting concept that was mentioned in Sect. 3 [26]. Also capacitive detection was employed in these studies on viruses and cells. An example of capacitive detection of a low molecular analyte was reported for creatinine that could be detected by a decrease in the electrode capacitance using a capacitive sensor based on MIP photografted onto a gold electrode. There was no response to the addition of sodium chloride, creatine, urea, or glucose and the detection limit for creatinine of 10 μM, should be sufficient for medical applications [47].

Conductometry can be used in many situations to detect binding events or reactions when there is an accompanying change in the distribution of ionic species. A conductometric sensor for atrazine using MIP as a recognition agent was prepared by using MAA and EDMA in the presence of atrazine as the template. The useful range was 0.01–0.5 mg/L, and the membrane was stable for up to 4 months without loss of sensitivity [48].

Another fairly general concept for detection is *holographic sensing*. This consists of a combination of a simple reflection hologram with an analyte-selective "smart polymer" with optical interrogation and a reporting transducer. Holographic sensors undergo visible optical changes in response to fluctuations in analyte concentration. The replay wavelength shifts as a consequence of the swelling/contraction induced in the base hydrogel, which alters the spacing between the embedded diffraction grating and thus varies the wavelength of the reflected light. A responsive gel for L-lactate could be produced through conjugation of a ligand that upon complex formation with the analyte charge change within the gel phase. Different boronic acid-based receptors were synthesized and it was found that the incorporation of 3-acrylamidophenyl boronic acid into an acrylamide hydrogel produced the largest response toward L-lactate [49].

Fluorescence-based sensors are highly sensitive and convenient for analysis. Wherever direct analyses of target molecules are possible and changes of fluorescence intensity appear in the presence of target molecules without any use of external reagents or conjugates it the most attractive situation [2]. The use of external fluorescence (conjugates) is presently a practiced method. Indicator-displacement assays including fluorescent-based assays have recently been reviewed [50]. Direct chemiluminescence may also be used. A chemiluminescence flow through sensor for 1,10-phenanthroline based on the combination of molecular imprinting and chemiluminescence was presented [51]. When analyte and H_2O_2 passed into the MIP column with the buffer stream, these were complexed by the pyridine–Cu(II) binding sites on the MIP and reacted with the H_2O_2 molecules. During the H_2O_2 decomposition, superoxide radical ions form and react with 1,10-phenanthraline and liberate a chemiluminescence signal. This liberates the binding site for another analyte molecule.

Chemiluminescence is also a popular detection method in competitive assays. In the first 2,4-D imaging assay analogous to a competitive enzyme immunoassay reported, microtiter plates were coated with polymer microspheres imprinted with 2,4-D using poly(vinyl alcohol) (PVA) as the glue. In a competitive mode the analyte-peroxidase conjugate was incubated with the free analyte in the microtiter plate; the bound fraction of the conjugate was then quantified using luminol. The light emission was measured by imaging with a cooled CCD camera. With this method, 2,4-D could be measured up to 34 nM with a useful range from 68 nM to 680 μM, and it was possible to measure a large number of samples simultaneously [52]. In a subsequent paper an imprinted polymer-based capillary assay using chemiluminescence and a PMT for 2,4-D detection was reported. A glass capillary was modified by covalently attaching 2,4-D MIP to the inner capillary wall. 2,4-D was labelled with tobacco peroxidase and used as a tracer in a competitive format in which the bound fraction of the conjugate was quantified in flow injection detection mode. The MIP capillary could be regenerated after each measurement, thereby allowing for consecutive measurements of large numbers of samples. The assay could be easily transformed into a FIA system and automated. A detection limit two orders of magnitude lower was achieved when detection was done in a discontinuous mode and the chemiluminescence light was conducted to the photomultiplier tube by an optical fiber bundle. A dynamic range of detection from 5 pg/mL–100 ng/mL (22.5 pM–450 nM) was obtained [53].

Aptamer-based detection methods are usually based on fluorescent measurements. In the overview on nucleic acid aptamers and enzymes as sensors previously cited [7] different detection methods including fluorescence spectroscopy are presented. A few other interesting reports on aptamers should be included. A new approach for generating unmodified DNA aptamers that can be immediately transformed into effective signaling probes without the need for further optimization was recently presented [54]. Signaling aptamers are aptamer probes that couple target binding to fluorescent-signal generation. The normal way to obtain signal generation capability is by post selection modifications. An alternative method is to create signaling aptamers directly by in vitro selection. The actual structure-switching idea takes advantage of the universal ability of any aptamer to adopt two different structures: a duplex structure with an antisense DNA and a complex structure with the cognate target. An aptamer can switch structures from duplex to complex upon target addition and, if the aptamer is labelled with a fluorophore and the antisense DNA is labelled with a quencher, the structure-switching process can be synchronized to fluorescence signaling. The new strategy permits the creation of standard aptamers that are encoded with a duplex-to-complex switching capability and can be converted into signaling probes immediately upon their isolation.

The fact that protein concentrations can vary over several orders of magnitude in many physiological and pathological processes presents an analytical problem. Affinity analysis of proteins with a very wide dynamic range can usually not be realized with a single affinity probe but requires multi-affinity probe analysis. In a study using kinetic capillary electrophoresis (KCE) for generation of smart DNA aptamers with high selectivity and a wide range of predefined K_d values a set of three aptamers with differ-

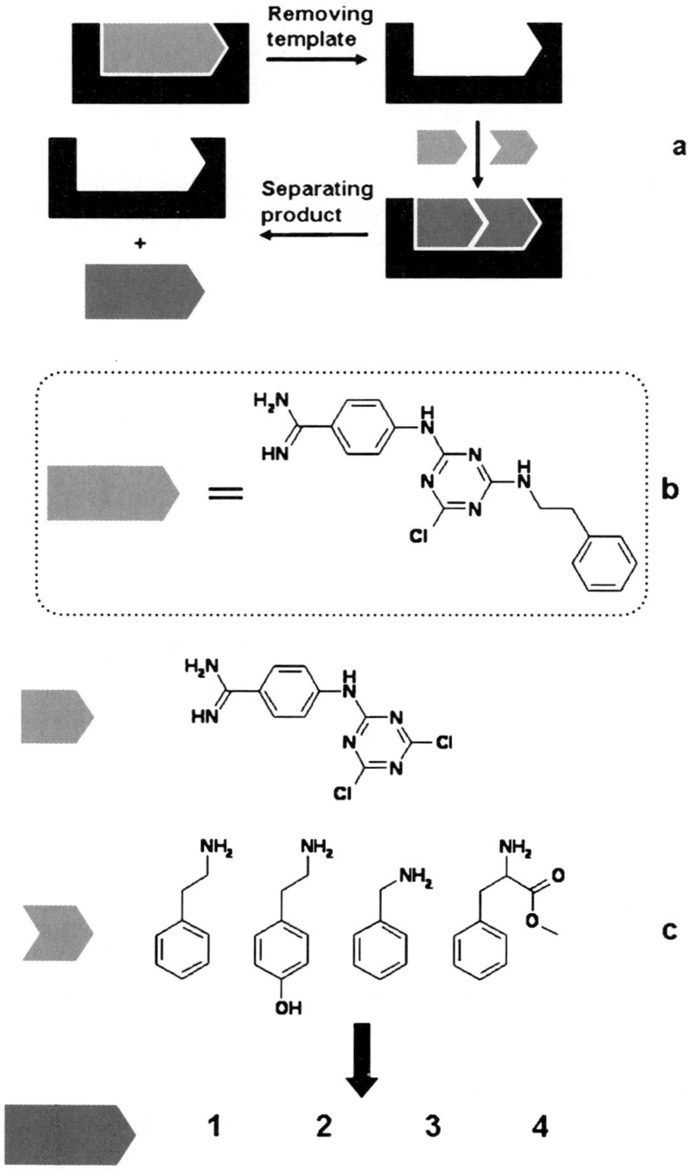

Fig. 7 **a** Schematic representation of the anti-idiotypic imprinting approach. After removal of the template (*light gray*) from the MIP (in *black*), the binding cavity was used to direct the assembly of the reactants (*gray*) to give products (*gray*) mimicking the original template (*light gray*). **b** Chemical structure of the template, 2-(4-amidinophenylamino)-4-chloro-6-phenylethylamino-s-triazine (1). **c** Synthetic reactions investigated. Reproduced from [61]

ent K_d-values was designed [55]. The dynamic range for a single-probe analysis is ca. 2 orders of magnitude. With the multi-probe set the dynamic range was more than 4 orders of magnitude. Finally, an interesting Quantum-Dot/Aptamer-Based Ultrasensitive Multi-Analyte Electrochemical Biosensor should be mentioned [56]. Nanocrystal tracers were used for designing multi-analyte electrochemical aptamer biosensors with subpicomolar (attomole) detection limits. Such quantum-dot (QD) semiconductor nanocrystals offer an electrodiverse population of electrical tags. Four encoding nanoparticles (cadmium sulfide, zinc sulfide, copper sulfide, and lead sulfide) can be used to differentiate the signals of four DNA targets, SNPs, or antigens in connection to stripping voltammetric measurements of the corresponding metals. This combination results in the extremely high sensitivity mentioned.

Thermometric monitoring. Another possibility for label-free detection of binding events and reactions is the use of calorimeters. In enzyme technology the enzyme thermistor is widely used for this purpose since most enzymatic reactions are accompanied by heat evolution in the range of 5–100 kJ/mol [57]. The combination of a calorimetric transducer – the thermistor – with analyte recognition by a catalytically active MIP was recently demonstrated for the first time [58]. Both effects, the binding/desorption and the catalytic conversion of the substrate, were reflected by concentration-dependant heat signals. This will be further discussed below (Fig. 7).

4.4 Carbohydrate Specific Receptors

Structural analysis of glycoconjugates is getting increasing attention due to the important functions that glycosylation have in many biological systems. Conventional structural analysis, although very efficient and exact today, is extremely laborious and requires sophisticated instrumentation. Analysis based on molecular recognition appears promising and carbohydrate-binding proteins such as lectins and also antibodies are very useful, but have the usual limitations of lack of robustness and availability. The design of artificial receptors that can compete with the natural systems in selectivity and affinity is an attractive area of research. Sugar recognition, especially in aqueous solutions at neutral pH has been particularly focused because of its potential applications in the development of glycomic tools, sensors, and therapeutic agents. A few examples to illustrate current activities in this field will be given below.

Sulfated sugars are a class of complex compounds with central biological roles in mammals. Among them, heparan sulfates are multifunctional cell regulators, whose biological activities are related to their sulfation pattern. Determination of fine structures of these sulfated sugars such as the distribution of the biologically relevant 2-*O*- and 6-*O*-sulfate substitutions on sugars in the heparin sulfate molecule is important for understanding the biological functions. The applicability of molecular imprinting technology for recognition of specific sequences was investigated using glucose-6-*O*-sulfate as a model [59]. It was found that MIPs can specifically recognize sulfated sugars by the introduction of primary amines at the polymer side. Imprinted polymers showed excellent selectivity with regard to the sulfate position, the sugar configuration, and the presence of *N*-acetyl groups. Studies of carbohydrate-binding proteins from a different angle, by probing with a library of inhibitors, was demonstrated in a study in which dynamic combinatorial libraries were generated from a pool of thiol components through reversible thiol-disulfide interchange and screened using QCM. Dimers based on 1-thio- and 6-thio-mannose analogues were found to be the most efficient concanavalin A inhibitors [5]. A cadmium-centered tris-boronic acid receptor was synthesized and its binding properties were screened towards various carboxy- and phospho-sugars with a quick 96-well plate assay [60]. The actual receptor showed high affinity toward specific sugar derivatives in protic media at neutral pH, especially for gluconic acid and to a lower degree to glucoronic acid and lactobionic acid.

4.5 MIP-Directed Synthesis

MIPs have been developed as artificial receptors in order to mimic the active sites or binding sites of enzymes, antibodies, and natural receptors, but with different properties such as higher robustness and ability to work in harsher environments. There is obviously an interest to take MIPs further than just to recognition. One application is to use MIPs for screening of new inhibitors or antagonists [16]. Another attractive application is to use the nano-cavities of MIPs as reactors for the synthesis of inhibitors or drugs. Recently, the "anti-idiotypic" approach was developed in which the MIP is assisting the synthesis and screening of inhibitors [61]. A polymer imprinted with a known inhibitor for the enzyme kallikrein was synthesized. Different inhibitor building blocks were then allowed to assemble and interconnect in the imprinted site, leading to a combinatorial library of new enzyme inhibitors (Fig. 7).

This in-cavity synthesis was developed further to exploring the feasibility of combinatorial drug or lead candidate "cloning" – a process that exploits properties (shape, size, and electronic features) embedded in the historical drugs or advanced drug candidates to identify novel biologically active entities as starting points. It was demonstrated, for the first time, that a MIP can act as a highly regioselective nano-reactor for the Huisgen 1,3-dipolar cycloaddition of azides and alkynes, which usually leads to a mixture of 1,4- and 1,5-regioisomers [62]. Unlike the previously reported regioselective MIP nano-reactors, these imprinted cavities bind two reactants by means of only noncovalent interactions, similar to what is utilized by enzymes.

The "direct molding" approach was developed for the direct generation of inhibitors inside the active sites of the enzymes as an extension of the "anti-idiotypic" approach and of the bio-imprinting method [63]. In the direct molding process, the binding sites of the enzymes are used instead of the cavities of the imprinted polymers. A library of building blocks for enzyme inhibitors was first incubated together with the enzyme, those of appropriate functionality and size spontaneously self-assembling at the binding sites of the enzyme, resulting in the in situ formation of enzyme inhibitors. This approach allows the direct generation of new inhibitor hits for the enzyme from a library of starting building blocks and does not need any known bioactive molecule to start with. There are, however, drawbacks in costs and the number of available reactions so further developments of both approaches can be expected.

Synthetic biomimetic catalysis by molecular imprints as an alternative to the use of biomolecules such as enzymes and catalytic antibodies is an appealing idea. Molecular imprinting with substrates or their transition state analogues can be a first step. At the same time, functional groups can be incorporated that act as binding sites, coenzyme analogs, or catalytic sites within the cavity and in a defined stereochemical manner [64]. These arti-

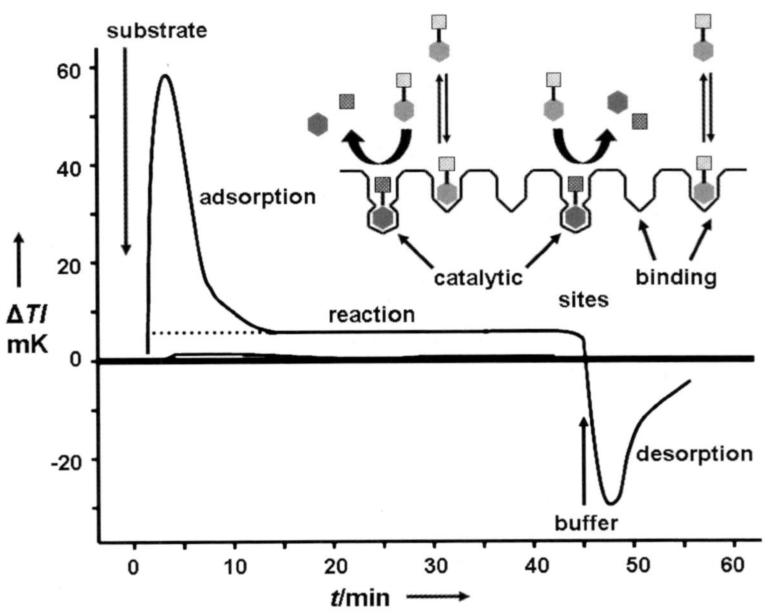

Fig. 8 Heat signals of the reactions of the bifunctional MIP and NIP with 5 mM phenylacetate. Start of substrate flow at 0 min and start of washing with buffer at 45 min. The curve close to the baseline represents the NIP. Inset shows two different interactions between the polymer and the substrate inside the MIP at times between 25 and 45 min. From [58]

ficial polymeric catalysts will hopefully be more durable and more resistant to harsh environments than biomolecules, which would be highly advantageous for industrial continuous transformation and/or conversion reactions. In order to improve the reaction rates of catalytic MIPs a new preparation method for an esterolytic imprinted polymer with catalytic sites on the surface was proposed. A template was prepared by immobilizing a transition state analogue (phosphoramidic acid derivative) of an esterolytic reaction within porous silica particles. Polymerization within the pores was carried out using 4-vinylimidazole as a functional monomer and DVB as a cross-linker. The polymer was thereafter released by dissolution of the silica support with hydrofluoric acid. The idea was to get a more flexible matrix and to generate the binding sites near the surface for improved reaction rate. The catalytic properties were studied by incubation with three different 4-nitrophenylesters and spectrophotometric determination of the released 4-nitrophenol [65]. This approach yielded a polymer that exhibited higher relative catalytic activity than conventional catalytic MIPs.

Further insights in the mechanisms involved in the binding and catalysis using this catalytic MIP were obtained with use of an enzyme thermistor loaded with the catalytic polymer [58].The enzyme-like catalysis and antibody-like binding of a bifunctional MIP could now be simultaneously resolved for the first time (Fig. 8). The bifunctionality is based on two different types of recognition sets in the MIP, which could be distinguished by the differences in heat generation using the flow-through thermistor. Thus, the MIP thermistor allows measurements of two events: substrate conversion in the catalytic sites and its adsorption at the binding sites. With this method it will be possible to construct label-free multianalyte detectors and gain a closer insight into the interactions between the polymer catalyst and the substrate which is necessary to optimize the polymer synthesis for an efficient tailor-made catalyst.

Acknowledgements The skillful assistance by MSc Maria Yakovleva at the preparation of the manuscript is gratefully acknowledged.

References

1. Haupt K, Mosbach K (2000) Chem Rev 100:2495–2504
2. Kandimalla VB, Ju H (2004) Anal Bioanal Chem 380:587–605
3. Schmuck C, Wich P (2006) New J Chem 30:1377–1385
4. Lehn JM (1999) Chem Eur J 5:2455–2463
5. Pei Z, Larsson R, Aastrup T, Andersson H, Lehn JM, Ramström O (2006) Biosens Bioelectron 22:42–48
6. Schalley CA, Liitzen A, Albrecht M (2004) Chem Eur J 10:1072–1080
7. Navani NK, Li Y (2006) Curr Opin Chem Biol 10:272–281
8. Peczuh MW, Hamilton AD (2000) Chem Rev 100:2479–2494
9. Schrader T, Koch S (2007) Mol Bio Syst 3:241–248

10. Favero G, Campanella L, Cavallo S, D'Annibale A, Perella M, Mattei E, Ferri T (2005) J Am Chem Soc 127:8103–8111
11. Trojanowicz M, Mulchandani A (2004) Anal Bioanal Chem 379:347–350
12. Nikolelis DP, Petropoulou S-SE, Theoharis G (2003) Electroanalysis 15:1616–1624
13. Umezawa Y, Aoki H (2004) Anal Chem 76:321A–326A
14. Di Natale C, Paolesse R, D'Amico A (2007) Sens Actuators B 121:238–246
15. Turiel E, Tadeo JL, Cormack PAG, Martin-Esteban A (2005) Analyst 130:1601–1607
16. Zhang H, Ye L, Mosbach K (2006) J Mol Recognit 19:248–259
17. Moral-Pérez N, Mayes AG (2006) Biosens Bioelectron 21:1798–1803
18. Spégel P, Schweitz L, Nilsson S (2003) Electrophoresis 24:3892–3899
19. Quaglia M, Sellergren B, De Lorenzi E (2004) J Chromatog A 1044:53–66
20. Biffis A, Braham NB, Siedlaczek G, Stalberg S, Wulff G (2001) Macromol Chem Phys 202:163–171
21. Chronakis IS, Milosevic B, Frenot A, Ye L (2006) Macromolecules 39:357–361
22. Xia Y, Whitesides GM (1998) Angew Chem Int Ed 37:550–575
23. Panasyuk TL, Mirsky VM, Piletsky SA, Wolfbeis OS (1999) Anal Chem 71:4609–4613
24. Piletsky SA et al (2000) Macromolecules 33:3092–3098
25. Shi H, Wei-Bor T, Michael D, Ferrari GS, Ratner BD (1999) Nature 398:593–597
26. Hayden O, Podlipna D, Chen X, Krassnig S, Leidl A, Dickert FL (2006) Mater Sci Eng C 26:924–928
27. Ramström O, Nicholls IA, Mosbach K (1994) Tetrahedron: Asymmetr 5:649–656
28. Kempe M (1996) Anal Chem 68:1948–1953
29. Zhou SN, Lai EPC, Miller JD (2004) Anal Bioanal Chem 378:1903–1906
30. Liu Y, Liu XL, Wang JD (2003) Anal Lett 36:1631–1645
31. Ye L, Weiss R, Mosbach K (2000) Macromolecules 33:8239–8245
32. Ye L, Yoshimatsu K, Kolodziej D, Da Cruz Francisco J, Dey ES (2006) J Appl Polym Sci 102:2863–2867
33. Glad M, Norrlöw O, Sellergren B, Siegbahn N, Mosbach K (1985) J Chromatogr 347:11–23
34. Kempe M, Glad M, Mosbach K (1995) J Mol Recogn 8:35–39
35. Liu X, Mosbach K (1998) Am Biotechnol Lab 16:90
36. Bossi A, Bonini F, Turner APF, Piletsky SA (2007) Biosens Bioelectron 22:1131–1137
37. Shiomi T, Matsui M, Mizukami F, Sakaguchi K (2005) Biomaterials 26:5564–5571
38. Pichon V (2007) J Chromatogr A 1152:41–53
39. Chapuis F, Pichon V, Lanza F, Sellergren B, Hennion C (2004) J Chromatogr B 804:93
40. Carabias-Martínez R, Rodríguez-Gonzalo E, Herrero-Hernández E, Díaz-García ME (2005) J Sep Sci 28:453
41. Karasová G, Lehotay J, Sádecká J, Skacáni I, Lachová M (2005) J Sep Sci 28:2468
42. Caro E, Marcé RM, Cormack PAG, Sherrington DC, Borrull F (2005) J Sep Sci 28:2080
43. Anslyn EV (2007) J Org Chem 72:687–699
44. Ji HS, McNiven S, Kyong-Hoon L, Saito T, Ikebukuro K, Karube I (2000) Biosens Bioelectron 15:403–409
45. Chianella I, Piletsky S, Tothill IE, Chen B, Turner APF (2003) Biosens Bioelectron 18:119–127
46. Dickert FL, Hayden O, Lieberzeit P, Haderspoeck C, Bindeus R, Palfinger C, Wirl B (2003) Synth Met 138:65–69
47. Delaney TP, Mirsky VM, Wolfbeis OS (2002) Electroanalysis 14:221–225
48. Piletsky SA, Piletskaya EV, Elgersma AV, Yano K, Karube I, Parhometz YP, ElSkaya AV (1995) Biosens Bioelectron 10:959–996
49. Sartain FK, Yang X, Lowe CR (2006) Anal Chem 78:5664–5670

50. Nguyen BT, Anslyn EV (2006) Coord Chem Rev 250:3118–3127
51. Lin JM, Yamada M (2001) Analyst 126:810–815
52. Surugiu I, Danielsson B, Ye L, Mosbach K, Haupt K (2001) Anal Chem 73:487–491
53. Surugiu I, Svitel J, Ye L, Haupt K, Danielsson B (2001) Anal Chem 73:4388–4392
54. Nutiu R, Li Y (2005) Angew Chem Int Ed 44:1061–1065
55. Drabovich AP, Okhonin V, Berezovski M, Krylov SN (2007) J Am Chem Soc 129:7260–7261
56. Hansen JA, Wang J, Kawde A-N, Xiang Y, Gothelf KV, Collins G (2006) J Am Chem Soc 128:2228–2229
57. Xie B, Ramanathan K, Danielsson B (2000) TRAC 19:340–349
58. Lettau K, Warsinke A, Katterle M, Danielsson B, Scheller FW (2006) Angew Chem Int Ed 45:6986–6990
59. Siñeriz F, Ikeda Y, Petit E, Bultel L, Haupt K, Kovensky J, Papy-Garcia D (2007) Tetrahedron 63:1857–1862
60. Zhang T, Anslyn EV (2006) Org Lett 8:1649–1652
61. Mosbach K, Yu Y, Andersch J, Ye L (2001) J Am Chem Soc 123:12420–12421
62. Zhang H, Piacham T, Drew M, Patek M, Mosbach K, Ye L (2006) J Am Chem Soc 128:4178–4179
63. Yu Y, Ye L, Haupt K, Mosbach K (2002) Angew Chem Int Ed 41:4459–4463
64. Wulff G (2002) Chem Rev 102:1–27
65. Lettau K, Warsinke A, Laschewsky A, Mosbach K, Yilmaz E, Scheller FW (2004) Chem Mater 16:2745–2749

Adv Biochem Engin/Biotechnol (2008) 109: 123–154
DOI 10.1007/10_2007_075

Published online: 14 September 2007

New Trends in Immunoassays

Cangel Pui-yee Chan[1] (✉) · Yiu-chi Cheung[1] · Reinhard Renneberg[1] · Matthias Seydack[2]

[1]Department of Chemistry, The Hong Kong University of Science and Technology, Clear Water Bay, Kowloon, SAR Hong Kong, China
cangel@ust.hk

[2]8sens.biognostic GmbH, Robert-Roessle-Str. 10, 13187 Berlin, Germany

Abstract This article takes a special focus on signal amplification technologies in immunoassays and new generations of lateral-flow assays. Novel signal amplification technologies based either on new classes of biofunctional nanocrystals consisting of releasable fluorophores or on aggregation-induced emission (AIE) can improve the sensitivity and the limits of detection in immunoassays. A bio-barcode assay also allows signal amplification by utilizing antibody-coated magnetic beads to concentrate the analytes and antibody-coated gold nanoparticle probes to carry with a large number of oligonucleotides. These innovative technologies boost the development of immunoassays.

Growth in rapid immunoassay is fueled by the increasing number of diabetics, the globalization of infectious diseases and the surge in cardiovascular and other chronic diseases as well as other chronic conditions. Rapid, near patient, decentralized, point-of-care (POC) tests are emerging as a tool for more efficient diagnosis and patient evaluation. Technological innovations in lateral-flow assays have enabled a move to bring testing closer to the patient. A novel "digital-style" lateral-flow assay provides semi-quantitative results by simply counting the number of red lines in the test without any expensive reading instrument. An immuno-threshold-based assay can give a signal directly proportional to the concentration of a hapten to prevent confusion on interpretation of the test results. In addition, POC tests become more meaningful to healthcare professionals by combining the benefits of new technologies to provide quantitative results. A molecular compact disc provides a high-resolution imaging capability that can identify and quantify many different antigens simultaneously in highly complex immunoassays.

Further advances in immunoassays will bring diagnostic testing even closer to the patient, and can help physicians to monitor diseases that require immediate test results, thereby enhancing the quality of patient care.

Keywords Aggregation-induced emission (AIE) · C-reactive protein (CRP) · Cortisol · Fatty acid-binding protein (FABP) · Fluorescein diacetate (FDA) · Lateral-flow assay · Nanoparticle

1 Introduction

Immunoassay, the most widely used analytical technology in biodiagnostics, includes the determination of antigens, hormones, drugs and antibodies. Immunoassays have been used in hospitals, laboratory medicine, and research since the mid-1960s. They are performed in central laboratories using a variety of instrument-based technologies or on-site via rapid test techniques, primarily lateral-flow assays. The rapid development of immunoassay technology was made possible by monoclonal antibodies or antibody-like molecules including novel fusion proteins, antibody mimics, and antibodies determining antigen-antibody complexes and by the application of recombinant antigens. There are assay formats for measuring proteins in solution [e.g. enzyme-linked immunosorbent assays (ELISA) and immunospot assays], on the surface of cells (e.g. flow cytometry), within cells (e.g. immunohistochemical and immunofluorescent microscopy) and in organs (e.g in vivo imaging with labeled antibodies).

A wide range of labeling and signal-enhancement strategies have been developed that allow ligand binding to be detected through association of the ligand with a "read-out" antibody that has particular fluorescent, colorimetric, histochemical or radioactive properties. Advances in nanoscience have a significant impact on these strategies, boosting the development of a variety of important technologies. The impact of these new technologies is particularly large in diagnostics, where a number of nanoparticle-based assays have been introduced for the detection of biomarkers. Given the variety of strategies afforded through nanoparticle technologies, a significant goal is to tailor nanoparticle surfaces to selectively bind a subset of biomarkers, either for direct detection and characterization or to sequester the target molecules for later study using other available techniques. To date, applications of nanoparticles have largely focused on DNA- or protein-functionalized nanoparticles used as the target-specific probes. Polymerase chain reaction (PCR) has enabled rapid advances in the development of powerful tools for detecting and quantifying DNA targets of interest for research, forensic, and clinical applications [1–3]. The development of comparable target amplification methods for proteins or DNA could substantially improve medical diagnostics and the developing field of proteomics [4–7]. The SuperNova® System is a novel signal amplification technology based on a new class of biofunctional fluorogenic nanocrystals to improve the sensitivity and the limits of detection in immunoassays [8–13]. It also allows the quantitative detection of pathogen DNA. Another novel class of biofunctional silole nanocrystals with aggregation-induced emission (AIE) feature has potential to create highly sensitive immunoassays [14].

Although it is impossible to chemically duplicate analytes yet, the analyte can be tagged with oligonucleotide markers that can be subsequently amplified with PCR and then identified by DNA detection [15–20]. This approach is often referred to as immuno-PCR that allows the detection of proteins with DNA markers. A bio-barcode assay utilizes antibody-coated magnetic beads to capture and concentrate the analytes, which are subsequently labeled with gold nanoparticle probes conjugated with specific antibodies and DNA barcodes [21–26]. The DNA barcodes are then released from the complex and detected via hybridization. This system allows signal amplification as the nanoparticle probe carries a large number of oligonucleotides per protein binding event and PCR can further enhance the sensitivity. This technology can also be applied to multiplexed assays by using different DNA barcode sequences conjugated with specific antibodies on each gold probe as indicators for different protein markers.

Recently, there has been increasing interest in multiplexed assays, especially its impact on genomics. Conventional immunoassay platforms have limited multiplexing capacity and high sample volume requirements. High-throughput multiplex immunoassays that measure hundreds of proteins in complex biological matrices in parallel have become significant tools for quantitative proteomics studies, diagnostic discovery and biomarker-assisted drug development [27–32]. Another aproach to multiplexing is a "molecular" compact disc, which provides a high-resolution imaging capability that can identify and quantify many different antigens simultaneously in highly complex immunoassays [33, 34].

At present there are a number of commercially available methods of determining haptens. However, such methods frequently suffer from the disadvantage of producing an increasing signal as the concentration of the hapten decreases (competitive assay format). This easily causes confusion. It would be desirable to provide a method that allows the determination to produce a response which increases with the concentration of the haptene. An immuno-threshold-based assay can give a signal directly proportional to the concentration of the hapten [35]. Also, a one-step, homogeneous non-competitive immunoassay for haptens using a highly specific antibody against the immune complex (IC) formed between an antibody and an analyte has been demonstrated [36]. The detection is based on Förster-type resonance energy transfer (FRET) between two fluorophores.

New generations of immunodiagnostic tests are suitable for application in point-of-care (POC) testing—the processing and analysis of samples at the patient's bedside or in a physician's office. Such testing enables the inclusion of real-time results into the patient care process. It allows a physician to quickly gain a more complete clinical impression, integrate the lab results into the diagnostic process, manage the patient's care based on timely lab data, and much more easily monitor the results of therapeutic intervention. The current generation of immunodiagnostic tests has benefited from industry improvements and breakthroughs in lateral-flow immunoassays [35, 37–41].

2 Nanoparticle-Based Signal Amplification Technologies

Advances in nanotechnology have had significant impacts on the field of biodiagnostics [42–44]. In our previous studies, we applied novel signal amplification technologies based on new classes of biofunctional fluorogenic nanocrystals to improve the sensitivity and the limits of detection in immunoassays [8–14].

2.1 SuperNova® System-Nanoencapsulated Microcrystalline Particles for Superamplified Biochemical Assays

A novel class of particulate labels based on nanoencapsulated organic microcrystals with the potential to create highly amplified biochemical assays was demonstrated [8]. Labels were constructed by encapsulating microcrystalline fluorescein diacetate (FDA; average size of 500 nm) within ultrathin polyelectrolyte layers of poly(allylamine hydrochloride) and poly(sodium 4-styrenesulfonate) via the layer-by-layer technique (Fig. 1). Subsequently, the polyelectrolyte coating was used as an "interface" for the attachment of anti-mouse antibodies through adsorption. A high molar ratio of fluorescent molecules present in the microcrystal core per biomolecule (on the particle surface) was achieved.

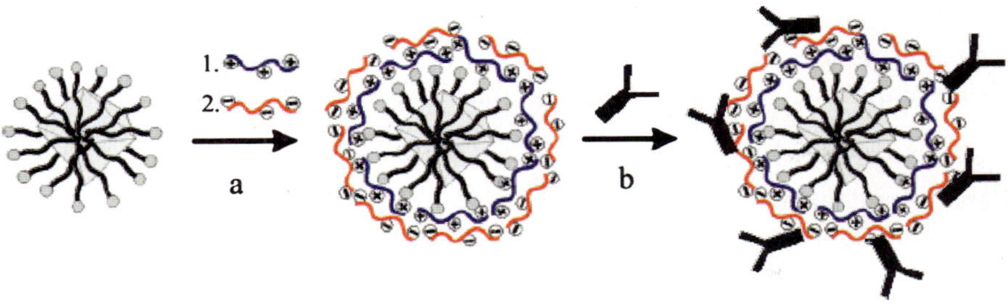

Fig. 1 Schematic illustration of the preparation of biolabeled, polyelectrolyte-encapsulated organic microcrystals. FDA was ball milled into micrometer-sized crystals in an aqueous surfactant (SDS) medium (introducing charges onto the originally uncharged microcrystal surface), followed by **a** encapsulation with polyelectrolyte multilayers of nanometer thickness, and **b** the attachment of a specific immunoreagent. Adapted from [8]

The applicability of the microcrystal-based label system was demonstrated in a model sandwich immunoassay for MIgG detection (Fig. 2). Following the immunoreaction, the FDA core was dissolved by exposure to organic solvent, leading to the release of the FDA molecules into the surrounding medium.

Amplification rates of 70- to 2000-fold (expressed as an increase in assay sensitivity) of the microcrystal label-based assay compared with the corresponding immunoassay performed with direct fluorescently labeled antibodies are reported. Our approach provides a general and facile means to prepare

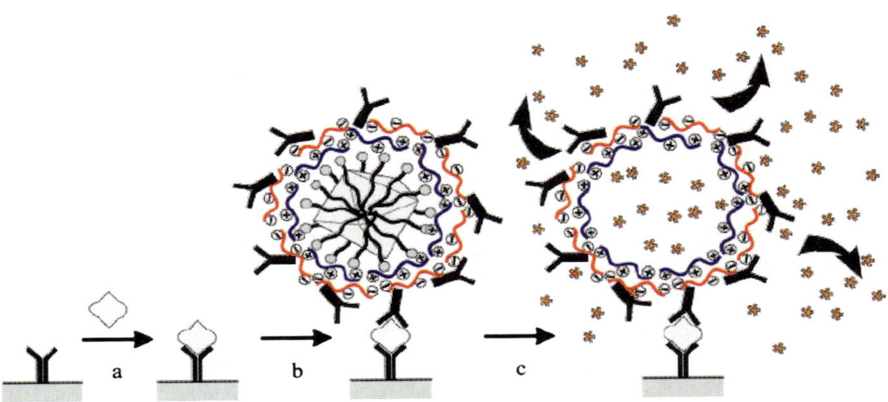

Fig. 2 Principle of a sandwich immunoassay using FDA particulate labels. **a** The analyte is first immobilized by the capture antibody preadsorbed on the solid phase and then **b** exposed to antibody-labeled microparticle detectors. Every microparticle contains 10^8 FDA molecules. **c** High signal amplification is achieved after solubilization, release, and conversion of the precursor FDA into fluorescein molecules by the addition of DMSO and NaOH. Fluorescence intensity is proportional to the analyte concentration. The surfactant is not shown in the last step; some surfactant may be adsorbed to the polyelectrolyte layers and some may be released with the fluorescein. Adapted from [8]

a novel class of biochemical assay labeling systems. The technology has the potential to compete with enzyme-based labels, as it does not require long incubation times and a stopping step, thus speeding up bioaffinity tests.

2.2 Nanocrystal Biolabels with Mega-Releasable Fluorophores for Immunoassays

We further simplify the signal amplification technology by using a two-step approach to encapsulate nanocrystalline FDA with an average size of 107 nm with ultrathin polyelectrolytic distearoylphosphatidylethanolamine (DSPE) lipids coupled with amino(poly(ethylene glycol)) (PEG2000-Amine) [9]. This

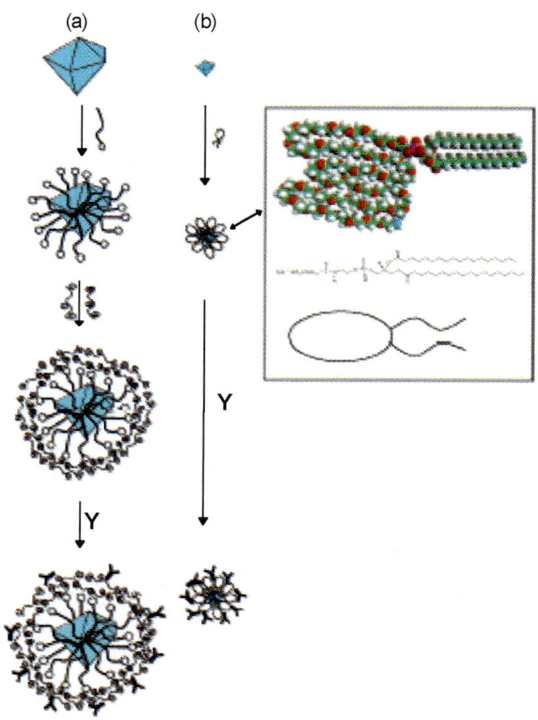

Fig. 3 Schematic illustration of the preparation of biofunctional fluorescent labels. In a previous study [8], **a** FDA was milled into microcrystals with an average size of 500 nm, stabilized with SDS surfactant, followed by encapsulation of polyelectrolyte multilayers, and the attachment of biorecognition molecules, e.g., antibodies. **b** A simple two-step approach is described in the current study: FDA was milled into nanometer-sized (107 nm on average) crystals in an aqueous surfactant (DSPE-PEG(2000)Amine) medium and subsequently biorecognition molecules, e.g., antibodies were deposited. The structure of DSPE-PEG(2000)Amine is shown in the *inset*. Adapted from [9]

polyelectrolyte coating was subsequently used as an interface for the attachment of biorecognition molecules through adsorption (Fig. 3). In brief, a suspension of 5% (w/w) FDA in 1.25% (w/w) DSPE-PEG(2000)Amine was milled for 48 to 72 h using a process developed by Elan Drug Delivery Inc. (King of Prussia, PA). The temperature was maintained at 20 °C to prevent the material from hydrolyzing. The pretreated particle suspension was diluted to 0.0626% (w/v) and incubated with 200 μg/mL biomolecules (e.g. antibodies, streptavidin) in 10 mM phosphate-buffered saline (PBS, pH 7.4) at room temperature for 1 h. The supernatant was removed after centrifugation at 16 000 g for 10 min. After three repeated centrifugation/washing cycles, the coated particles were finally separated from the mixture.

A high fluorescein equivalent (FE) value (2.6×10^6) and a high molar ratio of released fluorescent molecules per binding event (2.8×10^4) were achieved in this nanocrystal biolabel system.

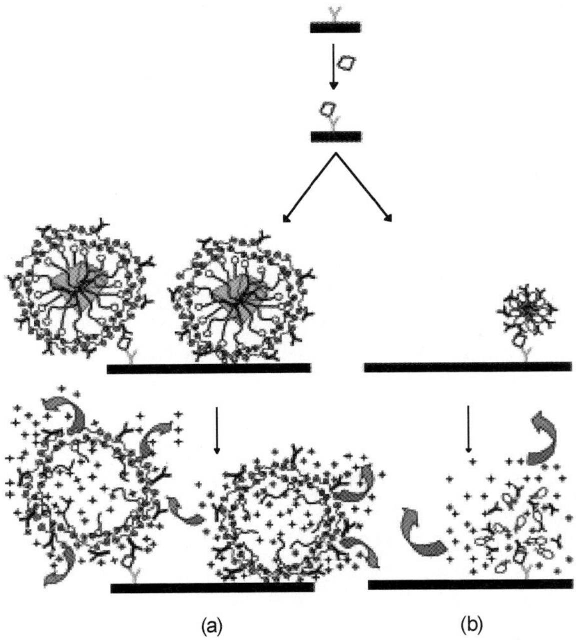

Fig. 4 Principle of a sandwich fluorescent immunoassay using nanocrystalline FDA conjugates. The analyte is first incubated with the capture antibody preadsorbed on the microtiter plate and then exposed to nanocrystalline FDA conjugates. High signal amplification is achieved after solubilization, release, and conversion of the precursor FDA into fluorescein molecules by the addition of DMSO and NaOH. **a** The previous microcrystal (500 nm in average size) contained 10^8 FDA molecules, and a limit of detection of 4 μg/L was achieved [8]. **b** A much lower limit of detection of 0.057 μg/L and minimized nonspecific interactions are demonstrated using the DSPE-PEG(2000)Amine-modified FDA as labels although each nanocrystal contains 2 orders of magnitude fewer fluorescein molecules. Adapted from [9]

The analytical performance of the nanocrystal-based label system is evaluated in a model sandwich immunoassay for the detection of MIgG (Fig. 4). After separation of the non-bound antibody nanocrystal labels, fluorophores are released by hydrolysis and dissolution of the nanocrystalline FDA. Due to the release of the fluorophores into a large volume of

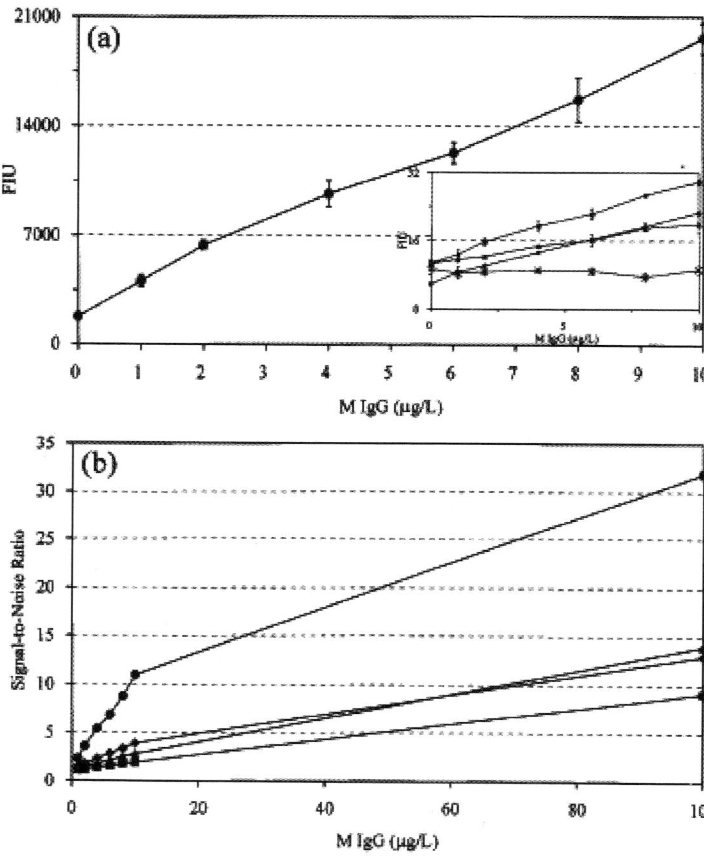

Fig. 5 **a** Sandwich fluorescence immunoassay using Gt α MIgG-FDA nanocrystals (•) and using Gt αMIgG-FITC from Arista (♦), Sigma-Aldrich (▲), and Molecular Probes (■) as labels. The *inset graph* shows the amplification of the fluorescent intensity (∗). Good linearity is observed in all cases using FDA nanocrystals as the label ($R^2 = 0.9954$) and direct FITC labels from Arista ($R^2 = 0.9962$), Sigma-Aldrich ($R^2 = 0.9915$), and Molecular Probes ($R^2 = 0.9947$). There is no increase in fluorescent intensity by increasing the MIgG concentrations using FDA-nanocrystal labels without addition of DMSO/NaOH as the releasing agent (∗). **b** Alternatively, use of the signal-to-noise ratio instead of the fluorescence intensity unit to compare the sensitivity of the assay using FDA-nanocrystals and Gt αMIgG-FITC as detector antibodies. *Error bars* correspond to standard deviations ((SD, n) 3). Adapted from [9]

Table 1 Signal-to-noise ratios (S/N) and sensitivity-to-sensitivity (S/S) of nanocrytalline FDA-labeled to FITC-labeled antibodies

Analyte (μg/L)	S/N of Gtα MIgG detector antibody FDA	FITC (Arista)	FITC (Sigma-Aldrich)	FITC (Molecular Probes)	S/S ofGtαMIgG-FDA/Gtα MIgG-FITC FDA/FITC (Arista)	FDA/FITC (Sigma-Aldrich)	FDA/FITC (Molecular Probes)
1	2.27	1.47	1.09	1.19	843.01	2276.13	1138.06
2	3.56	1.74	1.16	1.47	1065.63	2672.95	916.44
4	5.39	2.28	1.39	1.81	1061.92	1896.81	901.76
6	6.85	2.83	1.53	2.08	988.20	1880.11	905.24
8	8.77	3.34	1.80	2.49	1022.81	1650.37	869.39
10	10.96	3.90	1.89	2.79	1061.44	1891.30	931.53
100	31.87	12.93	9.11	13.95	798.84	644.93	398.52

Raw data were extracted from the measurements presented in Fig. 5.
Adapted from [9]

organic solvent/sodium hydroxide mixture, self-quenching is suppressed. The FDA[DSPE-PEG(2000)Amine]-modified biolabels form a highly stable colloidal suspension and, when coated with antibodies and used in an immunoassay, resulted in minimal non-specific interactions.

Figure 5a shows the calibration curves of Gt α MIgG-FDA nanocrystal labels in comparison with a direct FITC-labeled antibody conjugate. A limit of detection of 0.060 μg/L was achieved with the FDA-labeled antibodies. The detection limit of the FDA conjugate is lower by a factor of 5–28 than that of the direct FITC conjugate. Table 1 shows that the signal-to-noise ratio of the assays using nanocrystalline FDA-labeled antibodies was higher than the ratios of assays using the direct FITC-labeled antibodies at all analyte concentrations and was also higher than the ratios of assays using microcrystalline FDA conjugates in the previous study [8]. The greatest increase in the signal-to-noise ratio was observed in the low concentration range (Fig. 5b).

The FDA conjugates appear to be very useful in maximizing the sensitivity of a fluorescent assay. Using nanocrystalline FDA antibody conjugates allows for a low detection limit and a sensitivity that was 400- to 2700-fold higher compared with a state-of-the-art immunoassay using directly fluorescent-labeled antibodies.

2.3 A Highly Sensitive Fluorescent Immunoassay Based on Avidin-labeled Nanocrystals

The applicability of NeutrAvidin-labeled nanocrystals was demonstrated in an immunoassay using the avidin–biotin technique [11]. This can be applied to detect different kinds of analytes that are captured by corresponding biotinylated biomolecules in different bioanalytical applications. Biotinylated antibody and NeutrAvidin-labeled nanocrystals were applied to the assay sequentially. The performance was then compared with the traditional sandwich-type assay for MIgG detection. Following the immunoreaction, the nanocrystals were released by hydrolysis and dissolution instigated by adding a large volume of organic solvent/ sodium hydroxide mixture. The limit of detection was lowered by a factor of 2.5 to 21 while the sensitivity was found to be 3.5- to 30-fold higher than in immunoassays using commercial labeling systems [e.g. fluorescein 5-isothiocyanate (FITC) and horseradish peroxidase (HRP)]. This study shows that using fluorescent nanocrystals in combination with the avidin–biotin technique can act as a flexible probe to be applied in different kinds of immunoassays.

The biofunctional nanocrystals were also applied to detect different kinds of clinically interesting plasma proteins with high sensitivity, low limits of detection and short incubation times [12].

2.4 Biofunctional Organic Nanocrystals for Quantitative Detection of Pathogen Deoxyribonucleic Acid

A novel application of dissolvable, organic, and biofunctional nanocrystals for the quantitative detection of a PCR product was reported [13]. FDA was milled in a solution of a polymeric surfactant to create a stable, nanosized colloid with an interface for coupling streptavidin molecules (Fig. 6). The application of these particulate labels for the quantitative detection of biotinylated human papillomavirus (HPV) DNA, amplified in a standard PCR procedure, was demonstrated. A set of consensus primers, 5′ biotinylated PGMY09/11 [45], was used to amplify HPV-DNA from various concentrations of HPV 16, HPV 18 and HPV 45 plasmids. Biotinlyated HPV-DNA was amplified and hybridized specifically with DNA probes that had been immobilized on a NucleoLink™ microplate (Fig. 7). The biotin molecule of the

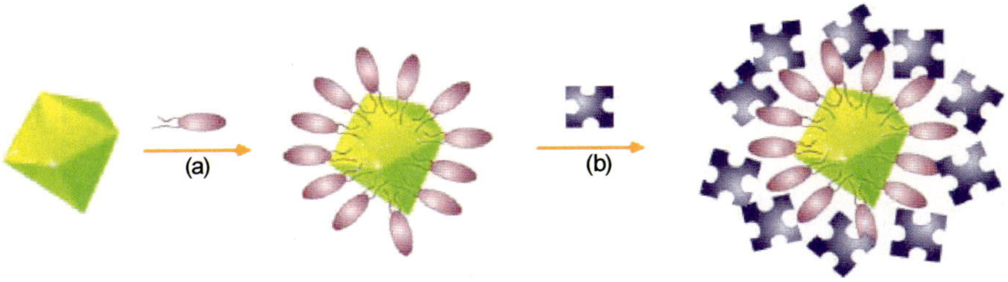

Fig. 6 Preparation of biofunctional fluorescent labels. A simple two-step approach is described in the current study: **a** FDA was milled into nanometer-sized (107 nm on average) crystals in an aqueous surfactant (DSPE-PEG(2000)Amine) medium and **b** subsequently adsorbed with streptavidin molecules. The FDA conjugated streptavidin can be used to capture any biotinylated biomolecules. Adapted from [13]

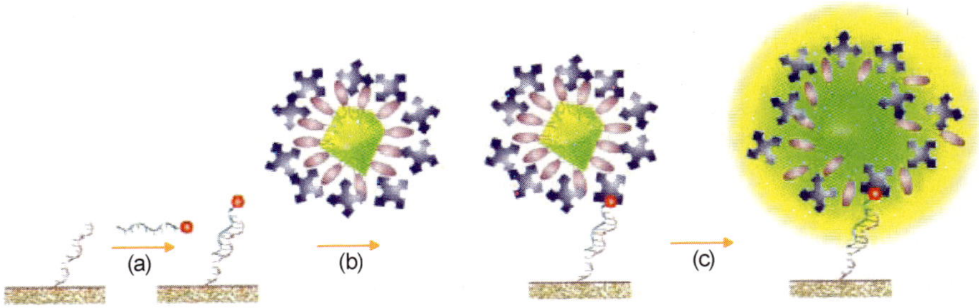

Fig. 7 Amplified HPV-DNA hybridization using nanocrystalline FDA as label. **a** The amplified HPV-DNA labeled with biotin was first hybridized with the immobilized probes in the microtiter plate and then **b** captured by the nanocrystalline FDA conjugated with streptavidin. **c** High signal amplification was achieved after solubilization, release and conversion of the precursor FDA into fluorescein molecules by the addition of DMSO and NaOH. Adapted from [13]

specific hybridized HPV-DNA was finally coupled with streptavidin that had been conjugated with FDA nanocrystals. After the affinity reaction, the FDA molecules were dissolved and concomitantly converted into fluorescein. The performance of the nanocrystal biolabel system was then compared with various directly labeled systems (i.e., FITC and peroxidase). This approach resulted in a high selectivity, short incubation times and a sensitivity up to

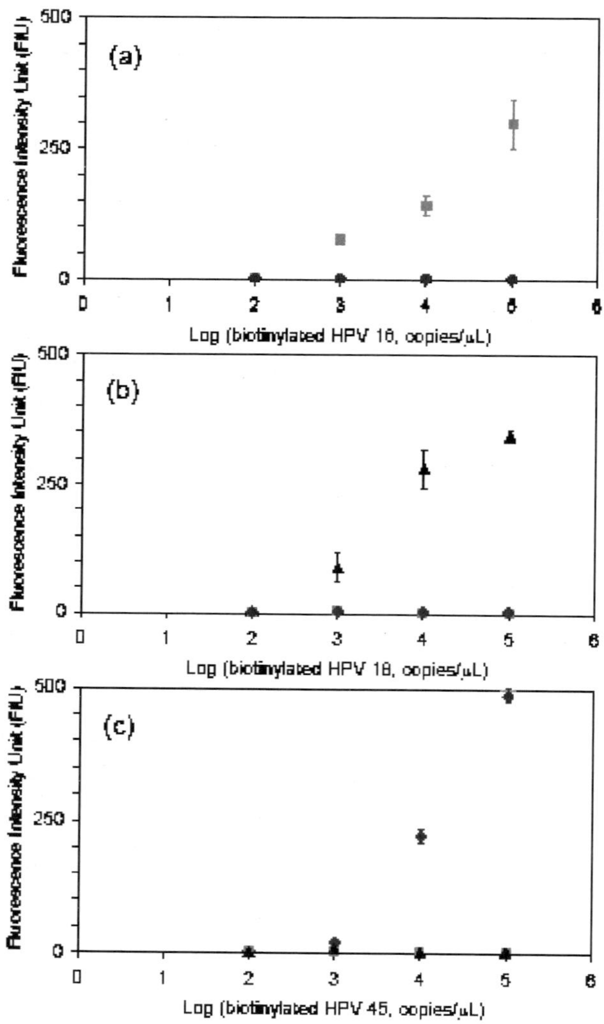

Fig. 8 Quantitative evaluation of the nanocrystalline FDA conjugates as shown in Fig. 7 by using different amplified HPV-DNA labeled with biotin that were separately hybridized with the immobilized probes of HPV 16 (■), 18 (▲) or 45 (♦) in the microtitre plate. *Error bars* correspond to standard deviations (±SD, $n = 3$). Adapted from [13]

147 times greater than obtained from state-of-the-art, directly fluorescent-labeled streptavidins. This innovative method offers a rapid detection of small amounts of nucleic acids because less target material and thus fewer PCR cycles are required.

The performance of the streptavidin-FDA nanocrystal label was compared with directly FITC-labeled and HRP-labeled streptavidin conjugates. Although the HRP system allowed ultra-sensitive detection of HPV-DNA (down to 100 copies/μL), the method did not provide quantitative information on the viral load. Neither did the FITC-labeled streptavidin conjugates. In contrast, the nanocrystalline FDA system did produce quantitative results without non-specific interactions with the other probes (Fig. 8). Moreover, depending on the amplicon concentration and the HPV genotype, the FDA system yielded much higher S/N ratios (Table 2) with a concomitant sensitivity enhancement by a factor of 2.8 to 147.0 compared with the directly FITC-labeled streptavidin (Table 3). The high sensitivity of the particle-based conjugates probably results from the boosting effect of the dissolving dye molecules, but the minimized quenching could also have contributed to the improved signal.

As well as detecting HPV, it is also important to discriminate between high-risk (HR) and low-risk (LR) HPV genotypes. The assay described above reliably detects 11 HR-HPV genotypes and yields the highest S/N ratios

Table 2 Signal-to-noise ratios (S/N) of FDA-labeled, FITC-labeled and HRP-labeled streptavidin

	S/N of streptavidin conjugate		
	FDA	FITC	HRP
Biotinylated HPV 16 (copies/μL)			
10^3	18.22	1.44	6.78
10^4	30.09	1.21	8.56
10^5	70.82	1.11	6.78
Biotinylated HPV 18 (copies/μL)			
10^3	20.33	1.33	4.20
10^4	45.50	1.21	9.91
10^5	78.34	1.22	6.45
Biotinylated HPV 45 (copies/μL)			
10^3	15.62	1.33	8.03
10^4	34.19	1.21	6.69
10^5	135.01	1.28	5.66

Adapted from [13]

Table 3 Sensitivity-to-sensitivity ratios (S/S) of FDA-labeled and FITC-labeled streptavidin

	S/S of Streptavidin-FDA/Streptavidin-FITC		
	16	18	45
Biotinylated HPV (copies/μL)			
10^3	9.06	14.08	2.79
10^4	34.46	68.34	54.47
10^5	146.96	84.50	96.96

Adapted from [13]

(Table 4). We immobilized the 11 HR-HPV genotypes in a single well and hybridized them with different amplicons by using FDA/FITC/HRP-labeled streptavidin for detection. The labeled amplicons and the FDA/FITC/HRP

Table 4 Signal-to-noise ratios (S/N) of FDA-labeled and HRP-labeled streptavidin for detection of multiple genotypes in a single well

	S/N of streptavidin conjugate	
	FDA	HRP
Biotinylated HPV 10^5 copies/μL		
16	43.28	8.36
18	58.06	9.29
31	56.72	9.96
33	50.11	8.81
35	58.65	9.25
39	97.34	9.47
45	92.96	8.83
51	69.74	8.31
56	68.16	9.14
58	106.93	9.60
68	108.02	8.95
Biotinylated HPV 10^4 copies/μL		
16	15.94	3.60
45	66.40	4.66
Biotinylated HPV 10^3 copies/μL		
16	13.62	2.85
45	8.37	3.36

Adapted from [13]

labeled streptavidin were added sequentially to coated microtitre plates to permit maximal binding. Compared with other reagents used to detect PCR products (e.g., ethidium bromide and radioisotopes), FDA is non-hazardous and fast because there is no color processing. After addition of the releasing reagent, the instant fluorescence intensity is over 90% of the maximum and reaches the maximum within 5 minutes.

We have demonstrated a novel signal amplification technology for a HPV-DNA hybridization assay based on FDA nanocrystals encapsulated within PEG-Amine-modified phospholipids (DSPE-PEG-(2000)Amine). The surface modification of the nanocrystalline biolabel system prevents non-specific binding and particle aggregation. It provides an interface for conjugation with streptavidin. The FDA nanocrystals are nearly water-insoluble precursors of fluorescein, which are dissolved and hydrolyzed by treatment with an organic solvent/hydroxide mixture. This technology allows easy quantification of PCR products and markedly increased sensitivities and S/N ratios over established systems. It should be noted that there is a clear potential for further improvement, to develop early-stage and instant detection systems for various diseases.

2.5 Silole Nanocrystals as Novel Biolabels

A novel class of biofunctional silole nanocrystals with the potential to create highly sensitive immunoassay was demonstrated for the first time [14]. Biolabels were constructed by encapsulating nanocrystalline hexaphenylsilole [$Ph_2Si(CPh)_4$; HPS] within ultrathin polyelectrolyte layers via the layer-by-layer technique that provided an “interface” for the attachment of antibodies (Fig. 9). A high ratio of fluorescent dyes to biomolecules (2.4×10^3) was achieved without self-quenching problem. The general concept of the application of the nanocrystalline silole biolabels as fluorescent labels in FIAs is depicted in Fig. 10.

The siloles are non-emissive (off) when molecularly dissolved in organic solvents at room temperature, while the silole molecules in poor solvents cluster into nanoaggregates, which are highly emissive and boost the photoluminescence quantum yields by up to 2 orders of magnitude [46]. A DMSO solution of HPS (22 μM) was virtually nonemissive. Only a noisy curve was obtained even in a 100 times magnified photoluminescent spectrum. However, after putting the same concentration of HPS in a poor solvent (e.g., water), an intense signal was recorded under identical measurement conditions. This intriguing and not yet completely understood aggregation-induced emission (AIE) feature is an invaluable property for the development of ultrasensitive FIA.

A 40- to 140-fold higher sensitivity with the assays using the nanocrystal biolabels was observed compared with the direct FITC-labeled antibod-

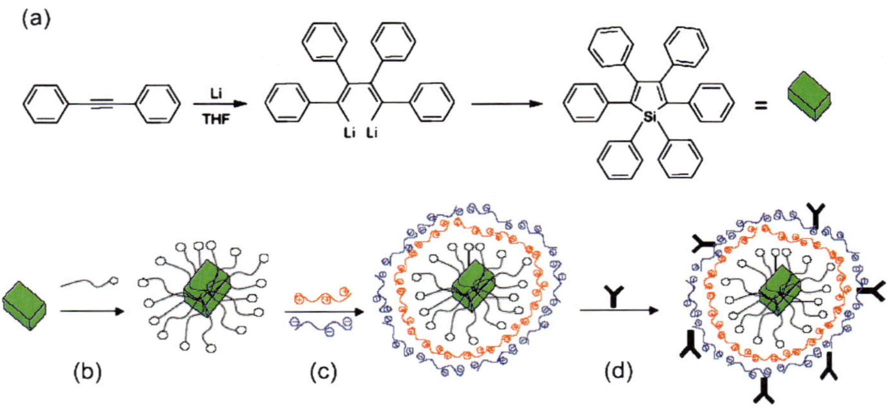

Fig. 9 Schematic illustration of the preparation of biofunctional silole nanocrystals. **a** HPS was synthesized according to a published experimental procedure [46] and **b** was ball-milled into nanocrystals in an aqueous surfactant, followed by **c** encapsulation with polyelectrolyte multilayers of nanometer thickness, and **d** the attachment of a specific immunoreagent. Adapted from [14]

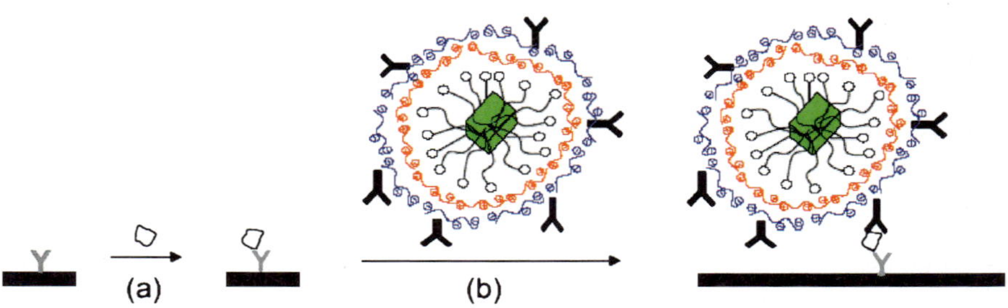

Fig. 10 Principle of a sandwich immunoassay using nanocrystalline silole biolabels. **a** The analyte is first immobilized by the capture antibody preadsorbed on the solid phase and then **b** exposed to antibody-labeled nanocrystal detectors. Fluorescence intensity is proportional to the analyte concentration. Adapted from [14]

ies, depending on the analyte concentration. The higher sensitivity of the nanocrystal biolabels compared with standard FITC conjugates may be explained by the boosting effect of the higher ratio of dye molecules to binding molecules, and the suppressed self-quenching by the AIE feature could also have contributed to the improved signal. Chemo- and photostabilities make siloles a useful tool for labeling purposes. The preparation of nanocrystalline HPS biolabels is straightforward to perform and controllable. By using a nanoscale polyelectrolyte coating as an interface for bioconjugation onto the nanocrystals, the siloles do not need to be water-soluble nor possess groups for bioconjugation, as this functionality is provided by the polyelectrolytes. The quenching problem normally arising from high F/P ratio labels can be prevented due to the AIE feature of siloles. Synthesis of water-soluble amphiphilic siloles through appropriate chemical modifications is now in progress. The soluble siloles may be used in homogenous assays, i.e., all reagents present in solution to achieve the measurement without any separation step.

We believe that, after optimization, the use of siloles provides us with a useful alternative to organic fluorophores.

2.6 Nanoparticle-Based Bio-Barcode Technology

An ultrasensitive method for detecting protein analytes has been developed. Mirkin et al. reported a bio-barcode assay as shown in Fig. 11 to detect analytes at levels 6 orders of magnitude below conventional diagnostic immunoassays [22]. The bio-barcode assay utilizes antibody-coated magnetic beads to capture and concentrate the analytes. The captured analytes are labeled with gold nanoparticle probes that are conjugated with specific antibodies and single-stranded DNA or double-stranded DNA barcodes. The resulting complexes are separated magnetically and washed to remove excess probe. The DNA barcodes are then released from the complex and detected via hybridization to a surface immobilized DNA probe and an oligonucleotide functionalized gold nanoparticle as shown in Fig. 12 [21]. The gold particles are enlarged through silver deposition, and the light scattered from the particles is detected [23]. Therefore, the detection of a specific DNA barcode sequence indicates the presence of a specific protein. The increased sensitivity is derived in part from the release of multiple barcodes per captured analyte [22]. The number of barcodes released is dependent on the size of the gold particle, with a 15 nm diameter gold particle containing up to 100 DNA barcodes [24]. Because the nanoparticle probe carries with a large number of oligonucleotides per protein binding event, there is substantial amplification. Using this system to detect prostate-specific antigen (PSA), the detection limit was 30 attomolar [21]. Alternatively, a polymerase chain reaction on the oligonucleotide barcodes can boost the sensitivity to 3 attomolar. Comparable clinically accepted conventional assays for detecting the same target have sensitivity limits of $\sim$3 picomolar, 6 orders of magnitude less sensitive than this system.

A more recent report describes an alternative approach to labeling the analyte after capture onto antibody coated magnetic beads [25]. This approach utilizes a biotinylated secondary antibody which is labeled with streptavidin-coated gold nanoparticles and biotinylated barcode DNA. Once the barcode DNA is released from the complex, the barcode DNA is detected as described above. Using the biotin-streptavidin approach, a 1000-fold improvement in detection limit was achieved with an excellent dose response compared with the best ELISA system.

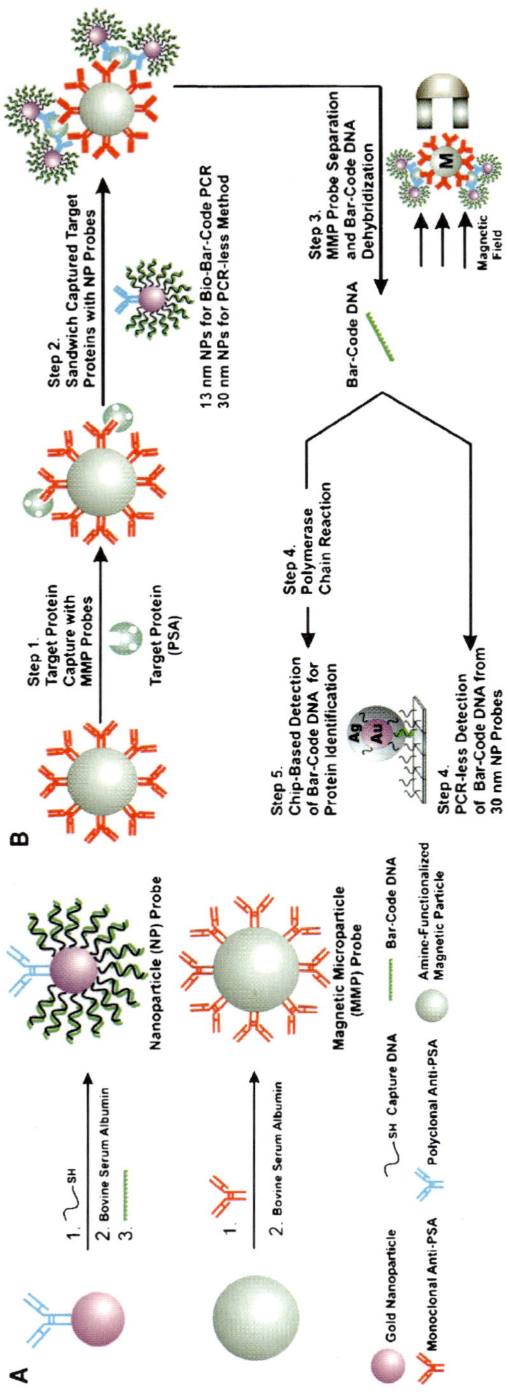
A
1. SH
2. Bovine Serum Albumin
3.
Nanoparticle (NP) Probe
1.
2. Bovine Serum Albumin
Magnetic Microparticle (MMP) Probe
Gold Nanoparticle
SH Capture DNA
Bar-Code DNA
Monoclonal Anti-PSA
Polyclonal Anti-PSA
Amine-Functionalized Magnetic Particle
B
Step 1. Target Protein Capture with MMP Probes
Target Protein (PSA)
Step 2. Sandwich Captured Target Proteins with NP Probes
13 nm NPs for Bio-Bar-Code PCR
30 nm NPs for PCR-less Method
Step 3. MMP Probe Separation and Bar-Code DNA Dehybridization
Bar-Code DNA
Magnetic Field
M
Step 4. Polymerase Chain Reaction
Step 5. Chip-Based Detection of Bar-Code DNA for Protein Identification
Ag
Au
Step 4. PCR-less Detection of Bar-Code DNA from 30 nm NP Probes

Fig. 11 Nanoparticle-based Bio-Barcode Technology. **A** Probe design and preparation. **B** PSA detection and barcode DNA amplification and identification. Adapted from [22]

This technology can also be applied to multiplexed assays by using different DNA barcode sequences conjugated with specific antibodies on each gold probe as indicators for different protein markers. This system is homogeneous, which makes this assay faster and simpler than heterogeneous immuno-PCR systems and can also increase the sensitivity because the capturing step is more efficient. In addition, it provides a high ratio of PCR-amplifiable DNA per labeling Ab, which results in a substantial increase in assay sensitivity. Moreover, direct detection or PCR is carried out on samples of barcode DNA that are free from analytes, most of the biological components, microparticles and nanoparticles. This step substantially reduces background signal. Furthermore, this system has the potential for massive multiplexing and the simultaneous detection of many analytes in one solution.

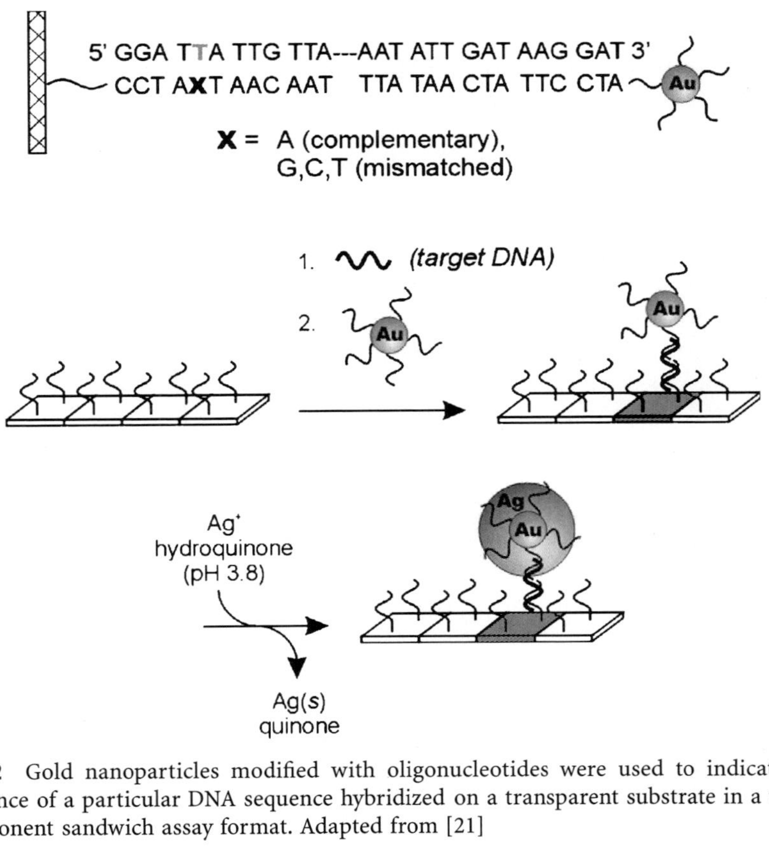

Fig. 12 Gold nanoparticles modified with oligonucleotides were used to indicate the presence of a particular DNA sequence hybridized on a transparent substrate in a three-component sandwich assay format. Adapted from [21]

3 New Generations of Immunodiagnostic Tests

Lateral-flow tests, also known as immunochromatographic tests, have been a popular platform for rapid immunoassays since their introduction in the mid-1980s. The features of lateral-flow tests shown below make them ideal for rapid point-of-care testing:

- User-friendly format that can be used by non-lab trained operators;
- Result available in short time (e.g. 5–30 minutes);
- Long-term stability over a wide climatic range (i.e., may not be necessarily be refrigerated).

3.1 A Novel Credit-Card Style Assay for Bedside Determination of Fatty Acid-Binding Protein

Heart-type fatty acid-binding protein (H-FABP), a low molecular mass cytoplasmic protein (15 kD) abundant in heart muscle cells, has a high potential as a sensitive biomarker for early diagnosis of acute myocardial infarction (AMI) [47–50]. This relatively small protein is involved in cellular long-chain fatty acid metabolism [51]. The concentration of H-FABP in the plasma of healthy persons is relatively low (2–6 μg/L) [52]. It is released rapidly from damaged cells into the circulation and is cleared from the circulation by the kidney with a plasma half-life of 20 min [47]. H-FABP levels rise as early as 1–3 hours after the onset of AMI, peak at 6–8 hours, and return to normal within 24–30 hours. Although FABP shows release characteristics from injured myocardium and elimination rates from plasma that are similar to those of myoglobin, several clinical studies have revealed a superior performance of H-FABP over myoglobin for early AMI detection as well as early estimation of infarct size [52–55]. In a multi-centre study (EUROCARDI), H-FABP was pointed as the most sensitive biochemical marker for early diagnosis of AMI [56]. H-FABP is also superior to CK-MB or cardiac troponins in the early detection of ischemic myocardial necrosis [57, 58]. The fact that it shows a ratio of cytoplasmic to vascular concentration one order of magnitude higher than any other cardiac protein makes H-FABP the most sensitive and specific marker for AMI diagnosis within 3 hours after the onset of infarction.

Using H-FABP as an early cardiac marker to confirm or exclude a diagnosis of AMI soon after the onset of symptoms, a one-step FABP immunotest, the so-called CardioDetect® shown in Fig. 13, has been successfully developed [37–39, 41].

- It is a commercially available FABP rapid test with a Conformité Européenne (CE) Mark approval.
- It is a rapid chromatographic immunoassay designed for qualitative determination of H-FABP in whole blood, plasma and serum samples.

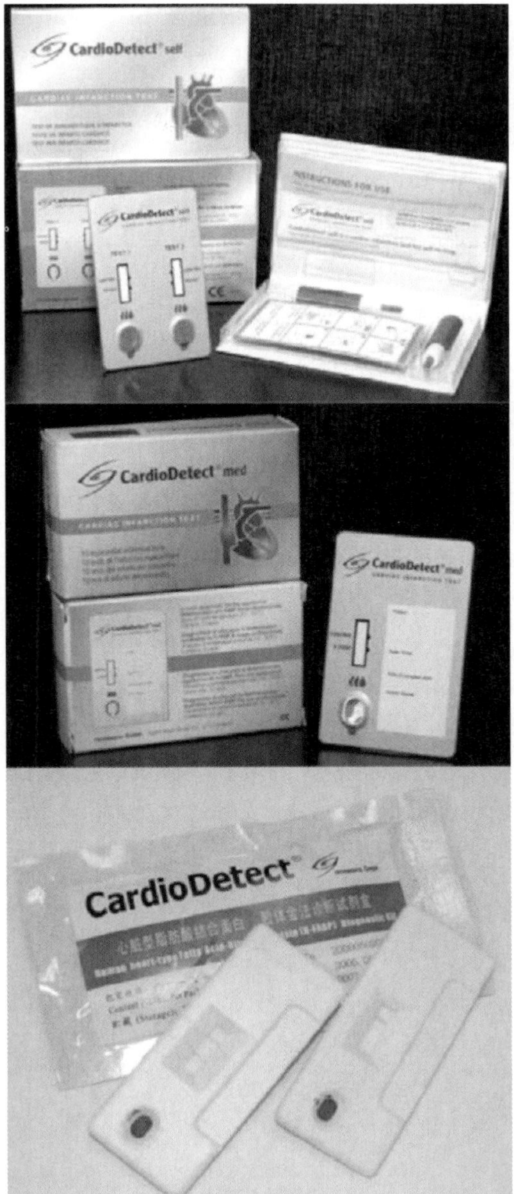

Fig. 13 CardioDetect® (rennesens GmbH, Berlin) is offered in CardioDetect® self version for self-testing (the *upper* one; its two separate test fields allow one repetition) and CardioDetect® med version for hospital use (the *middle* one; containing one test field and a label on which to note patient data). The test at the *bottom* with the State Food and Drug Administration (SFDA) of China approval is also now being manufactured in China by Shenzhen Kang Sheng Bao Bio-Technology Co., Ltd. (Shenzhen). Adapted from [41]

- The estimation of the test-strip results can be performed visually by the naked eye. The results can also be quantified by a test reader.
- After application of sample onto the test-strip, the result is available within 15 minutes.
- It can be stored at 4–8 °C for up to 1 year without any loss of activity.
- The whole blood immunotest requires no sample pretreatment and thus can be applied in emergency situation.
- Combined with the well-established markers, troponins, it may allow more accurate targeting of appropriate therapy and considerable cost savings than the current diagnostic tests, i.e. ECG, troponins, myoglobin and CK-MB tests.

In our previous study [41, 49], 434 plasma or serum samples from 218 patients admitted to the Coronary Care Unit (CCU) and Chest Pain Unit of the Prince of Wales Hospital in Hong Kong presenting with chest pain and suspected AMI were investigated. The diagnostic performance of the CardioDetect® self was evaluated according to the time interval from the onset of symptoms to analysis. The areas under the receiver operating characteristic (ROC) curves are shown in Table 5. Similar performance of measurements with the CardioDetect® and the FABP ELISA was observed. The areas for the CardioDetect® and the FABP ELISA were significantly greater than those for cardiac troponin I (cTnI) and creatine kinase (CK) within 6 hours after the onset of symptoms. Thus, FABP has great potential as an excellent cardiac marker for the diagnosis of AMI in the early phase.

Whole blood samples from 38 patients admitted to a hospital in Bernau (Brandenburg, Germany) and 171 patients at the Deutsches Diabetes Forschungsinstitut Düsseldorf (German Diabetes Research Institute) were also studied. The specificities of the CardioDetect® were 94% and 85.1% respectively. Both sensitivities and negative predictive values (NPV) were 100% implying that 100% of non-AMI patients could be excluded with no false negative results.

Table 5 Area under the ROC curve of different diagnostic tests for detection of myocardial infarction

Hours after onset of symptoms	Number of samples	Area under the ROC Curve CardioDetect®	FABP ELISA	cTnI ELISA	CK activity
0–3	71	0.80	0.83	0.50	0.61
> 3–6	156	0.93	0.95	0.83	0.86
> 6–12	181	0.98	0.98	0.97	0.98
> 12–24	148		0.90	0.98	0.99
> 24	227		0.57	1.00	0.90

Adapted from [41]

The potential impact of CardioDetect® is enormous. Its application to the detection of myocardial infarction is considerable, particularly if it leads medical doctors to more accurate and earlier AMI diagnosis, allowing better treatment to reduce patient morbidity and mortality, as well as avoiding unnecessary stay at hospital of non-AMI patients that would bring important economic and human advantages. Its simplicity of use could even make it a user-friendly test to be employed by chronic patients at home. This sophisticated but simple looking and highly effective diagnostic tool enables medical doctors and the entire population to prevent diseases of growing worldwide importance that cause extensive disability and take far too many lives.

3.2 Novel "Digital-Style" Rapid Test Simultaneously Detecting Heart Attack and Predicting Cardiovascular Disease Risk

An early cardiac marker, H-FABP, and an established risk marker for heart attack, C-reactive protein (CRP), were combined in a simple, rapid, and semi-quantitative "digital-style" lateral-flow assay as shown in Fig. 14 [40]. There is a vital need for developing a relatively simple and rapid test to confirm or exclude suspected AMI patients and simultaneously to identify apparent non-AMI patients at risk of developing cardiovascular events at an early stage. To interpret different clinical outcomes, monoclonal antibodies specific to H-FABP and CRP were employed to a "digital-style" rapid test with more than one line shown on the test zones (Fig. 15).

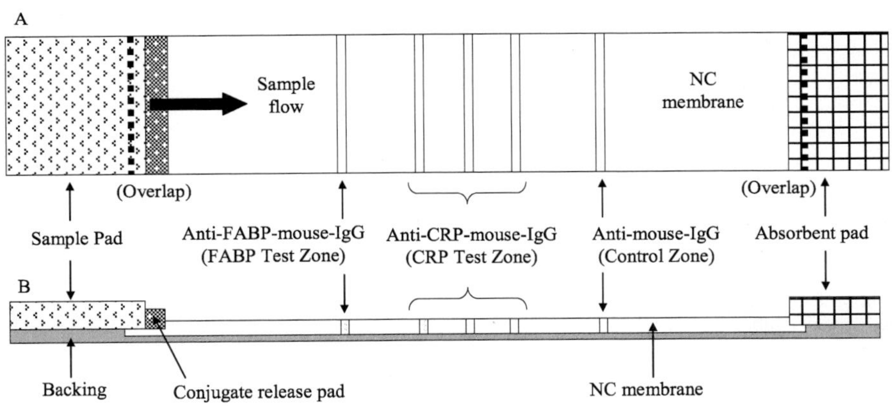

Fig. 14 Main components of a "digital-style" rapid test for prognosis of cardiovascular diseases and early diagnosis of heart attack. **A** Top view; **B** cross-section. Adapted from [40]

A total of 162 plasma samples from Country Brandenburg Hospital in Bernau (Germany) with CRP concentration between 0.03 and 283.2 mg/L

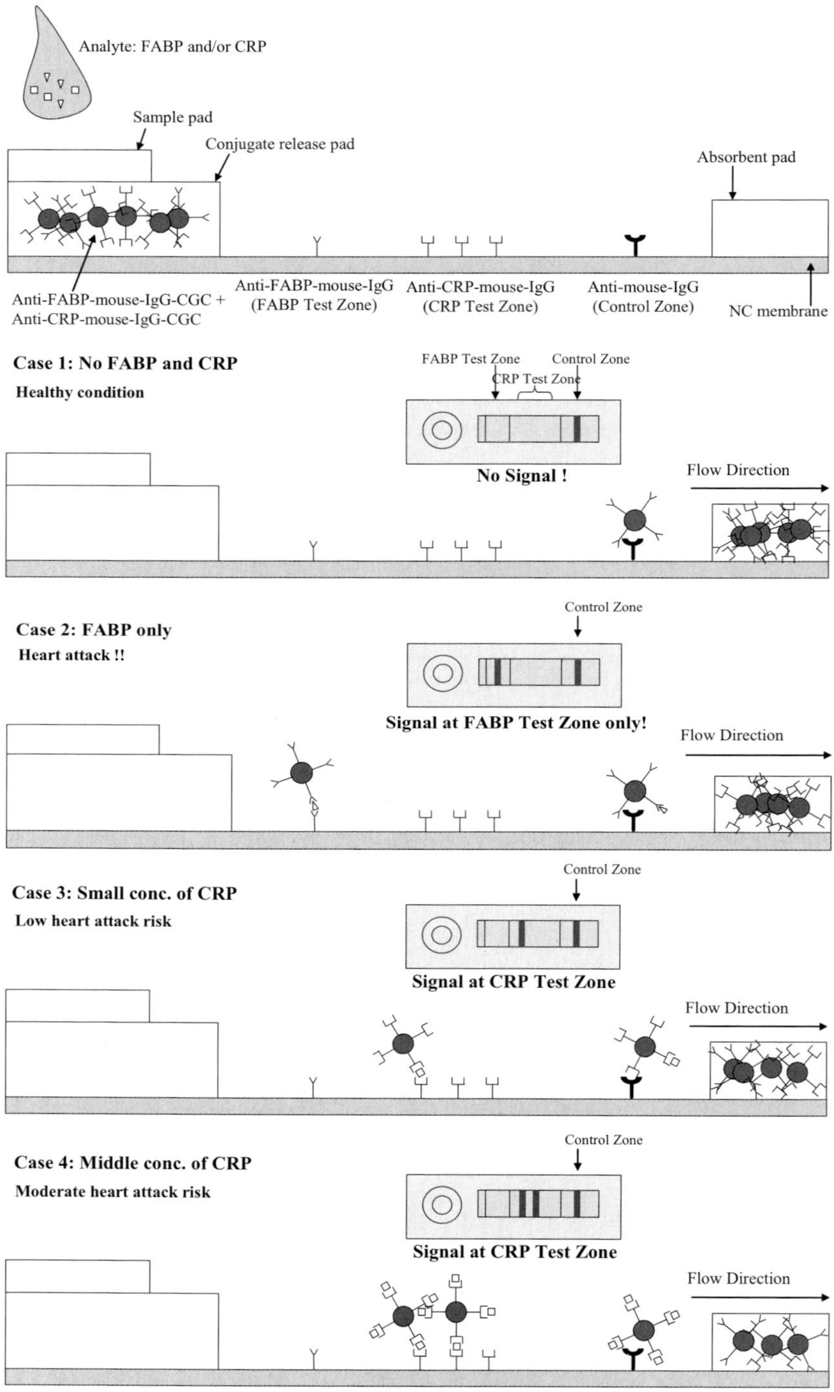
Analyte: FABP and/or CRP
Sample pad
Conjugate release pad
Absorbent pad
Anti-FABP-mouse-IgG-CGC +
Anti-CRP-mouse-IgG-CGC
Anti-FABP-mouse-IgG
(FABP Test Zone)
Anti-CRP-mouse-IgG
(CRP Test Zone)
Anti-mouse-IgG
(Control Zone)
NC membrane
Case 1: No FABP and CRP
Healthy condition
FABP Test Zone
Control Zone
CRP Test Zone
No Signal !
Flow Direction
Case 2: FABP only
Heart attack !!
Control Zone
Signal at FABP Test Zone only!
Flow Direction
Case 3: Small conc. of CRP
Low heart attack risk
Control Zone
Signal at CRP Test Zone
Flow Direction
Case 4: Middle conc. of CRP
Moderate heart attack risk
Control Zone
Signal at CRP Test Zone
Flow Direction

Fig. 15 Principle of the "digital-style" rapid test. An early cardiac marker, H-FABP, and an established risk marker, CRP, were combined in a lateral flow assay. H-FABP and CRP can also be detected simultaneously to give many different combinations. Adapted from [40]

were assessed using an in-house enzyme-linked immunosorbent assay (ELISA) and a newly developed "digital-style" assay. There was a good linearity ($r^2 = 0.9392$) and agreement by Bland and Altman statistic plot between these two methods. The intra- and inter-assay coefficients of variation (CV) were less than 15%. The "digital-style" assay with high stability provides long storage time and allows mass production and preparation of large batches.

A novel "digital-style" semi-quantitative lateral-flow assay simultaneously detecting heart attack and predicting early cardiovascular disease risk, just by simply counting the number of red lines in the test without any expensive reading instrument, has been successfully developed (Fig. 16). Medical doctors can early predict the extent of the risk and prescribe heart-attack-preventing therapy.

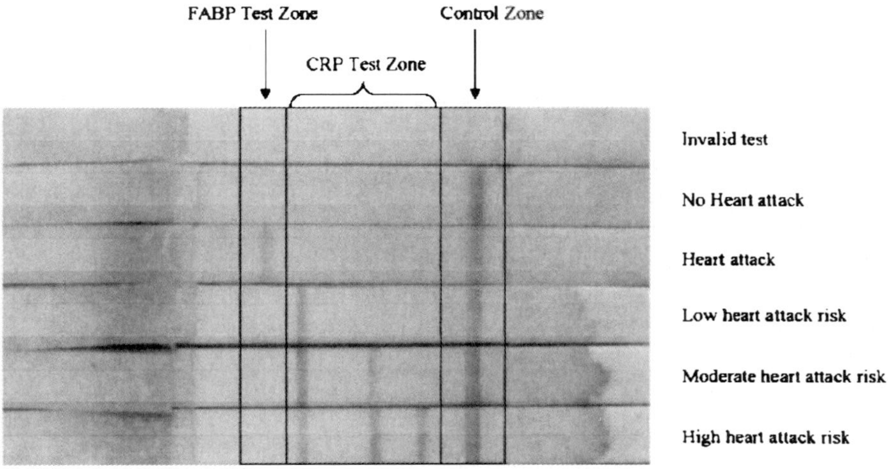

Fig. 16 Interpretation of the test results of the "digital-style" rapid test. There are many different combinations for H-FABP and CRP simultaneously detected that are not shown here. Adapted from [40]

3.3
One-step Quantitative Cortisol Immunodiagnostic Test with Proportional Reading

Rapid quantitative immuno-detection of haptens by lateral-flow assay without "typical" competitive inhibition results has been studied. An immunothreshold-based assay for quantitative detection of cortisol has been de-

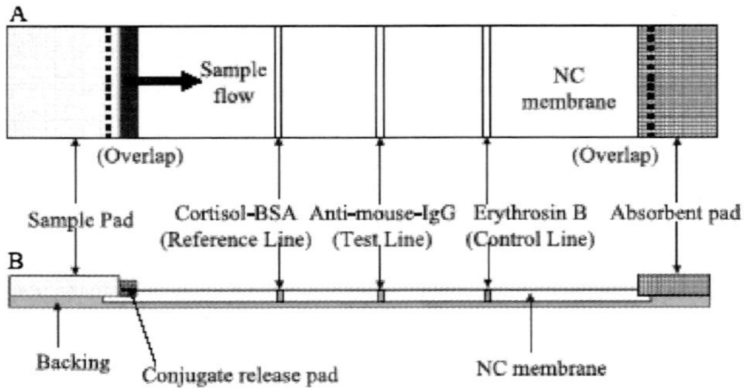

Fig. 17 Schematic diagram of a one-step quantitative immuno-threshold-based cortisol lateral-flow assay. **A** Top view; **B** cross-section. Adapted from [35]

veloped as shown in Fig. 17 [35]. It is a rapid chromatographic immunoassay designed for quantitative determination of cortisol in whole blood samples. It requires no sample pretreatment and gives result within 15 min. It gives a signal directly proportional to the cortisol concentration in samples (Fig. 18). This technique provides a practical calibration curve with detection limit of 3.5 μg/L. The intra- and inter-assay CVs are 6% and 10% respectively. Cross-reactivity with related steroids is acceptably low: corticosterone (3.4%), cortisone (2.1%), deoxycorticosterone (2.0%), 17a-hydroxyprogesterone (0.4%), and progesterone (0.05%). Furthermore, the test strips show the advantages of long shelf life and high stability that allow mass production and preparation of large batches.

This immuno-threshold-based principle can also be employed for detecting all low-molecular-weight haptens. It may be a useful and convenient test format for drug detection.

4 Other Ingenious Technologies

4.1 One-Step Homogeneous Non-Competitive Immunoassay for Small Analyte

Competitive immunoassays are usually performed for detection of small analytes. However, due to the non-specific nature of competitive immunoassays, often many false positives are generated. Also, the interpretation of the result can be confusing.

Pulli et al. described a one-step, homogeneous non-competitive immunoassay for small analytes using recombinant antibodies and morphine as the model analyte [36]. A highly specific antibody against the immune complex (IC) formed between an anti-morphine antibody and morphine was selected

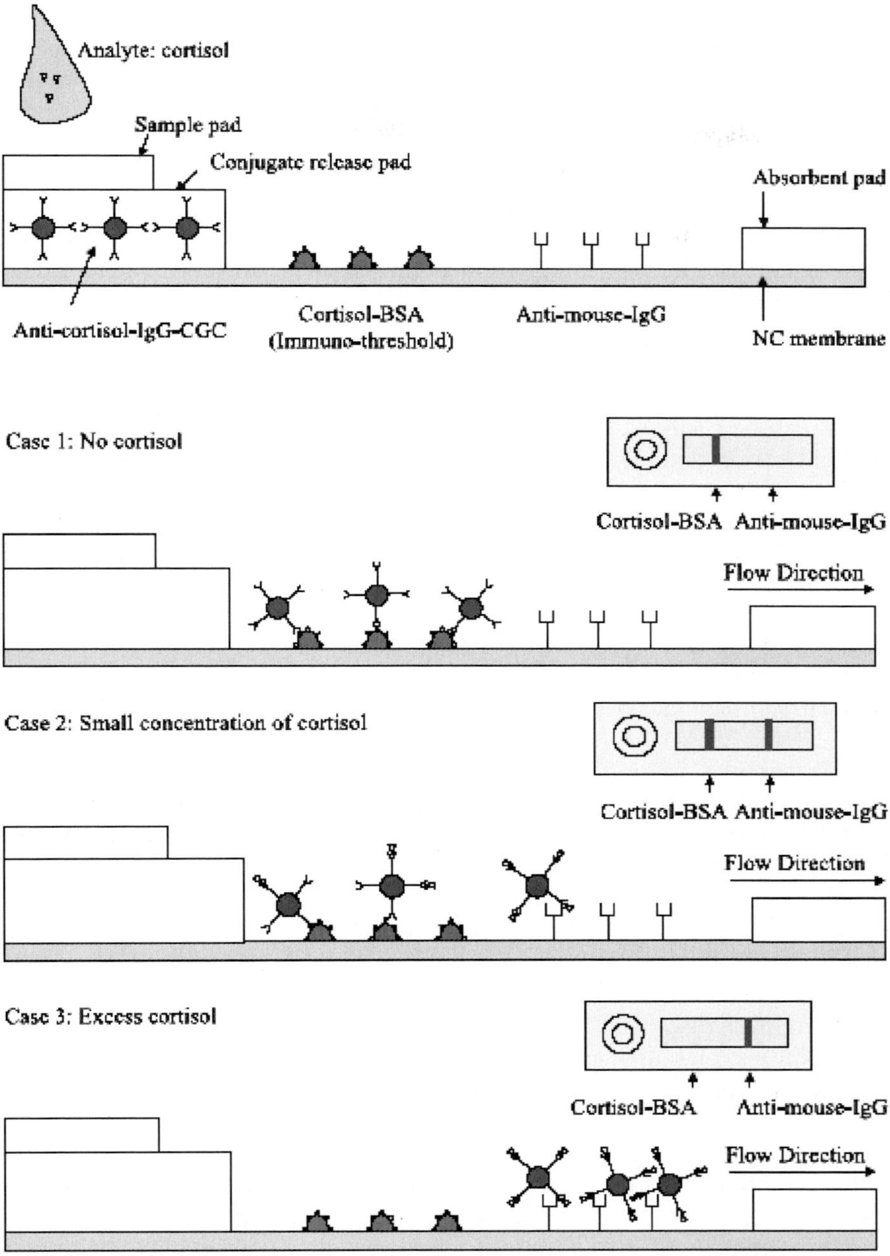

Fig. 18 Principle of the cortisol immunodiagnostic test with proportional reading. Control line is omitted for simplicity. Adapted from [35]

from a naïve scFv phage display library. The in vitro phage library selection procedure avoids the difficulties associated with the production of anti-IC antibodies by animal immunization. The anti-morphine and the anti-IC antibodies were labeled with a pair of fluorescence resonance energy transfer (FRET) fluorophores. In the FRET assay, 5 μg/L of morphine in saliva, which is clearly below the recommended cutoff level, can be detected without cross-reactivity to codeine or heroin. The assay time is 2 minutes.

This assay is fast and very simple to perform. To further simplify the use, the reagents can be kept in dry form in a vessel. The saliva sample is added, and the results can be read after a couple of minutes by a portable fluorometer. The method can be easily introduced in routine analysis. The only piece of equipment required is a fluorometer capable of FRET measurements. Furthermore, the production of recombinant antibody Fab fragments by fermentation can be easily scaled-up, enabling inexpensive manufacturing of large amounts of reagents. As a whole, there is a big demand for a sensitive, specific, and fast assay method for many small molecules, e.g., pharmaceuticals, drugs of abuse, steroids, and toxins. Because in principle the immunoassay described herein can be applied to any kind of small analytes, it offers a novel solution to this need.

4.2 A Compact Disc as Low-cost, High-sensitivity Readout System for Multiplex Immunoassays

Highly complex immunoassays that identify and quantify many different antigens simultaneously need high-resolution imaging capability. However, as microarray elements become smaller and smaller for larger and larger numbers of simultaneous tests, the state-of-the-art CMOS or CCD technologies cannot achieve sufficient imaging resolution at this moment. Confocal scanners using a microscope lens can provide the required resolution, but their use is limited by the cost, size and geometrical alignment of the optics. Lange et al. demonstrated an inexpensive classical silver staining and standard compact disc (CD) reader technology for array-based assays [33, 34]. It employs a pick-up head of a compact-disc player in a slightly modified form to provide a high efficient and sensitive readout in immunoassay (Fig. 19).

A conventional compact-disc pick-up reader works by focusing laser light onto the surface of the CD. As the CD rotates above the reader, information encoded as pits along a spiral track on its metal-coated surface can be read by means of light reflected back through a lens onto a photodiode. The system described by Lange et al. works in exactly the same way, but with a substrate covered with a regularly spaced array of capture antibodies taking the place of the CD. Using a sandwich immunoassay as an example, a capture antibody is firstly stamped in a 25-micrometre square pattern onto a solid substrate.

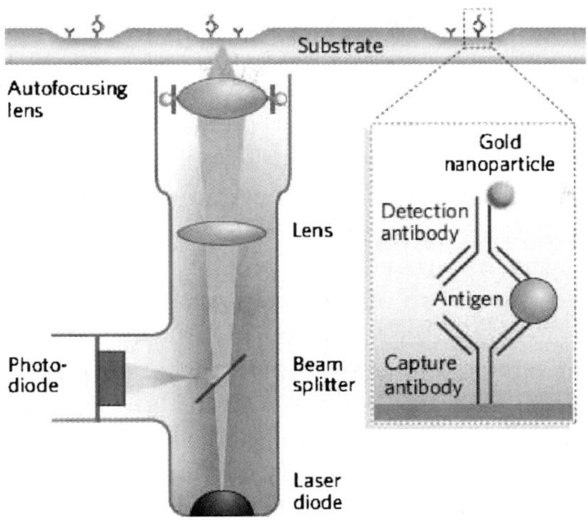

Fig. 19 Molecular compact disc. A substrate covered with a regularly spaced array of capture antibodies acts as a molecular CD. When an antigen binds to an antibody, its presence is signaled by a second, gold-containing detector antibody that catalyses the deposition of reflective silver particles onto the substrate. Adapted from [34]

Then the desired antigen binds to this antibody. After that, a detector antibody conjugated with gold nanoparticle binds to this antigen to make it visible to the CD pickup head. This gold nanoparticle catalyses the deposition of silver grains onto the substrate to which the capture antibody, antigen and detector antibody are all now attached. The reflections from the silver grains could be read by the CD pick-up head with a resolution of around 50 nm – considerably smaller than the diameter of most of the silver nanoparticles, which is of the order of several hundred nanometres. Using this system, the detection limit in serum could be reduced from 1 μg/L down to 100 pg/L. Lange et al. also demonstrated that different antigens were attached to their specific capture molecule, well apart from each other on the substrate. By correlating the density of the silver precipitate with the coverage of the antigen, it might be possible to detect single molecules with the system. The technology could be widely useful for low-cost, high-sensitivity readout systems for many different types of assay.

5
Prospects

Regarding new trends in immunoassays, the discussion often heard these days is an interest in multiplexing that help to define a patient's medical profile. The ability to evaluate one sample in one test for the presence or absence of multiple analytes is very appealing. Multiplexed protein measurement is a rapidly advancing field. Biomarker data from multiplexed protein measurement technologies will become increasingly the norm as we are entering an era of personalized medicine. Multiplexed protein measurement for biomarker-based diagnostic and prognostic testing will become the largest growth segment of the immunodiagnostics industry [59]. It has the capacity to identify surrogates that will be integrated into clinical indices, treatment algorithms, and, ultimately, into dynamic disease models that permit real-time, data-driven patient management.

Near patient testing is also one of the most rapidly growing segments as tests that were done in the central lab are now available as near patient tests. This has lead to an explosion of new tests and technologies. A key to effective point-of-care testing programs is the ability to link POC devices with centralized management stations and a healthcare system's primary information system. In this way, the POC test can be easily added to the patient's medical record. Very few lateral-flow tests provide quantitative results that can be downloaded on a computer or have the ability to send results to a receiving network. Improvements in all of these aspects will go a long way to increase the use of lateral-flow tests in the professional sector. In addition, more and more lateral-flow tests using saliva, urine and sweat for analysis can make them more patient-friendly.

References

1. Saiki RK et al. (1985) Science 230:1350
2. Gibbs RK (1991) Curr Opin Biotechnol 2:69
3. Bustin SA (2002) J Mol Endocrinol 29:23
4. MacBeath G, Schreiber SL (2000) Science 289:1760
5. Zhu H et al. (2001) Science 293:2101
6. Haab BB, Dunham MJ, Brown PO (2001) Genome Biol 2(2):research0004.1–0004.12
7. Nam M-J, Park S-J, Mirkin CA (2002) J Am Chem Soc 124:3820
8. Trau D, Yang W, Seydack M, Caruso F, Yu NT, Renneberg R (2002) Anal Chem 74:5480–5486
9. Chan CP, Bruemmel Y, Seydack M, Sin KK, Wong LW, Merisko-Liversidge E, Trau D, Renneberg R (2004) Anal Chem 76:3638–3645
10. Bruemmel Y, Chan CP, Renneberg R, Thuenemann A, Seydack M (2004) Langmuir 20(21):9371–9379
11. Sin KK, Chan CP, Pang TT, Seydack M, Renneberg R (2006) Anal Bioanal Chem 384:638–644
12. Sin KK, Chan CP, Leung WM, Seydack M, Renneberg R (2006) IEE Proc Nanobiotechnol 153:54–58
13. Chan CP, Tzang CH, Sin KK, Ji SL, Queenie Cheung KY, Ward Tam TK, Yang MS, Renneberg R, Seydack M (2007) Anal Chim Acta 584(1):7–11
14. Chan CP, Haeussler M, Tang BZ, Dong Y, Sin KK, Mak WC, Trau D, Seydack M, Renneberg R (2004) J Immunol Methods 295(1–2):111–118
15. Sano T, Smith CL, Cantor CR (1992) Science 258:120

16. Zhou H, Fisher RJ, Papas TS (1993) Nucl Acids Res 21:6038
17. Hendrickson ER, Hatfield-Truby TM, Joerger RD, Majarian WR, Ebersole RC (1995) Nucl Acids Res 23:522
18. Niemeyer CM et al. (1999) Nucl Acids Res 27:4553
19. Schweitzer B et al. (2000) Proc Natl Acad Sci USA 97:10113
20. McKie A, Samuel D, Cohen B, Saunders NA (2002) J Immunol Methods 270:135
21. Taton TA, Mirkin CA, Letsinger RL (2000) Science 289:1757–1760
22. Nam JM, Thaxton CS, Mirkin CA (2003) Science 301:1884–1886
23. Storhoff JJ, Marla SS, Bao P et al. (2004) Biosens Bioelectron 19:875–883
24. Bao YP, Wei TF, Lefebvre PA et al. (2006) Anal Chem 78:2055–2059
25. Hurst SJ, Lytton-Jean AKR, Mirkin CA (2006) Anal Chem 78:8313–8318
26. Stoeva SI, Lee JS, Smith JE, Rosen ST, Mirkin CA (2006) J Am Chem Soc 128:8378–8379
27. Hirabayashi J (2003) Trends Biotechnol 21:141–143
28. Bacarese-Hamilton T, Ardizzoni A, Gray J, Crisanti A (2004) Methods Mol Biol 264:271–283
29. Deinhofer K et al. (2004) Methods 32:249–254
30. Predki PF (2004) Curr Opin Chem Biol 8:8–13
31. Haab BB (2005) Mol Cell Proteomics 4:377–383
32. Utz PJ (2005) Immunol Rev 204:264–282
33. Lange SA et al. (2005) Angew Chem Int Edn 45:270–273
34. Ligler FS, Erickson JS (2006) Nature 440(7081):159–160
35. Leung W, Chan P, Bosgoed F, Lehmann K, Renneberg I, Lehmann M, Renneberg R (2003) J Immunol Methods 281(1–2):109–118
36. Pulli T, Höyhtyä M, Söderlund H, Takkinen K (2005) Anal Chem 77(8):2637–2642
37. Chan CP, Sum KW, Cheung KY, Glatz JF, Sanderson JE, Hempel A, Lehmann M, Renneberg I, Renneberg R (2003) J Immunol Methods 279(1–2):91–100
38. Chan CP, Cheng S, Glatz JFC, Sanderson E, Renneberg I, Lehmann M, Renneberg R (2004) J Anal Lett 36(9):1987–2004
39. Chan CP, Wan TS, Watkins KL, Pelsers MM, Van der Voort D, Tang FP, Lam KH, Mill J, Yuan Y, Lehmann M, Hempel A, Sanderson JE, Glatz JF, Renneberg R (2005) Biosens Bioelectron 20(12):2566–2580
40. Leung W, Chan CP, Leung MF, Lehmann K, Renneberg I, Lehmann M, Hempel A, Glatz JFC, Renneberg R (2005) Anal Lett 38:423–439
41. Chan CP, Lehmann M, Renneberg I, Ziegler J, George Cautherley WH, Jan Glatz FC, Renneberg R (2006) IVD Technology Magazine 12 2:51
42. Seydack M (2005) Biosens Bioelectron 20:2454–2469
43. Alivisatos P (2004) Nat Biotechnol 22:47–52
44. Penn SG, He L, Natan MJ (2003) Curr Opin Chem Biol 7:607–615
45. Gravitt PE, Peyton CL, Alessi TQ, Wheeler CM, Coutlee F, Hildesheim A, Schiffman MH, Scott DR, Apple RJ (2000) J Clin Microbiol 38:357–361
46. Chen J, Law CCW, Lam JWY, Dong Y, Lo SML, Williams ID, Zhu D, Tang BJ (2003) Chem Mater 15:1535–1546
47. Glatz JFC, Van der Voort D, Hermens WT (2002) J Clin Lig Assay 25:167–177
48. Nakata T, Hashimoto A, Hase M, Tsuchihashi K, Shimamoto K (2003) Cardiology 99:96–104
49. Chan CP, Sanderson JE, Glatz JF, Cheng WS, Hempel A, Renneberg R (2004) Z Kardiol 93(5):388–397
50. Ghani F, Wu AH, Graff L, Petry C, Armstrong G, Prigent F, Brown M (2000) Clin Chem 46:718–719

51. Schaap FG, Binas B, Danneberg H, van der Vusse GJ, Glatz JF (1999) Circ Res 85:329–337
52. Wodzig KW, Kragten JA, Hermens WT, Glatz JFC, van Dieijen-Visser MP (1997) Eur J Clin Chem Clin Biochem 35:191–198
53. Wodzig KWH, Pelsers MMAL, Van der Vusse GJ, Roos W, Glatz JFC (1997) Ann Clin Biochem 34:263–268
54. Pelsers MM, Hermens WT, Glatz JF (2005) Clin Chim Acta 352:15–35
55. Alhadi HA, Fox KA (2004) QJM 97:187–198
56. Glatz JFC, Haastrup B, Hermens WT et al. (1997) Circulation 96:I-215 (abstract)
57. Seino Y, Ogata K, Takano T, Ishii J, Hishida H, Morita H, Takeshita H, Takagi Y, Sugiyama H, Tanaka T, Kitaura Y (2003) Am J Med 115:185–190
58. Seino Y, Tomita Y, Takano T, Ohbayashi K et al. (2004) Circ J 68:144–148
59. Kingsmore SF (2006) Nature Reviews. Drug Discov 5(4):310–320

Adv Biochem Engin/Biotechnol (2008) 109: 155–193
DOI 10.1007/10_2007_079

Published online: 11 October 2007

Electrochemical Biochips for Protein Analysis

Axel Warsinke

University of Potsdam, Institute of Biochemistry and Biology, iPOC Research Group,
Karl-Liebknecht-Strasse 24–25, D-14476 Golm, Germany
warsinke@uni-potsdam.de

This chapter is dedicated in celebration of Professor Frieder Scheller's 65th birthday

Abstract Proteins bear important functions for most life processes. It is estimated that the human proteome comprises more than 250 000 proteins. Over the last years, highly sophisticated and powerful instruments have been developed that allow their detection and characterization with great precision and sensitivity. However, these instruments need well-equipped laboratories and a well-trained staff. For the determination of proteins in a hospital, in a doctor's office, or at home, low-budget protein analysis methods are needed that are easy to perform. In addition, for a proteomic approach, highly parallel measurements with small sample sizes are required. Biochips are considered as promising tools for such applications.

The following chapter describes electrochemical biochips for protein analysis that use antibodies or aptamers as recognition elements.

Keywords Antibodies · Aptamers · Biosensors · Immunosensors · Protein arrays · Protein chips · Proteomics

1 Protein Analysis in the Post-Genome Area

One of the driving forces in the development of new analytical methods for protein analysis is a better understanding of the molecular basis of diseases, the disposition for diseases, and the reactions of patients to certain drugs. For the pharmaceutical industry, the driving force is to find new drug targets and new therapeutical treatments. Proteins bear important functions for most of the life processes. Many proteins are enzymes catalyzing biochemical reactions, and are thereby vital to metabolism. About 4000 reactions are known to be catalyzed by enzymes and the rate acceleration can be up to 10^{17} in comparison to the uncatalyzed reaction. Other proteins have structural or mechanical functions, such as proteins in the cytoskeleton, which forms a system of scaffolding that maintains cell shape. Moreover, proteins are also important in cell signaling, immune responses, cell adhesion and the cell cycle.

The measurement of the levels of individual proteins in cells, tissues, extracts, and biological fluids is therefore of enormous importance for life science research and clinical practice.

Personalized medicine means the prescription of specific therapeutics best suited for an individual. It is usually based on pharmacogenetic, pharmacogenomic, and pharmacoproteomic information, but other individual variations in patients are also taken into consideration [1]. Although personalized medicine has started to become practice for the initial treatment of diseases, like cancer, it is recognized that the time scale for management of cancer and the treatment of other diseases will take much more time. One disease in which pharmacogenetics is already applied is acute lymphoblastic leukemia. Persons lacking the functional enzyme thiopurine methyl transferase are treated with a tailored chemotherapeutic cocktail of 6-mercaptopurine, 6-thioguanine, and azathiopurine [2].

A good indicator for the status of personalized medicine is how the drug labeling or package insert as a guide to how the drug should be used is realized. In 2005, there were only 22 approved drugs with such labeling. Some examples are shown in Table 1.

The main reasons for the delay in the realization of personalized medicine is the extreme complexity, the lack of large statistically and comparable approved analytical data, and the reservation of the pharmaceutical industry in the past which was more focussed on the developed of so-called "blockbusters". However, the picture is changing since pharmaceutical companies collaborate now more and more with biotech companies that develop the needed technologies but do not have the resources to develop the extensive practical applications. One example for a FDA-proven pharmacogenetic test is the AmpliChip CYP450, which was developed in a collaboration between Roche and Affymetrix. The microarray-based test is able to detect gene vari-

Table 1 Examples of drugs where predictive pharmacogenetic testing is proposed (adapted from [3])

Name of the drug	Manu-facturer	Disease	Biomarker	Label description
Somatotropin	Several	Prader Willi Syndrome	Chromosome 15 aberration	Use of drug is indicated for patients with the presence of the biomarker
Retinoid (Vesanoid)	Roche	Acute promyelocytic leukaemia	PML/RAR gene	Use of drug is indicated for patients with the presence of the biomarker
Cetuximab (Erbitux)	Imclone/ BMS	Colorectal cancer	EGFR	Use of drug is indicated for patients with the presence of the biomarker
Trastuzumab (Herceptin)	Roche/ Genentech	Breast cancer	HER 2 protein	Use of drug is indicated for patients overexpressing the biomarker
Alpha1-proteinase inhibitor (Prolastin)	Bayer	Congenital alpha1-proteinase inhibitor deficiency	PiMS or PiMZ alpha1-antitrypsin deficiency phenotypes	Use of drug is not indicated for patients with these phenotypes
Imatinib (Gleevec/ Glivec)	Novartis	Chronic myeloid leukaemia Gastrointestinal stromal tumours	Philadelphia Chromosome positive CD117 (*c-kit*) positive	Use of drug is indicated for patients with the presence of the biomarker

ations of the genes CYP2D6 and CYP2C19, known to play a major role in the metabolism of an estimated 25% of all prescription drugs.

Currently there is a debate regarding how long it will take that the clinical practise will benefit from technologies developed for the different fields of "-omics". A critical study sees the potential rather in the medium term (15–20 years) [3] whereas others rather in shorter term (5–10 years) [1, 2].

Besides new and improved genomic methods, a detailed analysis of proteins becomes more and more important. Compared to the genome, which is rather constant, a proteome differs from cell to cell, from different parts of a body, and from body to body. Moreover, organisms express different proteins in a life cycle and respond to environmental changes with different proteins.

One way to indicate if the synthesis of a protein is initiated or not is to determine the protein-coding m-RNA, also called as expression-profiling or transcriptomic approach. This can be done in a similar way as for genomic studies with well-developed high-throughput methods, e.g., by using nucleic-acid chip technologies.

However, since the produced m-RNAs correspond not inevitably with the amount of the mature proteins, the transcriptomic approach will give only some hints of which proteins could be present in the organism. As a consequence, it is unavoidable to indicate and measure the mature proteins directly in samples taken from the organism. In the past, there was a great hope that with the help of proteomics new biomarkers—substances whose detection indicates a disease or a particular disease state—can be found. There were several studies where the protein patterns of body fluids (blood, serum, or plasma) from sick and healthy persons were compared. Indeed, there have been a lot of differences and new biomarkers have been proposed. However, most of them did not withstand a critical validation. It was found that the differences between protein patterns of two persons are often greater than the differences due to diseases. In addition, the results often vary in dependence of the sample pretreatment and the analytical method used.

It is estimated that the human proteome comprises more than 250 000 proteins. This estimation is based on the assumption that from the ~25 000 human genes [4] about ten splice variants, post-translational modifications, and cleavage products could exist. However, if the immunoglobulin-class proteins also involved in this calculation, the number of proteins that could be present in the body exceeds 10×10^6. The length of the polypeptide chain in proteins can differ from a few amino acids to about 27 000. Moreover, many proteins carry modifications like phosphorylations, glycosylations, acylations, or contain additional compounds, like metal ions, heme, FAD, etc.

Due to the great variability of proteins, the analysis of proteins is much more challenging than the analysis of DNA. It is obvious that substantial efforts are needed to develop appropriate methods to analyze all the proteins in detail.

The analytical methods include protein sequencing (Edman sequencing), protein identification and analysis of the protein modifications (MALDI-TOF MS protein profiling, protein chips), localization of proteins on whole cells (fluorescence microscopy, X-ray tomography), identification of three dimensional structures (X-ray crystallography, NMR, circular dichroism, Fourier transform infrared spectroscopy and small angle X-ray scattering), protein–ligand interaction analysis (yeast two hybrid techniques, affinity chromatography, fluorescence resonance energy transfer and surface plasmon resonance) and quantification methods (immunoassays, protein chips, reverse-phase chromatography or 2D electrophoresis combined with tandem mass spectrometry).

For a proteomic approach where a snapshot-like analysis of many proteins at a given time under defined conditions in a cell, tissue or organ is needed a highly parallel methodology is required.

The value of multiplexed protein measurement is being established in applications such as comprehensive proteomic surveys, studies of protein networks and pathways, validation of genomic discoveries, and clinical biomarker development.

Examples for technologies that have already been developed for these applications are given in Table 2.

Whereas 2D electrophoresis or HPLC combined with mass spectrometry are mostly used for protein identification, immunoassays are seen as the gold standard for the quantitative determination of proteins, especially in complex media, like blood, serum, or food products. Although the development of new

Table 2 Applications and technologies for multiplexed protein measurements (adapted from [5])

Application category	Technology	Approximate multiplexing	Technology characteristics	Desired specification	Experimental design	Refs.
Surveys of most, large changes in protein abundance	"Shotgun" mass spectrometry	2500	Universal, open architecture	Semi-quantitative; CV 30%	Iterative, cross-sectional; 50 samples	[6]
Modelling networks, pathways, physiological and disease states	Bead and chip protein arrays; indirect immunoassays	250	Multiple sets of canonical optimized assays	Quantitative; CV 10%	Longitudinal or dose-response	[7]
Biomarker validation; outcome surrogates in clinical trials	Bead and chip protein arrays; sandwich immunoassays	25	Flexible platform with rapid assay development	Quantitative; CV 5%	Cross-sectional or longitudinal; 500 samples	[8]
Panel-based clinical diagnostics	Multiplexed immuno-diagnostic platform for rapid, point-of-care testing or automated central laboratory testing	5	Specific, highly optimized sets of assays	Quantitative; CV 3%; rapid format	Single sample comparison with reference range; >10 000 samples	[9]

CV coefficient of variation

methods in the field of mass spectrometry is in progress, quantification of proteins is still a problem.

In contrast to mass spectrometry where high abundant proteins have to be removed in a prepurfication step, immunoassays can handle such samples without any special pretreatments.

To increase the sample throughput, to reduce costs, and to reduce time for analysis protein arrays have been developed in different formats, by spotting target proteins, antibodies or other binding molecules. However, for proteomics, the chip technology is not as well developed as for nucleic-acid analysis, since (1) the number of different protein species is a multiple of the number of genes, (2) the physico-chemical diversity of proteins, e.g., the isoelectric point, is more complex than that of nucleic acids consisting of only four (five) unique building blocks, (3) up till now, no comparable methods to PCR for "protein amplification" have been found, (4) due to the highly complex 3D structures, interactions between proteins cannot be reduced to a general code of "complementary" amino acid—like it works for nucleic acid base pairing.

However, as long as the human proteome is not fully characterized and new biomarkers have not been found, the determination of key proteins is sufficient for clinical diagnostics. Here, factors like handling, price, and speed are at the center of interest.

Electrochemical biosensors have the potential to be fast, small, reliable, inexpensive and easy to handle. In this term, the commercially available electrochemical glucose and lactate sensors for patient self-control and point-of-care measurements have paved the way also for the development of other biosensors, e.g., sensors for protein analysis. Although it is obvious that new protein determination methods have a primary focus in the clinical area, it is clear that other areas like basic research in life sciences, food, forensics, and military research will also benefit from the developments of protein sensors.

The following gives an overview of important principles realized in electrochemical antibody and aptamer sensors for protein analysis.

2 Antibody-Based Electrochemical Sensors

2.1 Antibodies as Specific Recognition Elements for Protein Analysis

Antibodies are well-established recognition molecules in quantitative protein analysis. The high specificity and affinity of an antibody for its antigen allow the selective binding of the target protein in the presence of hundreds of other substances, even if they exceed the analyte concentration by 2–3 orders of magnitude. Antibodies can be produced against most proteins. In addition to the standard methods for antibody generation like hybridoma technology or polyclonal antibody recovery from the serum, nowadays also recombinant technologies combined with in vitro evolutive methods, e.g., phage display [10] or ribosome display [11, 12] are used. These technologies can be applied to modify an antibody molecule to the analytical requirements, e.g., by increasing the specificity, affinity, or stability [13, 14].

Moreover, recombinant antibody technology has now been developed to a level that allows the expression of antibody fragments like Fab (fragment antigen-binding) or scFv (single chain fragment variable) (Fig. 1) in *E. coli* in large quantities and at an acceptable cost. In mammalian cell culture, the yield is about 0.5–1 g/l with costs of 300 /g whereas in bacteria the yield can be as high as 3 g/l with costs of 1 /g [16]. With recombinant antibody technology, additional peptide sequences or functional proteins can be attached as tags that can be used for immobilization on a chip surface (Ingvarsson et al. 2006) or for coupling of labeling compounds, e.g., alkaline phosphatase (Wang et al. 2006).

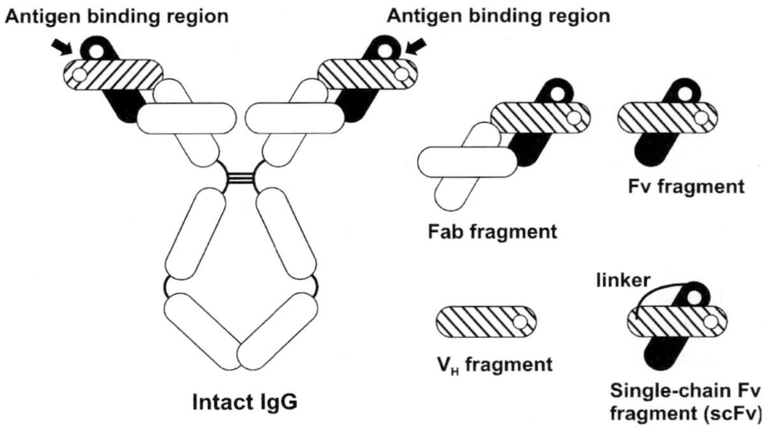

Fig. 1 Schematic representation of an intact IgG antibody and important antibody fragments that are often used in recombinant antibody technology. The *small circles* represent the N-termini of the polypeptide chains (adapted from [15])

Analytical methods based on antibodies can be divided into direct and indirect methods. Direct methods, like surface plasmon resonance (SPR), reflectometric interference spectroscopy (RIfS), quartz crystal microbalance (QCM), cantilever-based methods or isothermal titration calorimetry (ITC), can usually record the antigen antibody binding in real time and without any labeling compound [19–21].

For an indirect indication of analyte binding additional compounds or helper reactions are needed. To date, a huge number of different labels, including radioactive compounds, enzymes, fluorophores, redox compounds, cofactors, fluorescence quenchers, chemiluminescence metals, latex particles and liposomes are used in immunoassays [22].

In principle, immunoassays with labeled compounds are carried out in two different ways: (i) in competitive assays a labeled analyte competes with the analyte for the antigen binding sites of the antibody. The key feature of a competitive assay is that maximum assay sensitivity is attained using an amount of antibody tending to zero. The sensitivity of a competitive immunoassay is determined by the affinity of the antibody for its antigen. Since the affinity of an antibody can be in the range of 10^5–10^{12} M^{-1} [23], competitive immunoassays using antibodies with a $K_d = 10^{-12}$ M reach their highest sensitivity in the picomolar range. However, the lower detection limit can be much lower when appropriate assay conditions are chosen. (ii) In non-competitive assays (e.g., sandwich or sequential saturation assays) a significant excess of antibodies over the antigen is used. Here, diffusion processes are much more important than the affinity of the antibody used. While the lowest analyte concentration detectable using competitive design is on the order of 10^7 molecules per milliliter, a non-competitive method is potentially capable of measuring concentrations lower by several orders of magnitude [24]. In addition, due to the excess of one immunoreactant, non-competitive assays can be performed much faster than their competitive counterparts. Besides, a classification due to amount of used antibodies immunoassays can also be classified in heterogenous and homogenous immunoassays. In heterogenous assays, a separation of the immunocomplex from the free immunoreactants is necessary.

The most successful heterogenous immunoassay type is the enzyme-linked immunosorbent assay (ELISA), which is performed in a 96-well microtiter plate with immobilized antibodies or antigens, allowing the determination of many of samples at once. After sequential incubations of sample, antibody- or antigen-enzyme conjugate and enzyme substrate reagents with washing steps in between the indication takes place by measuring the absorbance or fluorescence of each well by a standard microtiter plate reader. In contrast, no washing steps are necessary in homogenous immunoassays. Most of the homogeneous immunoassays make use of the modulation of the enzymatic activity or the fluorescence intensity due to the antigen antibody binding. For homogenous enzyme immunoassays, one immunoreactant is coupled with an enzyme (EMIT), substrate (SLFIA), inhibitor (EMMIA) or cofactor (ARIS), or immunoreactants are coupled with two cooperating enzymes (Channeling-EIA). For homogenous fluorescence immunoassays one (e.g., quenching IA, polarization IA, confocal IA) or two (e.g., Förster Resonance Energy Transfer, FRET) fluorophor- labeled immunoreactants are often used. Immunoassays have been adapted to highly sophisticated automated analyzers facilitating the analysis of a very large number of samples within a short time. Such analyzers can handle homogeneous or heterogeneous test formats which are performed in plastic tubes or in wells of 96-, 384- or 1536-microtiter well plates. Seidel et al. [25] described the use of gold-coated nanotiter plates (75 nl/well) for a phase-separation fluorescence immunoassay (PSFIA). Recently, Ghatnekar-Nilsson et al. [26] described atto-vial-based recombinant antibody arrays containing vials with a volume of 6–4000 aL. In combination with a planar waive-guide detection system, these arrays were able to detect proteins within subzeptomole range. In addition to multi-well plate assays, microspot assays are currently becoming more and more popular [27]. Here many different antibodies or antigens are spotted, e.g., with a piezoelectrical pipetting system on a very small area on a planar chip. Since one immunoreactive spot on a chip usually has an area of $<10\,\mu m^2$, the benefit of this technology could be seen in (i) a dramatic reduction of reagent consumption, (ii) in its use with cross-reactive antibodies for pattern recognition or (iii) in providing more exact results by highly multiple determinations.

In contrast to DNA analysis, where different specificities can easily be predetermined by simple chemical synthesis of the nucleotide sequence, the generation of antibodies for each analyte needs many more efforts. For microarrays, fluorophores were usually used as the labeling compound. However, to increase the sensitivity, enzymes have also been applied as labels.

In 1999 Weller et al. [28] described an optical immunoassay with 1600 spots per $1.8\,cm^2$ which can be used in environmental control for parallel determination of different pesticides. With a volume of 2 nl per spot of an antibody solution of 10 g/ml and 1000 spots per chip the total amount of antibody per chip was calculated to be only 20 ng. Peroxidase was used as the label and the chemiluminescence was indicated with a CCD camera. A similar approach was recently used for the screening of allergen-specific protein IgE [29].

The company Molecular Staging in New Haven, Connecticut, is using another method to increase the sensitivity for protein chips, the rolling circle amplification technology. Thereby the proteins on a chip are detected by antibodies that are conjugated to a single-stranded DNA. The end hybridizes on a defined region of a circular DNA which elongates the antibody-coupled DNA by the rolling circle principle. To the antibody-coupled elongated DNA, many fluorophor-labeled oligonucleotide molecules can then be hybridized, which gives a high fluorophor to antibody ratio [30].

2.2 Electrochemical Immunoassays

Electrochemical immunoassays are currently not as widely used for protein analysis as optical immunoassays. However, since no microtiter plate reader or other sophisticated optical detection systems are needed, the costs can be reduced, and since no high voltage supply is necessary the measurements can be done on-site. Other advantages of an electrochemical detection are that it can be easily miniaturized, permitting detection in small volumes or integration into microfluidic systems and the lack of interferences caused by turbid or colored samples [31].

Although most of the electrochemical immunoassays use electrochemical measuring cells, flow through detectors or microchips with potentiometric or amperometric transducers (for immunoassays based on impedimetric transducers see [32]) the integration of amperometric electrodes directly in the wells of microtiter plates has also been described [33, 34]. Thus the pipetting and washing equipment from conventional ELISA methods can be used.

For electrochemical immunoassays, in principle the same assay formats have been used as for optical immunoassays, like competitive, non-competitive, heterogenous, and homogenous formats.

Instead of fluorophor-labeled immunoreactants, redox-labeled immunoreactants are used in electrochemical immunoassays. The redox label should have the following properties:

(i) It should be electroactive in a potential range of 0 mV to –200 mV [35].
(ii) It should not cause electrode fouling.
(iii) There should be no side reactions with the matrix.
(iv) It should be stable in buffer.
(v) Chemical groups for coupling should be available.

In non-amplified redox-labeled electrochemical immunoassays the indication of one antigen or antibody molecule will generate only one signal equivalent. Since the sensitivity of an amperometric sensor for the redox label is in the lower micromolar to submicromolar range, this kind of assay can be used if the analyte has to be determined also in that range, e.g., creatinine. In the so-called size exclusion redox-labeled immunoassay (SERI), creatinine from the sample competes with naphthoquinone-labeled creatinine conjugate for the antigen-binding sites of the antibody. Unbound conjugate passes through a membrane and is indicated at the amperometric electrode ($E = -195$ mV vs. Ag/AgCl) whereas antibody-bound conjugate is size excluded. This principle allows a wash-free and therefore convenient determination of creatinine within the submicromolar to submillimolar range [36]. The SERI principle should also be applicable for protein analysis if small redox-labeled peptides are used as competitor to the complete target protein for antibody binding. Instead of using a membrane to separate the redox-labeled immunocomplex from the free redox-labeled immunocomponent, Wang et al. [37] used capillary electrophoresis on a chip (Fig. 2). The electrophoretic separation was carried out in a 78-mm long separation channel (50 μm wide, and 20 μm deep) on a glass chip and the indication of the ferrocene conjugate took place on a gold-plated screen-printed carbon working electrode (+600 mV vs. Ag/AgCl). It was shown that this principle can be used for low-molecular-weight analytes by using a competitive principle but also for proteins by using a non-competitive principle. 3,3′,5-triiodo-L-thyronine (T3) and mouse IgG were determined with a detection limit of 2.5×10^{-6} μg/ml and 1 μg/ml, respectively.

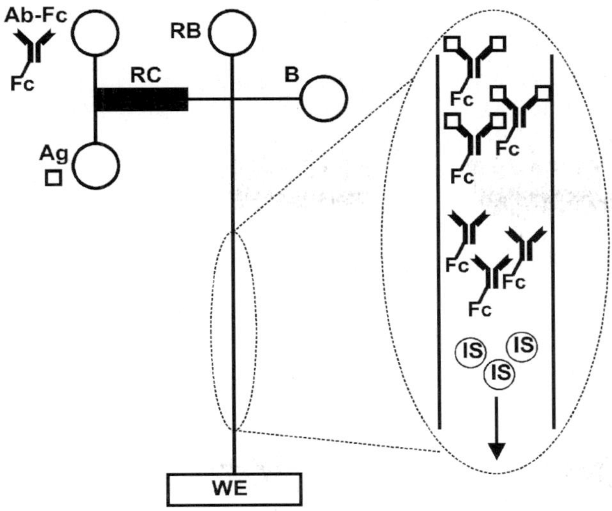

Fig. 2 Scheme of a non-competitive electrochemical immunoassay based on electrophoretic separation of the redox-labeled components. *Ag* mouse IgG; *Ab-Fc* anti-mouse-IgG-ferrocene conjugate; *RC* reaction chamber; *RB* running buffer; *B* unused buffer reservoir; *IS* internal standard; *WE* screen-printed working electrode (adapted from [37])

Hemoglobin A1c (HbA1c)—a long-term diabetes marker has been determined with a flow immunoassay by using ferrocene-labeled anti-hemoglobin antibodies (anti-Hb-IgG-Fc) and a boronate affinity column [38]. In a 60-min preincubation step anti-Hb-IgG-Fc binds to HbA1c and hemoglobin. Subsequently, the HbA1c-anti-Hb-IgG-Fc complex is captured on a boronate gel. After elution of the complex with 50 mM sorbitol, it is indicated within a electrochemical flow trough cell containing a glassy carbon electrode (+600 mV vs. Ag/AgCl). The linear measuring range was described to be 0–500 μg/ml HbA1c.

For more sensitive immunoassays analyte accumulation or signal amplification principles are applied. Bordes et al. [39, 40] used in a multi-analyte immunoassay a 5-min preconcentration step of the cationic labels used (ferrocene ammonium salt with $E_p = +260$ mV, cobaltocenium salt with $E_p = -1.05$ V) at a negatively charged nafion-loaded carbon paste electrode resulting in a linear response down to 75 nM. However, to regenerate the sensor, the nafion-loaded carbon paste surface has to be renewed after each measurement.

Another possibility for amplification is to use immunocomponents that are multi-labeled. Multi-labeling can be done by covalent coupling of a redox compound to more than one amino or carboxylic group of the protein. However, the amount of redox molecules that can be coupled in that way is limited.

To get a much higher labeling rate, Mak et al. [41] used encapsulated microcrystals of a redox molecule. Therefore they encapsulated ferrocene microcrystals (1 μm in diameter) by using layer-by-layer (LbL) technology. An anti-mouse-antibody was coupled to the encapsulated particles by adsorption which was used as indicating antibody in a sandwich immunoassay for mouse IgG. The indication was carried out after a releasing agent was added to mediate the dissolvation of the ferrocene microcrystals within and across the capsule. With the so-called "supernova effect" many ferrocene molecules were released and were detected by an amperometric sensor. The detection limit for mouse IgG was 2.82 μg/l.

High sensitivities are also achieved when redox recycling is applied. Thereby the redox compound is oxidized and reduced in a cyclic manner so that the indication of one labeled antigen or antibody molecule will generate multiple signal equivalents. Redox recycling can be done by using electrode–electrode coupling.

It was used by Niwa et al. [42], Wollenberger et al. [43], and Hintsche et al. [44] to enhance the sensitivity by one order of magnitude and to thereby achieve detection limits in the lower nanomolar range. Therefore interdigitated array electrodes (IDA) with an electrode width of 1.5 mm and an interelectrode space of 0.8 mm were used. As the authors described, redox recycling with IDA can be used for the design of an electrode-supported ELISA or for a separation-free displacement assay. Anti-analyte antibodies that have been coupled to small polymeric beads are saturated with a ferrocene-labeled analyte conjugate. The analyte displaces the conjugates, which are then detected by the IDA. The advantage of this kind of redox recycling is that a sensitive amperometric determination is reached without any transducer-immobilized biomolecule, which makes the system more stable. However, high-quality IDA and a bipotentiostat are also needed. Furthermore, the reliability of measurements in serum or whole blood has not been shown until now.

Enzymes were used as a labeling compound to increase the sensitivity of electrochemical immunoassays much further. Table 3 shows some examples for enzymes that have been used as labels in electrochemical immunoassays.

Catalase is known to be one of the most active enzymes with a catalytic reaction rate constant of approximately 10^5 sec^{-1}. This enzyme was used as labeling compound to improve the sensitivity of an immunoassay. After the addition of hydrogen peroxide, the produced oxygen was measured with an oxygen electrode as the transducer [68, 69].

One of the most popular enzymes in electrochemical immunoanalysis is alkaline phosphatase (aP). This enzyme catalyzes a dephosphorylation reaction of different organic phosphates. Some of the products formed as a result of the reaction can be detected amperometrically in significantly low concentrations. This approach was introduced by J. Kulys et al. in 1980 [70]. Successful developments of the electrochemical immunoassays based on aP determination was achieved by W.R. Heineman and his coworkers [71–74].

Table 3 Representative examples of enzymes used in electrochemical immunoassays. Indicated substances are underlined

Enzymes	Substrates	Products	Transducer	Refs.
Glucose-6-phosphate-Dehydrogenase	glucose-6-phosphate + NAD^+	gluconate-6-phosphate + NADH	Platinized carbon (+150 mV)	[45]
Catalase	H_2O_2	O_2	Clark electrode (–600 mV)	[46]
Peroxidase	H_2O_2 + iodide	iodine	Highly dispersed carbon electrode (+127 mV)	[47]
			Graphite (0 mV)	
	H_2O_2 + 5-aminosalicylic acid	5-aminosalicylic acid quinoneimine	Graphite (–20 mV)	[49]
	H_2O_2 + hydroquinone	benzoquinone	Screen-printed platinum (–600 mV)	[50]
			Graphite methacrylate (–100 mV)	[51]
	H_2O_2 + tetrathiafulvalene (TTF)	TTF^+	Glassy carbon (+20 mV)	[52]
	H_2O_2 + *o*-phenylendiamine (OPD)	OP imine	(–300 mV)	[53, 54]
	H_2O_2 + tetramethylbenzidine (TMB red.)	TMB ox.	(+100 mV)	[55]
Microperoxidase	H_2O_2 + electrode	$2e^-$	Gold	[56]
Alkaline phosphatase	Glucose-6-phosphate	Glucose + phosphate	Glucose oxidase enzyme sensor (+600 mV)	[57]
	Phenylphosphate	Phenol + phosphate	(+870 mV)	[58, 59]
	4-Aminophenylphosphate	4-Aminophenol + phosphate	Gold (+190 mV)	[60]
	3-Indoxyl phosphate	Indigo blue + phosphate	Screen printed carbon (CV –250 to 200 mV)	[61]
	1-Naphthyl phosphate	1-Naphthol	(+550 mV)	[62]

Table 3 (continued)

Enzymes	Substrates	Products	Transducer	Refs.
Glucose oxidase	Glucose + ferrocene (Oxid.)	Gluconolactone + ferrocene (Red.)	Glassy carbon (+250 mV)	[63]
β-Galactosidase	4-Aminophenyl-β-galactopyranoside	4-Aminophenol + galactose	Laccase/ODH enzyme sensor (–600 mV)	[64]
	4-Aminophenyl-β-galactopyranoside	4-Aminophenol + galactose	Glassy carbon (+200 mV)	[65]
Urease	Urea	NH_4^+ + CO_2	Ammonia gas electrode (Ingold)	[66]
β-Lactamase	Benzyl penicillin	H^+	pH-FET	[67]

Here, instead of *p*-nitrophenylphosphate, which is routinely used as an aP, substrate for optical ELISA *p*-aminophenylphosphate (*p*-APP) and the more stable phenylphosphate is used as substrates for aP. The enzymatically produced *p*-aminophenol (*p*-AP) or phenol can be indicated in micro- to submicromolar concentrations at a glassy carbon electrode [58, 72, 75–77]. To improve the sensitivity further, substrate recycling can be applied. Here, the product of the enzyme label is not only measured once but is converted back by an additional enzyme or on a second electrode to be measured again.

The formation of NAD+ from NADP+ [78], pyruvate from phosphoenolpyruvate [79], phenol from phenol phosphate [80, 81] and aminophenol from aminophenylphosphate by the aP label [82, 83] have been followed by using enzymatic substrate recycling with electrochemical indication. The commercially available amplified optical ELISA (AmpliQ from Dako, NovoClone AELIA from Novo BioLabs) uses Iodonitrotetrazolium violet (INT-violet) as diaphorase substrate [84, 85]. Instead of INT-violet for the electrochemical amplified immunoassay, ferricyanide was used as diaphorase substrate [86]. The ferrocyanide that was produced by the diaphorase reaction was indicated amperometrically, and thereby recycled back to ferricyanide (Fig. 3). In similar configurations, ADH was replaced by formiate dehydrogenase, glucose dehydrogenase, carnitine dehydrogenase [87], or glycerol dehydrogenase [88].

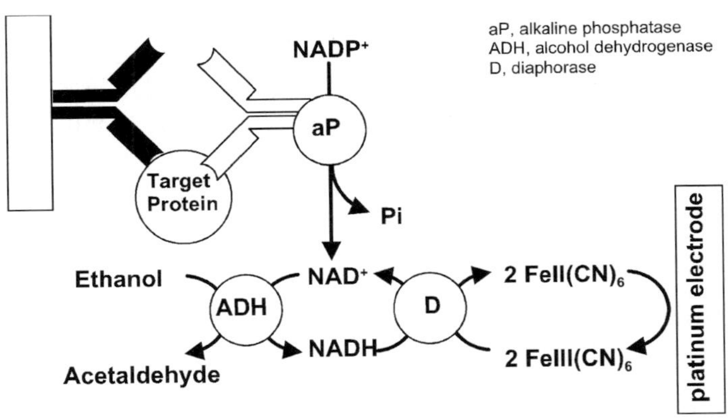

Fig. 3 Schematic representation of an amplified electrochemical sandwich immunoassay by using substrate recycling (adapted from [86])

Examples of protein determination with such a type of electrochemical immunoassays was described for the determination of goat IgG and human thyroid stimulating hormone (hTSH) by using a sandwich-type immunoassay in combination with an oligosaccharide dehydrogenase/laccase bi-enzyme sensor [82]. In a first investigation alkaline phosphatase was used for the model compound IgG, and the liberation of *p*-aminophenol from *p*-aminophenyl phosphate was followed. This reaction, however, is known to suffer from drawbacks related to the limited stability of both *p*-aminophenyl phosphate and *p*-aminophenol in alkaline solution [74]. Therefore a large blank signal, which in some cases exceeds the dynamic measuring range of the electrode, is obtained and the incubation time is limited. Furthermore, the use of alkaline phosphatase requires a change of the pH between the immunoassay and the electrode reaction. Therefore, β-galactosidase was used as labeling enzyme. Since the optimum pH of β-galactosidase is close to that of the bienzyme sensor (pH 6.5), the whole assay could be performed under the same conditions. The sensitivity of the total assay was comparable to that of the photometric test. For the determination of hTSH, the sandwich-type assay has been performed using biotinylated tracer antibody and streptavidin-β-galactosidase conjugate. The measuring range extends from 0.5 pg/ml to 20 ng/ml with a detection limit of 0.3 pg/ml. The electrochemical immunoassays mentioned were using amperometric transducers. A special application of electrode supported EIA with a potentiometric transducer is the commercially available Threshold Immuno-Ligand Assay System (Molecular Devices, USA). Here, urease is used as the labeling enzyme. After incubation of the analyte-containing sample with urease- and avidin-labeled immunoreactants, the incubation mixture is passed through a filtration membrane with immobilized biotin. After washing, the membrane-bound urease is indicated by a light addressable potentiometric sensor [89].

Homogenous immunoassays are very attractive for use since no washing procedures are necessary. Athey et al. [45] have adapted a commercially available EMIT (from Syva)–based on the modulation of the enzyme activity—for the competitive determination of theophylline in whole blood with theophylline-glucose-6-phosphate dehydrogenase conjugate. The produced NADH was determined by an amperometric detector at a potential of +150 mV vs. Ag/AgCl.

2.3 Electrochemical Immunosensors

2.3.1 Direct Immunosensors

In contrast to electrochemical immunoassays, in immunosensors the antibody or its complementary binding partner is immobilized in direct spatial contact with the transducer. Amperometric transducers are mostly used, which are made from metals (e.g., gold, platinum) or carbon materials (e.g., graphite, carbon paste, glassy carbon). The immunocomponent can be immobilized directly on the surface for example by adsorption, by thiol-gold interactions, or by chemical coupling to carboxyl groups generated on the surface of carbon electrodes. Indirect immobilization methods have been developed by using membranes, polymers, gels, or paramagnetic microbeads in front of the electrode [31].

Direct immunosensors can measure the antigen–antibody reaction without any labeled immuno components, e.g., by using changes in charge densities or conductivities for transduction.

Thirty years ago Janata observed in an affinity sensor a potential change when he incubated mannan with a potentiometric electrode covered with PVC membrane-immobilized ConA [90]. Thereby, it was possible to follow the binding process directly in real-time without any labeling.

In 1984, Keating and Rechnitz [91] used a potassium-sensitive electrode with a dioxin-ionophore conjugate within a PVC membrane for the determination of anti-dioxin-antibodies. Although many more publications have been published describing similar effects [92–94], it is not yet completely clear whether the potential change is a result of disturbed ion transport or of charge changes at the protein surface during binding (for a critical review see [95]).

In addition to potentiometric techniques, amperometric techniques have also been used for the construction of direct immunosensors. Sadik and van Emon [96] described a polymer-modified antibody electrode in combination with pulsed amperometric detection (PAD). An antibody or a ligand that binds the analyte of interest is immobilized on the surface of a membrane electrode. The analytical signal was generated by applying a pulsed waveform between +0.6 and –0.6 V with a pulse duration of 120 and 480 ms. When the current is sampled at the end of the pulse, the component, due to charging, is significantly reduced or close to zero, and the level of current due to charging and discharging induced by pulsing can be treated as a constant. Under these circumstances it has been postulated that the change in the response level reflects only the binding of the analyte to the antibody and that the pulse amperometric technique can be used to repulse the antigen from the immobilized antibody and thereby regenerate the sensor. Recently, Grant et al. [97] have used PAD for the determination of the protein bovine serum albumin. They described that the actual source of the signal is as yet unclear, but that the currents observed were due in some way to the antibody–antigen interaction. Further investigations and controls are necessary to resolve the background of the signal generation.

Since electrochemical transducers detect only species that reach the electrode surface, it is possible to differentiate between molecules that are near the transducer and molecules that are within the bulk. One principle is to hinder the diffusion of a redox active substance, e.g., potassium hexacyanoferrate(II) [98] or ferrocene-labeled glucose oxidase [99] to the electrode surface as result of the antigen–antibody complexation. Analogous to this principle, the diffusion of an enzyme to a mediator-modified electrode can be hindered. Both principles are applicable especially for high-molecular-weight analytes, like microorganisms and proteins. By hindrance of the ferrocene-labeled glucose oxidase anti-dinitrophenyl-antibodies were determined within a concentration range of 1–50 μg/ml [99]. However, since the principle is sensitive to defects within the layer, high-quality monolayers with a high reproducibility have to be built up on the electrode surface.

Direct electrochemical immunosensors based on impedance spectroscopy are described in Pänke et al. [32].

2.3.2 Redox-Labeled Immunosensors

Although direct electrochemical immunosensors for protein analysis without any labeling compounds have been developed, most of the electrochemical immunosensors are indirect ones and use redox-labeled or enzyme-labeled immunocomponents.

In contrast to the electrochemical immunoassays where the redox-labeled compound can diffuse to the sensor surface, in immunosensors the redox-labeled immunocomponent is bound within the antigen–antibody complex. Therefore, to enable an electrochemical indication, it is necessary that the redox moiety is very close to the sensor surface. Often ferrocene is used as the redox-labeling compound. The labeling can be done covalently via the amino groups of the antibody molecule by using amino reactive ferrocene derivatives. Akram et al. [100] used such a conjugate for the construction of a sandwich immunosensor for the determination of human chorionic gonadotropin (hCG). A detection limit of 2.2 IU/L was described. For the determination of the ratio of HbA1c to total hemoglobin, Halamek et al. [101] used another way of ferrocene antibody conjugation. They used ferrocene boronic acid, which binds to the cis-diols of the sugar moiety of the antibody. For the measurement, in a first step total hemoglobin was captured on a surfactant-modified gold electrode. After incubation with an anti-HbA1c antibody and subsequently with the ferrocene boronic acid the sensor bound ferrocene boronic acid was indicated at a potential of +300 mV vs. Ag/AgCl.

Dai et al. [102] described an immunosensor for carcinoma antigen-125 (CA125) with horseradish peroxidase (POD) as the redox label. Serum CA125 has been widely considered to be an especially valuable serum marker with a threshold value of 35 units/mL expressed by >80% of patients with non-mucinous epithelial ovarian cancer. Since POD was not indicated via its biocatalytic activity, this sensor works substrateless and mediatorless. For the construction of the sensor, the antigen (CA125) was immobilized with titanium sol-gel on a glassy carbon electrode by vapor deposition. The CA125 to be determined competes with the immobilized CA for the binding sites of the anti-CA125-POD conjugate molecules. By application of differential pulse voltammetry (DPV) a peak current decrease was observed in the presence of CA125. The detection limit for CA125 was 1.29 U/ml. However, a washing step after anti-CA125-POD conjugate incubation and oxygen exclusion during DPV measurements are necessary.

2.3.3 Enzyme Immunosensors

Enzymes are the most prominent labels in electrochemical immunosensor development. In principle, all types of electrochemical immunoassays with enzymes described in Sect. 2.2 have been realized with the recognition element immobilized at the transducer surface.

Alkaline phosphatase was used for the immunochemical determination of fatty acid binding protein (FABP), which is an early diagnostic marker for acute myocardial infarction (AMI). Disposable electrodes were constructed using graphite working electrodes that were coated with anti-FABP-IgG as capture antibody. A sandwich principle was applied using anti-FABP-alkaline phosphatase conjugate [103].

The advantages of using glucose oxidase are the high stability of the enzyme and substrate in solution, the moderate price, and above all the easy and reliable indication (without electrode fouling) at a conventional platinum electrode, which can be done via the consumed oxygen at −600 mV or via the produced hydrogen peroxide at +600 mV vs. Ag/AgCl. The first publication describing GOD as label for the development of immunosensors dates back to 1977 when Mathiasson and Nilsson developed an albumin immunosensor based on immobilized anti-insulin antibodies [104]. GOD was indicated by oxygen consumption measurements.

Many proteins that need to be analyzed are enzymatically or non-enzymatically modified. Often it is important to know the percentage of this protein modification. For this purpose, two sensors are normally used. Recently, we have developed a general principle by using only one sensor [105].

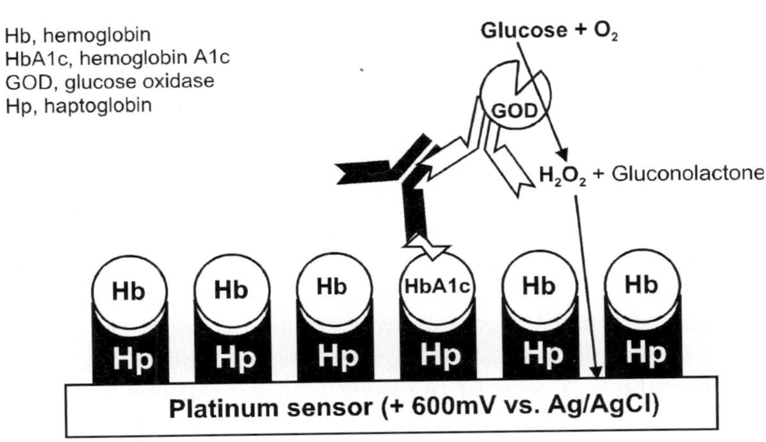

Fig. 4 Scheme for an enzyme immunosensor for the determination of the percentage of protein modification exemplified on glycated hemoglobin (HbA1c) to total hemoglobin (Hb) (adapted from [105])

In a first step the binding sites of sensor immobilized capture molecules are saturated with both the unmodified and the modified protein molecules. Then the captured modified protein is indicated via an antibody that binds specifically to the modified protein. In this way, an immunosensor is used for direct determination of the HbA1c value (percentage of glycated to total hemoglobin). GOD was used as the labeling enzyme and the enzymatically produced hydrogen peroxide was indicated with the amperometric sensor. In an optical test it was shown that anti-hemoglobin-antibodies as well as haptoglobin can be used as capture molecules. However, since the ratio of specific to unspecific binding was superior in the case of using haptoglobin, the sensor was constructed by using haptoglobin as capture the molecule (Fig. 4). With this configuration it was possible to determine HbA1c in the clinically relevant range of 5–20%.

2.3.4 Wash-Free Enzyme Immunosensors

For the enzyme immunosensors described above, several washing steps are still necessary. To shorten the time of analysis and to improve the convenience of handling wash-free enzyme immunosensors are highly desirable and in the focus of many researchers of the biosensor community. For this purpose, mostly proximity effects are utilized.

For a number of redox enzymes, the ability to catalyze electrode reactions by a mediatorless mechanism was established, which is also known as direct electron transfer [106]. In such cases, the electrons are transferred directly from the electrode to the substrate molecule via the active site of the enzyme. Therefore, electrons act as the second substrate for an enzymatic reaction and the potential is shifted in the presence of the corresponding substrate. This phenomenon results in catalytic elimination of the over-voltage.

The electrocatalytic properties of several redox enzymes permit their application as labels for separation-free potentiometric or amperometric immunosensors. Binding of the electrocatalytically active enzyme label to the electrode surface initiates an electrocatalytic reaction resulting in a potential shift or a catalytic current. In the assay procedure, the antigen immobilized on the electrode surface interacts with the enzyme-labeled antibody, resulting in fixation of the enzyme to the electrode surface. Therefore, the formation of antigen-labeled antibody complex on the electrode surface is accompanied by a potential shift or the generation of a catalytic current. The presence of free antigen in the solution leads to a competition for labeled antibodies. The competition results in a decrease of the electrode signal. The decrease in the rate of the potential shift in this case is proportional to free antigen concentration in solution.

Laccase is known to catalyze the reaction of oxygen electroreduction via a direct mechanism [107]. This property of the enzyme allows the detection of the biospecific interaction of laccase-labeled antibodies with an antigen-modified electrode. Formation of a complex between the laccase-labeled antibody and antigen on the electrode surface results in a considerable shift in electrode potential. Immunosensors based on the laccase-labeled immunoagents and direct potentiometric detection of antigen–antibody interaction have been developed by Ghindilis et al. [108–111]. The laccase near the electrode surface, which catalyzes the oxygen electroreduction, leads to an increase in the electrode potential due to the catalytic removal of the reaction overvoltage. Since only oxygen is needed as a laccase substrate, the sensor works substrate-free. The analysis can be performed in a competitive or in a sandwich scheme. A single measurement requires 20 min. Insulin and mouse immunoglobulin were used in a model assay. For insulin, a concentration range of 10–1000 ng/ml was covered.

Peroxidase is known to catalyze direct (mediatorless) electroreduction of hydrogen peroxide [106]. The electrocatalytic properties of peroxidase permit its application as a label for immunosensors. Attachment of peroxidase to the electrode surface in the presence of H_2O_2 initiates an electrocatalytic reaction, resulting in a potential shift or a catalytic current. The direct electron transfer of a peroxidase label was applied by McNeil et al. [112] for the development of amperometric immunosensors using amperometric detection of hydrogen peroxide electroreduction.

If an antibody is immobilized on a mediator-modified electrode, the antigen-enzyme conjugate can only be determined if it is in direct contact with the mediator.

The so-called "electrically wired" amperometric immunosensors normally use peroxidase as label and sensor immobilized osmium redox polymers [113, 114] or tetrathiafulvalene [52] as mediators (Fig. 5). Only if the peroxidase-antibody-conjugate is near the tetrathiavalene modified electrode an electron transfer can take place. With this immunosensor rabbit IgG was monitored in the range 5–100 ng/ml [52].

Another principle that has been used for wash-free enzyme immunosensors is substrate channeling. Keay and McNeil [115] used an electrode with coimmobilized antibodies and peroxidase. If the GOD-labeled antigen binds to the antibody H_2O_2 "channels" to the peroxidase if glucose is added. The H_2O_2 that is produced in the bulk is destroyed by catalase and cannot reach the electrode. Since the peroxidase activity is measured at the electrode (+50 mV), the bound GOD-labeled antigen can be quantified. This principle should be applicable for competitive and sandwich formats (Fig. 6A). A similar principle but with peroxidase as labeling enzyme and with GOD immobilized on the electrode was used by the Risphon group [48, 49]. Here, the H_2O_2 is produced at the electrode interface and not in homogeneous solution. This has the advantage that no catalase for H_2O_2 destruction has to be used. The indication takes place by amperometric measurement of the peroxidase reaction products (Fig. 6B).

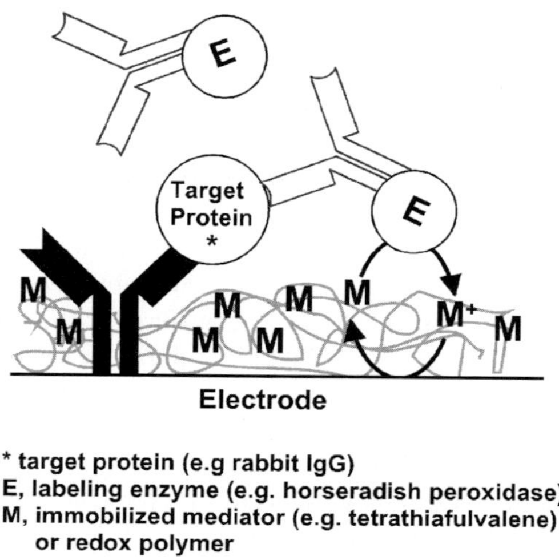

Fig. 5 Schematic representation of an "electrically wired" immunosensor based on Wendzinski et al. [52]

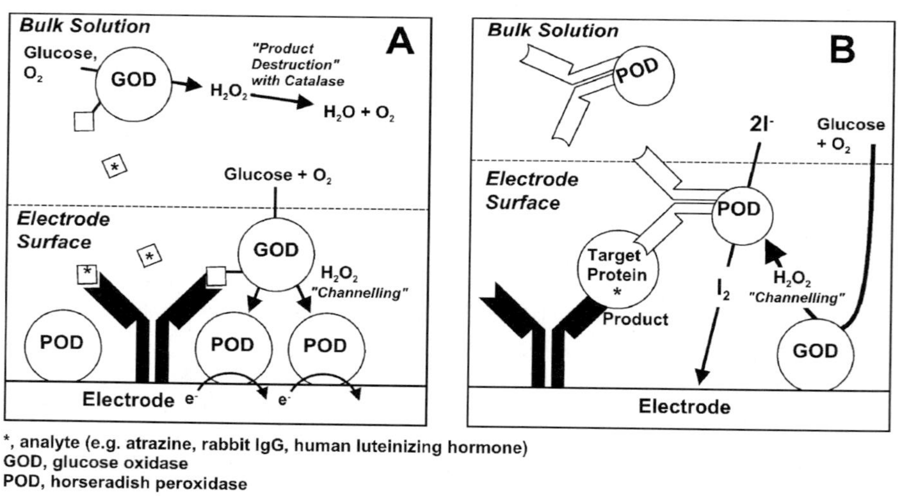

Fig. 6 Two principles of wash-free enzyme immunosensors based on substrate channelling: **A** with and **B** without destruction reaction in the bulk [48, 49, 115]

However, due to the immobilized enzymes, the regeneration of these immunosensors is problematic. To avoid this problem, Ducey et al. [60] let the substrate (*p*-APP) for the label enzyme (aP) diffuse through the back of the microporous gold electrode so that the substrate concentration near the electrode was high. Since no immobilized enzymes are used, this sensor can be regenerated.

3 Aptamers in Analytics

3.1 Aptamers as Specific Recognition Elements for Protein Analysis

Although most of the protein microarrays are being developed with antibodies as recognition molecule [116, 117], alternative binders have already been explored that could overcome certain limitations of antibodies. Compared to antibodies, aptamers offer several advantages, e.g., ease of manufacture, possibility of regeneration, high stability under elevated temperatures and many possibilities for labeling [118].

It is well known that DNAs and RNAs bind in vivo not only complementary nucleic acids but also small molecules or proteins. By using the SELEX method (Systematic Evolution of Ligands by Exponential Enrichment, Fig. 7) [119, 120], binders for proteins (Table 4), amino acids, nucleotides, drugs, vitamins, and other organic and inorganic compounds have been generated.

The preparation of these aptamers takes advantage of the well-established nucleic-acid chemistry, polymeric chain reaction, and modern separation techniques. Aptamers are selected from mainly random pools of RNA or DNA

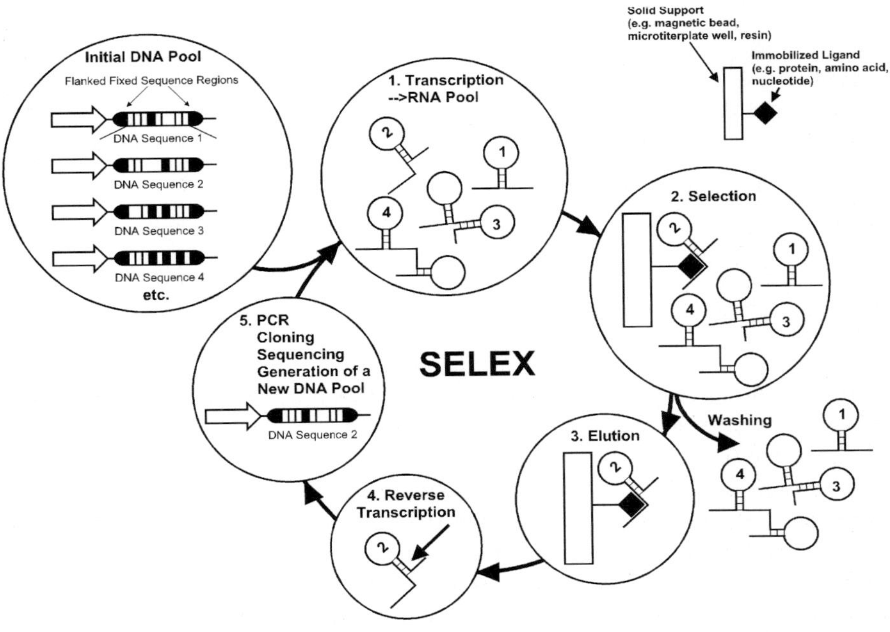

Fig. 7 Principle of SELEX (Systematic Evolution of Ligands by Exponential Enrichment)

Table 4 Examples of proteins against which aptamers have been developed

Target protein (Type of aptamer)	Refs.
NS5B (RNA-Pol of Hepatitis C Virus) (DNA) [121]	Vascular endothelial growth factor (RNA) [147]
Complement C5 (RNA) [122]	SelB (RNA) [148]
Neutrophil elastase (DNA) [123]	HIV-1 Rev (RNA) [149]
α-Thrombin (DNA) [124]	Interferon-γ (RNA) [150]
HIV-2 Tat (RNA) [125]	Hepatitis C NS3 (RNA) [151]
Cytochrome c (DNA) [126]	Anti-insulin receptor antibody MA 20 (RNA) [152]
NFAT (RNA) [127]	PSMA prostate-specific antigen (RNA) [153]
HIV-1 rev peptide (RNA) [128]	L-Selectin (DNA) [154]
Bacteriophage MS2 coat protein (RNA) [128]	Human RNase H1 (DNA) [155]
Transcription factor E2F (DNA) [129]	Human Oncostatin M (RNA) [156]
Neuropeptide nociceptin/orphanin FQ (RNA) [130]	Factor VIIa (RNA) [157]
NS3 prot. domain hepatitis C virus (RNA) [131]	Factor IXa (RNA) [158]
Human activated protein C (RNA) [132]	Cytotoxic T cell antigen 4 (RNA) [159]
Platelet-derived growth factor (PDGF)-AB (DNA) [133]	HIV-1 gp120 (RNA) [160]
Ghrelin (RNA) [134]	Streptavidin (RNA) [161]
Ricin A-chain (RNA) [135]	HIV-1 Tat (RNA) [162]
Tenascin-C [136]	Subtilisin (RNA) [163]
Colicin E3 (RNA) [137]	Thrombin (DNA) [164]
Pepocin (ribosome inactiv. protein) (RNA) [138]	Yeast RNA polymerase II (RNA) [165]
Anti-acetylcholine receptor antibody (RNA) [139]	Nuclear factor κB (DNA) [166]
Tenascin-C (RNA) [140]	Neuropeptide calcitonin gene-related peptide 1 (RNA) [167]
Basic fibroblastic groth factor (RNA) [141]	Prion Protein PrP^{Sc} (RNA) [168, 169]
P-Selectin (RNA) [142]	α-Thrombin (RNA) [170]
Influenza A virus haemagglutinin (DNA) [143]	Angiopoietin-2 (RNA) [171]
HIV-1 integrase (DNA) [144]	IgE (DNA, RNA) [172]
HIV RT (RNA) [145]	Gonadotropin-releasing hormone (DNA) [173]
HIV-1 RT (RNA) [146]	HIV-1 Tat protein (RNA) [174]

by affinity chromatography, or other selection techniques suitable for the enrichment of a desired property. The complexity of a typical combinatorial oligonucleotide library obtained from 1-µmol scale solid-phase DNA synthesis is limited to 10^{14}–10^{15} individual sequences. After selection the enriched sequences can be amplified and mutated and can be used as a new pool for another selection step. After 8–15 cycles, aptamers have been obtained that show affinities and specificities often described as good as antibodies. Including cloning and sequencing, a typical SELEX experiment may take 2–3 months. Full-length aptamers are generally 70–80 nucleotides long but

can be truncated to eliminate nucleotide sequences which are not necessary for target binding. Thereby, functional aptamers with less than 40 nucleotides have been obtained.

On the therapeutic front, aptamers have made tremendous progress and are already in clinical trials only 8 years after establishing the technology. The application of aptamers for in vivo diagnostics has already been explored. In comparison to antibodies, aptamers have also some features that make them very attractive for protein chips.

In contrast to antibodies, aptamers can easily be produced by chemical synthesis. Thereby reporter molecules as biotin or functional group for further conjugation can be attached at precise locations for labeling or immobilization. Aptamers containing 5-halo-uracil analogs (e.g., 5-bromodeoxyuridine or 5-iodouracil) could be used to lock the bound proteins irreversibly. Therefore, UV light is briefly shone on the mixture. The photoaptamer–target pairs that have the bromodeoxyuracil adjacent to a tyrosine, or a similar electron-rich amino acid, at the time the UV light is turned on become covalently cross-linked to each other. It was claimed that the special requirements for photocrosslinking can increase the specificity by a factor of about 1000 [175].

3.2 Aptamer-Based Assays

Although aptamers were discovered 16 years ago, they were for a long time viewed as novelty molecules with only a few applications in analytics. Now, a great variety of optical analytical methods have been developed using mostly fluorophores as label molecules.

In analogy to antibody arrays, aptamer arrays have been developed for the parallel determination of proteins. Since it is often emphasized that the development of aptamers for a target could be faster and easier than the development of an antibody, aptamers are often seen as recognition elements of choice for the development of chips for proteomic applications. The company Archemix for instance already uses aptamers to produce small arrays with scanning fluorescence anisotropy readout. McCauley et al. [176] described an optical sensor array for the determination of thrombin, bFGF, VEGF, and IMPDH II and Bock et al. [177] a photoaptamer array for the parallel determination of 17 different proteins. Photoaptamers are a relatively new class of capture reagents that augment specificity by a photochemical cross-link after protein binding. Detection limits below 10 fM for several proteins were reached. The company Somalogic in Boulder, Colorado, uses photoaptamers containing photoreactive 5-bromodeoxyuridine (BrdU) for the production of protein microarrays.

One great advantage of using photoaptamers is that the cross-over to other spots can be avoided even under rigorous washing conditions.

In the area of homogeneous assays based on fluorescence intensity changes, an adenosine aptamer labeled with fluorescein or acridine has for instance been utilized to detect ATP [178]. Aptamers which couple target binding to fluorescent-signal generation were termed by the same authors "signaling aptamers" (Fig. 8).

In 2001, Hamaguchi et al. [180] described an aptamer beacon for the quantification of thrombin. The 5′ and 3′ ends of the aptamer carry a fluorophore and a quencher. Due to the thrombin binding, the 3D structure of the aptamer (and therefore the distance of the fluorophore and the quencher) changes. This was detected by a change in fluorescence intensity.

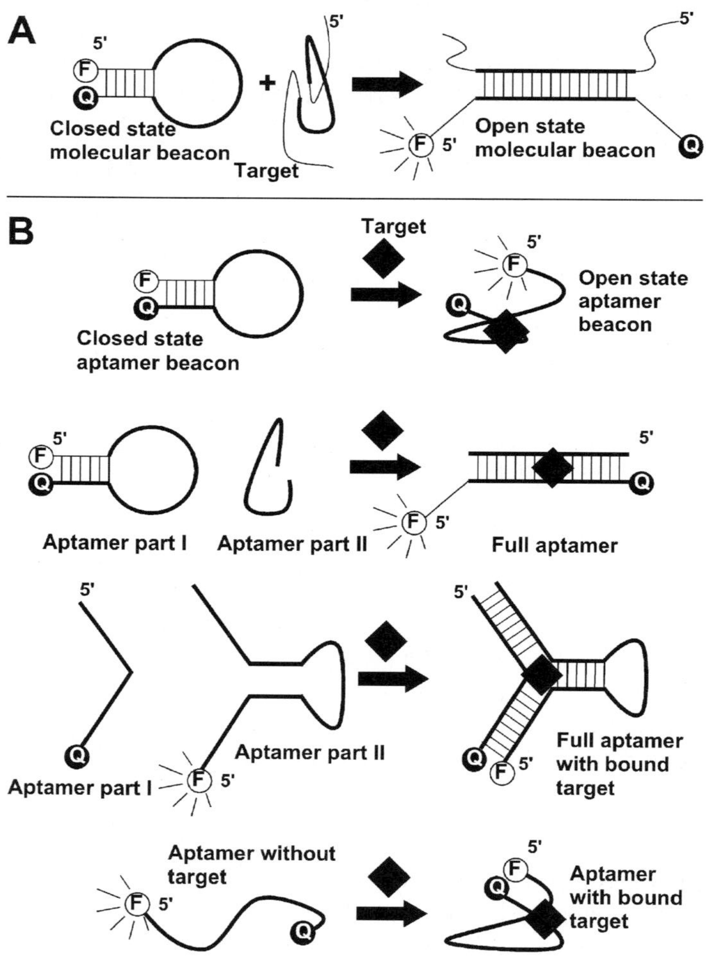

Fig. 8 "Signaling aptamers" based on the molecular beacon principle. **A** Principle of the signaling mechanism of a standard molecular beacon. **B** Examples of signaling mechanisms with aptamers. *F* Fluorophore; *Q* Fluorescence quencher (adapted from [179])

The group of Wilson has described a homogeneous DNA aptamer-based bioanalytical method for IgE with a detection limit of 350 pM by using fluorescence anisotropy [181].

A very low detection limit for a protein was reached by using the proximity ligation assay combined with real-time PCR indication. Therefore two different aptamers carrying additional DNA sequence extensions bind to separated areas of the protein. Due to the proximity of both extensions, an oligonucleotide sequence can hybridize to both ends, which are then subsequently ligated. The ligation product is then amplified and quantified by real-time PCR. Fredriksson et al. [182] were able to detect 40×10^{-21} mol cytokine platelet-derived growth factor (PDGF) without any washing and separation steps. Another technique, the so-called "proximity extension" technique, is also based on the close orientation of two different aptamers on the target protein (e.g., thrombin). One aptamer has a circular DNA architecture and the other one has an extension which, if the protein is present, binds to a defined sequence of the circular DNA. Finally, the rolling circle principle is applied for a highly sensitive detection [183] (Fig. 9).

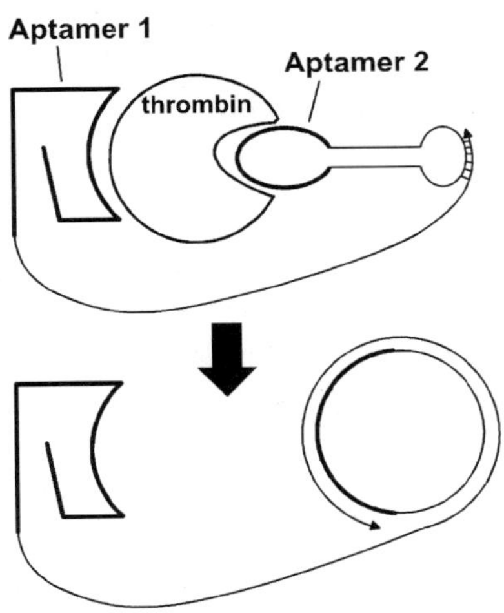

Fig. 9 Proximity extension principle: Simultaneous binding of two aptamers to thrombin primes DNA polymerase-mediated rolling circle amplification (adapted from [183])

3.3 Aptamer-Based Electrochemical Sensors

Aptamer-based sensors have already been described in the form of optical sensors [184–186] electronic tongue, and optical sensor arrays [176, 187, 188]. In addition to optical transducers, other types of transducers have also been used in combination with aptamers. Quartz crystal aptamer sensors that detect mass changes due to the analyte binding have been described for the detection of 0.5 nM human IgE [189], for the arginine-rich motif Rev peptide, which is involved in HIV-1 infection [190], for the HIV-1 Tat protein [191] and for thrombin and the anticoagulants heparin and antithrombin III [192].

One of the obvious limitations of immunosensors is their poor capacity for regeneration. Therefore one of the most attractive features of aptamers in biosensor applications is their ability to be regenerable. Nucleic acids can withstand repeated cycles of denaturation and renaturation. Several methods can be used for regeneration: heat, salt, pH and chelating agents. However, although aptamers seem to be potent antibody mimics for a general application in analytics, a systematic analysis of aptamer structures and behavior is needed to understand whether any overarching rules govern aptamer function in assays, as in the case with antibodies [175].

In the past, only a few aptamer sensors with electrochemical transducers were described, but now they are starting to appear in the scientific community.

Recently, a very sensitive displacement aptamer sensor for two proteins has been described; by using quantum dot semiconductor nanocrystals as label and electrochemical stripping detection for sensor readout [193]. Two different aptamers for the binding of thrombin and lysozyme were co-immobilised on a gold sensor surface. The protein sample was applied to the sensor that had been preincubated with CDs-labeled thrombin and PbS-labeled lysozyme. After displacement of the labeled proteins from the aptamers, the remaining captured quantum dots were electrochemically measured. Since CDs and PbS can be indicated at different potentials, lysozyme and thrombin can be detected simultaneously. A detection limit of 54.5 amol thrombin in 100 μl sample was described. Unfortunately, after the displacement step, some additional incubation steps are necessary before electrochemical indication can take place.

Ikebukuro et al. [194, 195] described electrochemical aptamer sensors for thrombin based on a sandwich principle. They used two different aptamers which recognize two separated areas on the protein molecule. One aptamer was coupled to the gold sensor surface and the other one was labeled with glucose dehydrogenase. For the determination of the sensor bound glucose dehydrogenase m-PMS and glucose were used as substrates and the enzymatically reduced PMS was electrochemically indicated at a potential of +100 mV (vs. Ag/AgCl). A linear measuring range of 40–100 nM and a detection limit of 10 nm were described [194].

The principles described in 2.3 for wash-free antibody based sensors can also be used for the development of wash-free aptamer sensors. In addition, very interesting electrochemical aptamer sensors have been developed that work in a similar way as the known beacon aptamers. Here, instead of the fluorophor/quencher couple, a redox-labeled compound, e.g., ferrocene or

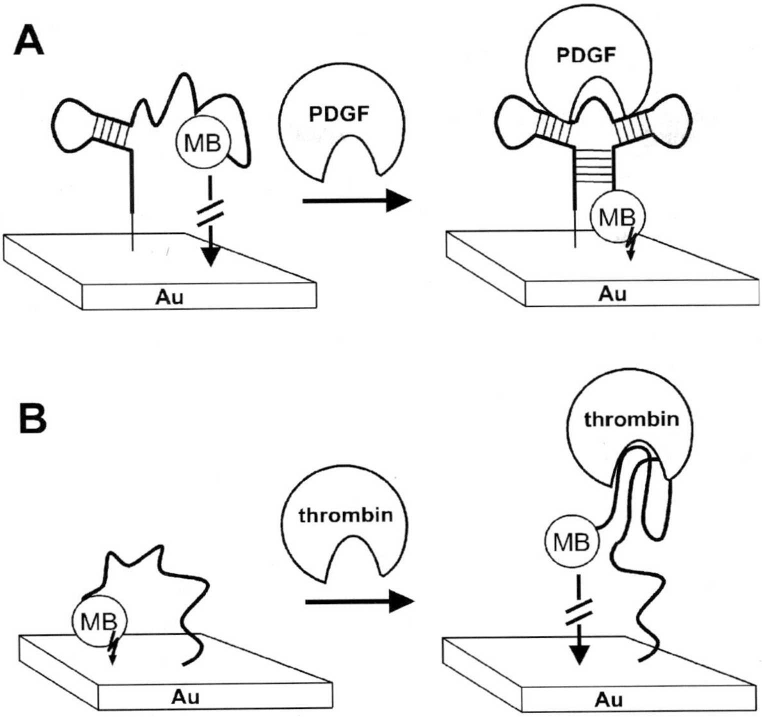

Fig. 10 Principle of electrochemical aptamer-based sensors for the reagentless and reusable determination of proteins. **A** Determination of platelet-derived growth factor (PDGF) (adapted from [198]). **B** Determination of thrombin (adapted from [199])

methylene blue is attached to one end of the aptamer nucleic acid sequence. The other end of the sequence is attached to the surface of a gold sensor via thiol chemistry. Due to the analyte binding, the aptamer changes its conformation and thereby brings the redox label in close proximity to the sensor surface. This can then be indicated, e.g., by differential pulse voltammetry. These sensors work without any washing steps and are reusable. Such sensors have been developed for thrombin [196, 197] with detection limits in the μM to nM range. Recently, Lai et al. [198] used this principle for the determination of platelet-derived growth factor (PDGF) with detection limits of 1 nM in undiluted blood and 50 pM in blood serum (Fig. 10a).

In contrast, Xiao et al. [199] have shown that the binding of the target protein (thrombin) to an aptamer sequences stabilizes the binding motif of the aptamer and increases thereby the distances between methylene blue label and the sensor surface. As a consequence, the electron transfer is stopped, which is indicated electrochemically (Fig. 10b).

4
Parallelization of Protein Analysis with Electrochemical Biochips

For application of electrochemical biochips in proteomics or protein analysis, principles are needed that allow a high degree of parallelization and miniaturization. High parallel analysis procedure on a chip surface is not possible by dispensing reagents to each spot—as it is realized for the microliter and nanoliter plates.

(i) The binding partners should possess very high specificity for the target molecule in order to prevent cross over to the compounds immobilized at the other spots. Unspecific interactions with the "carrier" material should be avoided. In order to prevent the cross talk between neighboring spots, the signal generation must be restricted to the immediate vicinity of each spot.

(ii) All electrodes or ISFETs of the chip should be readable individually.

Two different principles are feasible to avoid cross talk. Micromechanical generation of a "wall" between the spots effectively prevents the migration of products which are generated by an enzyme label. This principle was applied to create a low-density electrochemical DNA chip with 28 or 128 individual positions [200, 201]. Therefore polymeric ring structures of 10 μm height, 15 μm width and 10 μm distance were built up around each array position by photolithography.

The second way to avoid cross talk between the spots is to use assay techniques that indicate the binding partner only at the point where the immunochemical complexation takes place. With this principle, additional mechanical barriers are not necessary. Substrate channeling assays or formats using the direct electron transfer between the marker enzyme or redox label and the electrode or label-free immunosensor principles could fulfil these demands.

Due to a multiplicity of unsolved problems, highly parallel electrochemical chips for protein analysis are at the very beginning of development. On the other hand, different technologies of DNA chip production may be transferred to the development of electrochemical biochips for protein analysis.

Electrochemical biochips for DNA sequences allow for the integration of the signal generation using individually addressable microelectrodes or field effect transistors by highly parallel amperometric or potentiometric measurements. Recently, a fully processed 16×8 CMOS sensor array was described with interdigitated gold electrodes arranged within a circular polymeric ring structure described above [200, 201]. The interdigitated gold structures with 1-μm-wide fingers, gaps of 0.8 μm, and 17 000 μm^2 were generated by the lift-off technique from a deposited 120-nm-thick gold electrode. By using a micro spotter, single-stranded DNA molecules were immobilized on the gold sensor surfaces. The biotinylated target from the sample and an alkaline phosphatase-avidin conjugate was bound subsequently. After adding *p*-aminophenylphosphate, the released *p*-aminophenol was detected by electrode–electrode recycling. The amplification factor was eight. The cross-over of the enzymatically released *p*-aminophenol from one to the other sensor positions has been avoided due the polymeric ring structure around the electrode and the time-dependent measurement. This approach should

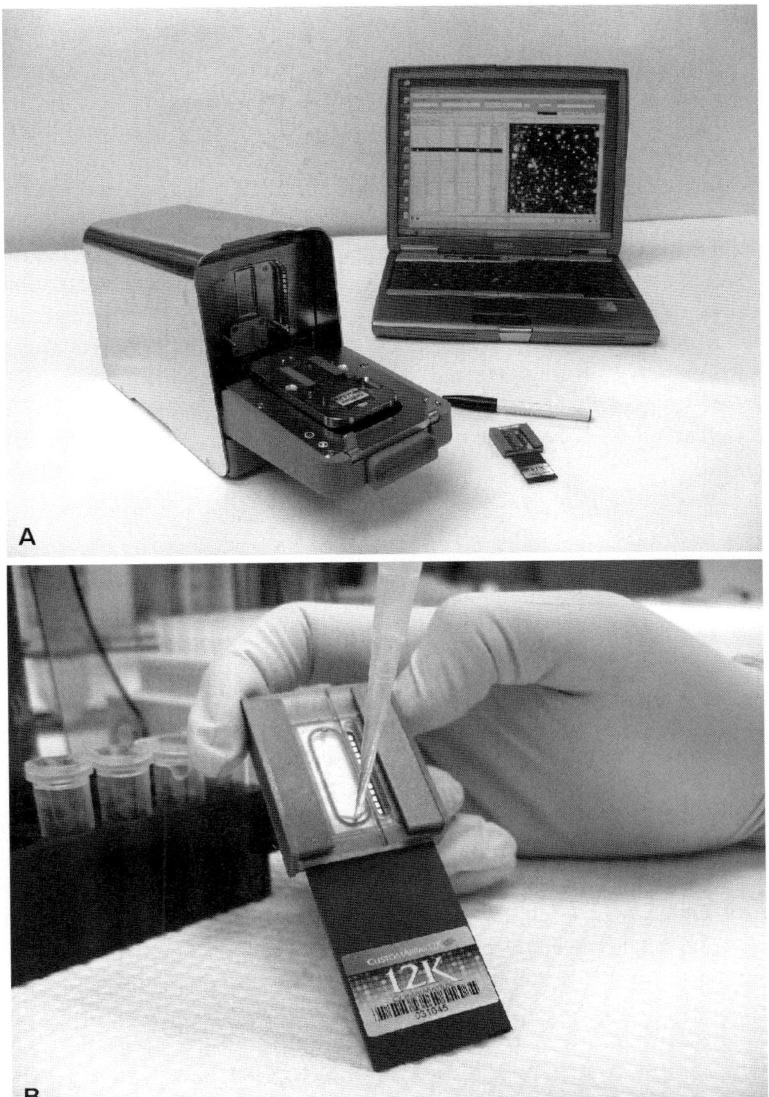

Fig. 11 ElectraSense Device (CombiMatrix Corp.) (**A**) for the electrochemical readout of a 12 k DNA-Chip (**B**)

principally be applicable also for a highly parallel determination of proteins. Therefore alkaline phosphatase-labeled antibodies could be used.

The problem of how to get the huge number of the required antibodies immobilised on a chip with spatial resolution could be solved by the principle of "refunctionalization" of a DNA array [202, 203]. Therefore, the array is incubated with proteins that are conjugated with the respective complementary oligonucleotide sequence. Due to the high specificity of hybridization, the protein-DNA conjugates will be guided and immobilized on the spots due to their DNA sequences.

A CMOS-fabricated silicon microchip was used as a platform with more than 1000 electrode spots [53]. For site-directed DNA synthesis, the authors used an electrochemical method. Protons were generated electrochemically on an addressed electrode spot, which caused a local pH change. Thereby, the chemical synthesis of the oligonucleotides (phosphoramidite chemistry) was controlled. The capture antibodies were labeled with oligonucleotides and hybridized with the complementary oligonucleotides on the electrode spots (refunctionalization approach). A sandwich-type multi-step principle with electrochemical indication and peroxidase as label was applied for the parallel determination of human alpha 1 acid protein, ricin, M13 phage, *Bacillus globigii* spores and fluorescein [54]. Very recently, this technology was further enhanced to create a electrochemical readout device called ElectraSense (CombiMatrix Corp.), which uses a chip with 12 544 features (12 kChip) and reaches a detection limit of 0.75 pM [204] (Fig. 11).

5 Conclusion and Outlook

Electrochemical biochips for protein analysis have been developed in a great variety, mostly by using redox or enzyme-labeled antibodies. The detection limits are usually within the nanomolar range, which is sufficient for many analytes. For much lower detection limits, amplification techniques, e.g., recycling have been applied, shifting the detection limits into the picomolar to femtomolar range.

As for conventional immunoassays like ELISA, biochips often need several incubation and washing steps. To make use of the full potential of chip and sensor technology, like fast, convenient, inexpensive and sensitive detection, the assay configurations needs to be simplified. First-wash and reagent-free biochips have already been developed for protein analysis.

The development of new antibody-like recognitions elements, a chip-integrated sample pretreatment, and a further miniaturization will pave the way of electrochemical biochips into the doctor's office and for point-of-care measurements, making in future protein determinations as simple as blood glucose measurements today.

Acknowledgements I would like to thank Mark Humble (CombiMatrix Corp.) for the photos in Fig. 5.

References

1. Jain KK (2005) Drug Discov Today 10(21):1435
2. Jain KK (2006) Drug Discov Today 11(13–14):573
3. Anonymous (2005) Personalized medicines: hopes and realities. The Royal Society, London, UK
4. Pennisi E (2003) Science 300(5625):1484
5. Kingsmore SF (2006) Nat Rev Drug Discov 5(4):310–320
6. Shevchenko A, Jensen ON, Podtelejnikov AV, Sagliocco F, Wilm M, Vorm O, Mortensen P, Shevchenko A, Boucherie H, Mann M (1996) PNAS 93(25):14440
7. Petricoin EF, Bichsel VE, Calvert VS, Espina V, Winters M, Young L, Belluco C, Trock BJ, Lippman M, Fishman DA, Sgroi DC, Munson PJ, Esserman LJ, Liotta LA (2005) J Clin Oncol 23(15):3614
8. Frank R, Hargreaves R (2003) Nat Rev Drug Discov 2(7):566
9. Buechler KF, Moi S, Noar B, McGrath D, Villela J, Clancy M, Shenhav A, Colleymore A, Valkirs G, Lee T, Bruni JF, Walsh M, Hoffman R, Ahmuty F, Nowakowski M, Buechler J, Mitchell M, Boyd D, Stiso N, Anderson R (1992) Clin Chem 38(9):1678
10. Barbas III CF, Burton DR, Scott JK (2004) Phage Display: A Laboratory Manual. Cold Spring Harbor Laboratory Press, Cold Spring Harbor, NY, USA
11. He MY, Taussig MJ (1997) Nucleic Acids Res 25(24):5132
12. Hanes J, Plueckthun A (1997) PNAS 94(10):4937
13. Kramer K, Hock B (2003) Anal Bioanal Chem 377(3):417
14. Hock B, Seifert M, Kramer K (2002) Biosens Bioelectron 17(3):239
15. Hudson PJ (1999) Curr Opin Immunol 11(5):548
16. Harris B (1999) Trends Biotechnol 17:290
17. Ingvarsson J, Lindstedt M Borrebaeck CAK, Wingren C (2006) J Proteome Res 5(1):170
18. Wang SH, Zhang JB, Zhang ZP, Zhou YF, Yang RF, Chen J, Guo YC, You F, Zhang XE (2006) Anal Chem 78(4):997
19. Lechuga LM (2006) Anal Bioanal Chem 384(1):44
20. Gauglitz G (2005) Anal Bioanal Chem 381(1):141
21. Cooper MA (2006) Drug Discov Today 11(23/34):1061
22. Blake C, Gould BJ (1984) Analyst 109:533
23. Harlow E, Lane D (1988) Antibodies: A Laboratory Manual. Cold Spring Harbor Laboratory Press, Cold Spring Harbor, NY, USA
24. Ekins R (1989) Nature 340(6231):256
25. Seidel M, Dankbar DM, Gauglitz G (2004) Anal Bioanal Chem 379(7–8):904
26. Ghatnekar-Nilsson S, Dexlin L, Wingren C, Montelius L, Borrebaeck CAK (2007) Proteomics 7(4):540
27. Ekins RP, Chu FW (1991) Clin Chem 37(11):1955
28. Weller MG, Schuetz AJ, Winklmair M, Niessner R (1999) Anal Chim Acta 393:29
29. Fall BI, Eberlein-Konig B, Behrendt H, Niessner R, Ring J, Weller MG (2003) Anal Chem 75(3):556
30. Schweitzer B, Wiltshire S, Lambert J, O'Malley S, Kukanskis K, Zhu ZR, Kingsmore SF, Lizardi PM, Ward DC (2000) PNAS 97(18):10113
31. Diaz-Gonzalez M, Gonzalez-Garcia MB, Costa-Garcia A (2005) Electroanalysis 17(21):1901

32. Pänke O, Balkenhohl T, Kafka J, Schäfer D, Lisdat F (2008) Impedance Spectroscopy and Biosensing (in this volume)
33. Tang TC, Deng AP, Huang HJ (2002) Anal Chem 74(11):2617
34. Bagel O, Degrand C, Limoges B, Joannes M, Azek F, Brossier P (2000) Electroanalysis 12(18):1447
35. Marko-Varga G, Emneus J, Gorton L, Ruzgas T (1995) Trends Anal Chem 14:319
36. Benkert A, Scheller FW, Schoessler W, Micheel B, Warsinke A (2000) Electroanalysis 12:1318
37. Wang J, Ibanez A, Chatrathi MP (2002) Electrophoresis 23:3744
38. Tanaka T, Tsukube S, Izawa K, Okochi M, Lim TK, Watanabe S, Harada M, Matsunaga T (2007) Biosens Bioelectron 22(9–10):2051
39. Bordes AL, Schöllhorn B, Limoges B, Degrand C (1999) Talanta 48:201
40. Bordes AL, Limoges B, Brossier P, Degrand C (1997) Anal Chim Acta 356:195
41. Mak WC, Cheung KY, Trau D, Warsinke A, Scheller F, Renneberg R (2005) Anal Chem 77(9):2835
42. Niwa O, Morita M, Tabei H (1991) Electroanalysis 3:163
43. Wollenberger U, Paeschke M, Hintsche R (1994) Analyst 119:1245
44. Hintsche R, Paeschke M, Uhlig A, Seitz R (1997) Microbiosensors using electrodes made in Si-technology. In: Scheller FW, Schubert F, Fedrowitz J (eds) Frontiers in Biosensorics I. Birkhäuser, Basel, pp 267–283
45. Athey D, McNeil CJ, Bailey WR, Hager HJ, Mullen WH, Russell LJ (1993) Biosens Bioelectron 8:415
46. Aizawa M, Morioka A, Suzuki S (1980) Anal Chim Acta 115:61
47. Abdel-Hamid I, Ghindilis AL, Atanasov P, Wilkins E (1999) Anal Lett 32:1081
48. Rishpon J, Ivnitski D (1997) Biosens Bioelectron 12:195
49. Ivnitski D, Wolf T, Solomon B, Fleminger G, Rishpon J (1998) Bioelectrochem Bioenerg 45:27
50. Romero FJM, Stiene M, Kast R, Luque de Castro MD, Bilitewski U (1998) Biosens Bioelectron 13:1107
51. Santandreu M, Alegret A, Fabregas E (1999) Anal Chim Acta 396:181
52. Wendzinski F, Gründig B, Renneberg R, Spener F (1997) Biosens Bioelectron 12:43
53. Dill K, Montgomery DD, Ghindilis AL, Schwarzkopf KR (2004) J Biochem Biophys Meth 59(2):181
54. Dill K, Montgomery DD, Ghindilis AL, Schwarzkopf KR, Ragsdale SR, Oleinikov AV (2004) Biosens Bioelectron 20(4):736
55. Volpe G, Compagnone D, Draisci R, Palleschi G (1998) Analyst 123(6):1303
56. Padeste C, Grubelnik A, Tiefenauer L (1998) Anal Chim Acta 374:167
57. Renneberg R, Schoessler W, Scheller F (1983) Anal Lett 16:1279
58. Wehmeyer KR, Halsall HB, Heineman WR, Volle CD, Chen IW (1986) Anal Chem 58:135
59. Jenkins SH, Heineman WR, Halsall HB (1988) Anal Biochem 168:292
60. Ducey MW, Smith AM, Guo X, Meyerhoff ME (1997) Anal Chim Acta 357:5
61. Diaz-Gonzalez M, Gonzalez-Garcia MB, Costa-Garcia A (2005) Biosens Bioelectron 20:2035
62. Volpe G, Fares G, Quadri FD, Draisci R, Ferretti G, Marchiafava C, Moscone D, Palleschi G (2006) Anal Chim Acta 572(1):11
63. Robinson GA, Cole VM, Forrest GC (1987/88) Biosensors 3:147
64. Bier FF, Ehrentreich-Förster E, Bauer CH, Scheller FW (1996) Fresensius J Anal Chem 354:861
65. Masson M, Liu Z, Haruyama T, Kobatake E, Ikariyama Y, Aizawa M (1995) Anal Chim Acta 304(3):353

66. Campanella L, Attioli R, Colapicchioni C, Tomassetti M (1999) Sens Actuat B 55:23
67. Sergeyeva TA, Soldatkin AP, Rachkov AE, Tereschenko MI, Piletsky SA, El'skaya AV (1999) Anal Chim Acta 390:73
68. Mirhabibollahi B, Brooks JL, Kroll RG (1990) J Appl Bacteriol 68:577
69. Ikariyama Y, Aizawa M (1988) Bioaffinity sensors. In: Methods in Enzymology, vol 137. Academic Press, New York, pp 111–124
70. Kulys J, Razumas V, Malinauskas A (1980) Bioelectrochem Bioenerg 7:11
71. Kaneki N, Xu Y, Kumari A, Halsall HB, Heineman WR, Kissinger PT (1994) Anal Chim Acta 287:253
72. Niwa O, Xy Y, Halsall BH, Heineman WR (1993) Anal Chem 65:1559
73. Buckley E, Smyth MR, Heineman W, Halsall HB (1989) Anal Proc 26:5
74. Tang HT, Lunte CF, Halsall HB, Heineman WR (1988) Anal Chim Acta 214:187
75. Meusel M, Renneberg R, Spener F, Schmitz G (1995) Biosens Bioelectron 10:577
76. Mirhabibollahi B, Brooks JL, Kroll RG (1990) Lett Appl Microbiol 11:119
77. Frew JE, Foulds NC, Wilshere JM, Forrow NJ, Green MJ (1989) Electroanal Chem 266:309
78. Athey D, McNeil C (1994) J Immunol Meth 176:153
79. Makower A, Bauer CG, Ghindilis A, Scheller FW (1994) Biosensors 94, Proceedings of the 3rd World Congress on Biosensors, edited by T. Turner. Oxford: Elsevier Advanced Technology
80. Scheller FW, Bauer CG, Makower A, Wollenberger U, Warsinke A, Bier FF (2002) Immunoassays using enzymatic amplification electrodes. pp. 207- 238. In: Biomolecular Sensors, edited by E Gizeli and CR Lowe. Taylor and Francis,
81. Bauer CG, Eremenko AV, Ehrentreich-Förster E, Bier FF, Makower A, Halsall HB, Heineman WR, Scheller FW (1996) Anal Chem 68:2453
82. Bier FF, Förster E, Makower A, Scheller FW (1996) Anal Chim Acta 328:27
83. Ghindilis AL, Makower A, Bauer CG, Bier FF, Scheller FW (1995) Anal Chim Acta 304:25
84. Johansson A, Ellis DH, Bates DL, Plumb AM, Stanley CJ (1986) J Immunol Meth 87(1):7
85. Self CH (1981) EU Patent Appl 80303478.4
86. Stanley CJ, Cox RB, Cardosi MF, Turner APF (1988) J Immunol Meth 112(2):153
87. Bergel A, Souppe J, Comtant M (1989) Anal Biochem 179:382
88. Tang X, Johansson G (1995) Anal Lett 28:2595
89. Olson JD, Panfili PR, Armenta R, Fremmel MB, Merrick H, Gumperz J, Goltz M, Zuk RF (1990) J Immunol Meth 134:71
90. Janata J (1975) Immunoelectrode. J Am Chem Soc 97:2914
91. Keating MY, Rechnitz GA (1984) Anal Chem 56:801
92. Engel L, Baumann W (1994) Fresenius J Anal Chem 349:447
93. Pfeifer U, Baumann W (1992) Fresenius J Anal Chem 343:541
94. Gotoh M, Tamiya E, Momoi M, Kagawa Y, Karube I (1987) Anal Lett 20:857
95. Bergveld P (1991) Biosens Bioelectron 6:55
96. Sadik OA, van Emon JM (1997) Chemtech 27(6):38
97. Grant S, Davis F, Pritchard JA, Law KA, Higson SPJ, Gibson TD (2003) Anal Chim Acta 495:21
98. Susmel S, Guilbault GG, O'Sullivan CK (2003) Biosens Bioelectron 18:881
99. Blonder R, Katz E, Cohen Y, Itzhak N, Riklin A, Willner I (1996) Anal Chem 68:3151
100. Akram M, Stuart MC, Wong DKY (2006) Electroanalysis 18(3):237
101. Halamek J, Wollenberger U, Stöcklein WFM, Warsinke A, Scheller FW (2007) Anal Lett 40:1434

102. Dai Z, Yan F, Chen J, Ju H (2003) Anal Chem 75:5429
103. Key G, Schreiber A, Feldbrügge R, McNeil CJ, Joergensen P, Pelsers MMAL, Glatz JFC, Spener F (1999) Clin Biochem 32:229
104. Mattiasson B, Nilsson H (1977) FEBS Lett 78:251
105. Stöllner D, Stocklein W, Scheller F, Warsinke A (2002) Anal Chim Acta 470(2):111
106. Tarasewich MR (1985) Chapter 4: Bioelectrocatalysis, p. 275 In: Comprehensive Treatise of Electrochemistry. 10th Bioelectrochemistry, Plenum Press, NY
107. Berezin IV, Bogdanovskaya VA, Tarasevich MR, Varfolomeev SD, Yaropolov AI (1978) Doklady Akademii Nauk SSSR 240:615
108. Ghindilis AL, Skorobogat'ko OV, Yaropolov AI (1991) Biomed Sci 2(5):520
109. Ghindilis AL, Skorobogat'ko OV, Gavrilova VP, Yaropolov AI (1992) Biosens Bioelectron 7:301
110. Ghindilis AL, Atanasov P, Wilkins E (1996) Sens Actuat B 34(1–3):528
111. Milligan C, Ghindilis A (2002) Electroanalysis 14:415
112. McNeil CJ, Athey D, Ho WO (1995) Biosens Bioelectron 10:75
113. Lu B, Iwuoha EI, Smyth MR, O'Kennedy R (1997) Anal Chim Acta 345:59
114. Lu B, Iwuoha EI, Smyth MR, O'Kennedy R (1997) Anal Comm 34:21
115. Keay RW, McNeil CJ (1998) Biosens Bioelectron 13:963
116. Wingren C, Ingvarsson J, Lindstedt M, Borrebaeck CAK (2003) Nat Biotechnol 21:223
117. Dove A (1999) Nat Biotechnol 17:233
118. O'Sullivan CK (2002) Anal Bioanal Chem 372(1):44
119. Tuerk C, Gold L (1990) Science 249:505
120. Ellington AD, Szostak J (1990) Nature 346(6287):818
121. Bellecave P, Andreola ML, Ventura M, Tarrago-Litvak L, Litvak S, Astier-Gin T (2003) Oligonucleotides 13:455
122. Biesecker G, Dihel L, Enney K, Bendele RA (1999) Immunopharmacology 42(1–3):219
123. Bless NM, Smith D, Charlton J, Czermak BJ, Schmal H, Friedl HP, Ward PA (1997) Curr Biol 7:877
124. Bock LC, Griffin LC, Latham JA, Vermaas EH, Toole JJ (1992) Nature 355(6360):564
125. Browning CM, Cagnon L, Good PD, Rossi J, Engelke DR, Markovitz DM (1999) J Virol 73(6):5191
126. Chinnapen DJ, Sen D (2002) Biochemistry 41:5202
127. Cho JS, Lee YJ, Shin KS, Jeong S, Park J, Lee SW (2004) Mol Cells 18:17
128. Convery MA, Rowsell S, Stonehouse NJ, Ellington AD, Hirao I, Murray JB, Peabody DS, Phillips SE, Stockley PG (1998) Nat Struct Biol 5:133
129. Ehsan A, Mann MJ, Dell'Acqua G, Dzau VJ (2001) J Thorac Cardiovasc Surg 121(4):714
130. Faulhammer D, Eschgfaller B, Stark S, Burgstaller P, Englberger W, Erfurth J, Kleinjung F, Rupp J, Vulcu SD, Schroder W, Vonhoff S, Nawrath H, Gillen C, Klussmann S (2004) RNA 10(3):516
131. Fukuda K, Vishnuvardhan D, Sekiya S, Hwang J, Kakiuchi N, Taira K, Shimotohno K, Kumar PK, Nishikawa S (2000) Eur J Biochem 267:3685
132. Gal SW, Amontov S, Urvil PT, Vishnuvardhan D, Nishikawa F, Kumar PK, Nishikawa S (1998) Eur J Biochem 252:553
133. Green LS, Jellinek D, Jenison R, Ostman A, Heldin CH, Janjic N (1996) Biochemistry 35:14413
134. Helmling S, Maasch C, Eulberg D, Buchner K, Schroder W, Lange C, Vonhoff S, Wlotzka B, Tschop MH, Rosewicz S, Klussmann S (2004) PNAS 101(36):13174
135. Hesselberth JR, Miller D, Robertus J, Ellington AD (2000) J Biol Chem 275:4937

136. Hicke BJ, Marion C, Chang YF, Gould T, Lynott CK, Parma D, Schmidt PG, Warren S (2001) J Biol Chem 276:48644
137. Hirao I, Harada Y, Nojima T, Osawa Y, Masaki H, Yokoyama S (2004) Biochemistry 43:3214
138. Hirao I, Madin K, Endo Y, Yokoyama S, Ellington AD (2000) J Biol Chem 275:4943
139. Hwang B, Lee SW (2002) Biochem Biophys Res Commun 290(2):656
140. Ishizaki J, Nevins JR, Sullenger BA (1996) Nat Med 2(12):1386
141. Jellinek D, Green LS, Bell C, Lynott CK, Gill N, Vargeese C, Kirschenheuter G, McGee DP, Abesinghe P, Pieken WA, Shapiro R, Rifin DB, Moscatelli D, Janjic N (1995) Biochemistry 34(36):11363
142. Jenison RD, Jennings SD, Walker DW, Bargatze RF, Parma D (1998) Antisense Nucleic Acid Drug Dev 8(4):265
143. Jeon SH, Kayhan B, Ben-Yedidia T, Arnon R (2004) J Biol Chem 279:48410
144. Jing N, Rando RF, Pommier Y, Hogan ME (1997) Biochemistry 36:12498
145. Joshi P, Prasad VR (2002) J Virol 76(13):6545
146. Kensch O, Connolly BA, Steinhoff HJ, McGregor A, Goody RS, Restle T (2000) Biol Chem 275:18271
147. Kim ES, Serur A, Huang JZ, Manley CA, McCrudden KW, Frischer JS, Soffer SZ, Ring L, New T, Zabski S, Rudge JS, Holash J, Yancopoulos GD, Kandel JJ, Yamashiro DJ (2002) PNAS 99(17):11399
148. Klug SJ, Huttenhofer A, Famulok M (1999) RNA 5:1180
149. Kohn DB, Bauer G, Rice CR, Rothschild JR, Carbonaro DA, Valdez P, Hao QL, Zhou C, Bahner I, Kearns K, Brody K, Fox S, Haden E, Wilson K, Salata C, Dolan C, Wetter C, Aguilar-Cordova E, Church J (1999) Blood 94(1):368
150. Kubik MF, Bell C, Fitzwater T, Watson SR, Tasset DM (1997) J Immunol 159(1):259
151. Kumar PK, Machida K, Urvil PT, Kakiuchi N, Vishnuvardhan D, Shimotohno K, Taira K, Nishikawa S (1997) Virology 237(2):270
152. Lee SW, Sullenger BA (1996) J Exp Med 184(2):315
153. Lupold SE, Hicke BJ, Lin Y, Coffey DS (2002) Cancer Res 62:4029
154. O'Connell D, Koenig A, Jennings S, Hicke B, Han HL, Fitzwater T, Chang YF, Varki N, Parma D, Varki A (1996) PNAS 93(12):5883
155. Pileur F, Andreola ML, Dausse E, Michel J, Moreau S, Yamada H, Gaidamakov SA, Crouch RJ, Toulme JJ, Cazenave C (2003) Nucleic Acids Res 31:5776
156. Rhodes A, Deakin A, Spaull J, Coomber B, Aitken A, Life P, Rees S (2000) J Biol Chem 275:28555
157. Rusconi CP, Yeh A, Lyerly HK, Lawson JH, Sullenger BA (2000) Thromb Haemost 84(5):841
158. Rusconi CP, Scardino E, Layzer J, Pitoc GA, Ortel TL, Monroe D, Sullenger BA (2002) Nature 419(6902):90
159. Santulli-Marotto S, Nair SK, Rusconi C, Sullenger B, Gilboa E (2003) Cancer Res 63(21):7483
160. Sayer N, Ibrahim J, Turner K, Tahiri-Alaoui A, James W (2002) Biochem Biophys Res Commun 293:924
161. Srisawat C, Engelke DR (2001) RNA 7:632
162. Sullenger BA, Gallardo HF, Ungers GE, Gilboa E (1990) Cell 63(3):601
163. Takeno H, Yamamoto S, Tanaka T, Sakano Y, Kikuchi Y (1999) J Biochem (Tokyo) 125:1115
164. Tasset DM, Kubik MF, Steiner W (1997) J Mol Biol 272:688
165. Thomas M, Chedin S, Carles C, Riva M, Famulok M, Sentenac A (1997) J Biol Chem 272:27980

166. Ueno T, Sawa Y, Kitagawa-Sakakida S, Nishimura M, Morishita R, Kaneda Y, Kohmura E, Yoshimine T, Matsuda H (2001) J Thorac Cardiovasc Surg 122(4):720
167. Vater A, Jarosch F, Buchner K, Klussmann S (2003) Nucleic Acids Res 31(21):e130
168. Proske D, Gilch S, Wopfner F, Schätzl HM, Winnacker EL, Famulok M (2002) Chembiochem 3(8):717
169. Weiss S, Proske D, Neumann M, Groschup MH, Kretzschmar HA, Famulok M, Winnacker EL (1997) J Virol 71:8790
170. White R, Rusconi C, Scardino E, Wolberg A, Lawson J, Hoffman M, Sullenger B (2001) Mol Ther 4(6):567
171. White RR, Shan S, Rusconi CP, Shetty G, Dewhirst MW, Kontos CD, Sullenger BA (2003) PNAS 100(9):5028
172. Wiegand TW, Williams PB, Dreskin SC, Jouvin MH, Kinet JP, Tasset D (1996) J Immunol 157:221
173. Wlotzka B, Leva S, Eschgfaller B, Burmeister J, Kleinjung F, Kaduk C, Muhn P, Hess-Stump H, Klussmann S (2002) PNAS 99(13):8898
174. Yamamoto R, Baba T, Kumar PK (2000) Genes Cells 5:389
175. Mukhopadhyay R (2005) Anal Chem 77:115A
176. McCauley TG, Hamaguchi N, Stanton M (2003) Anal Biochem 319(2):244
177. Bock C, Coleman M, Collins B, Davis J, Foulds G, Gold L, Greef C, Heil J, Heilig JS, Hicke B, Hurst MN, Husar GM, Miller D, Ostroff R, Petach H, Schneider D, Vant-Hull B, Waugh S, Weiss A, Wilcox SK, Zichi D (2004) Proteomics 4(3):609
178. Jhaveri SD, Kirby R, Conrad R, Maglott E, Bowser M, Kennedy RT, Glick G, Ellington AD (2000) J Am Chem Soc 122:2469
179. Nutiu R, Li YF (2004) Chem Eur J 10(8):1868
180. Hamaguchi N, Ellington A, Stanton M (2001) Anal Biochem 294(2):126
181. Gokulrangan G, Unruh JR, Holub DF, Ingram B, Johnson CK, Wilson GS (2005) Anal Chem 77(7):1963
182. Fredriksson S, Gullberg M, Jarvius J, Olsson C, Pietras K, Gustafsdottir SM, Ostman A, Landegren U (2002) Nat Biotechnol 20(5):473
183. Di Giusto DA, Wlassoff WA, Gooding JJ, Messerle BA, King GC (2005) Nucleic Acids Res 33(6):e64
184. Lee M, Walt DR (2000) Anal Biochem 282:142
185. Kleinjung F, Klussmann S, Erdmann VA, Scheller FW, Fuerste JP, Bier FF (1998) Anal Chem 70:328
186. Potyrailo RA, Conrad RC, Ellington AD, Hieftje GM (1998) Anal Chem 70:3419
187. Kirby R, Cho EJ, Gehrke B, Bayer T, Park YS, Neikirk DP, McDevitt JT, Ellington AD (2004) Anal Chem 76:4066
188. Savran CA, Knudsen SM, Ellington AD, Manalis SR (2004) Anal Chem 76(11):3194
189. Liss M, Petersen B, Wolf H, Prohaska E (2002) Anal Chem 74:4488
190. Fukusho S, Furusawa H, Okahata Y (2002) Chem Commun 1:88
191. Minunni M, Tombelli S, Gullotto A, Luzi E, Mascini M (2004) Biosens Bioelectron 20:1149
192. Gronewold TMA, Glass S, Quandt E, Famulok M (2005) Biosens Bioelectron 20:2044
193. Hansen JA, Wang J, Kawde AN, Xiang Y, Gothelf KV, Collins G (2006) J Am Chem Soc 128(7):2228
194. Ikebukuro K, Kiyohara C, Sode K (2005) Biosens Bioelectron 20(10):2168
195. Ikebukuro K, Kiyohara C, Sode K (2004) Anal Lett 37(14):2901
196. Radi AE, Sanchez JLA, Baldrich E, O'Sullivan CK (2006) J Am Chem Soc 128(1):117
197. Xiao Y, Piorek BD, Plaxco KW, Heeger AJ (2005) J Am Chem Soc 127(51):17990
198. Lai RY, Plaxco KW, Heeger AJ (2007) Anal Chem 79(1):229

199. Xiao Y, Lubin AA, Heeger AJ, Plaxco KW (2005) Angew Chem Int Ed 44(34):5456
200. Nebling E, Grunwald T, Albers J, Schäfer P, Hintsche R (2004) Anal Chem 76:689
201. Frey A, Hofmann F, Peters R, Holzapfl B, Schienle M, Paulus C, Schindler-Bauer P, Kuhlmeier D, Krause J, Eckstein G, Thewes R (2002) Microelectron Reliabil 42:1801
202. Bier FF, Kleinjung F, Ehrentreich-Förster E, Scheller FW (1999) Biotechniques 27:752
203. Niemeyer CM, Sano T, Smith CL, Cantor CR (1994) Nucleic Acids Res 22:5530
204. Ghindilis AL, Smith MW, Schwarzkopf KR, Roth KM, Peyvan K, Munro SB, Lodes MJ, Stöver AG, Bernards K, Dill K, McShea A (2006) CombiMatrix Oligonucleotide Arrays: Genotyping and Gene Expression Assays Employing Electrochemical Detection, presented at The Ninth World Congress on Biosensors, May 10–12, Toronto

Adv Biochem Engin/Biotechnol (2008) 109: 195–237
DOI 10.1007/10_2007_081

Published online: 9 November 2007

Impedance Spectroscopy and Biosensing

O. Pänke · T. Balkenhohl · J. Kafka · D. Schäfer · F. Lisdat (✉)

Biosystems Technology, Wildau University of Applied Sciences,
Bahnhofstrasse 1, 15745 Wildau, Germany
flisdat@igw.tfh-wildau.de

Abstract This chapter introduces the basic terms of impedance and the technique of impedance measurements. Furthermore, an overview of the application of this transduction method for analytical purposes will be given. Examples for combination with enzymes, antibodies, DNA but also for the analysis of living cells will be described. Special attention is devoted to the different electrode design and amplification schemes developed for sensitivity enhancement. Finally, the last two sections will show examples from the label-free determination of DNA and the sensorial detection of autoantibodies involved in celiac disease.

Keywords Impedance · Sensor · DNA · Antibodies · Cells · Enzymes

Abbreviations

AEC	3-Amino-9-ethylcarbazole
ac	Alternating current
C_{dl}	Double layer capacitance
CPE	Constant phase element
dc	Direct current
EIS	Electrochemical impedance spectroscopy

EIS structure	Electrolyte-insulator-semiconductor structure
ELISA	Enzyme linked immunosorbent assay
f	Frequency
ϕ	Phase shift (between voltage and current time functions)
FET	Field effect transistor
FRA	Frequency response analyser
HRP	Horseradish peroxidase
I	Current
IDEs	Interdigitated electrodes
IgG	Immunoglobulin of type G
IgA	Immunoglobulin of type A
ISFET	Ion-sensitive field effect transistor
ITO	Indium tin oxide
LB	Langmuir–Blodgett
MCB	Mercaptobutanol
MIP	Molecular imprinted polymers
OCP	Open circuit potential
POD	Peroxidase
PNA	Peptide nucleic acid
PSS	Polystyrene sulfonic acid
R_{ct}	Charge transfer resistance
R_s	Solution resistance
SAM	Self-assembling monolayers
SNP	Single nucleotide polymorphism
τ	Time constant for a system which can be described by a parallel circuit of R and C ($\tau = RC$)
t	Time
TTG	Tissue transglutaminase
V	Voltage
V_G	Gate voltage (EIS structure)
ω	Angular frequency
W	Warburg impedance
Y	Admittance
Z	Impedance
$\|Z\|$	Modulus (or absolute value) of impedance
Z_R	Real part of impedance
Z_I	Imaginary part of impedance

Impedance spectroscopy is a powerful method to analyse the complex electrical resistance of a system and is sensitive to surface phenomena and changes of bulk properties. Thus, in the field of biosensors, it is particularly suited to the detection of binding events. In addition, it is a valuable tool in characterising surface modifications, e.g. on transducer devices during the immobilisation procedure of recognition elements. In this chapter, a short introduction into the measuring principles will be given followed by an overview on the usage of the technique in the area of biosensors. Finally two examples from the field of immunosensing and DNA detection will be presented and discussed.

1
Introduction to Impedance Spectroscopy

The impedance Z of a system is determined by applying a voltage perturbation of small amplitude and detecting the current response. From the definition impedance is the quotient of the voltage-time function $V(t)$ and the resulting current-time function $I(t)$:

$$Z = \frac{V(t)}{I(t)} = \frac{V_0 \sin(2\Pi ft)}{I_0 \sin(2\Pi ft + \phi)} = \frac{1}{Y} \tag{1}$$

In analogy to the conductance definition in a dc circuit one can define the admittance Y as the reciprocal impedance or complex conductance. The impedance is a complex value since the current can not only differ in the amplitude (as in the case for a pure ohmic resistance) but can also show a phase shift ϕ compared to the voltage-time function (as in the case of capacitive or inductive resistances). Thus, the value can be described either by the modulus $|Z|$ and the phase shift ϕ or alternatively by the real part Z_R and the imaginary part Z_I of the impedance. This is illustrated in Fig. 1. Therefore the result of an impedance measurement can be illustrated in two different ways, the Bode plot using $|Z|$ and ϕ as a function of $\log f$ or the Nyquist plot using Z_R and Z_I.

The name impedance "spectroscopy" is derived from the fact that usually not only a single frequency is measured but impedance is determined at different frequencies. Thus, an impedance spectrum is obtained allowing the characterisation of surfaces, layers or membranes as well as exchange and diffusion processes. For this purpose the impedance spectrum is often analysed by means of an equivalent circuit. The circuit, commonly consisting of resistances and capacitances, represents the different physico-chemical properties of the system under investigation [1–3]. Alternatively the description of the system can be performed on the basis of transfer functions, which are derived from the basic laws of the processes involved such as electrokinetics, diffusion, partition, etc.

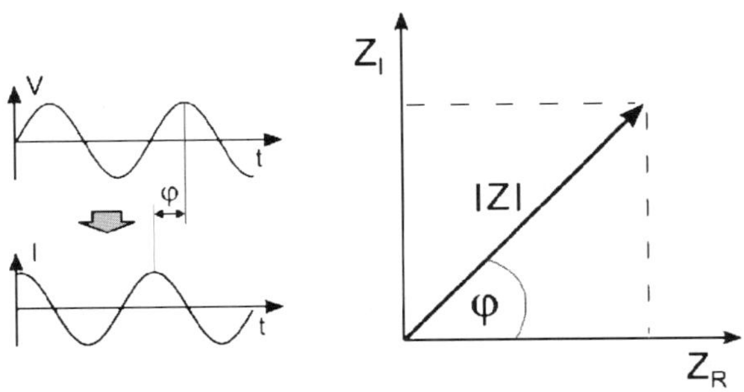

Fig. 1 Impedance is a complex value and defined as quotient of the voltage(time) and current(time) function. It can be expressed as modulus $|Z|$ and phase angle ϕ or can be given by the real part (Z_R) and the imaginary part (Z_I) of impedance

However, it is not only possible to describe a system, but the measurement can be also used for analytical purposes. In this case the change of one impedance element, a resistance or a capacitance, depending on the solution composition is evaluated. In some cases it is also suitable to correlate the overall impedance to a concentration change. This can result in a more simple measurement since here it is often sufficient to determine the impedance at one selected frequency or within a limited frequency window (where the relative changes are largest).

1.1 Impedance Elements and Simple Circuits

In electrochemical impedance spectroscopy, where the electrolyte solution is one component of the system to be investigated, usually four elements are used for the description of the impedance behaviour: ohmic resistance, capacitance, constant phase element and Warburg impedance. These elements and their definitions are summarised in Table 1.

An ohmic resistance R appears in the impedance spectrum when charge carriers are transported within the bulk of a material, through thin layers or across interfaces. An ohmic resistance causes no phase shift between the ac

Table 1 Impedance definition, frequency dependence and phase shift of different impedance elements most often used for the description of (bio)electrochemical systems

Impedance element	Definition	Phase angle	Frequency dependence
R	$Z = R$	0°	No
C	$Z_C = \frac{1}{j \cdot \omega \cdot C}$	90°	Yes
CPE	$Z_{CPE} = \frac{1}{A(j \cdot \omega)^\alpha}$	0–90°	Yes
W (infinite)	$Z_W = \frac{\sigma}{\sqrt{\omega}}(1 - j)$	45°	Yes
	$\sigma = \frac{R \cdot T}{n^2 \cdot F^2 \cdot \sqrt{2}} \left(\frac{1}{\sqrt{D_O} \cdot c_O} + \frac{1}{\sqrt{D_R} \cdot c_R} \right)$		
W (finite)	$Z_W = R_0 \frac{\tanh l \cdot \sqrt{j \cdot \omega / D}}{l \cdot \sqrt{j \cdot \omega / D}}$	0–45°	Yes

ω angular frequency
l length of diffusion region
D diffusion coefficient
R_0 diffusion resistance for $\omega = 0$
c_O,c_R concentration of oxidized and reduced species

voltage and the ac current and the impedance value is independent of the frequency used for the measurement. This means the impedance is constant over the whole frequency range.

A capacitance C appears when charge carriers are separated at interfaces (e.g. electrochemical double layer) or at membranes or films. In a simplified model this can be interpreted as a plate capacitor where the two "plates" are separated by a non-conducting medium. A capacitance results in a phase shift of $-90°$ of the current-time function compared to the applied voltage function and the impedance is strongly dependent on the frequency of measurement (as also seen from the impedance definition in Table 1). Particularly at low frequencies the capacitive impedance becomes very high.

Besides these two "basic" elements, often so-called distributed impedance elements have to be included for an appropriate description of the impedance behaviour of a real system. A constant phase element (CPE) is used when an ideal capacitive behaviour with a phase shift of $-90°$ is not observed. In electrical engineering this corresponds to the behaviour of a "loss capacitor". The physico-chemical background can be of different origin but is very often a heterogeneity of the surface (e.g. microroughness) or a heterogeneity of the bulk material (e.g. water content in a polymeric membrane). The degree of frequency dispersion is indicated by the α-value: When α is 1, a pure capacitive behaviour is observed, with decreasing α the properties show also resistive behaviour and when α is 0 a pure ohmic resistance appears.

A second distributed impedance element often used in equivalent circuits is the Warburg impedance W. Here, the system is under diffusion control; that means that the diffusion of substances is limiting the current. This impedance element is normally relevant at small frequencies and results in a linear behaviour with a slope of 45° in the Nyquist plot. From this impedance behaviour the diffusion coefficient of the relevant species can be calculated (see also Table 1). However, the so-called infinite Warburg impedance can only be observed when the region available for diffusion is not limiting. For example in a stirred cell the diffusion layer is finite. In this case the "finite" Warburg impedance appears resulting in a bending of the impedance curve towards the real axis at very low frequencies. This produces a distorted semicircle in the Nyquist plot. In contrast finite diffusion in a thin layer cell (blocked ion transfer at the end of diffusion layer) can result in increasing capacitive behaviour (turn of the 45° line parallel to the imaginary axis).

The individual impedance elements can be connected in two ways: in series and in parallel. For the resistance and the capacitance this is illustrated in Fig. 2. A series circuit means that the overall impedance is the sum of the individual elements. Since the capacitive impedance is frequency dependent, the whole impedance is also dependent on the frequency. This behaviour is shown as Bode and Nyquist plots in Fig. 2. From both plots the resistance in series can be easily determined. In the Bode plot the transition can be clearly seen from the frequency range where the resistance is dominating to

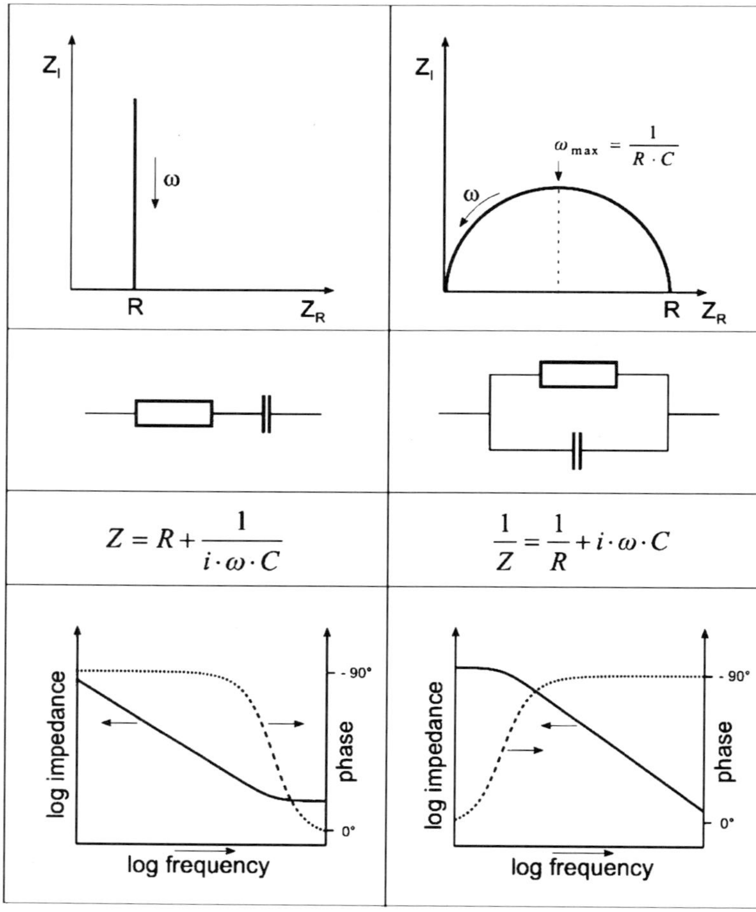

Fig. 2 The figure illustrates a series (*left*) and a parallel circuit (*right*) of a capacitance and a resistance. The frequency dependence of the impedance is given as a Nyquist plot (*above*) and Bode plot (*below*). The overall impedance is the sum of the individual impedance elements in a series circuit whereas the reciprocal impedance (admittance) is the sum of the reciprocal impedances of the individual elements for a parallel circuit

the range where the capacitance is controlling the impedance behaviour (by the course of the phase shift).

In a parallel circuit the current has two ways to flow through the system and will preferably choose the way of low impedance. Here the overall impedance is the sum of the reciprocal impedance of the individual elements (or in other words, the reciprocal impedance is the sum of the individual admittances). Thus, here a very different impedance spectrum is obtained. In the Nyquist plot, a typical semicircle appears, the diameter of which represents the ohmic resistance. The Bode plot shows that at high frequencies the impedance is dominated by the capacitance whereas at low frequencies the resistance can be determined.

1.2 Equivalent Circuits

Equivalent circuits are used in order to approximate the experimental impedance data with the above mentioned ideal or distributed impedance elements, both arranged in series and/or in parallel. Many electrochemical systems have been analysed according to this procedure. For an individual system under investigation one can probably find models in the literature for the description of the impedance behaviour – at least as a starting point for the analysis. In this chapter two examples will be illustrated and a general conclusion for an electrochemical cell comprising a biological component will be drawn.

Electrode in Electrolyte Contact

When an electrode is immersed into an electrolyte solution the ac current has to flow through the solution and has to pass the interface between the electron- and ion-conducting phase. The equivalent circuit for this situation is shown in Fig. 3 and is often referred to as Randles circuit. According to the electrochemical theory a current flow is possible by charging and discharging the double layer at the interface or by oxidising/reducing substances from the electrode or in solution. Subsequently, the model of the interface is a parallel circuit of the double layer capacitance C_{dl} and the charge transfer resistance R_{ct}. At low frequencies where the charge transfer might be faster than the dif-

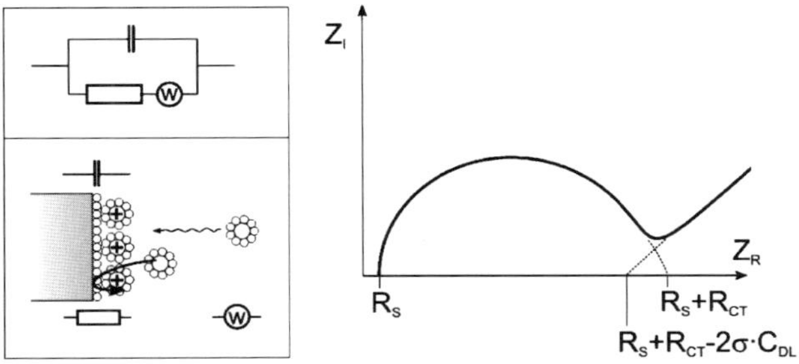

Fig. 3 Randles equivalent circuit for an electrode in electrolyte contact. The double layer capacitance C_{dl} results from the possibility of charge storage in the double layer at the interface. The charge transfer resistance R_{Ct} denotes the possibility of current flow by redox reactions at the interface and the Warburg impedance results from the current limitation by diffusion from the bulk of the solution to the interface. R_s is the solution resistance given by the ion concentration and the cell geometry. From the Nyquist plot R_s and R_{Ct} can be easily determined, but also the σ value from the Warburg impedance can be evaluated (see Table 1)

fusion of the substances to the electrode a Warburg impedance needs to be introduced in series to R_{ct} for an appropriate impedance description. In series to this parallel circuit the solution resistance R_s is added. In the Nyquist plot shown in Fig. 3 the values for R_s and R_{ct} can be easily determined. From the frequency at the maximum of the semicircle the double layer capacitance can be calculated ($\omega = 2\Pi f = 1/R_{ct}C_{dl}$). The product of R_{ct} and C_{dl} is often termed as time constant τ of the electrochemical process. The 45° line indicating Warburg limited behaviour can be extrapolated to the real axis. The intercept is equal to $R_s + R_{ct} - 2\sigma C_{dl}$ from which σ and subsequently diffusion coefficients can be calculated (see Table 1).

For analytical applications, however, the equivalent circuit is often simplified by neglecting the Warburg impedance. This can be done by choosing a frequency range where no 45° line in the Nyquist plot is observed and the interfacial or bulk impedance is dominating.

Polymer Membranes in Electrolyte Contact

Polymeric membranes are often used for separation purposes and are applied for the construction of sensors as immobilisation matrix for the recognition element or as filters improving the selectivity of the sensor by rejecting disturbing substances. In general, the impedance behaviour of such membranes can be described by a series circuit of two parallel RC combinations describing the bulk properties of the membrane and the properties of the membrane/electrolyte interface and a Warburg impedance for the diffusion limitation [4]. Under the assumption of three well-separated impedance contributions Fig. 4 illustrates the behaviour in the Nyquist plot. From the first RC term the membrane resistance and the geometrical membrane capacitance can determined. These values are governed by the chemical nature of the polymer, the pore size and distribution, the water content, the electrolyte uptake and the membrane thickness. From the second RC term (second semi-

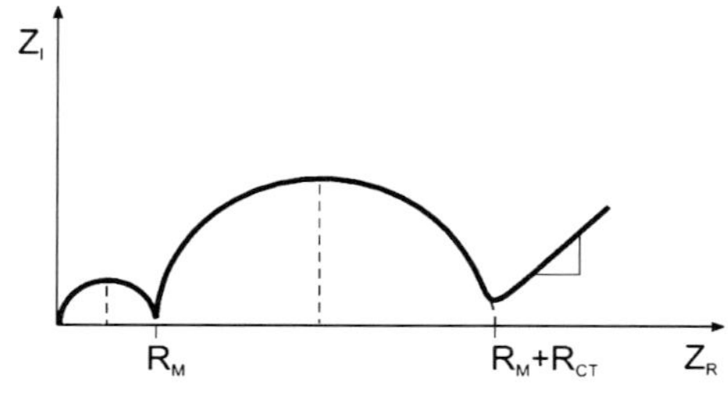

Fig. 4 The figure illustrates schematically the principle impedance spectrum of a polymer membrane in electrolyte contact. The first semicircle describes the impedance of the bulk of the membrane, the second semicircle represents the membrane-solution interface and the 45° line shows the infinite Warburg impedance. It has to be noted that depending on the nature of the membrane not all impedance contributions can always be seen. One impedance may dominate over the others

circle in the Nyquist plot) the values for the exchange resistance and the double layer capacitance of the interface can be determined. Both values are also influenced by the above mentioned membrane parameters with the exception of the membrane thickness. At lower frequencies the Warburg impedance dominates the impedance spectrum.

However, this general behaviour will not be necessarily observed for all membranes. The values for the different impedance values might differ significantly, so that for example only one semicircle occurs in the impedance spectrum when the bulk impedance of the membrane is strongly dominating. This is an important aspect in impedance spectroscopy: In a series of impedance contributions the largest impedance determines the overall impedance behaviour to a large extent so that minor components cannot be well-characterised. In the literature one can find examples concentrating on the one hand on analysis of the bulk properties and on the other hand on ion exchange at the interface – mainly because of membrane applications in separation (e.g. [5, 6]), sensor construction [7–9], fuel cell electrodes [10–12] or applications in chemical ion sensors (ion-selective electrodes) [13–15].

Electrochemical Cell with a Biocomponent

For the characterisation of a biological material, e.g. antibodies or cells, electrodes have to be introduced into the system forming an electrochemical cell. By applying an ac voltage perturbation the current has to flow through all the components of the systems – the working electrode, the biological material, the solution and the counter electrode. The measured impedance is a sum of all the individual contributions. Figure 5 gives a simplified equivalent circuit for this situation neglecting the potential contribution of the Warburg impedance. For analytical applications two different situations can occur:

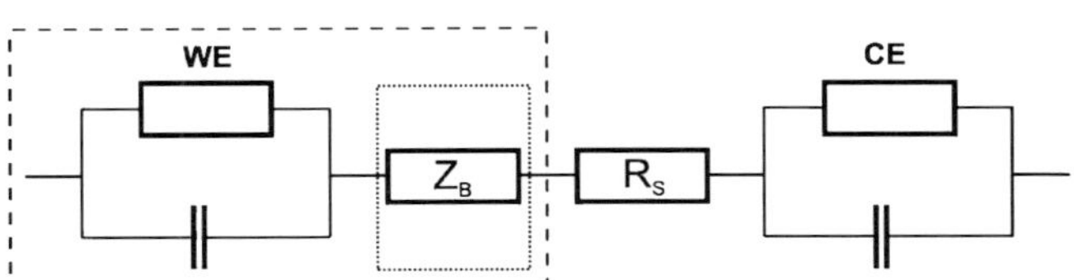

Fig. 5 Equivalent circuit for an electrochemical cell using two electrodes (working electrode -WE- and counter electrode -CE-) and a biological component. Both electrodes are exemplified by a parallel circuit of the double layer capacitance and a charge transfer resistance. In series to the electrode impedance is the solution resistance R_S and the impedance of the biological component Z_B. The working electrode can be modified with the biocomponent and thus appears as the sensing electrode. This is indicated by the *dashed* box. For this kind of measurement the other impedance elements should be comparatively small. When the biological component needs to be investigated in solution, its value should determine the overall impedance to a large extent and the impedance of the electrodes should be small. This kind of investigation is indicated by the *dotted* box

1. The impedance of the biological material is investigated as a function of the concentration of a certain analyte in solution or as a function of time. This is analogous to the analysis of the solution resistance as a function of ion concentration in solution. In both cases the impedance of the two electrodes has to be small compared to the impedance to be determined. This can be achieved using electrodes with a rather large surface area. In addition non-specific adsorption of biological material from solution has to be avoided since this would also increase the interfacial impedance.
   An example for this kind of measurement is the determination of the cell mass during a fermentation process [16, 17]. The increasing density of cells within the solution results in an impedance increase which can be easily monitored.
2. Alternatively, the biological component is immobilised on the working electrode and the interaction with an analyte molecule is detected. Here the impedance of the sensing electrode (working electrode modified with the biological component) controls the overall impedance. Thus, the impedance of the counter electrode has to be significantly smaller. Usually this is achieved by use of an electrode with an area which is at least 10 times larger than the sensing electrode. In addition, the surface can be roughened by electrochemical metal deposition or metal dissolution. By use of a supporting electrolyte (0.01–1 M) the solution resistance can be adjusted to not limit the current response. For measurements with such surface-modified sensing electrodes redox-active compounds are commonly added to the solution, which results in a well-defined charge transfer resistance R_{ct}. If the redox-active compound is omitted a rather capacitive impedance behaviour will be observed (since R_{ct} will become extremely large). Thus, a binding event at the electrode can be detected by following the change in R_{ct} in the first case or the change in the capacitance in the second case.
   By analysing the impedance behaviour over a broad frequency range the respective region where the impedance is dominated by the impedance element under investigation can be identified. Thus, for analytical applications it is often possible to limit the frequency window for the impedance analysis. In some applications also a single frequency measurement is sufficient. For example, this can be used in the measurement of chemical sensitive semiconductor structures which show a very capacitive behaviour [18, 19].

1.3 Impedance Measurement

The impedance can be measured in the frequency and in the time domain. In both cases a perturbating signal is applied to the system and the response is recorded. In the frequency domain this is an ac excitation voltage of small amplitude (usually within the range 5–50 mV) and the corresponding ac current which will be detected as a function of the frequency of the applied voltage. Since the impedance is a complex value the modulus/phase shift or the real/imaginary part couple have to be determined. Nowadays this can be easily performed with a frequency response analyser (FRA). Alternatively a lock-in amplifier can be used. In the time domain the perturbation is a potential step function, and the response the corresponding current time function. To obtain the impedance in the frequency domain from this kind of measurement a Fourier transformation is necessary. Another technique which can be termed as multisine technique applies different frequencies at the same time to the system. In this case impedance data can be obtained rather fast, however also here a Fourier transformation is necessary to transfer the data into the normal format of the frequency domain.

In general some requirements have to be fulfilled in order to get a reliable impedimetric description of the system:

1. The response of the system should only be originated by the perturbation applied. This means other conditions have to be controlled to be constant (e.g. temperature).
2. The system should react linearly on the perturbating signal. This means that the impedance should be independent on the amplitude of the perturbation applied which has to be verified experimentally.
3. The system has to be stable during the measurement and should not change because of the perturbation applied. (This can be tested by repeated measurements. For changeable systems the faster multisine technique might be advantageous.)
4. The impedance has to be a continuous and finite-valued function in the frequency range of the measurement.

These are important features which have to be controlled during the impedance analysis of a system. More accurately, the measured impedance data should follow the Kramers–Kroning relation which allows the calculation of the $Z_I = f(\omega)$ function from the $Z_R = f(\omega)$ function [1].

The frequency range which has to be analysed varies with the system under investigation. However, the range in general is from several tens of kilohertz, where the capacitive impedance becomes rather small and often the impedance is determined by the solution resistance, to 10^{-1} or 10^{-3} Hz, where often the diffusion is limiting. Commercial instruments easily cover a range from 10^{-5} to 10^{6} Hz.

For electrochemical impedance spectroscopy it is often necessary to control the situation at the working electrode: (1) In some cases impedance can be measured at the so-called open circuit potential (OCP). This means a situation which has equilibrated at the electrode depending on the solution composition. However, this situation has to be stable during the measurement as mentioned above. Here a simple two electrode arrangement can often be used. (2) Alternatively, a dc potential has to be fixed at the working electrode in order to define a situation at the electrode or to enforce a process to occur (e.g. an oxidation process). This is a typical potentiostatic mode where a three electrode arrangement is necessary. However, ac voltage and current appear only in the working/counter electrode circuit. (3) Another possibility is to define a current to flow over the working electrode. This can be achieved in a galvanostatic mode again with a three electrode arrangement.

Modern impedance analyzers also contain data evaluation software and fitting programmes which allow the calculation of transfer functions according to an equivalent circuit. These equivalent circuits are based on a physicochemical model of the system under investigation, as described, e.g. in Sect. 1.2. As a starting point for the description of an individual system one can find models in the literature or one can choose an equivalent circuit from the implemented software. In order to verify the validity of the chosen circuit not only does the fit quality in relation to the experimental curve need to be evaluated, but also a variation of experimental parameters such as electrolyte or analyte concentration, layer thickness or electrode area. From such an analysis of the impedance data the frequency range necessary to determine a parameter of interest can also be deduced. For analytical application, it is also possible not only to reduce the frequency range, but to limit the measurement to a few or even one selected frequency. Here the impedance parameter of interest (e.g. resistance or capacitance as a function of a concentration to be determined) can be obtained from a fit of the measured impedance data by a verified equivalent circuit or the overall impedance can be directly correlated to a concentration.

2 Overview on Biosensorial Applications

In analytical science, the term biosensor is used when the biochemical recognition event is integrated with the signal transduction into a single device. Thus, an electrochemical biosensor represents some kind of "bioelectronic device" [20] which is based on the interaction of biomolecules with a conductive or semiconductive support and the electronic transduction of a specific biological reaction associated with the biological function. The biological components might be proteins, including enzymes, receptors and antibodies, nucleic acids, or low molecular weight molecules such as cofactors, inhibitors or antigens. More recently, artificial recognition elements such as molecular imprinted polymers (MIPs), peptide nucleic acids (PNA), aptamers, abzymes, a.o., have been developed for sensorial applications. The detection process involves the formation of a recognition complex between the sensing biomolecule and the analyte at the interface of the electronic transducer which directly or indirectly alters the electrical properties of the recognition surface. Electrochemical biosensors for direct and often label-free measurement of various analytes are especially attractive because of their fast performance and less demanding instrumentation (e.g. [21–23]). In this context, impedance spectroscopy is perfectly qualified for the read-out of biosensors since it provides detailed information on the resistive and capacitive properties of the recognition surface. Because of the affordable availability of impedance instrumentation nowadays, a trend towards the development of impedimetric biosensors has currently been observed. This can be seen from the increasing number of publications on this topic over the past few years. A simple inquiry with the search engine from the ISI Web of Science using the catchword "impedance spectroscopy" yielded a remarkable number of 1400 hits for the year 2006. If the term "biosensor" is added to the search still 50 publications are returned. Figure 6 shows the result of such searches for the last 10 years.

Impedance spectroscopy is applied in bioanalytics both to monitor the fabrication of biosensors and the biosensorial recognition process. Applications have been demonstrated for all types of (bio)analytes such as proteins, nucleic acids, whole cells, microorganisms, antibodies and antigens. A detailed and complete description of the different approaches is difficult; thus, the present chapter tries to give an overview on the various strategies that have been attempted to design impedimetric biosensors without claiming to be complete. The presentation is focused on general concepts, some selected examples and discusses novel designs which were published. Alternative reviews on impedimetric biosensors are available [24–26].

The various efforts for specific and sensitive detection can either be classified according to the electrode material (metals, metal oxides, glassy car-

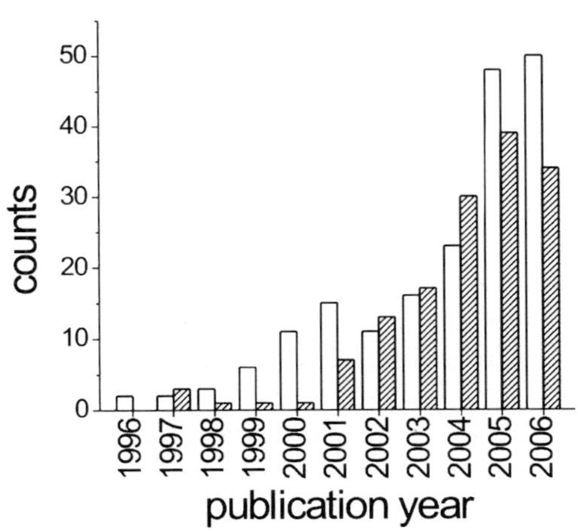

Fig. 6 Number of publications on impedance spectroscopy and biosensors for the last 10 years. *Open* and *hatched* bars correspond to bibliographic searches using the keyword combination "impedance spectroscopy AND biosensor" with ISI Web of Science and Pubmed, respectively

Table 2 Elements of impedimetric biosensors. The various approaches in literature consist of these main elements in nearly any combination

Electrode material	Electrode geometry	Recognition elements for selectivity and specificity and analytes	Elements for signal amplification
• Gold, platinium, and other metals • Glassy carbon • Metal oxides (e.g. ITO, etc.) • Nanoparticles • Semiconducting materials • Field-effect devices	• Single working electrodes of different shape • Interdigitated electrodes for in-plane impedance measurements	• Antibodies and their antigens • Nucleic acids • Proteins, enzymes, receptors, and their substrates, inhibitors, cofactors • Molecular imprinted polymers (MIP)	• Enzyme labels • Redox probes, labels and mediators • Nanoparticles • Liposomes

bon, semiconductors, etc.), the electrode geometry (conventional electrode arrangement or "in-plane" measurement), the analyte (proteins, antibodies, nucleic acids, etc.), or according to the mediator or amplification protocol used (enzyme labels, redox labels, antibodies, polyelectrolytes, conductive polymer films, carbon nanotubes, nanoparticles, etc.). Table 2 summarises the typical elements of such a biosensor. The numerous approaches in literature apply these main elements in nearly any combination. In the following, the impedimetric biosensors will be divided into two categories. The first approach uses the electric properties at the surface of a single working electrode ("one sensor electrode" approach). The second method is based on two shortly spaced electrodes ("in-plane" geometry). In the latter case, the recognition reaction can also occur in the gap between the two electrodes.

2.1 Impedimetric Detection Using a Single Sensor Electrode

Biosensors of this type follow the idea that the impedance at a single working electrode should determine the overall impedance. Thus after modification with the recognition element, this electrode becomes the "sensing" electrode. This approach is sometimes also referred to as "interfacial" measurement. Figure 7a shows the schematic structure of such an impedimetric biosensor. Different options exist for the evaluation of the recognition event at the sensor surface; besides the simple detection of the overall impedance at selected frequencies one can analyse the changes in electrode capacitance or resistance.

Capacitance and Resistance Measurements

When the biosensor electrode is immersed in an aqueous solution an electrochemical double layer is formed. The capacitance C_{dl} depends on all compounds present at the interface, which are predominantly solvent molecules, ions, immobilised biomolecules, and optional films which promote the immobilisation and detection processes. A capacitance change is induced when the dielectric constant or the thickness of the double layer on the transducer surface changes. Both options have been a matter of examination [26, 27]. When constructing a capacitive biosensor, the electrode surface is usually covered with an additional insulating layer to reduce Faradaic currents (i.e. to increase the charge transfer resistance in parallel to the capacity). The biorecognition element is immobilised on top of this layer.

Capacitance measurements in the absence of Faradaic currents also allow for the investigation of the influence of the electrical field on the biological recognition event by applying different dc working potentials [28]. The interaction between the recognition element and the analyte as well as their orientation might be influenced by the applied dc potential, particularly when the components of the interaction are charged.

The alternative approach for impedimetric detection of analyte binding is the determination of the charge transfer resistance R_{ct}. These Faradaic impedance measurements are usually performed in the presence of a redox couple in solution and rely on the change in the barrier for the redox conversion by the formation of the recognition complex itself or a subsequent complex. The ferri-/ferrocyanide system is often chosen for this purpose. For sensitive detection, impedance spectra are generally recorded for dc potentials equal to the open circuit potential (OCP) of the employed redox couple. Although an appropriate redox probe is necessary, the Faradaic impedance

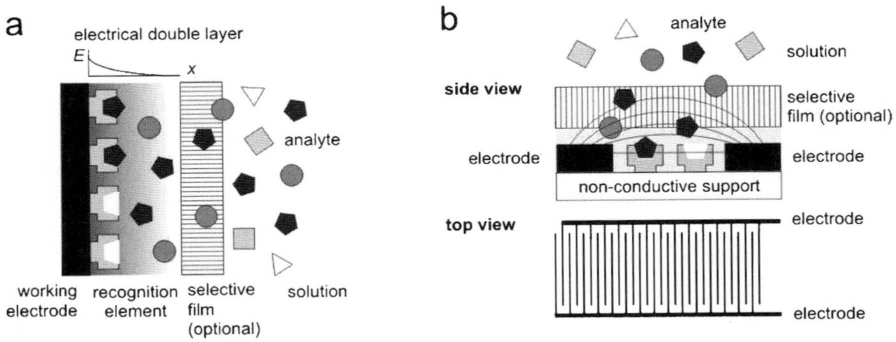

Fig. 7 Schemes of impedimetric biosensors using a modified working electrode as sensor (**a**) or an in-plane arrangement of two electrodes (interdigitated electrodes IDE) (**b**). Immobilisation occurs in **a** on the electrode and in **b** between and/or on the electrodes

measurement is considered to be more sensitive since the changes in R_{ct} are usually larger than the respective capacitance changes.

Although very often used, the presence of a redox couple is not absolutely necessary. The measurement in an inert electrolyte solution results usually in a much higher impedance, but as long as the impedance changes upon the biorecognition process are reproducible as well this kind of measurement can result in reliable data. Particularly interesting is this approach when the conductivity of the recognition layer is changed by the specific binding event (as shown for example with proteins embedded in a lipid layer).

Sensor and Sensor Preparation

Typical transducer elements are carbon and metal electrodes. Preferentially gold is chosen, because it allows the application of thiol chemistry for immobilisation [29–32]. Additionally, thin oxide layers on the electrode surface like indium tin oxide (ITO) or platinum/platinum hydroxide electrodes provide hydroxyl groups for silanisation chemistry [33]. Metal oxides of platinum [34, 35], tantalum [22, 36] and titanium [37] have also been used for capacitance-based detection. Semiconductor-based systems appear as a special group of devices which will be treated separately below.

Self-assembled monolayers (SAM) are frequently used to promote the immobilisation of the biorecognition element. Such films are formed spontaneously by chemisorption and self-organisation from solution [30, 38–40]. Long-chain thiols can effectively block the electrode surface from unwanted reactions and thus are often used in capacitive sensors. Since the SAM films do not have specific recognition properties, the biocompound is normally coupled to the surface with functionalised SAMs [41]. Alternatively the SAM layer has to be structured in order to create recognition sites. This can be termed as a kind of surface imprinting [42, 43]. An alternative method to obtain well-ordered recognition layers is the transfer of a lipid bilayer (from liposomes) or a Langmuir–Blodgett (LB) film onto the electrode surface [44, 45]. Other immobilisation protocols rely on the preparation of thin polymer films, electropolymers or bioaffinity layers (e.g. avidin films capable for binding biotinylated biomolecules). More recently polyelectrolyte films have been introduced for biomolecule immobilisation. This technique is based on the layer-by-layer adsorption of oppositely charged polymers on the transducer surface and is particularly suited for the preparation of multilayers [46, 47].

Impedimetric Immuno- and DNA Sensing

As already implied, impedimetric analyte detection is frequently applied for immunosensing. This originates from the fact that classical immunoassays (ELISA) are wide-spread and used in clinical diagnostics, food quality control, environmental analysis, for the detection of pathogens, toxins, and explosives or drugs. Even though immunosensing is an established technique, background for the sensorial developments with impedimetric detection are the demands of point-of-care diagnostics with the need of rather simple equipment and the potential for a label-free analysis.

For immunosensing two immobilisation strategies are possible. The sensor chip can either be modified with antibodies binding the respective antigen molecule or the antigen itself is immobilised, which binds then the complementary antibody. In both cases the binding event results in a change of the electrical surface properties, although in the latter case larger changes can be expected because of the high molecular weight and the low dielectric constant of antibodies.

Initially capacitive sensors were preferably used for the detection of immune reactions [30, 31, 48, 49]. Charge transfer can be diminished, e.g. by long-chain SAM layers. In combination with a potential-jump method very sensitive measurements have been reported [23]. In order to reduce the influence of non-specific binding, which would influence the capacitance in a similar way compared to the specific interaction, a differential mode of measurement was proposed [35]. Antibodies can be incorporated into Langmuir–Blodgett films while retaining their binding properties. This can result in very ordered recognition structures [44, 50].

Resistance-based sensors have also been developed [51–53]. For example, the human mammary tumor-associated glycoprotein was detected with specific antibodies immobilised on gold by spontaneous adsorption. The binding of the complementary antigen resulted in a change of the charge transfer resistance R_{ct} [52]. Another system used ultrathin platinum layers and evaluated the conductivity changes based on an impedance model analysis [53]. But, the change in ion conductivity of a lipid bilayer with incorporated ion channels can also be used for the antibody-antigen binding detection [54, 55].

A special approach for the immobilisation can be seen in the use of conducting polymers such as polypyrrole. This can help in amplifying the response since the conductivity of the polymeric network is strongly influenced by conformational changes induced by binding events [56, 57]. Biotinylated polypyrrole films were used for the immobilisation of biotinylated antibodies via avidin [58]. Antigen binding was connected with an increase in the charge transfer resistance. The respective detection limit was 10 pg mL^{-1} of human IgG.

Besides immunosensing, the impedimetric detection of nucleic acids is also a matter of research. ssDNA and dsDNA have been discriminated using a hanging mercury drop electrode [59, 60]. Desorption of denaturated ssDNA was accompanied by a larger dielectric loss than desorption of native dsDNA. This was explained with the higher flexibility of the ssDNA compared to the rigid dsDNA double strand.

For a sensorial application more interesting are investigations on gold electrodes. Several sensor systems have been reported for a voltammetric analysis of the hybridisation [61–65]. But here the addition or the coupling of a redox indicator is necessary. The presence of ss- or dsDNA on the electrode surface was found to result in different molecular structures which can be evaluated by different spectroscopic and microscopic methods [66]. This provides the basis for impedimetric detection of oligonucleotide concentrations and single base pair mismatches, as will be shown in Sect. 3 [67].

In analogy to the immune detection using conductive polymers, probe DNA can also be linked to conducting polymer films to evaluate the change in interface properties after target DNA binding. Here an increase in resistance was observed which could be used for a sensorial detection of suitable sensitivity [68–70]. Amplification of the response by the use of nanoparticles coupled to an oligonucleotide has been reported [71]. As an alternative to polypyrrole a quinone-containing polymer was developed [72] or polythiophene was used [73]. The impedance technique can be used not only for the detection of DNA-DNA binding but also for the analysis of DNA-analyte interactions. Several small organic molecules but also large proteins show a specific interaction with DNA duplexes. This can be used for an analytical determination of these substances. Examples are cisplatin [74] or mythramycin, a.o. [75], but also specific DNA-binding proteins [76].

Molecular imprinted polymers are a rather new group within the recognition elements. They provide a surface which acts as a "negative" of the analyte and thus can recognise it within a mixture. Because of its chemical nature it can be reproducibly fabricated in different formats and is rather long term stable in different environments. Besides its application in chromatographic separation interesting examples have been reported for applications in sensors (see the chapter by Danielsson). For an impedimetric transduction of the binding event, particularly thin surface films are suited. Thus, capacitive sensors have been developed for organic molecules [42, 77] but also for whole cells and viruses [78, 79].

Impedimetric Enzyme- and Cell-based Sensors

In the area of enzyme sensors amperometric electrodes are by far the most attractive transducers. This is because of the ease of use and their wide dynamic range. However, the change in redox conversion of a redox-active compound by the turnover of an enzyme (in the presence of the respective substrate) can also be detected by following the charge transfer resistance. This has been demonstrated for glucose detection using glucose oxidase and parabenzoquinone as mediator [80].

In contrast to the transduction process itself impedance measurements have found increasing applications in optimising biosensors mainly by following the different steps of surface modification during sensor preparation. This includes formation of SAMs, Langmuir–Blodgett films, adsorption layers, a.o., but also the binding of the recognition element to the transducer surface (e.g. [81–83, 169]).

Another direction in using impedance for sensing can be seen in the incorporation of a "signaling" protein in a lipid layer on an electrode and evaluating the conformational changes upon interaction with a specific analyte. Often this is accompanied by changes in ion mobility through the surface-confined film. Thus, conductivity changes can be analysed. One example of sensing substrates and inhibitors is given by Naumann et al. [84]. The authors used peptide-tethered lipid membranes on a gold support with cytochrome c oxidase. Impedance spectroscopy was used to study proton transport across the membrane. Impedance spectra could be correlated to the concentration of inhibitor and substrate of the enzyme. Other examples are the detection of paraoxone [85], quinacrine [86], or the bacterial toxin streptolysin [87], but also proteins or DNA can be analysed [88].

In analytics not only substrates and inhibitors of enzymes have been determined, but also the enzyme activity itself has often been analysed. Here an interesting approach was introduced by using degradable polymer films. These polymers can be coated onto an electrode and impedimetrically characterised. The degradation by a biocatalytic reaction can occur directly by the enzyme acting on the polymer chains or by products of the enzymatic conversion. The diminished film thickness is easily followed by impedance measurements. This has been shown for urea [89], glucose [90] and enzymes such as chymotrypsin or lipase [91]. This scheme can be nicely combined with immunoassays using enzymes as labels [89].

Recent developments are cell-based assays for the detection of specific analytes or for the screening of large libraries of substances for their biolog-

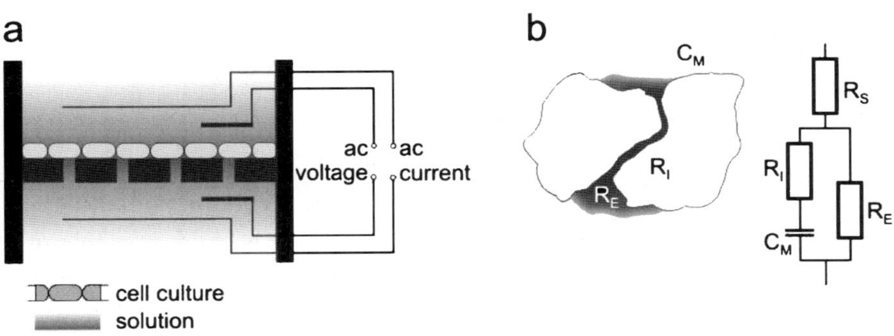

Fig. 8 **a** Experimental setup for the impedimetric characterisation of a tissue culture (four-electrode measurement: ac current flow in the outer circuit and ac voltage detection between the two other electrodes) **b** Simplified equivalent circuit for the impedance description consisting of the solution resistance R_S, the membrane capacity C_M and the resistances for the internal and external charge carrier transport R_I and R_E (adapted from [97])

ical effect. This relies on the concept of effect monitoring. Here, parameters are needed that report the response of the whole biological system with respect to the stimulus under investigation. Impedance is one of the suitable parameters because it can be used for the indirect detection of metabolic activity, cell adhesion on surfaces, response to potential drugs and for cytotoxicity tests [92, 93]. First examples have been reported analysing the status of a cell culture [94–97]. Figure 8 illustrates a typical experimental setup for cell-based measurements and a simplified equivalent circuit describing the impedance of a cell layer as a parallel circuit of intracellular and extracellular pathways. It has to be mentioned that cells can also be cultivated directly on the working electrode with an appropriate surface modification.

Impedance is not only suited to give information on the status of surface-fixed cells, but can also be used for the detection of the number of cells in solution. This has been intensively investigated in microbiology and was used for the detection, quantification, and identification of bacteria [98, 99], and resulted also in micromachined devices for cell counting [100] or cell differentiation [101].

Amplification of Impedimetric Signals

Despite the potential for a direct binding detection many interfacial impedimetric assays suffer from the fact that the generated signals (changes in C_{dl} and/or R_{ct}) are relatively small. This problem is intensified when the surface coverage with the recognition complex is low and small amounts of the analyte have to be detected. This has led to the development of several amplification strategies. In order to increase the capacitance change due to an altered dielectric constant, polymeric membranes were used for fixation of the recognition element. Indeed, the incorporation of antibodies into a membrane induced a change of C_{dl}. As an example Maupas et al. [102] have modified platinum electrodes with different polymeric layers and used them in a flow injection system for real-time immune detection. As mentioned above a conducting polymer layer can be nicely used for generating an enhanced signal [56, 68, 70].

Bresler et al. developed a strategy for sensitivity enhancement by using a capacitive sensor structure (based on passivated interdigitated electrodes) and combining the antigen-antibody reaction with a second binding event and thus coupling catalase to the surface. The enzymatic activity of catalase converts H_2O_2 to oxygen. The gas bubbles formed near the surface change the dielectric constant drastically and thus allow a very sensitive detection, which was shown, e.g. for HIV antibodies [103].

A very simple amplification route has been introduced by the Willner group [104, 105]. Binding of negatively charged liposomes to the recognition surface can drastically alter the charge transfer of a redox couple, particularly when it posses the same charge as, e.g. ferri-/ferrocyanide. Figure 9 illustrates

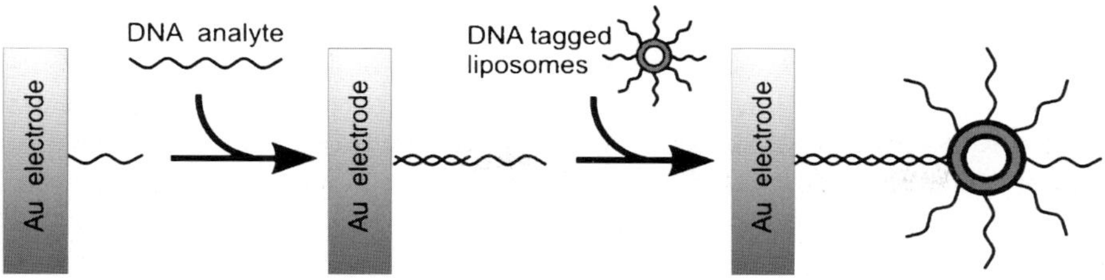

Fig. 9 Amplification of the impedimetric detection of a binding event using secondary binding reactions coupling liposomes to the electrode surface. The change in surface properties by the DNA binding is much enhanced by negatively charged liposomes which can repel the ferri-/ferrocyanide redox couple much more effectively from the electrode surface resulting in a significant impedance increase

the general assay procedure. In the scheme, the DNA hybridisation is followed by one or several secondary binding steps resulting in an accumulation of highly charged liposomes on the electrode.

A smart amplification approach was taken over from classical immunoassays and includes the application of enzyme labels bound to a component of the sensing system. The enzyme label catalyses the formation of an insoluble product which precipitates onto the electrode surface and amplifies the changes of R_{ct} and C_{dl} [106–111]. The principle was introduced for the detection of DNA [106] and antibodies against dinitrophenyl at an electrode surface functionalised with the self-assembled monolayer (SAM) technique [107]. Horseradish peroxidase (HRP) was used as the enzyme label. The technique can also be applied for the detection of autoantibodies in physiological samples [112]. This will be illustrated in Sect. 4.

The HRP label was also used for DNA detection of the Tay–Sachs genetic disorder with high sensitivity [106]. Further applications of this amplification method, but with alkaline phosphatase as the label, have been demonstrated for the analysis of viral RNA and single nucleotide polymorphisms [113, 114]. Amplified DNA detection was achieved using liposomes labelled with multiple HRP tags, which resulted in an amplification factor of approximately 10^5 [110].

An enzyme label was also used for the impedimetric detection of bacteria cells of *Escherichia coli* [115]. The electrode surface was initially modified with an *E. coli* antibody. The binding of the cells were subsequently detected with Faradaic impedance measurements. Signal amplification was achieved using a secondary antibody against *E. coli*, which was modified with alkaline phosphatase catalysing the oxidation of 5-bromo-4-chloro-3-indolyl phosphate yielding the insoluble indigo derivate. The detection limit was 6000 cells mL^{-1}.

This amplification technique enables the detection of very low surface coverage of antibody-antigen complexes. However, a regeneration of the sensor surface is difficult, if not almost impossible, and impedes a repeated usage of the sensor. A way out of this problem might be the application of nanoparticles which have been also applied for signal generation and amplification [71, 75, 116–118]. Examples are gold nanoparticles which were used for electrode modification [75, 119] or quantum dots [116] which were coupled to oligonucleotides and thus able to bind to surface-confined DNA. Also in this case the interfacial charge situation is altered and thus the negatively charged redox compound ferri-/ferrocyanide is repelled from the electrode surface.

Semiconductor Devices

A special field in impedance related sensors are semiconductor-based structures. Mainly these are semiconductor/insulator structures which detect chemical changes at the insulator surface in contact with the solution by analysing the semiconductor/insulator interface. The classical ion-sensitive field effect transistor (ISFET) which can be combined with enzymes or antibodies yielding enzyme FETs (ENFETs) or immunoFETs (IMFETs) is outside the focus of this review [120]. Although this device has an inherent impedance conversion from the high impedance line between the semiconductor bulk and the gate to the low impedance line between source and drain, the output is a simple source-drain current and thus, the device does not follow the general idea of impedance measurement discussed within this chapter. However, the FET-derived, but simplified structure, electrolyte-insulator-semiconductor (EIS structure) is characterised by capacitance measurements and therefore will be discussed here. A general scheme is given in Fig. 10.

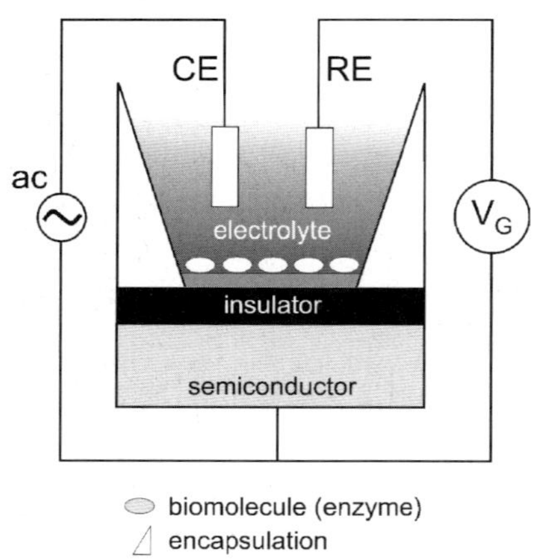

Fig. 10 Schematic view on an EIS structure (electrolyte-insulator-semiconductor). The gate voltage V_G is varied in order to establish different states at the semiconductor-insulator interface such as accumulation, depletion, and inversion. The ac voltage is applied to measure the capacitance changes caused by V_G variation and the biosensing event (CE – counter electrode, RE – reference electrode)

First attempts to measure capacitance changes on Si/SiO_2 surfaces take advantage of the fact that silicon is commercially available, easy to handle and biocompatible. Structures can be miniaturised and produced in large quantities. Long time EIS structures have been only seen as an intermediate step towards ISFET development, but meanwhile they have gained attention as a separate sensor platform mainly because of their ease of preparation and the progress in instruments detecting capacitance changes.

Enzymes can be effectively coupled to ion-sensitive semiconductor structures when the enzyme reaction changes the ion concentration near the surface as shown with protons [121, 122] and fluoride ions [123]. When the semiconductor structure carries the respective ion-sensitive layer, an interfacial potential is generated affecting the space charge region in the semiconductor which can be analysed by following the capacitance of the structure. For EIS structures, examples are the detection of penicillin using penicillinase [124] or organophosphorous pesticides using the respective hydrolase [125]. Very often the shift of the capacitance-voltage curve along the voltage axis is evaluated rather than the impedance change by the biocatalysis. Also in this area impedance measurements are used for the characterisation of the immobilisation steps (e.g. [126, 127]).

There are also several attempts to combine immune reactions with semiconductor-based sensors. However, the formation of the immune complex often does not enhance the overall impedance or change the charge situation on the gate. Thus, a limited number of systems have been reported. Moreover, for the detection of charged proteins the measurements are additionally sensitive to the ionic strength of the solution.

Using immobilised antibodies on Si/SiO_2 surfaces the antigen α-fetoprotein was detected down to $1\ \mathrm{ng\,mL^{-1}}$ (15 pM) [128]. A similar approach was used to detect enterotoxins. A capacitance change was observed after incubation with $10\ \mathrm{\mu g\,mL^{-1}}$ (0.4 μM) [129].

Because of the high charge in a DNA strand semiconductive heterostructures have been successfully used to detect nucleic acids. The charge accumulation during hybridisation results in a change of the flat band potential which can be easily detected by capacitance measurements [130–134]. However, different sensitivities have been reported and not all factors determining the signal output are fully understood. Thus, recent investigations concentrate on elucidating these parameters and modelling the response of the device [135, 136].

In the field of semiconductor devices a lot of effort is still going on in order to overcome practical limitations as for example the limited sensitivity. One direction can be seen in the use of modified device structures and the application of nanowires [137, 138]. Although high sensitivities have been reported, the reproducible production and operational stability are still a problem.

2.2 Impedimetric Sensors Using Interdigitated Electrodes

Instead of using the surface of a single working electrode an in-plane geometry can be applied, which makes use of two closely spaced electrodes. Two electrodes are prepared on an insulating substrate in a finger-like structure leaving a small non-conductive gap between them. These electrodes are termed as interdigitated electrodes (IDEs) and can be formed from different metals such as palladium [139], titanium [140], platinum [141] and gold [142]. The geometry takes advantage of the changed conditions for the current flow which occurs mainly very near to the surface and thus shows a much higher sensitivity towards surface changes compared to the conventional design. The biorecognition element can be immobilised on both electrodes, in the gap between the electrodes or on top of both. This electrode design has an additional advantage since by preparing the electrodes on a chip the whole electrochemical setup is already integrated and no external electrodes are needed. The basic scheme is shown in Fig. 7b.

Immune detection using IDEs takes advantage of the change in resistance close to the surface due to protein accumulation and the change in the dielectric properties due to the presence of substances with a lower dielectric constant compared to water. Thus, a label-free detection becomes feasible [143–145]. This kind of measurement can be also applied to the detection of DNA hybridisation without the need of a label. This relies mainly on the changed charge accumulation after double strand formation [141–143, 146].

Receptor-ligand binding is often accompanied by structural changes of the complex. This can be also used for a sensitive ligand detection particularly when a receptor is used, which is a chemically gated transport protein such as the nicotinic acetylcholine receptor. Incorporated in a lipid membrane, acetylcholine but also receptor antagonists can be measured by capacitance changes [147].

IDE-based sensors have also found application in enzyme detection. For example, an enzyme-based impedimetric sensor was developed for the detection of collagenase. Impedance changes were induced by the proteolytic digestion of gelatine-coated interdigitated electrodes [148]. In this case, enzyme degradation of the layer produced a rapid increase in the impedance due to a changed thickness of the layer. The change in impedance with protease digestion was correlated to the solubilisation of the gelatine layer from the sensor surface. Other examples based on the enzyme catalysed polymer transformation, explained in Sect. 2.1, are sensors for urea [149, 150] or creatinine [150].

However, for the direct binding detection the sensitivity is often limited for practical applications. Thus, several amplification protocols have been developed (see also Fig. 11). First of all, the transducer itself provides the potential for further sensitivity enhancement. A highly sensitive response of

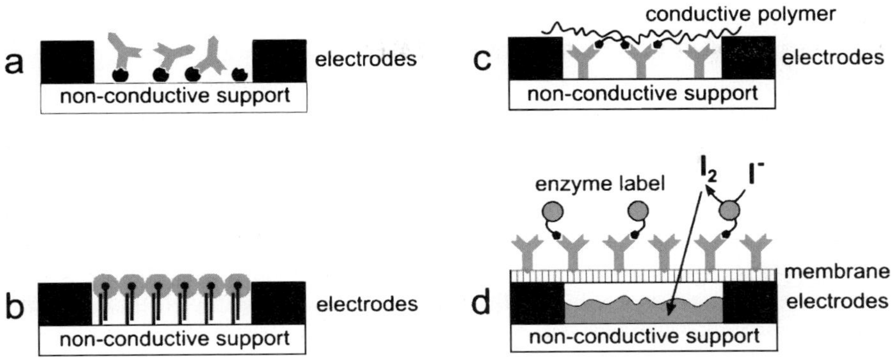

Fig. 11 Examples for sensor systems using in-plane impedance detection. **a** Direct binding detection by evaluating the change in conductivity and dielectric constant near the surface by the binding event. **b** Amplification of the binding event by use of gold nanoparticle labelled molecules. The catalytic activity of the gold surface is used for a silver precipitation enhancing the diameter of the gold particle. This results in the formation of a conductive bridge between the electrodes. **c** Antigens labelled with a conductive polymer decrease the resistance of the gap. **d** An enzyme label (HRP) converts iodide to iodine, which can permeate a membrane and enhance the conductivity of a polymer film between the electrodes

the sensor requires strong electrical fields due to short distances of the electrodes [151]. Typical distances between the electrode fingers are in the range of 1–10 μm and are limited by the fabrication procedure. However, electrode widths from 500 nm [143] to less than 200 nm have meanwhile been achieved. A theoretical analysis demonstrates that 80% of the current between the electrodes with a spacing of about 250 nm will flow in a layer not higher than 250 nm above the surface. This may illustrate that a reproducible production of interdigitated electrodes with dimensions in the nanometer range will significantly increase the sensitivity for a direct binding analysis.

In analogy to amplification schemes developed for impedimetric sensors using a single sensor electrode also on IDEs additional films, enzyme labels, or nanoparticles can be applied. For example the DNA hybridisation can be extremely sensitively detected by use of Au nanoparticles coupled to an oligonucleotide. The catalytic properties of the nanoparticle surface can be used for a silver deposition and thus an enlargement of the particles. When this enlargement proceeds to an extent that the surface-bound particles face each other, then a conductive pathway is formed between two electrodes [152, 153]. This is schematically shown in Fig. 11b and results in an ultrasensitive detection scheme.

An alternative approach for sensitivity enhancement in immunosensing uses conducting polymer chains coupled to the antigen (see Fig. 11c) [154]. After binding of the modified antigen to the recognition surface, the conductance of the gap increases and detection becomes possible. The same effect can be induced, if an enzyme label is coupled to the antigen. In this case, the sensor surface can be additionally covered with a membrane permeable for iodine, which is produced by the enzyme label HRP. After binding of the antigen HRP conjugate iodine is generated from iodide and hydrogen peroxide, which leads to an increased conductance of the promoting film consisting of tetra-*tert*-butyl copper phthalocyanine within the gap [155] (see Fig. 11d). Another example relies on the production of a gas on the sensor surface by the activity of catalase which was labelled to a secondary binding protein [103].

The various examples in the literature may illustrate the potential of the impedance technique for analytical purposes. However, reproducibility and reliability of the sensors is often a problem particularly when repeatedly used or applied in real samples. In the following sections two examples will be demonstrated showing the useful sensorial properties of a reusable DNA sensor and a disposable antibody sensor applicable in serum samples.

3 Impedimetric DNA Detection

Based on recent progress in genomics the interest in DNA detection devices has increased drastically. Various fields of application have been opened up for devices which allow a highly efficient analysis of DNA sequences. Examples are the identification of pathogenic species in medicine or food industry and paternity tests in criminology. Also the need for an early identification and therapy of genetic diseases plays a key role in the development of new diagnostical tools for a sensitive, selective and quantitative DNA detection.

From the analytical point of view there are mainly two different approaches to identify the sequence of bases within a DNA strand. On the one hand there are biochemical sequencing techniques, and on the other hand methods allowing the detection of specific hybridisation events between two complementary strands. The high costs, time consumption and the complexity of nucleotide sequencing resulted in an increasing usage of hybridisation-based approaches. Here, one DNA strand (probe DNA) is immobilised on a surface. By contact with a solution containing the complementary DNA (target) both DNA strands hybridise forming a double strand. The detection of the hybridisation event can be achieved by different transduction principles making use of changes in the interfacial properties, mass accumulation, changes in optical density or simply exploiting the detection of a label (mostly an optical). Examples are shown with fluorescence [156–158], quartz crystal microbalance [159, 160], surface plasmon resonance [161], a.o. A more detailed overview will be given in the chapter by Bier.

Recently there has been an increasing interest in electrochemical devices for hybridisation detection since they allow a simple, fast and sensitive analysis without usage of expensive equipment. Very often the hybridisation event is detected by voltammetric measurements of a redox label which interacts with or is directly bound to the DNA [162–165]. The elimination of the redox label simplifies and speeds up the whole process and reduces the cost for each analysis. Currently label-free detection of hybridisation is mainly achieved by oxidation of the DNA bases [166–168] or by monitoring the conductance/resistance or capacitance of the surface after the DNA binding event (see Chap. 2).

3.1 Sensor Concept

Impedance spectroscopy has the potential to monitor DNA hybridisation on electrodes directly, without any label. This is reasonable if no trace amounts of DNA have to be analysed. A fixed ssDNA layer acts as the recognition element and a gold chip electrode can be used as transducer (see Fig. 12). Probe DNA is immobilised on the gold surface via a thiol modification at the 5′-end of the DNA. Unspecific binding between the electrode and other components of the test solution is reduced by blocking the surface with a hydrophilic, short thiol as, e.g. mercaptobutanol (MCB). The treatment of the surface with the thiol also ensures that the immobilised ssDNA is exposed to

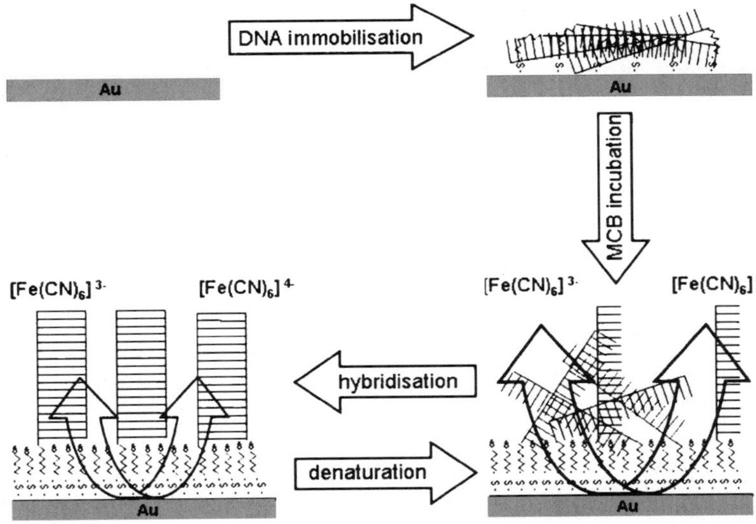

Fig. 12 DNA sensor. Principle of electrode construction and DNA detection. Probe DNA is immobilised on cleaned gold chip electrodes. Blocking of the surface with MCB prevents unspecific binding. Hybridisation changes the surface properties and thereby hinders the redox conversion of ferri-/ferrocyanide. This is detected by impedance spectroscopy. The change of surface properties is reversible by denaturation of the DNA duplex

the solution and thus available for hybridisation. The surface concentration of ssDNA can be detected for example by hybridisation with methylenblue-labelled DNA and a voltammetric read out [65]. The surface coverage of about 5–9 pmol/cm^2 is sufficiently high for an impedimetric detection and sufficiently low to avoid steric hindrance during the binding of the complementary strands.

The impedance is analysed in the presence of a redox component in solution, here ferri-/ferrocyanide. The measurement is thus based on the change of the interfacial impedance for the redox conversion of the hexacyanoferrate ion due to the hybridisation process. Binding of the target DNA causes an increased barrier for the redox reaction because of charge accumulation and therefore should result in a reduced charge transfer depending on the amount of bound DNA. In contrast, denaturation of the DNA duplexes on the electrode surface induces a higher charge transfer and therefore a decrease of the impedance [67].

Impedance spectra of a denaturated and hybridised surface are shown in Fig. 13a as Nyquist plot and demonstrate that the impedance increases after hybridisation. From an analysis of these spectra using a simplified Randles equivalent circuit without a component for the Warburg impedance (Fig. 13b) the values of charge transfer resistance and capacitance are available. This analysis shows that after hybridisation the charge transfer resistance, which reflects the diameter of the half circle in the Nyquist plot, rises up several times. In contrast, the changes of the capacitance are in range of the measuring error.

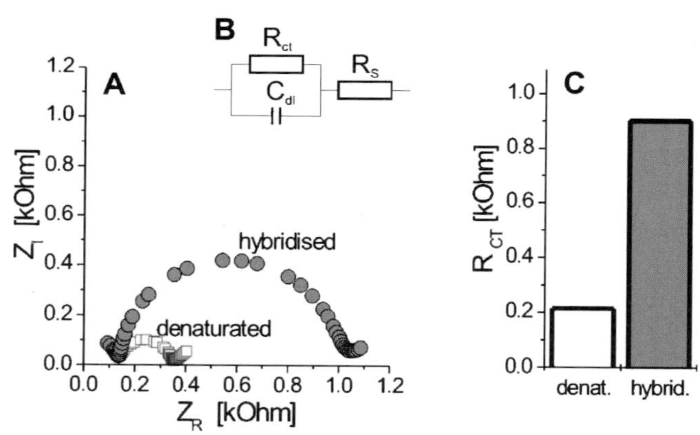

Fig. 13 **A** Impedance spectra of a denaturated and hybridised sensor surface displayed as a Nyquist plot (conditions: sodium phosphate 0.1 M; pH 7.0; 10 mM ferri/ferrocyanide; OCP; frequency range 0.3 to 10 kHz). **B** Equivalent circuit which is used for analysing the impedance spectra with solution resistance (R_S), charge transfer resistance (R_{ct}) and electrode capacitance (C_{dl}). **C** Charge transfer resistance of a denaturated and hybridised sensor surface calculated from a fit of the measured impedance curve to the equivalent circuit in (**B**)

For sensor applications the change of the charge transfer resistance can be analysed in different ways: (1) First, the difference between the charge transfer resistance before and after hybridisation can be taken as sensor signal. (2) Alternatively, the ratio between both R_{ct} values can be calculated. The resulting factor describes the relative increase in the impedance after contacting the sensor electrode with the solution under investigation. In the following the latter approach is used for the description of the sensing properties of the electrode.

3.2
Sensor Properties

Reusability. For repeated usage of the sensor the denaturation procedure is very important. In general, various compounds such as isopropanol, guanidine hydrochloride, sodium carbonate and urea are used for denaturation of duplex DNA. For the choice of denaturation procedure using impedance measurements as transduction principle a few aspects have to be considered. First a stable impedance signal for the ssDNA modified (denatured) surface has to be achieved. Second a high hybridisation efficiency has to be guaranteed even after several denaturation cycles. This means that the surface properties of the electrode should not be altered by the denaturation process while effectively breaking the double strand structure on the electrode. Thus, not all denaturation procedures are suitable for an impedimetric assay. For example guanidinium hydrochloride and isopropanol change the surface properties in such a way that the impedance changes are diminished after a few detection cycles. With urea, on the other hand, the sensor surface becomes reusable and only a small increase of R_{ct} for the denatured state can be observed after each cycle. Additionally, a high relative signal increase after hybridisation can be retained.

Selectivity. Alternating incubations of the sensor chip with complementary and non-complementary ssDNA strands demonstrate the specificity of the recognition interface. After hybridisation of the complementary strand the charge transfer resistance increases about three times in relation to the denaturated state. In contrast, the incubation with a non-complementary strand has a negligible effect on the charge transfer resistance ($< 5\%$). This is illustrated in Fig. 14. The redox reaction of ferri-/ferrocyanide is only altered after hybridisation with complementary target DNA but it is not influenced in the presence of non-complementary DNA. This indicates low unspecific binding of oligonucleotides to the surface.

Sensitivity. When the sensor surface is exposed to different concentrations of complementary DNA the resulting impedance changes accordingly. The curve obtained for an 18-mer oligonucleotide is shown in Fig. 15b. The curve has

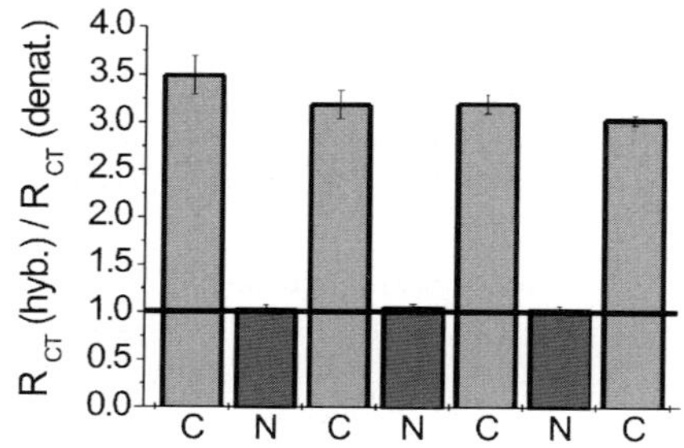

Fig. 14 Change of the charge transfer resistance R_{Ct} after incubation of the sensor surface with complementary (C) and non-complementary (N) target DNA illustrated as relative change compared to the R_{ct} value of the denatured sensor

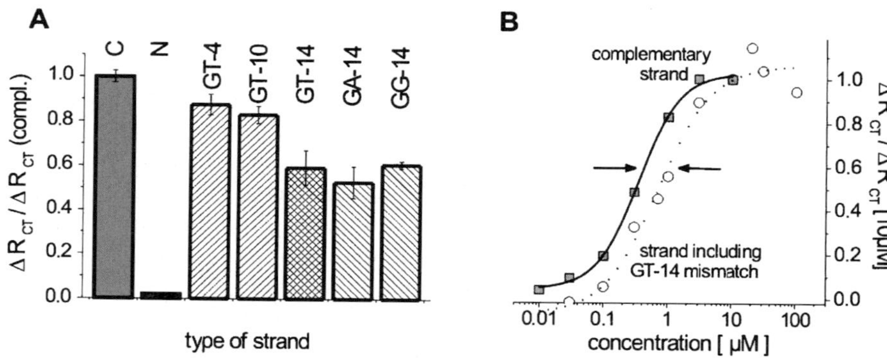

Fig. 15 **A** Impedimetric results of a mismatch detection expressed as relative impedance changes compared to a hybridisation with a complementary strand. The mismatch strands vary in the type of mismatch (GT, GA, GG) and in the position within the duplex counted from the solution end of the strand (4,10,14) (conditions: 18-mer modified gold electrode; 0.1 M phosphate buffer pH 7; 10 mM ferri-/ferrocyanide). **B** Affinity curves obtained for a fully complementary 18-mer and a strand containing a GT mismatch at position 14. The lower affinity can be deduced from the shift of the curve to higher concentrations

a sigmoidal shape and describes the affinity between the sensor surface and the respective ssDNA strand in solution. The half maximum signal for the complementary DNA is observed at 350 nM. The linear range is between 0.1 and 1 μM and marks the concentration range applicable for DNA quantification.

Mismatch detection. Detection of single base pair mismatches within DNA duplexes (single nucleotide polymorphism) is of importance in medical therapy and diagnostics. Therefore, sensor constructions which can discriminate single base pair mismatches are of particular interest. For the impedimetric sensor presented here a smaller R_{ct} value can be found for all strands containing a single mismatch compared to the situation of a fullmatch strand binding. In addition, different mismatches (variation of position in the strand and/or type of base pair mismatch) result in different impedance changes. This is shown in Fig. 15a.

For example, the GT mismatch causes decreasing changes of the charge transfer resistance when the distance of the mismatch position in the immobilised DNA duplex from the electrode surface is reduced (GT4 $\rightarrow$ GT14). On the other hand, all possible types of mismatches in position 14 from the 5′-end of the target strand yield nearly the same change of charge transfer resistance signal, which is lower than a GT mismatch signal in position four or ten.

To exclude that the observed effects are caused by reduced affinities, binding curves are recorded for the mismatched ssDNA targets as well. Figure 15b shows the affinity curve for the GT-14 mismatch in comparison to the fullmatch strand. As expected, the mismatch reduces the affinity between the immobilised probe and the target DNA. However, saturated DNA duplex coverage of the electrode is observed for 10 μM target DNA, which is also the working concentration for the mismatch detection measurements. This proves that the variation in R_{ct} for the different mismatches is not caused by different affinities. Thus, structural differences are behind this effect, which can be used for an analytical detection of SNPs.

In conclusion, it can be stated that based on ssDNA modified Au-chip electrodes and an impedimetric surface analysis a label-free and repeated DNA detection is possible in the nanomolar concentration range with low unspecific binding of non-complementary DNA.

4
Impedimetric Detection of Autoantibodies Involved in Celiac Disease

For the detection and quantification of antibodies, various immunological methods such as ELISAs (enzyme-linked immunosorbent assay), western blot procedures or line assays are available. A common feature of these methods is that they require either expensive technical equipment or are characterised by rather extensive work demanding a skilled person. Therefore these common biochemical immunoassay methods are restricted to the laboratory.

For the development of small-scale devices, which can be used by untrained persons and which are suited for point-of-care measurements, electrochemical immunosensors are promising alternatives. In contrast to the above mentioned standard biochemical immunoassays, where the antigen is immobilised onto a microplate or a membrane, electrodes serve as solid supports. Here, the formation of the antigen-antibody complex on the electrode surface is detected by electrochemical means making use of the changed interfacial properties. Besides amperometric and potentiometric detection, impedance spectroscopy has gained increasing attention. However, as already indicated in Sect. 2.1, amplification of the impedance signal is often necessary in order to detect substances in the relevant concentration range. Herein one approach is the formation of a precipitate on the electrode surface by an enzymatic reaction which results in an increase of the interfacial impedance.

One example for such an impedimetric immunosensor for the diagnosis of the autoimmune disorder celiac disease is described in this section. Autoimmune diseases result from disturbances in the immune defence reaction which triggers the immune system to the production of antibodies directed against structures of the own body. Therefore these antibodies are termed autoantibodies. Celiac disease is caused by an incompatibility against cereal proteins, especially gliadins. Consuming wheat, rye or barley, a.o., leads the immune system to the production of antibodies mainly directed against gliadins and tissue transglutaminase (TTG). These two types of autoantibodies are responsible for an inflammation reaction in the small intestine that prevents the absorption of essential nutrients. For the diagnosis of this disease, the concentrations of antigliadin and antitransglutaminase autoantibodies in human serum are valuable clinical markers [83].

4.1 Sensor Preparation and Assay Procedure

For the preparation of a suitable immunosensor which enables an electrochemical detection of the relevant autoantibodies, different approaches for immobilisation of the antigens onto the gold electrode surfaces can be applied. Besides polymer films, covalent coupling or bioaffinity approaches, more recently the polyelectrolyte adsorption technique (or layer-by-layer technique [46, 47]) has gained some interest. The method relies on the adsorption of oppositely charged macromolecules to the surface and thus also allows the immobilisation of proteins. The method is rather simple and particularly suited for when disposable sensors are developed. For the construction of sensors for antigliadin antibodies but also for antiTTG antibodies poly-(styrene sulfonic acid) (PSS) can be used. This polyelectrolyte forms a negatively charged layer on the electrode surface which enables the electrostatic binding of gliadins and TTG. Residual and unspecific binding sites are finally blocked with bovine serum albumin.

After exposing this ready-to-use immunosensor to a solution containing the relevant autoantibody, two further incubation steps have to be performed: First with a secondary, peroxidase-labelled antibody and second with a substrate solution consisting of 3-amino-9-ethylcarbazole (AEC) and H_2O_2. The peroxidase catalyses the conversion of AEC to 3-azo-9-ethylcarbazole, which forms a precipitate on the sensor electrode surface. The resulting interfacial

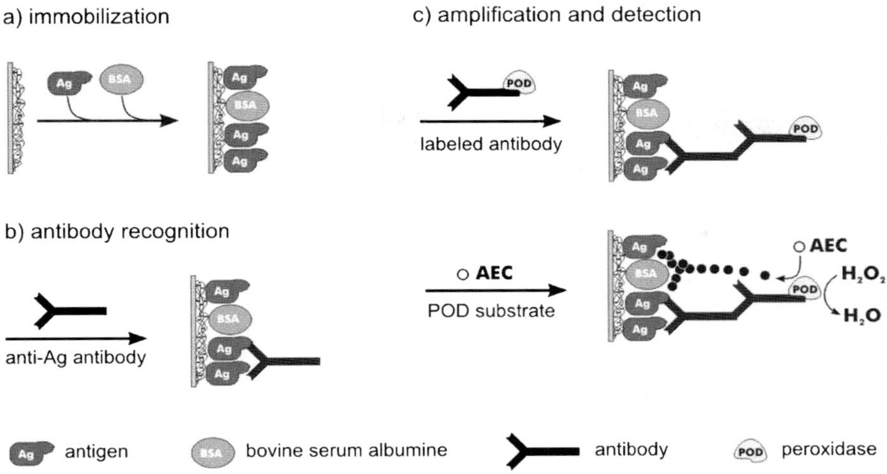

Fig. 16 Schematic presentation of the antigen (gliadin or TTG) immobilisation, immune reaction and signal amplification on gold electrodes modified with the polyelectrolyte poly-styrenesulfonic acid (PSS). Binding of proteins is achieved by electrostatic interaction with the negatively charged PSS layer. Blocking of residual and unspecific binding sites is performed with bovine serum albumin (BSA). Signal amplification of the antibody-antigen interaction is achieved by a second POD-labelled antibody with subsequent conversion of the enzyme substrates (AEC + H_2O_2)

change of the electrode is finally measured by EIS in the presence of a redox mediator $[Fe(CN)_6]^{3-/4-}$ [112]. The procedure is illustrated in Fig. 16.

4.2 Immunosensors Based on Gold Electrodes

After preparation of the immunosensor using gold wire electrodes and performing the immunoassay procedure as described in Fig. 16, the impedance of the sensor electrode can be measured after the different incubation steps of the assay. In the course of the immunoassay procedure, an increase in the interfacial impedance is detected with each successive step (Fig. 17a). The increase from the ready-to-use immunosensor (a) to binding of antigliadin antibodies (b) and incubation with secondary POD-labelled antibodies (c) is relatively low. However, a drastic impedance increase can be detected after POD-catalysed oxidation of its AEC substrate (d). The amplification step generates a signal allowing the quantification of different antibody concentrations.

Fitting of the impedance spectra according to a simplified Randles equivalent circuit shows that changes in the solution resistance R_s and the double layer capacitance C_{dl} are only small, but a clear increase in the charge transfer resistance R_{ct} can be observed. Therefore R_{ct} can be elucidated as the sensor

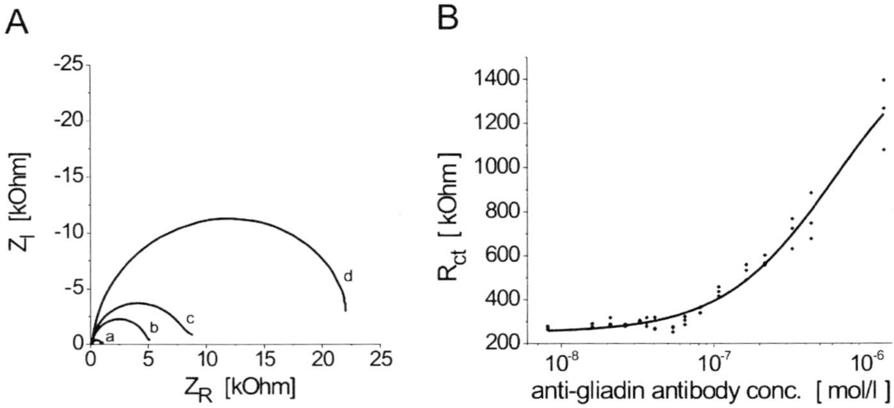

Fig. 17 **A** Impedance spectra of the gliadin-immunosensor after the different steps of the assay procedure: *a* ready-to-use immunosensor; *b* binding of antigliadin antibodies; *c* incubation with secondary POD-labelled antibodies; *d* POD-catalysed oxidation of AEC (impedance measurement: 5 mM phosphate buffer pH 7; 5 mM ferri-/ferrocyanide; OCP). **B** Calibration graph established by incubating the immunosensor with solutions of different concentrations of antigliadin antibodies. Fitting of the R_{CT} values according to a sigmoidal equation resulted in a half-maximal saturation of $c_{1/2} = 613$ nM and a correlation of $r^2 = 0.96$

parameter. By investigating solutions of different antigliadin antibody concentrations, it can be seen that increasing R_{ct} values correspond to increasing amounts of the autoantibody. The correlation can be described by a calibration graph in the range from approximately 10 nM to 1 μM and is shown in Fig. 17b. The experiments may illustrate that modified gold electrodes in combination with a sandwich-type of assay and an impedimetric detection are suitable for the detection of physiologically relevant autoantibody concentrations.

4.3 Immunosensors Based on Screen-printed Electrodes

The impedance analysis of modified metal electrodes provides sufficient sensitivity for autoantibody detection. However, disposable sensors are needed when the analysis is performed close to the patient, e.g. at the physician's site. With this respect screen-printing technology opens the door to a large number of sensor electrodes with reproducible properties and a reasonable price.

In this section the application of screen-printed gold electrodes as immunosensors is exemplified for the detection of antitransglutaminase antibodies – the other important diagnostic marker for celiac disease. For the immobilisation of the antigen (transglutaminase) onto the surface of screen-printed electrodes, the same immobilisation strategy applied for the fixation of gliadins onto gold wire electrodes can be used. The assay procedure and the impedance analysis can be performed in a planar chip cell with several electrodes (Fig. 18). The measurement also results in increasing interfacial impedance values after the successive incubation steps and enables a reliable detection of antitransglutaminase antibodies after the peroxidase-catalysed AEC oxidation. In contrast to the gold wire-based immunosensor, no complete semicircles can be obtained within the frequency range from 0.5 to 1000 Hz. Furthermore, only little differences in the impedance values are detected between the ready-to-use transglutaminase immunosensor, the immunosensor after binding of antitransglutaminase antibodies and after the incubation of the sensor with secondary POD-labelled antibodies. Only the amplification step by the peroxidase reaction results in a clear signal increase, although the impedance differences between the ready-to-use immunosensor and the sensor after the assay procedure is smaller compared to the gold wire-based gliadin sensor [169].

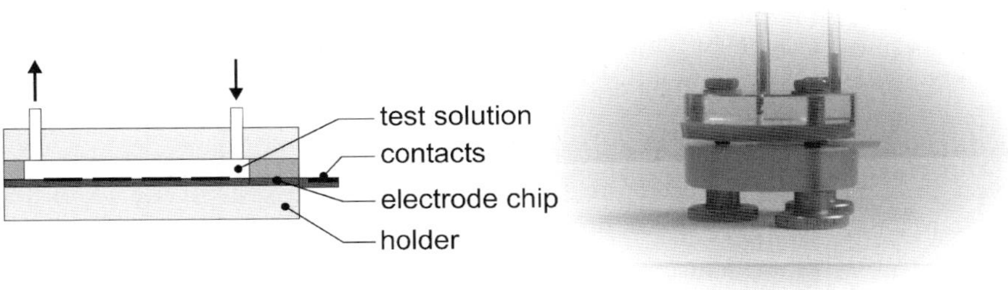

Fig. 18 Schematic view of the measuring cell used for the impedimetric detection of autoantibodies based on a screen-printed electrode chip

Fitting of the impedance spectra of the ready-to-use immunosensor and the immunosensor after POD-catalysed AEC oxidation, according to a simplified Randles equivalent circuit, revealed that the C_{dl} element has to be replaced by a constant phase element (CPE) (see inset Fig. 19). This can be explained by the relatively rough surface of the screen-printed electrodes, in comparison to gold metal electrodes, which leads to an increase in the heterogeneity of the electrode surface. However, the very low variation in the α values ($\alpha = 0.90 \pm 0.02$) between different immunosensors results in a highly reproducible determination of antitransglutaminase antibody concentrations by evaluating the R_{ct} value. This enables the establishment of a calibration graph with a very high correlation of $r^2 = 0.98$.

Measurement of impedance data in the whole frequency range is technically expending. This procedure can be simplified by performing impedance measurements at only two selected frequencies which guarantee reliable fitting results. From the Bode plot in Fig. 19 it can be seen that fitting of impedance data obtained at only 0.5 and 100 Hz results in a curve approx-

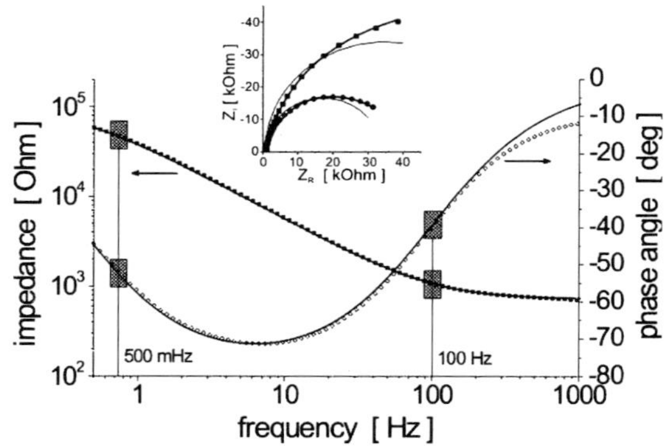

Fig. 19 Bode plot of measured impedance data and fitted impedance curves of a TTG immunosensor after the assay procedure (after AEC oxidation): (■) impedance data; (•) phase angle; (—) fitting results of the impedance curve and the phase angle shift using a Randles equivalent circuit containing a CPE element. The fit is only based on the measured data at 0.5 Hz and 1000 Hz. *Insert*: Fitting of impedance spectra of the ready-to-use immunosensor and the immunosensor after the AEC oxidation step according to a Randles equivalent circuit containing a C_{dl} element (—) or alternatively a CPE (—)

imation nearly close to the measured data points over the whole frequency range. Only at higher frequencies do small deviations appear in the absolute impedance, however this is not important for the determination of R_{ct} as the sensor parameter.

4.4 Analysis of Human Sera

For demonstrating the suitability of the developed immunosensor for clinical purposes, different human sera can be analysed for their autoantibody concentrations. The results are compared with a commercially available line assay (Seramun Diagnostica Ltd., Germany). In order to get a reliable diagnostic indication for celiac disease, human sera have to be analysed for their autoantibody concentrations from the IgG and the IgA subtype of immunoglobulins. IgA analysis is performed similar to the IgG detection, only the second POD-labelled antibody is replaced by a POD-labelled antiIgA antibody.

Table 3 collects the data obtained with the impedimetric immunosensor and a commercial immunoassay. It can be seen that sera which are classified as positive in the line assay with respect to a distinct autoantibody (P1–P4), also give in the measurement with the immunosensor high autoantibody concentrations. Negative sera (N1–N4) are explicitly identified by both assay procedures. Even discrepant samples (e.g. P1 and N4 with respect to the antitransglutaminase antibody concentration from the IgG and IgA subtype, respectively) are recognised by both immunoassay procedures.

Table 3 Assay results of different (clinical unambiguously characterised) human sera analysed with the developed immunosensor and with a commercially available line assay. P – serum which is positive for celiac disease; N – negative serum; (+) positive tested serum; (–) negative tested serum; (+ / –) weak positive serum. Antitransglutaminase antibody concentrations are given in kΩ because of the lack of concentration values in the calibration graph (dilution was used), n.d. – not determined

Sample	Immunosensor Antigliadin IgG (µM)		IgA (µM)		AntiTTG IgG (kΩ)		IgA (kΩ)		Line assay Antigliadin IgG	IgA	AntiTTG IgG	IgA
P1	840	+	140	+	74	+ / –	93	+ / –	+	+	+ / –	+
P2	396	+	56	–	60	–	20	–	+	–	–	–
P3	n.d.		n.d.		198	+	144	+	n.d.	n.d.	+	+
P4	n.d.		n.d.		109	+	137	+	n.d.	n.d.	+	+
N1	48	–	77	–	n.d.		n.d.		–	–	n.d.	n.d.
N2	54	–	67	–	n.d.		n.d.		–	–	n.d.	n.d.
N3	n.d.		n.d.		60	–	21	–	n.d.	n.d.	–	–
N4	n.d.		n.d.		66	–	44	+ / –	n.d.	n.d.	+ / –	+ / –

In summary, it can be concluded that the impedimetric immunosensor offers comparable results to a commercially available assay. Furthermore, the immunosensor can clearly distinguish between the different sera. That means a differentiation of positive sera in weak and strong reacting specimens is possible.

The detection of individual autoantibodies against gliadin or TTG by impedance measurements can serve as a basis for the development of a multianalyte assay procedure. Future work will focus on the production of multiparameter immunosensor chips which are able to detect several antibodies simultaneously. Hereby the use of industrially manufactured screen-printed electrodes facilitates a reproducible, large-scale production of disposable immunosensors. Furthermore, the electrochemical evaluation of the sensor signal by impedance spectroscopy at evaluated frequencies enables the development of small scale devices which are applicable for point-of-care measurements.

5 Conclusions

In conclusion it can be stated that impedimetric detection can achieve the necessary analytical sensitivity for practical applications in medicine, gene analysis, food industry or environmental analysis. Interdigitated electrodes and amplification protocols contribute enormously to the present state-of-the-art. However, many sensors described in the literature have not been tested with real samples. Additional problems are expected when interfering compounds are present and non-specific binding occurs. This is much more severe in impedance analysis compared to optical or amperometric detection and will increase the reported detection limits. Furthermore the recording of a full impedance spectrum within a broad range of frequencies is time-consuming but not always needed. Thus, until now impedimetric techniques are preferentially used as a characterisation method rather than as analytical method for daily routine applications. However, the label-free and direct binding detection is so attractive that the door to the analytical applications has been widely opened. Therefore, it can be expected that further research will be directed to simplify measurement schemes and devices (e.g. single or dual frequency measurement), robust amplification cycles and applications under real conditions.

References

1. Macdonald JR (1987) Impedance Spectroscopy. Wiley, New York
2. Rubinstein I (1995) Physical Electrochemistry. Dekker, New York

3. Krause S (2003) Encyclopedia of Electrochemistry, vol 3. Instrumentation and Electroanalytical Chemistry. Wiley-VCH, Weinheim
4. Buck RP (1975) Crit Rev Anal Chem 5:323
5. Malmgreen-Hansen B, Sorensen TS, Jensen JB, Henneberg M (1989) J Colloid Interface Sci 130:359
6. Canas A, Ariza MJ, Benavente J (2001) J Membrane Sci 183:135
7. Lisdat F, Moritz W (1994) Thin Solid Films 248:126
8. Wiegand G, Arribas-Layton N, Hillebrandt H, Sackmann E, Wagner P (2002) J Phys Chem B 106:4245
9. Pejcic B, de Marco R (2006) Electrochim Acta 51:6217
10. Wagner N (2002) J Appl Electrochem 32:859
11. Himanen O, Hottinen T (2006) Electrochim Acta 52:581
12. Agel E, Bouet J, Fauvarque JF (2001) J Power Sour 101:267
13. Xie SL, Camman K (1987) J Electroanal Chem 229:249
14. Nagarale RK, Gohil GS, Shahi VK (2006) Adv Coll Interface Sci 119:97
15. Touzi H, Chevalier Y, Kalfat R, Ben Ouada H, Zarrouk H, Chapel JP, Jaffrezic-Renault N (2006) Mater Sci Eng C 26:462
16. Krommenhoek EE, Gardeniers JGE, Bomer JG, van den Berg A, Li X, Ottens M, van der Wielen LAM, van Dedem GWK, van Leeuwen M, van Gulik WM, Heijnen JJ (2006) Sensor Actuator B Chem 115:384
17. Cheung K, Gawad S, Renaud P (2005) Cytometry Part A 65A:124
18. Moritz W, Szeponik J, Lisdat F, Friebe A, Krause S, Hintsche R, Scheller F (1992) Sensor Actuator B7:497
19. Friebe A, Lisdat F, Moritz W (1993) Sensor Mater 5:65
20. Willner I (2002) Science 298:2407
21. Gardies F, Martelet C, Colin B, Mandrand B (1989) Sensor Actuator 17:461
22. Gebbert A, Alvarezicaza M, Stocklein W, Schmid RD (1992) Anal Chem 64:997
23. Berggren C, Johansson G (1997) Anal Chem 693:651
24. Katz E, Willner I (2003) Electroanalysis 15:913
25. Guan JG, Miao YQ, Zhang QJ (2004) J Biosci Bioeng 97:219
26. Berggren C, Bjarnason B, Johansson G (2001) Electroanalysis 13:173
27. Saby C, Jaffrezicrenault N, Martelet C, Colin B, Charles MH, Delair T, Mandrand B (1993) Sensor Actuator B Chem 16:458
28. Bart M, Stigter ECA, Stapert HR, de Jong GJ, van Bennekom WP (2005) Biosens Bioelectron 21:49
29. Knichel M, Heiduschka P, Beck W, Jung G, Gopel W (1995) Sensor Actuator B Chem 28:85
30. Rickert R, Gopel W, Beck W, Jung G, Heiduschka P (1996) Biosens Bioelectron 11:757
31. Taira H, Nakano K, Maeda M, Takagi M (1993) Anal Sci 9:199
32. Ameur S, Martelet C, Jaffrezic-Renault N, Chovelon JM (2000) Appl Biochem Biotechnol 89:161
33. Sadik OA, Xu H, Gheorghiu E, Andreescu D, Balut C, Gheorghiu M, Bratu D (2002) Anal Chem 74:3142
34. Ameur S, Maupas H, Martelet C, Jaffrezic-Renault N, Ben Ouada H, Cosnier S, Labbe P (1997) Mater Sci Eng C 5:111
35. Maupas H, Soldatkin AP, Martelet C, Jaffrezicrenault N, Mandrand B (1997) J Electroanal Chem 421:165
36. Gebbert A, Alvarezicaza M, Peters H, Jager V, Bilitewski U, Schmid RD (1994) J Biotechnol 32:213

37. Varlan AR, Suls J, Sansen W, Veelaert D, de Loof A (1997) Sensor Actuator B Chem 44:334
38. Bain CD, Troughton EB, Tao YT, Evall J, Whitesides GM, Nuzzo RG (1989) J Am Chem Soc 111:321
39. Anderson JL, Bowden EF, Pickup PG (1996) Anal Chem 68:R379
40. Mirsky VM, Riepl M, Wolfbeis OS (1997) Biosens Bioelectron 12:977
41. Wink T, Zuilen SJ, Bult A, van Bennekom WP (1997) Analyst 122:R43
42. Prodromidis MI, Hirsch T, Mirsky VM, Wolfbeis OS (2003) Electroanalysis 15:1795
43. Mirsky VM (2002) TRAC Trends Anal Chem 21:439
44. Barraud A, Perrot H, Billard V, Martelet C, Therasse J (1993) Biosens Bioelectron 8:39
45. Wang LG, Li YH, Tien HT (1995) Bioelectrochem Bioenerg 36:145
46. Cassier T, Lowack K, Decher G (1998) Supramol Sci 5:309
47. Calvo EJ, Battaglini F, Danilowicz C, Wolosiuk A, Otero M (2000) Faraday Discuss 47:75
48. Berggren C, Bjarnason B, Johansson G (1998) Biosens Bioelectron 13:1061
49. Bataillard P, Gardies F, Jaffrezicrenault N, Martelet C, Colin B, Mandrand B (1988) Anal Chem 60:2374
50. Hou YX, Tlili C, Faffrezic-Renault N, Zhang AD, Martelet C, Ponsonnet L, Errachid A, Samitier J, Bausells J (2004) Biosens Bioelectron 20:1126
51. Hleli S, Martelet C, Abdelghani A, Burais N, Jaffrezic-Renault N (2006) Sensor Actuator B Chem 113:711
52. Jie M, Ming CY, Jing D, Cheng LS, Na LH, Jun F, Xiang CY (1999) Electrochem Commun 1:425
53. Pak SC, Penrose W, Hesketh PJ (2001) Biosens Bioelectron 16:371
54. Woodhouse GE, King LG, Wieczorek L, Cornell BA (1998) Faraday Discuss 111:247
55. Cornell BA, Braach-Maksvytis VLB, King LG, Osman PDJ, Raguse B, Wieczorek L, Pace RJ (1997) Nature 387:580
56. Sargent A, Loi T, Gal S, Sadik OA (1999) J Electroanal Chem 470:144
57. Ouerghi O, Senillou A, Jaffrezic-Renault N, Martelet C, Ben Ouada H, Cosnier S (2001) J Electroanal Chem 501:62
58. Ouerghi O, Touhami A, Jaffrezic-Renault N, Martelet C, Ben Ouada H, Cosnier S (2002) Bioelectrochemistry 56:131
59. Hason S, Dvorak J, Jelen F, Vetterl V (2002) Talanta 56:905
60. Strasak L, Dvorak J, Hason S, Vetterl V (2002) Bioelectrochemistry 56:37
61. Drummond TG, Hill MG, Barton JK (2003) Nat Biotechnol 21:1192
62. Kelley SO, Boon EM, Barton JK, Jackson NM, Hill MG (1999) Nucleic Acids Res 27:4830
63. Fojta M, Havran L, Billova S, Kostecka P, Masarik M, Kizek R (2003) Electroanalysis 15:431
64. Hashimoto K, Ito K, Ishimore Y (1994) Anal Chim Acta 286:219
65. Pänke O, Kirbs A, Lisdat F (2007) Biosensor Bioelectron 22:2656
66. Zhao YD, Pang DW, Hu S, Wang ZL, Cheng JK, Qi YP, Dai HP, Mao BW, Tian ZQ, Luo J, Lin ZH (1999) Anal Chim Acta 388:93
67. Kafka J, Pänke O, Abendroth B, Lisdat F (2007) submitted
68. Korri-Youssoufi H, Garnier F, Srivtava P, Godillot P, Yassar A (1997) J Am Chem Soc 119:7388
69. Li CM, Sun CQ, Song S, Choong VE, Maracas G, Zhang XJ (2005) Front Biosci 10:180
70. Wang J, Jiang M, Fortes A, Mukherjee B (1999) Anal Chim Acta 402:7
71. Peng H, Soeller C, Cannell MB, Bowmaker GA, Cooney RP, Travas-Seijdic J (2006) Biosens Bioelectron 21:1727

72. Piro B, Haccoun J, Pham MC, Tran LD, Rubin A, Perrot H, Gabrielli C (2005) J Electroanal Chem 577:155
73. Gautier C, Cougnon C, Pilard JF, Casse N (2006) J Electroanal Chem 587:276
74. Yan F, Sadik OA (2001) Anal Chem 73:5272
75. Li CZ, Liu YL, Luong JHT (2005) Anal Chem 77:478
76. Li CZ, Long YT, Lee JS, Kraatz HB (2004) Chem Commun 5:574
77. Liu XY, Li CY, Wang CF, Li T, Hu SS (2006) J Appl Polym Sci 101:2222
78. Hayden O, Lieberzeit PA, Blaas D, Dickert FL (2006) Adv Funct Mater 16:1269
79. Dickert FL, Hayden O, Lieberzeit PA, Haderspoeck C, Bindeus R, Palfinger C, Wirl B (2003) Synth Metal 138:65
80. Shervedani RK, Mehrjardi AH, Zamiri N (2006) Bioelectrochemistry 69:201
81. Ye JS, Ottova A, Tien HT, Sheu FS (2003) Bioelectrochemistry 59:65
82. Jeuken LJC, Connell SD, Henderson PJF, Gennis RB, Evans SD, Bushby RJ (2006) J Am Chem Soc 128:1711
83. Alaedini A, Green PHR (2005) Ann Internal Med 142:289
84. Naumann R, Schmidt EK, Jonczyk A, Fendler K, Kadenbach B, Liebermann T, Offenhausser A, Knoll W (1999) Biosens Bioelectron 14:651
85. Schasfoort RBM, Niedziela T (1994) Sensor Actuator B Chem 18:175
86. Tong YH, Han XJ, Song YH, Jiang JG, Wang EK (2003) Biophys Chem 105:1
87. Wilkop T, Xu DK, Cheng Q (2007) Langmuir 23:1403
88. Cornell BA, Braach-Maksvytis VLB, King LG, Osman PDJ, Raguse B, Wieczorek L, Pace RJ (1997) Nature 387:580
89. McNeil CJ, Athey D, Ball M, Ho WO, Krause S, Armstrong RD, Wright JD, Rawson K (1995) Anal Chem 67:3928
90. Gooding JJ, Hall EA (1996) Biosens Bioelectron 11:1031
91. Sumner C, Krause S, Sabot A, Turner K, McNeil CJ (2001) Biosens Bioelectron 16:709
92. Hug TS (2003) Assay Drug Devel Technol 1:479
93. Yotter RA, Wilson DM (2004) IEEE Sens J 4:412
94. Yeon JH, Park JK (2005) Anal Biochem 341:308
95. Xiao C, Luong JHT (2003) Biotechnol Prog 19:1000
96. K'Owino IO, Sadik OA (2005) Electroanalysis 17:2101
97. Kyle AH, Chan CTO, Minchinton AI (1999) Biophys J 76:2640
98. Silley P, Forsythe S (1996) J Appl Bacteriol 80:233
99. K'Owino IO, Sadik OA (2005) Electroanalysis 17:2101
100. Jonsson M, Welch K, Hamp S, Stromme M (2006) J Phys Chem B 110:10165
101. Cheung K, Gawad S, Renaud P (2005) Cytometry Part A 65A:124
102. Maupas H, Soldatkin AP, Martelet C, Jaffrezicrenault N, Mandrand B (1997) J Electroanal Chem 421:165
103. Bresler HS, Lenkevich MJ, Murdock JF, Newmann AL, Roblin RO (1992) In: Mathewson PR, Finley JW (eds) Biosensor Design and Application. American Chemical Society, Washington DC
104. Patolsky F, Lichtenstein A, Willner I (2001) J Am Chem Soc 123:5194
105. Patolsky F, Lichtenstein A, Willner I (2000) Angew Chem Int Ed 39:940
106. Patolsky F, Katz E, Bardea A, Willner I (1999) Langmuir 15:3703
107. Bardea A, Katz E, Willner I (2000) Electroanalysis 12:1097
108. Alfonta L, Bardea A, Khersonsky O, Katz E, Willner I (2001) Biosens Bioelectron 16:675
109. Katz E, Alfonta L, Willner I (2001) Sensor Actuator B Chem 76:134
110. Alfonta L, Singh AK, Willner I (2001) Anal Chem 73:91
111. Yoon HC, Yang HS, Kim YT (2002) Analyst 127:1082

112. Balkenhohl T, Lisdat F (2007) Analyst 132:314
113. Patolsky F, Lichtenstein A, Kotler M, Willner I (2001) Angew Chem Int Ed 40:2261
114. Patolsky F, Lichtenstein A, Willner I (2001) Nat Biotechnol 19:253
115. Ruan CM, Yang LJ, Li YB (2002) Anal Chem 74:4814
116. Xu Y, Cai H, He PG, Fang YZ (2004) Electroanalysis 16:150
117. Peng H, Soeller C, Cannell MB, Bowmaker GA, Cooney RP, Travas-Sejdic J (2006) Biosens Bioelectron 21:1727
118. Kerman K, Morita Y, Takamura Y, Ozsoz M, Tamiya E (2004) Electroanalysis 16:1667
119. Fu YZ, Yuan R, Xu L, Chai YQ, Zhong X, Tang DP (2005) Biochem Eng J 23:37
120. Göpel W, Hesse J, Zemel JN (eds) (1998) Sensors: A Comprehensive Survey, vol 2. Wiley-VCH, Weinheim
121. Caras S, Janata J (1980) Anal Chem 52:1935
122. Kharitonov AB, Zayats M, Lichtenstein A, Katz E, Willner I (2000) Sensor Actuator B Chem 70:222
123. Hintsche R, Dransfeld I, Scheller F, Pham MT, Hoffmann W, Hueller J, Moritz W (1990) Biosens Bioelectron 5:327
124. Thust M, Schoning MJ, Vetter J, Kordos P, Luth H (1996) Anal Chim Acta 323:115
125. Schoning MJ, Arzdorf M, Mulchandani P, Chen W, Mulchandani A (2003) Sensors 3:119
126. Abdelmalek F, Shadaram M, Boushriha H (2001) Sensor Actuator B Chem 72:208
127. Kharitonov AB, Zayats M, Alfonta L, Katz E, Willner I (2001) Sensor Actuator B Chem 76:203
128. Maupas H, Saby C, Martelet C, Jaffrezic-Renault N, Soldatkin AP, Charles MH, Delair T, Mandrand B (1996) J Electroanal Chem 406:53
129. Billard V, Martelet C, Binder P, Therasse J (1991) Anal Chim Acta 249:367
130. Souteyrand E, Cloarec JP, Martin JR, Wilson C, Lawrence I, Mikkelsen S, Lawrence MF (1997) J Phys Chem B 101:2980
131. Souteyrand E, Chen C, Cloarec JP, Nesme X, Simonet P, Navarro I, Martin JR (2000) Appl Biochem Biotechnol 89:195
132. Berney H, West J, Haefele E, Alderman J, Lane W, Collins JK (2000) Sensor Actuator B Chem 68:100
133. Estrela P, Stewart AG, Yan F, Migliorato P (2005) Electrochim Acta 50:4995
134. Ingebrandt S, Offenhausser A (2006) Phys Status Solidi A 203:3399
135. Poghossian A, Ingebrandt S, Abouzar MH, Schoning MJ (2007) Appl Phys A Mater 87:517
136. Lillis B, Manning M, Hurley E, Berney H, Duane R, Mathewson A, Sheehan MM (2007) Biosens Bioelectron 22:1289
137. Hahm J, Lieber CM (2004) Nano Lett 4:51
138. Gruner G (2006) Anal Bioanal Chem 384:322
139. van Gerwen P, Laureyn W, Laureys W, Huyberechts G, de Beeck MO, Baert K, Suls J, Sansen W, Jacobs P, Hermans L, Mertens R (1998) Sensor Actuator B Chem 49:73
140. Laureyn W, Nelis D, van Gerwen P, Baert K, Hermans L, Magnee R, Pireaux JJ, Maes G (2000) Sensor Actuator B Chem 68:360
141. Gheorghe M, Guiseppi-Elie A (2003) Biosens Bioelectron 19:95
142. Dharuman V, Grunwald I, Nebling E, Albers J, Blohm L, Hintsche R (2005) Biosens Bioelectron 21:645
143. van Gerwen PV, Laureyn W, Laureys W, Huyberechts H, Op De Beeck M, Baert K, Suls J, Sansen W, Jacobs P, Hermans L, Martens R (1998) Sensor Actuator B 49:73
144. Taylor RF, Marenchic IG, Spencer RH (1991) Anal Chim Acta 249:67
145. Reichl H, Heuberger A (eds) (1996) Microsystem Technologies. VDE-Verlag, Frankfurt am Main

146. Hang TC, Elie AG (2004) Biosensor Bioelectron 19:1537
147. Eldefrawi ME, Sherby SM, Andreou AG, Mansour NA, Annau Z, Blum NA, Valdes JJ (1988) Anal Lett 21:1665
148. Saum AEG, Cumming RH, Rowell FJ (1998) Biosens Bioelectron 13:511
149. Sheppard NF, Mears DJ, Guiseppi-Elie A (1996) Biosens Bioelectron 11:967
150. Ho WO, Krause S, McNeil CJ, Pritchard JA, Armstrong RD, Athey D, Rawson K (1999) Anal Chem 71:1940
151. Montelius L, Tegenfeldt JO, Ling TGI (1995) J Vac Sci Technol A 13:1755
152. Möller R, Csáki A, Köhler JM, Fritsche W (2001) Langmuir 17:5426
153. Park SJ, Taton TA, Mirkin CA (2002) Science 295:1503
154. Sergeyeva TA, Lavrik NV, Piletsky SA, Rachkov AE, El'skaya AV (1996) Sensor Actuator 34:283
155. Sergeyeva TA, Lavrik NV, Rachkov AE, Kazantseva ZI, El'skaya AV (1998) Biosens Bioelectron 13:359
156. Puinno P, Krull U, Hudson R, Damha M, Cohen H (1994) Anal Chim Acta 288:205
157. Puinno P, Krull U (1995) Anal Chem 67:2635
158. Taton T, Lu G, Mirkin C (2001) J Am Chem Soc 123:5164
159. Caruso F, Rodda E, Furlong D (1997) Anal Chem 69:2043
160. Patolsky F, Lichtenstein A, Willner I (2001) J Am Chem Soc 123:5194
161. Nielson B, Grimsrud T, Liles M, Goodman R, Corn R (2001) Anal Chem 73:1
162. Steel A, Herne T, Tarlov M (1998) Anal Chem 70:4670
163. Kelley S, Boon E, Barton J, Jackson N, Hill M (1999) Nucleic Acid Res 27:24
164. Boon E, Ceres D, Drummond T, Hill M, Barton J (2000) Nat Biotechnol 18:1096
165. Umek R, Lin S, Vielmetter J, Terbrueggen R, Irvine B, Yu C, Kayyem J, Yowanto H, Blackburn G, Farkas D, Chen Y (2001) J Mol Diag 374:84
166. Singhal P, Kuhr W (1997) Anal Chem 69:4828
167. Wang J, Rivas G, Fernandes J, Paz J, Jiang M, Waymire R (1998) Anal Chim Acta 375:197
168. Wang J, Kawde A (2001) Anal Chim Acta 431:219
169. Balkenhohl T, Lisdat F (2007) Anal Chim Acta 597:50–57

Adv Biochem Engin/Biotechnol (2008) 109: 239–254
DOI 10.1007/10_2007_074

Published online: 7 November 2007

Amplified Transduction of Biomolecular Interactions Based on the Use of Nanomaterials

Joseph Wang

Biodesign Institute,
Departments of Chemical Engineering and Chemistry and Biochemistry,
Arizona State University, 85287-5801, Tempe, Arizona USA
joseph.wang@asu.edu

Abstract The achievement of very high sensitivity is a major goal in biological assays such as the monitoring of DNA hybridization or protein interactions. This chapter reviews progress in the development of nanomaterials for amplified biosensing and discusses different nanomaterial-based bioamplification strategies. The emergence of nanotechnology is opening new horizons for highly sensitive bioassays and for novel biosensor protocols that employ electronic, optical, or microgravimetric signal transduction. Antibodies or nucleic acids functionalized with metal or semiconductor nanoparticles have been employed as amplifying tags for the biodetection of proteins and DNA. The coupling of different nanomaterial-based amplification platforms and amplification processes dramatically enhances the intensity of the analytical readout and leads to ultrasensitive bioassays. The remarkable sensitivity of the new nanomaterial-based sensing protocols opens up the possibility of detecting disease markers, biothreat agents, or infectious agents that cannot be measured by conventional methods.

Keywords Amplification · Bioelectronic detection · Biosensors · DNA · Nanomaterials · Nanoparticles

1 Why Amplification?

Amplification of analytical signals is of considerable research interest in detecting trace levels of analytes of clinical, security, or environmental significance. Highly sensitive methods are needed for measuring disease biomarkers present at ultra-low levels during early stages of the disease progress. Such early detection should facilitate the treatment of diseases. While the polymerase chain reaction (PCR) amplification has revolutionized genetic testing, it is somewhat restricted owing to its complexity, potential contamination and cost. Highly sensitive monitoring of protein interactions is particularly challenging owing to the absence of PCR-like amplification protocols. Signal amplification has also attracted considerable attention for ultrasensitive detection of infectious agents or biothreat agents. New highly sensitive bioagent detection protocols could prevent outbreaks of foodborne illnesses or provide an early warning of their release, and hence minimizing human casualties. Conventional (optical and electrochemical) sandwich bioaffinity assays have the drawback of capturing a small number of tracers per binding event.

The achievement of highly sensitive biodetection requires novel approaches that combine different amplification platforms and amplification processes. Enzymes have been used traditionally as amplifying labels for detecting biomolecular interactions. Nanotechnology offers unique opportunities for developing novel highly sensitive biosensing devices and ultrasensitive bioassays. This chapter reviews recent advances in the use of nanomaterials for amplified biosensing and discusses different nanomaterial-based amplification pathways aimed at addressing the high sensitivity demands of modern biosensing protocols.

2 Nanoparticle-Based Amplification of Biomolecular Interactions

Nanotechnology offers highly innovative approaches for the monitoring of biomolecular interactions. The attractive properties of nanomaterials offer excellent prospects for increasing the sensitivity of bioaffinity assays of proteins and nucleic acids by several orders of magnitude. The creation of designer nanomaterials for specific sensing applications greatly benefits from the ability to vary the size, composition, and shape of such materials, and hence tailoring their physical properties. Owing to the tiny size of nanomaterials, their properties are strongly influenced by the binding of target biomolecules. Nanoparticles of different dimensions and compositions have been widely used in recent years as sensitive and versatile tracers for the optical, electronic, and microgravimetric transduction of different biomolecular recognition events [1–5]. The dramatic signal enhancement associated with the use of nanoparticle-amplifying tags and with the formation of nanoparticle-biomolecule networks provides the basis for ultrasensitive optical and electrical detection with PCR-like sensitivity. Such procedures combine the amplification properties of nanoparticle-biomolecule assemblies with highly sensitive optical or electrochemical transduction schemes. Multi-amplification schemes, coupling several nanomaterial-based amplification units and processes, can also be designed for addressing further the high sensitivity demands of modern bioassays. The catalytic properties of metal nanoparticles stimulate their enlargement by the same metal or another one to offer dramatic signal amplification. An enormous amplification can be achieved by maximizing the number of tracers captured per binding event, e.g., by encapsulating numerous signal-generating molecules within a microsphere host ("carrier"). These nanomaterial-based amplification schemes can be coupled with additional amplification processes, such as enzymatic recycling or surface accumulation. The following sections discuss modern nanomaterial-based amplification routes for ultrasensitive optical, electronic, or microgravimetric bioassays.

3 Nanoparticle-Amplified Optical Bioaffinity Assays

The driving force behind the use of nanoparticle tracers in optical bioassays has been to address the drawbacks of organic fluorophores. Such ability of nanoparticles to transduce biomolecular binding events was demonstrated first by Mirkin's laboratory [6, 7]. These studies demonstrated that the aggregation of gold nanoparticles, induced by DNA hybridization, leads to materials with remarkable optical properties. For example, the distance-dependent optical properties of aggregated gold nanoparticles were exploited for developing a simple and fast colorimetric protocol for detecting polynucleotides [7]. Such hybridization-induced aggregation of nanoparticle-modified DNA led to a rapid change of the solution color from red to blue. The resulting gold-nanoparticle/DNA networks displayed remarkably sharp DNA melting curves that allowed convenient differentiation of oligonucleotides with single base mismatches [6, 8].

Mirkin [9] reported on a highly sensitive scanometric DNA array detection based on the use of oligonucleotide targets, labelled with gold nanoparticles, for recognizing DNA segments on a chip. When coupled with a signal-amplification method based on nanoparticle-promoted (hydroquinone-induced) reduction of silver(I), the sensitivity of this scanometric array detection system exceeds that of the analogous fluorophore system by two orders of magnitude. The silver precipitation relies on the chemical reduction of silver ions by hydroquinone to silver metal on the surface of the gold nanoparticles. Such silver enhancement facilitated visualization of the nanoparticle tag and enabled quantitation of the hybridized target based on the imaged gray-scale values. In addition, the use of nanoparticle tags altered the melting profiles to offer effective discrimination against single-base mismatches.

Willner's group [10] reported on the use of gold nanoparticles for amplified optical transduction of aptamer–protein interactions. The gold nanoparticle tags were functionalized with the thiolated aptamer, with about 80 aptamers per particle. Binding of the target thrombin protein to the functionalized aptamer resulted in aggregation of the gold nanoparticles and led to a decrease in their plasmon absorbance spectra.

Surface-enhanced Raman scattering (SERS) is another spectroscopic transduction mode that can greatly benefits from the use of gold nanoparticles. Cao et al. [11] used nanoparticles functionalized with oligonucleotides and Raman-active dyes for detecting DNA hybridization. The gold nanoparticles facilitated the formation of a silver coating that acted as a promoter for the Raman scattering of the dyes. High sensitivity down to the 20 femtomolar DNA level was reported. Multiplexed detection was accomplished by using different Raman dyes.

The high-fluorescence intensity of quantum dots (QD) nanocrystals can also be exploited for highly sensitive bioassays. Hahn et al. [12] developed a highly sensitive protocol for detecting a single bacterial pathogen *E. coli* O157 cell using CdSe/ZnS core-shell QD conjugated to streptavidin. This system displayed two orders of magnitude higher sensitivity (as well as higher stability) compared to commonly used fluorescent dyes.

3.1 Nanoparticle-Based DNA-Barcode

The bio-barcode assay (Fig. 1), developed by Mirkin's group, is a powerful amplification method for detecting proteins down to the low attomolar level [13]. This powerful method relies on magnetic spheres functionalized with an antibody that binds specifically the target protein and a secondary antibody conjugated to gold nanoparticles that are encoded with 15–20 mer oligonucleotide strands that are unique to the target protein. Magnetic separation of the complexed probes and target followed by thermal dehybridization of the oligonucleotides on the nanoparticle surface allowed highly sensitive scanometric measurements of the target protein by identifying the oligonucleotide sequence released from the nanoparticle probe. Substantial amplification was achieved since each nanoparticle probe carried a large number of oligonucleotides per protein binding event. Since the extent of signal amplification correlated with the amount of DNA on each nanoparticle carrier (i.e., a high ratio of barcode to target recognition element), efforts have been made to maximize the DNA loading [14]. "Large" gold nanoparticles of up to 250 nm diameter offered ca. two orders of magnitude higher

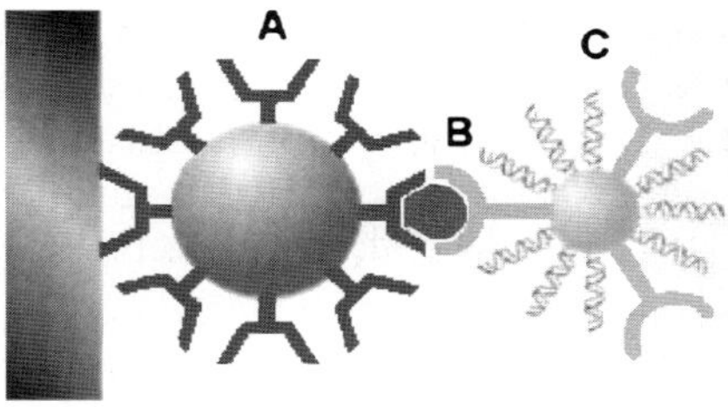

Fig. 1 Nanoparticle-based DNA barcode method for detecting proteins. The system relies on magnetic spheres loaded with antibodies (**A**) that specifically bind to the target antigen (**B**) and gold nanoparticles that are encoded with DNA (**C**) that is unique to the target protein target and antibodies that can sandwich the target captured by the microparticle probes (Reproduced from [13] with permission.)

DNA loading than smaller (13–30 nm) particles. A multi-analyte immunoassay could be accomplished by using different oligonucleotide sequences for encoding different target antigens. Note that with 20-mer oligonucleotides there are 4^{20} unique coding combinations.

The nanoparticle-based bio-barcode assay was applied for detecting the prostate-specific antigen (PSA) at the low attomolar level, i.e., a sensitivity that is six orders of magnitude greater than the corresponding ELISA assay, and for measuring picomolar levels the pathogenic Alzheimer's disease marker ADDL in cerebral spinal fluid [15]. The bio-barcode concept was extended also to PCR-less amplified detection of DNA hybridization down to the 500 zeptomolar level (corresponding to 30 copies in 30 μL samples) [16]. Niemeyer's team recently demonstrated a sensitive optical sandwich immunoassay of proteins using difunctional DNA-gold nanoparticles [17]. In this case, the sandwich immunoassay resulted in the formation of multilayers of DNA-linked gold nanoparticles that were detected by UV/Vis spectroscopy.

3.2 Tag-Embedded Particle Carriers

An attractive protocol for amplifying optical detection of biomolecular interactions involves the encapsulation of a huge amount of a fluorescent tag within a nanoscale carrier [18, 19]. Trau et al. [19] developed on a highly sensitive immunoassay of proteins based on polyelectrolyte-encapsulated microcrystalline fluorescent material interfaced to the antibody. A dramatically (∼2000-fold) amplified immunoassay was obtained in connection to the release of the fluorescent molecules to the detection medium following the antibody–antigen interaction. An analogous route for maximizing the number of fluorescent molecules per binding event, based on encapsulation of a fluorescent dye within silica nanoparticles functionalized with an oligonucleotide probe, was reported by Zhao et al. [19]. This protocol offered a very low detection limit of 0.8 fM, along with effective discrimination against mismatched DNA. The silica matrix also provided good protection against bleaching of the fluorophore. The fluorescence-bioconjugated silica nanoparticles were also applied for ultrasensitive bacterial detection, down to a single *E. coli* O157 cell [20]. A similar encapsulation of electroactive markers for amplified bioelectronic detection is described below.

3.3 Nanoparticle-Based Amplification of Surface Plasmon Resonance and Chemiluminescence Detection

Surface plasmon resonance (SPR) is attractive method for real-time label-free monitoring of biomolecular interactions. The technique relies on measuring changes in the refractive index of a prism coated with a thin metal film associated with the binding event. The sensitivity of SPR-based bioassays is commonly limited by the inability to measure small changes in the refractive index. Keating's team [21] demonstrated the utility of gold nanoparticle tags for dramatically amplifying the SPR detection of sandwich DNA hybridization. Such use of gold nanoparticle labels led to a ten-fold increase in the angle shift, that corresponded to a 1000-fold sensitivity enhancement (compared to the unamplified binding event) and to a detection limit of ~10 pM.

Chemiluminescence and electrochemiluminescence (ECL) are powerful analytical techniques involving the formation of light-emitting excited-state species. Zhang et al. [22] reported recently on the gold-nanoparticle-catalyzed chemiluminescent detection of the luminol/hydrogen-peroxide system. Such enhancement has been attributed to the ability of gold nanoparticles to facilitate the radical generation and electron transfer processes taking place on the surface of the nanoparticles. The catalytic behavior of luminal on gold nanoparticles was exploited by the same group for dramatically enhancing its ECL response in alkaline and neutral media [23]. Willner's group demonstrated the use of platinum nanoparticles as catalytic labels for enhanced chemiluminescent detection of DNA hybridization [24]. The Pt-nanoparticles were modified with thiolated DNA complementary to the captured target, and catalyzed the generation of chemiluminescence in the presence of the luminol/hydrogen-peroxide system.

3.4 Enzyme-Stimulated Growth of Nanoparticles for Amplified Optical Assays

The integration of metal nanoparticles with enzymatic reactions is a relatively new field that offers great promise for designing sensitive biocatalytic assays. For example, the catalytic properties of metal nanoparticles have recently been exploited by Willner's team for amplified biocatalytic optical assays in connection to enzymatically induced particle growth processes [25–29]. Such biocatalytic enlargement of gold-nanoparticles involves the reduction and deposition of a metal onto the nanoparticle by an enzymatically generated reducing agent. It has been exploited for the quantitative optical monitoring of several enzymatic reactions. For example, the NADH-stimulated enlargement of gold-nanoparticle seeds by the biocatalytic deposition of copper shells was exploited for the optical detection of ethanol in the presence of alcohol dehydrogenase [25]. Such NADH-mediated enlargement of gold nanoparticles by the biocatalytic deposition of copper shells is displayed in Fig. 2. Similarly, an optical detection of glucose was accomplished through a hydrogen-peroxide induced growth of gold nanoparticles with gold in the presence of glucose oxidase [26]. Biometallization-based sensing of nerve agents was also reported [27]. In this case, the nerve agent blocks the biocatalytic enlargement of gold nanoparticles by the thiocholine product of the acetylcholine–esterase enzymatic reaction. The neurotransmitter-mediated growth of gold nanoparticles was used for probing the activity of tyrosinase in connection to the biocatalytic conversion of tyrosine to L-DOPA [28].

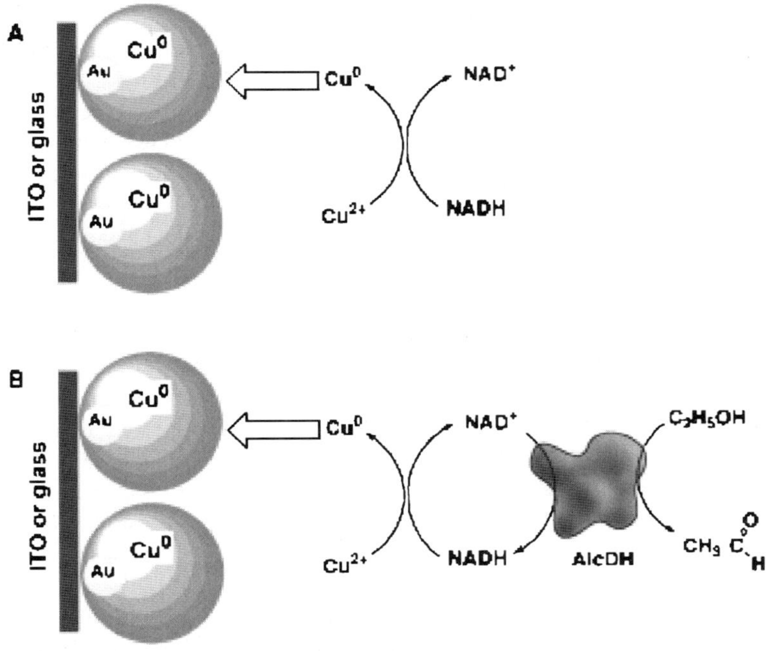

Fig. 2 Enlargement of Au-nanoparticle seeds by the biocatalyzed deposition of copper shells (**A**) Using NADH; (**B**) using enzymatically-generated NADH. (Reproduced from [25] with permission.)

4 Nanoparticle-Enhanced Electrochemical Bioassays

Electrochemical biosensors are devices that intimately combine a biological recognition element to an electrode transducer. Such devices have received a major share of the attention in biosensor development [29]. As desired for numerous decentralized applications, electrochemical transduction is extremely useful for delivering the diagnostic information in a simple, fast, and cost-effective fashion in connection to compact (hand-held) instrumentation.

Several electrochemical protocols offer great promise for ultrasensitive nanoparticle-based transduction of biomolecular interactions. Such metal nanoparticle-tagging is an attractive labelling method in bioaffinity assays systems since it enhances the signal by generating thousands of detectable metal ions for each binding event. Anodic stripping voltammetry (ASV) has been particularly useful for detecting metal nanoparticle tracers, owing to its "built-in" electrolytic preconcentration step that leads to a remarkable sensitivity [30]. About 7 years ago, Limoges's group [31, 32] and our group [33] demonstrated the use of gold nanoparticle tags for stripping-based electrochemical detection of DNA hybridization and antibody–antigen interactions. Such protocols relied on capturing the gold nanoparticles to the hybridized target or captured antibody, followed by their dissolution in acidic bromine-bromide solution and ASV measurement of the metal tracer. A huge number of gold(III) ions are thus released from each captured gold nanoparticle. (A 20-nm spherical gold nanoparticle theoretically contains 2.3×10^5 gold atoms.) Solid-state measurements of the metal nanoparticle tag (without dissolving it) have also been realized through a "magnetic" collection of the DNA-linked particle assembly onto an electrode transducer [35]. Such bioassay involved the hybridization of a target oligonucleotide to probe-coated magnetic beads, followed by binding of the streptavidin-coated gold nanoparticles to the captured target, catalytic silver precipitation on the gold-particle tags, a magnetic attraction of the DNA-linked particle assembly and a chronopotentiometric detection.

Since the sensitivity of such stripping-based electronic bioassays depends on the size of the metallic tag, a dramatic amplification of the hybridization signals is expected in connection to larger nanomaterial tags. For example, a substantial sensitivity enhancement can be achieved by using the metal nanosphere tracers as catalytic labels for subsequent enlargement and further amplification. This was demonstrated for autocatalytic enlargement of the gold tag in connection to nanoparticle-promoted reductive deposition of gold [33, 34] or hydroquinone-induced silver [36, 37]. Combining such enlargement of the metal-particle tracers with the effective "built-in" amplification of ASV paved the way to sub-femtomolar detection limits [34]. The silver or gold reduction/deposition process must be controlled as a trade-off between larger enhancement and a background contribution (due to non-specific metal deposition on the solid-phase support or on the negatively charged DNA). A triple-amplification bioassay, coupling polymeric carrier-sphere amplifying units (loaded with numerous gold nanoparticles tags) with the "built-in" preconcentration of the electrochemical stripping detection and catalytic electroless enlargement of the multiple gold-particle tags has also been demonstrated [38]. Such enlargement of numerous gold nanoparticles tags (on a carrier sphere) represents the fourth generation of amplification (Fig. 3), starting with the early use of single gold nanoparticle tags [31–33]. In addition to ASV measurements of the metal tags, it is now possible to employ new highly sensitive ion selective electrodes (ISE) as transducers. For example, it was shown recently that such potentiometric silver sensors offer highly sensitive monitoring of antibody–antigen interactions in connection to silver nanoparticle-conjugated secondary antibodies [39].

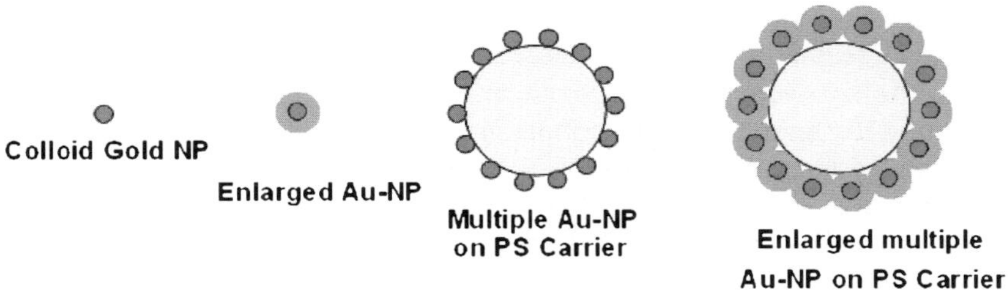

Fig. 3 Gold nanoparticle tracers for bioelectronic detection: different generations of amplification platforms: (*from left to right*) A single nanoparticle tag; catalytic enlargement of the nanoparticle tag; polymer carrier bead loaded with numerous gold nanoparticle tags; catalytic enlargement of multiple tags on the carrier bead

Instead of enlarging the metal nanoparticle tracer, it is possible to increase the amount of the metal marker and hence the sensitivity of such electrochemical detection by using micrometer-long metal nanowires [40]. Indium nanowires have been shown to be particularly useful for such sensitive DNA detection in view of the favorable stripping-voltammetric behavior of indium. Such nanowires were prepared using a porous-membrane template electrodeposition route in connection to a short gold segment for conjugating the thiolated DNA probe. The use of micrometer-long wire tags results in greatly enhanced sensitivity (compared to spherical nanoparticle tracers) and with fM detection limits. DNA-templated metallization represents another approach to increase the amount of the metal tracer and hence for amplifying the electrochemical response. This relies on the vectorial electrostatic "collection" of silver ions along the captured DNA target, followed by hydroquinone-induced reductive formation of silver aggregates along the DNA skeleton, along with stripping detection of the dissolved nanoscale silver cluster [41]. Enzymatically deposited silver along the DNA backbone can also lead to an amplified electronic detection of DNA hybridization [42]. This protocol involves the capture of an alkaline-phosphatase enzyme tracer, and biocatalytic generation of aminophenol that reduces silver ions in the solution and deposits them onto the DNA duplex. Such a biometallization route offers high sensitivity, down to the 100 aM (10 zmol) of the target DNA.

It is possible to use metal nanoparticle labels as catalytic labels in a manner analogous to the common use of enzyme tags in sandwich assays. Polsky et al. [43] recently reported on the use of nucleic-acid functionalized Pt nanoparticles as catalytic labels for amplified ampeonetric detection of DNA hybridization and of aptamer–protein interactions. Capture of the Pt-DNA conjugate was followed by electrocatalytic reduction of hydrogen peroxide as the readout signal (Fig. 2). Similarly, binding of the Pt-particle-aptamer conjugate to the aptamer-thrombin complex offered highly sensitive detection of thrombin down to the 1 nM level. Yang's group recently described a similar use of gold-nanoparticles labels for amplified electrochemical detection of protein immunoassays in connection to electrocatalytic reduction of nitrophenol [44]. Highly sensitive measurements of prostate specific antigen (PSA) down to 1 fg/mL were reported. Signal amplification of impedance biosensors by labelling the analyte with nanoparticles [45] or nanocrystals [46] has also been reported for protein or DNA biosensors.

Nanoparticle-induced generation of conductivity paths across a microelectrode gap can also be exploited for highly sensitive and selective detection of DNA hybridization. Mirkin's team developed an array-based electronic detection system utilizing oligonucleotide-functionalized gold nanoparticles and closely spaced interdigitated microelectrodes [47]. The oligonucleotide probe was anchored in the gap between the two-band microelectrodes. The hybridization event localized the gold nanoparticles in the electrode gap, and along with subsequent silver deposition led to measurable conductivity signals. Such hybridization-induced generation of a "conductive bridge" offered high sensitivity with a 0.5-picomolar detection limit. Control of the salt concentration allowed high point-mutation selectivity without thermal stringency. Changes in the resistance across a microelectrode gap, resulting from the hybridization of nanoparticle-labelled DNA, were exploited also for a paralleled array readout system [48]. A self-contained microanalyzer was used for parallel readout of the entire array, indicating great promise for point-of-care applications. Applicability of these resistive measurements for the detection of single nucleotide polymorphism (SNP) was illustrated by Diessel's group [49]. Factors affecting such silver-enhanced bioanalyte detection were discussed by Moller et al. [50].

4.1 Internal Encapsulation or External Loading of Redox Markers

Another attractive route for amplifying bioelectronic signals relies on maximizing the number of signal-generating tags per biorecognition event. For example, dramatic signal amplification can be achieved by linking the biorecognition units to polymeric microspheres carrying multiple redox tracers. For example, ultrasensitive electrical detection of DNA hybridization was reported on the basis of polystyrene beads impregnated with the ferrocenecarboxaldehyde redox marker [51]. The hybridized system with the captured "electroactive beads" was transferred into an acetonitrile solution where the beads were dissolved and the tags were released. This allowed the chronopotentiometric detection of the target DNA down to the 5×10^{-21} mol (31 000 molecules) level. Similar polystyrene carrier spheres loaded with a large number of electrogenerated chemiluminescence (ECL) labels [$Ru(bpy)_3^{+2}$] were shown useful as carriers for ultrasensitive ECL detection of DNA hybridization [52] and protein immunoassays [53] down to the 1 fM and 10 ng/mL levels, respectively. The [$Ru(bpy)_3^{+2}$] markers were released following the binding event in an acetonitrile solution containing the tri-*n*-propylamine ECL co-reactant. Another approach for internal loading of numerous electroactive markers relies on the cavity structure of apoferritin [54]. Liu et al. reported on the encapsulation of ferricyanide within such apoferritin "carrier" in connection to electrochemical immunoassay of IgG down to the 0.52 pM level (Fig. 4).

Polymeric beads carrying numerous DNA tags were used for ultrasensitive electrochemical immunoassays of proteins [55]. Such amplification feature of polymeric carrier beads was combined with the highly sensitive stripping response of the guanine and adenine nucleobases to create distinct identifiable oligonucleotide barcodes for electrochemical immunoassays. Oligomer tags with different predetermined A/G ratios yielded recognizable electrochemical signatures useful for simultaneous electrochemical immunoassays of several proteins. Similarly, it is possible to greatly amplify the coulometric detection of DNA hybridization by capturing a gold nanoparticle loaded with hundreds of DNA oligonucleotides to which numerous $Ru(NH_3)_6^{3+}$ tags are bound (Fig. 5). This DNA sensor could detect as low as femtomolar (zeptomoles) DNA targets. Lin's team reported on the use of silica nanoparticle carriers functionalized with numerous poly(guanine) for highly sensitive voltammetric immunoassays of the tumor nomarker TNF-α in connection to a $Ru(bpy)_3^{+2}$-induced catalytic oxidation of guanine [57].

Instead of using polymeric carrier spheres, it is possible to encapsulate a huge amount of the signal-generating redox marker as a microcrystal in combination with a layer-by-layer (LBL) technology. Mak et al. [58] reported on an amplified electrochemical sandwich immunoassay based on such polyelectrolyte-coated ferrocene microcrystal (Fig. 6). Capturing of the antibody-conjugated ferrocene microcrystal was followed by release of a large amount of the redox marker through the capsule wall (by a releasing agent), and led to a highly sensitive amperometric biodetection.

Carbon nanotubes can also serve as amplification platforms in connection with their loading with numerous signal-generating molecules. For example,

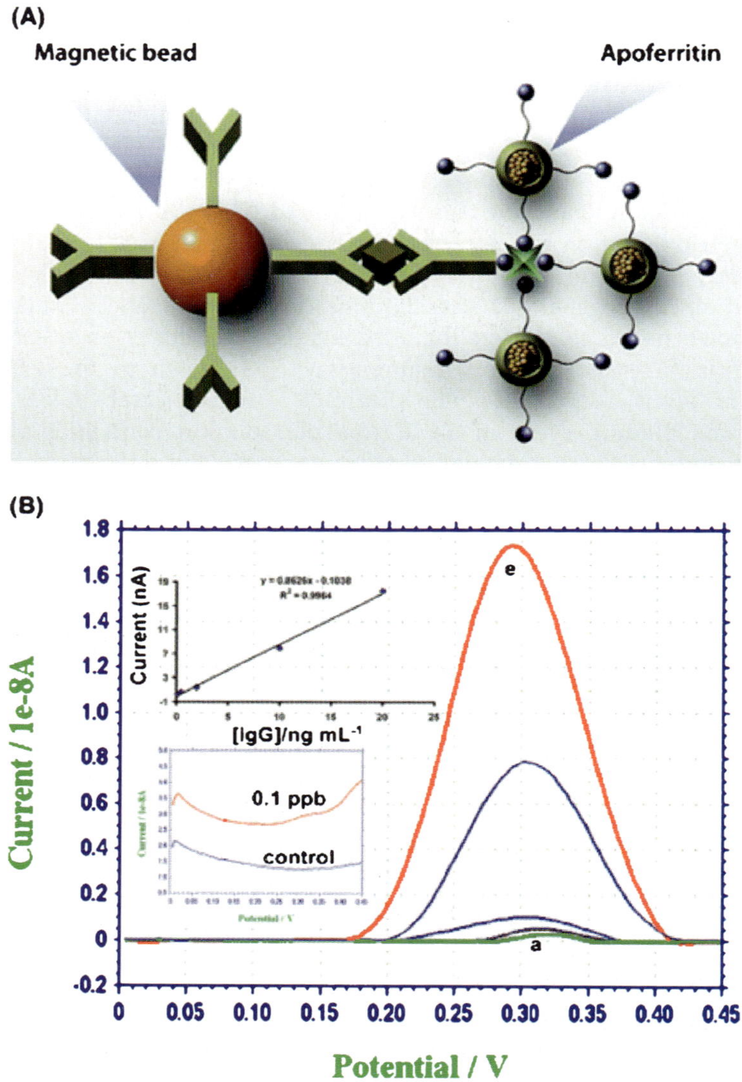

Fig. 4 **A** Magnetic beads and electrochemical sandwich immunoassay protocol based on biotin-functionalized hexacyanoferrate MLAN labels. **B** Typical square wave voltammograms of electrochemical immunoassay with increasing concentration of the IgG (from a to e, 0.1, 0.5, 2, 10, and 20 ng mL^{-1} IgG, respectively). (Reproduced from [54] with permission.)

Wang et al. developed a highly sensitive bioelectronic protocol for detecting proteins and DNA [59]. In this procedure, CNT played a dual amplification role in both the recognition and transduction events, namely as carriers for

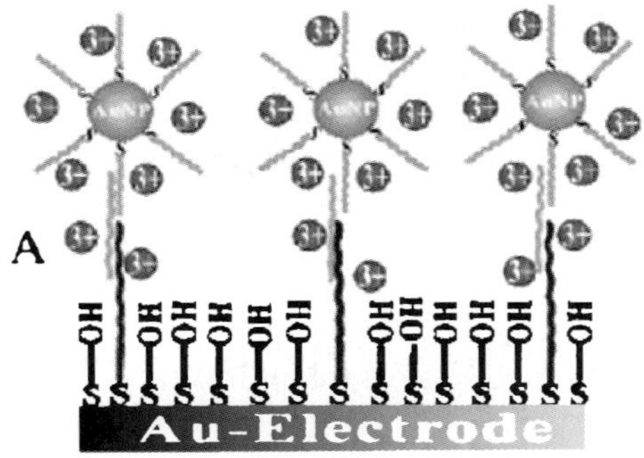

Fig. 5 Amplified chronocoulometric DNA detection. Electrochemical signals are generated by chronocoulometric interrogation of $[Ru(NH_3)_6]^{3+}$ that quantitatively binds to surface-confined capture probe DNA via electrostatic interactions (Reproduced from [56] with permission.)

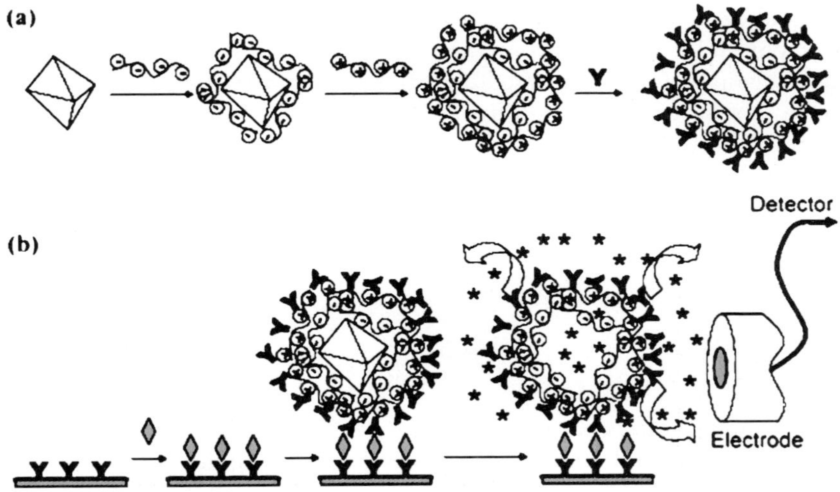

Fig. 6 Schematic diagram illustrating **a** the preparation of microcrystal-labelled antibody conjugate and **b** the principle of microcrystal-labelled electrochemical immunoassay (Reproduced from [58] with permission.)

numerous enzyme tags and for accumulating the α-naphtol product of the enzymatic reaction onto the electrode. A coverage of around 9600 enzyme molecules per a CNT (i.e., binding event) was estimated. Such CNT-derived double-step amplification pathway (of both the recognition and transduction events) allows the detection of DNA and proteins down to 1.3 and 160 zmol, respectively, in 25–50 μL samples and indicates great promise for PCR-free DNA analysis. The concept was used recently by Rusling's team for highly sensitive immunodetection of the PSA cancer biomarker in serum and tissue lysates in connection with HRP-loaded CNT [60]. Further sensitivity enhancement was achieved by a LBL assembly of multilayer enzyme films on the CNT template [61]. The coupling of CNT carriers and protein multilayer architectures was shown to maximize the ratio of enzyme tags per binding event and hence can offer a remarkably high amplification factor.

5
Amplified Microgravimetric Bioassays Monitoring

Microgravimetric transducers monitor mass changes that occur during the binding of target analytes to the surface-confined recognition layer. Metal nanoparticles have been used by Patolsky et al. [62] as "weight" labels for amplified microgravimetric monitoring of bioaffinity events (particularly DNA hybridization). For example, dendritic amplified detection of DNA hybridization was accomplished in connection with a nucleic-acid functionalized gold nanoparticles. Amplified detection of single-base mismatches was also reported by the same group [63]. This protocol involved a polymerase-I-induced coupling of biotinylated base complementary to the point-mutation site, and subsequent detection of the biotin label using an avidin-gold nanoparticle conjugate that stimulated the catalytic electroless deposition of gold. Iron-oxide (Fe_3O_4) nanoparticles (145 nm) "mass enhancers" were also used for amplifying the microgravimetric DNA-hybridization detection of *E. coli* 0157 : H7 [64].

6
Summary

Nanotechnology offers innovative approaches for the monitoring of biomolecular interactions and unique opportunities for designing highly sensitive bioassays based on different amplifying labels. The studies described in this chapter demonstrate the broad scope of bioconjugated nanoparticles for amplified transduction of biomolecular interactions. Given the huge amplification afforded by nanoparticle tags, such nanomaterials provide the basis for ultrasensitive assays of proteins and nucleic acids, approaching PCR sensitivity. The remarkable sensitivity of the new nanomaterial-based sensing protocols opens up the possibility of detecting disease markers, infectious agents, or biothreat agents that cannot be measured by conventional methods. Such highly sensitive biodetection schemes could provide early detection of diseases or warning of a terrorist attack. The use of nanoparticle tags for detecting proteins is still in its infancy, but the lessons learned in ultrasensitive DNA detection should provide useful starting points. The successful realization of the new signal-amplification strategies requires proper attention to non-specific adsorption issues that commonly control the detectability of bioaffinity assays. Proper surface blocking and washing steps should thus be employed to avoid amplification of background signals associated with non-specific adsorption of the nanoparticle amplifiers. Newly introduced nanomaterials are expected to further expand the scope of nanomaterial-based biosensors.

Acknowledgements This research was supported by grants from the National Science Foundation (Grant Number CHE 0209707) and the NIH (Awards Number R01A 1056047-01 and EB002189).

References

1. Niemeyer CM (2001) Angew Chem Int Ed 40:4129
2. Alivisatos P (2002) Nat Biotechnol 22:47
3. Katz E, Willner I, Wang J (2004) Electroanalysis 16:19
4. Willner I, Katz E (2004) Angew Chem Int Ed 43:6042
5. Rosi NL, Mirkin CA (2005) Chem Rev 105:1547
6. Letsinger RL, Mucic RC, Storhoff JJ (1996) Nature 382:607
7. Storhoff JJ, Elghanian R, Mucic RC, Mirkin CA, Letsinger RL (1998) J Am Chem Soc 120:1959
8. Elghanian R, Storhoff JJ, Mucic RC, Letsinger RL, Mirkin CA (1977) Science 277:1078
9. Taton TA, Mirkin CA, Letsinger RL (2000) Science 289:1757
10. Pavlov V, Shyahovsky B, Willner I (2004) J Am Chem Soc 126:11768
11. Cao YC, Jin R, Mirkin CA (2002) Science 297:1536
12. Hahn MA, Tabb JS, Krauss TD (2005) Anal Chem 77:4861
13. Nam JM, Thaxton CS, Mirkin CA (2003) Science 301:1884
14. Hurst S, Lytton Jean AK, Mirkin CA (2006) Anal Chem 78:8313
15. Georganopoulou DG, Chang L, Nam JM, Thaxton CS, Mufson EJ, Klein WL, Mirkin CA (2005) Proc Natl Acad Sci USA 102:2273
16. Nam JM, Stoeva SI, Mirkin CA (2004) J Am Chem Soc 126:5932
17. Hazarika P, Ceyhan B, Niemeyer CM (2005) Small 1:844
18. Trau D, Yang W, Seydack M, Caruso F, Yu NT, Renneberg R (2002) Anal Chem 74:5480
19. Zhao X, Tapec-Dytioco R, Tan W (2003) J Am Chem Soc 125:11474
20. Zhao X, Hilliard LR, Mechery SJ, Wang Y, Bagwe RP, Jin S, Tan W (2004) Proc Natl Acad Sci USA 101:15027
21. He L, Musick M, Nicewarner S, Salinas F, Benkovic S, Nathan M, Keating C (2000) J Am Chem Soc 122:9071
22. Zhang ZF, Cui H, Lai CZ, Liu LJ (2005) Anal Chem 77:3324
23. Cui H, Xu Y, Zhang ZF (2004) Anal Chem 76:4002
24. Gill R, Polsky R, Willner I (2006) Small 2:1037
25. Shlyahovsky B, Katz E, Xiao Y, Pavlov V, Willner I (2005) Small 1:213
26. Zayats M, Baron R, Popov I, Willner I (2005) Nano Lett 5:21
27. Parlov V, Xiao Y, Willner I (2005) Nano Lett 5:649
28. Baron R, Zoyats M, Willner I (2005) Anal Chem 77:1566
29. Frew J, Hill HA (1987) Anal Chem 59:933A
30. Wang J (1985) Stripping Analysis: Principles, Instrumentation, and Applications. VCH Publishing, New York

31. Dequaire M, Degrand C, Limoges B (2000) Anal Chem 72:5521
32. Authier L, Grossiord C, Brossier P, Limoges B (2001) Anal Chem 73:4450
33. Wang J, Xu D, Kawde A, Polsky R (2001) Anal Chem 73:5576
34. Rochelet-Dequaire M, Limoges B, Brossier P (2006) Analyst 131:923
35. Wang J, Xu D, Polsky R (2002) J Am Chem Soc 124:4208
36. Wang J, Polsky R, Danke X (2001) Langmuir 17:5739
37. Cai H, Wang Y, He P, Fang Y (2002) Anal Chim Acta 469:165
38. Kawde A, Wang J (2004) Electroanalysis 16:101
39. Chumbimuni-Torres KY, Dai Z, Rubinova N, Xiang Y, Pretsch E, Wang J, Bakker E (2006) J Am Chem Soc 128:13676
40. Wang J, Liu G, Zhou J (2003) Anal Chem 75:6218
41. Wang J, Rincon O, Polsky R, Dominguez E (2003) Electrochem Commun 5:83
42. Hwang S, Kim E, Kwak J (2005) Anal Chem 77:579
43. Polsky R, Gill R, Kaganovsky L, Willner I (2006) Anal Chem 78:2268
44. Das J, Abdul Aziz M, Yang H (2006) J Am Chem Soc 128:16022
45. Wang J, Profitt JA, Pugia MJ, Suni II (2006) Anal Chem 78:1769
46. Xu Y, Cai H, He PG, Fang YZ (2004) Electroanalysis 16:150
47. Park S, Taton TA, Mirkin CA (2002) Science 295:1503
48. Urban M, Moller R, Fritzsche W (2003) Rev Sci Instr 74:1077
49. Diessel E, Grothe K, Siebert HM, Warner B, Burmeister J (2004) Biosens Bioelectron 19:1229
50. Moller R, Csaki A, Kohler JM, Fritzsche W (2001) Langmuir 17:5426
51. Wang J, Polsky R, Merkoci A, Turner K (2003) Langmuir 19:989
52. Miao W, Bard AJ (2004) Anal Chem 76:5379
53. Miao W, Bard AJ (2004) Anal Chem 76:7109
54. Liu G, Wang J, Wu H, Lin Y (2006) Anal Chem 78:7417
55. Wang J, Liu G, Munge B, Lin L, Zhu Q (2004) Angew Chem Int Ed 43:2158
56. Zhang J, Song S, Zhang L, Wang L, Wu H, Pan D, Fan C (2006) J Am Chem Soc 128:8575
57. Wang J, Liu G, Engelhard ME, Lin Y (2006) Anal Chem 78:6974
58. Mak WC, Cheung KY, Trau D, Warsinke A, Scheller F, Rennenberg R (2005) Anal Chem 77:2835
59. Wang J, Liu G, Rasul M (2003) J Am Chem Soc 126:3010
60. Yu X, Munge B, Patel V, Jensen G, Bhirde A, Gong JD, Kim SN, Gillespie J, Gutkind JS, Papadimitrakopoulos F, Rusling J (2006) J Am Chem Soc 128:11199
61. Munge B, Liu G, Collins G, Wang J (2005) Anal Chem 77:4662
62. Patolsky F, Ranjit KT, Lichtenstein A, Willner I (2000) Chem Commun, p 1025
63. Willner I, Patolsky F, Weizmann Y, Willner B (2002) Talanta 56:847
64. Mao X, Yang L, Su XS, Li Y (2006) Biosens Bioelectron 21:1178

Adv Biochem Engin/Biotechnol (2008) 109: 255–283
DOI 10.1007/10_2007_077

Published online: 15 November 2007

Photoelectrochemical and Optical Applications of Semiconductor Quantum Dots for Bioanalysis

Maya Zayats · Itamar Willner (✉)

Institute of Chemistry, The Hebrew University of Jerusalem, 91904 Jerusalem, Israel
willnea@vms.huji.ac.il

Abstract Semiconductor nanoparticles (NPs) or quantum dots (QDs) exhibit unique photophysical properties reflected by size-controlled fluorescence, high fluorescence quantum yields, and stability against photobleaching. These properties are utilized by applying the QDs as optical labels for the multiplexed analysis of immunocomplexes and DNA hybridization. Also, semiconductor QDs are used to probe biocatalytic transformations. The time-dependent replication or telomerization of nucleic acids, the oxidation of phenol derivatives by tyrosinase, and the hydrolytic cleavage of peptides by proteases are probed by using fluorescence resonance energy transfer or photoinduced electron transfer. The photoexcitation of semiconductor NP–biomolecule hybrids associated with electrodes enables the photoelectrochemical transduction of biorecognition events or biocatalytic transformations. This is exemplified with the generation of photocurrents by duplex DNA assemblies bridging CdS NPs to electrodes, and by the formation of photocurrents as a result of biocatalyzed transformations, or redox protein-mediated electron transfer in the presence of the NPs.

Keywords Biosensor · DNA · Photocurrent · Protein · Quantum dots · Semiconductor nanoparticles

1 Introduction

Semiconductor nanoparticles (NPs) or quantum dots (QDs) reveal unique photophysical properties that offer significant advantages in their application as optical labels for sensing and, specifically, biosensing. The semiconductor QDs are characterized by high fluorescence quantum yield stability against photobleaching, and size-controlled luminescence properties [1–5]. The efficient fluorescence and stability of the QDs add improved sensitivities and prolonged lifetime to the use of the QDs as optical labels. The size-controlled luminescence functions of the QDs demonstrates, however, the major advantages of the QDs as optical labels, since particles of the same material with variable sizes may be used as fluorescent labels for the parallel multiplexed analysis of different analytes. Furthermore, unlike molecular fluorophores that reveal narrow excitation spectra, semiconductor QDs absorb light over a broad spectral region. This allows the simultaneous excitation of many different-sized QDs at the same excitation wavelength while imaging in parallel the fluorescence of the different QDs. Also, efficient methods to synthesize semiconductor QDs and to chemically modify the surfaces of the QDs by functional capping monolayers or thin films have been developed in recent years [6, 7]. These functional QDs allow the secondary tethering of ligands or receptor units to the surface of the QDs, thus yielding QD–ligand conjugates where the NP acts as an optical transducer for recognition of sensing events occurring on the surfaces of the NPs.

Quantum dots reveal also photoelectrochemical properties [8]. The photoexcitation of the semiconductor QDs results in the transfer of an electron from the valence band to the conduction band, thus yielding an electron-hole pair. While the luminescence properties of the QDs originate from the radiative electron-hole recombination, the trapping of conduction-band electrons in surface traps yields sufficiently long-lived electron-hole pairs that permit the ejection of the conduction-band electrons to electrodes (or solution-solubilized electron acceptors) giving rise to photoelectrochemical current generation. The ejection of the conduction-band electrons to the electrode, with the concomitant transfer of electrons from a solution-solubilized electron donor yields an anodic photocurrent (Fig. 1A), whereas the transfer of the conduction-band electrons to a solution-solubilized electron acceptor followed by the supply of electrons from the electrode to neutralize the valence-bound holes yields a cathodic photocurrent (Fig. 1B). Enormous research activities are directed towards the utilization of semi-

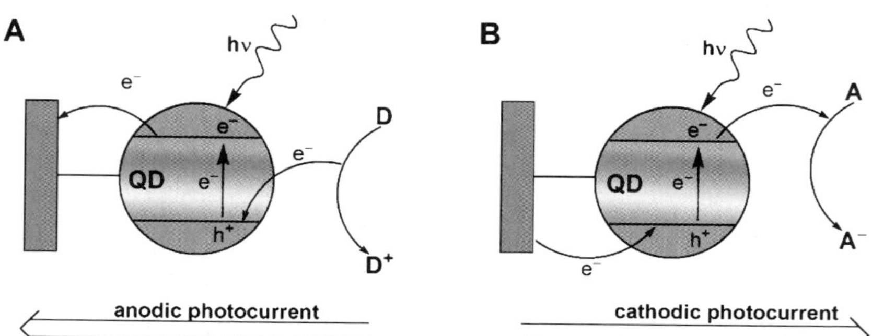

Fig. 1 Photocurrent generated by semiconductor NPs associated with electrodes. **A** Anodic photocurrent. **B** Cathodic photocurrent

conductor QDs associated with electrodes for developing light-to-electrical energy conversion systems. The coupling of semiconductor QDs–biomolecule conjugates to electrode surfaces might control the photocurrent generation through biorecognition or biocatalytic processes occurring on the electrode surface.

The present article summarizes some recent research activities directed to the utilization of semiconductor QDs as optical labels or photoelectrochemical transduction elements for the sensing of biorecognition and biocatalytic transformations. Specifically, we will emphasize the use of the QDs as optical labels that follow dynamic biochemical transformations, or as photoelectrochemical transduction means for biorecognition events or biocatalytic transformations.

2 Semiconductor Quantum Dots as Optical Labels for Bioanalysis

Functionalized semiconductor QDs have been used as fluorescence labels for numerous biorecognition events [9–12] including their use in immunoassays for protein detection (Fig. 2A) and nucleic acid detection (Fig. 2B). For example, CdSe/ZnS QDs were functionalized with avidin, and these were used as fluorescent labels for biotinylated antibodies. Fluoroimmunoassays utilizing these antibody-conjugated NPs were successfully used in the detection of protein toxins (staphylococcal enterotoxin B and cholera toxin) [13, 14]. Similarly, CdSe/ZnS QDs conjugated to appropriate antibodies were applied for the multiplexed fluouroimmunoassay of toxins using a sandwich configuration (Fig. 3A). The sandwich immunoassays for the simultaneous detection of the four toxins (cholera toxin, ricin, shiga-like toxin 1, and staphylococcal enterotoxin B) by using different-sized QDs were performed in single wells

Fig. 2 Fluorescence analysis of **A** antigen by antibody-functionalized QDs, and **B** DNA by a nucleic acid-functionalized QDs

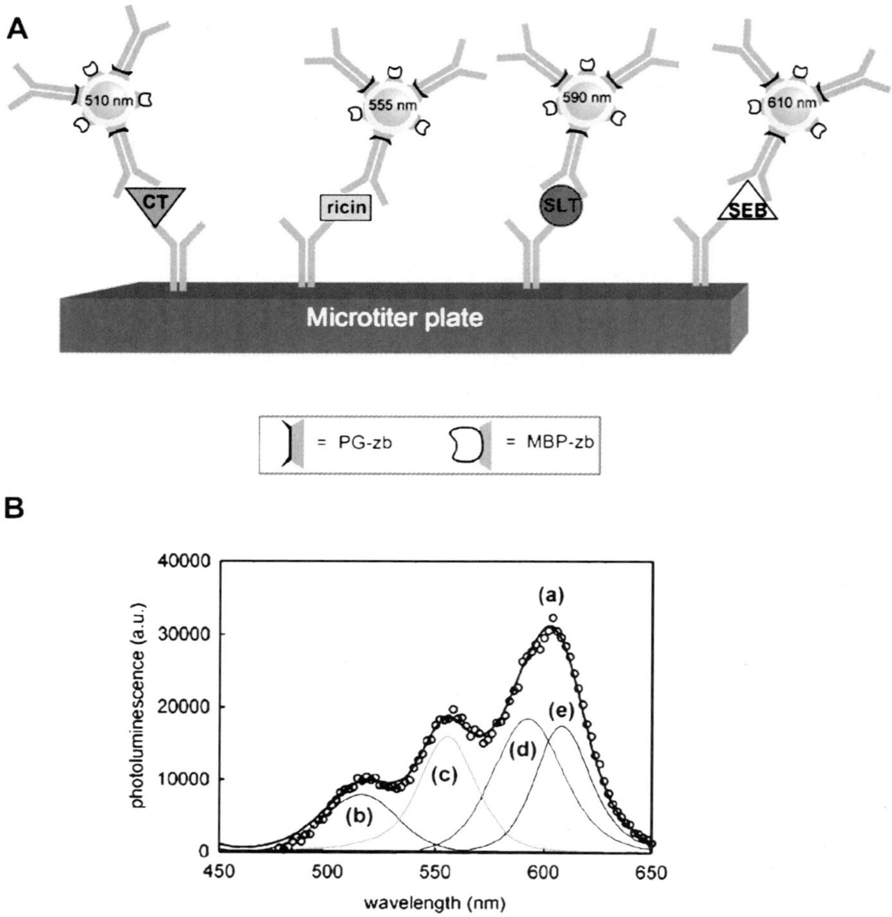

Fig. 3 **A** Parallel optical analysis of different antigens in a well-array format using the fluorescence of different sized quantum dots. **B** Fluorescence spectrum observed upon analyzing the four analytes, 1 $\mu g\,mL^{-1}$, by the different sized QDs (*a*) and deconvoluted spectra of individual toxins: CT (*b*), ricin (*c*), SLT (*d*), and SEB (*e*). (Reprinted in part with permission from [15], © 2004, American Chemical Society)

of a microtiter plate in the presence of a mixture of all four QD–antibody conjugates (Fig. 3B) [15], thus leading to the fluorescence that encodes for the toxin. In another example, multiplexed immunoassay formats based on antibody-functionalized QDs were used for simultaneous detection of *Escherichia coli* O157:H7 and *Salmonella Typhimurium* bacteria [16] and for the discrimination between the diphtheria toxin and the tetanus toxin proteins [17].

Fluorescent QDs were used for the detection of single-nucleotide polymorphism in human oncogene p53 and for the multiallele detection of the hepatitis B and hepatitis C virus in microarray configurations [18]. DNA-functionalized CdSe/ZnS QDs of different sizes were used to probe hepatitis B and C genotypes in the presence of a background of human genes. The discrimination of a perfectly matched sequence of p53 gene in the presence of background oligonucleotides that included different single-nucleotide polymorphism sequences was detected with true-to-false signal ratios higher than ten (under stringent buffer conditions) at room temperature within minutes. Also, QDs were used for microRNA detection in microarray configuration [19]. The streptavidin-conjugated QDs were used to label miRNA targets biotinylated at their 3′ termini, and hybridized with the corresponding complementary DNA probes immobilized on glass slides. The resulting fluorescence from the QDs that labeled the miRNAs was then used as readout signal (Fig. 4A). Analysis of a model system indicated that the detection limit for miRNA was ~0.4 fmol and that the detection dynamic range spanned about two orders of magnitude, from 156 to 20 000 pM (Fig. 4B). The method was applied to develop an assay for profiling 11 miRNA targets from rice (Fig. 4C).

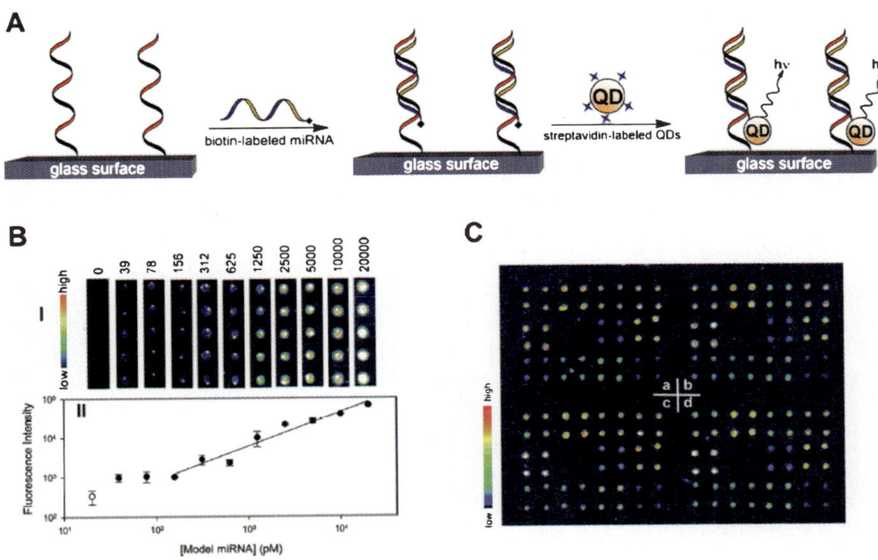

Fig. 4 Analysis of miRNA by means of fluorescent QDs. **A** Organization of the streptavidin-labeled QDs on a DNA/miRNA duplex bound to a glass support. **B** Fluorescence intensities observed upon analyzing different concentrations of a target miRNA (*panel I*), and derived calibration curve (*panel II*). **C** Analysis of 11 target miRNAs from rice in an array format. (Reprinted by permission in part from [19], © 2005, Oxford University Press)

Also, DNA/QDs conjugates were used as fluorescence probes for in situ hybridization assays (FISH). For example, the QD-based FISH labeling method was used for the detection of the Y chromosome in human sperm cells [20]. A QDs-based FISH method to analyze human metaphase chromosomes was reported [21] by using QD-conjugated total genomic DNA as a probe for the detection of EBRB2/HER2/neu gene. Also, the FISH technique was used for the multiplexed cellular detection of different mRNA targets [22].

Extensive efforts are directed to the incorporation of semiconductor QDs into cells [6, 23–25]. The fluorescence properties of the QDs were applied to image intracellular regions and, specifically, to follow intracellular mobility of receptors in living cells. For example, the early stages of the transmembrane receptor tyrosine kinases (RTKs)-dependent signaling in living cells were imaged by following the binding of the QDs bearing the epidermal growth factor (EGF–QDs), and the subsequent activation of the EGF receptor by erbB1 [26]. By using a combination of the EGF–QDs and the fluorescent protein-tagged receptor (erbB1–eGFP), the binding and endocytic uptake of EGF–QDs was followed (Fig. 5A). Endocytosis of both components of the EGF–QDs/erbB1–eGFP complex proceeded rapidly, and the colocalization was mainly inside the cell, in endosome vesicles (Fig. 5B). It was shown that upon incorporation of the QDs into the cell, endocytic vesicles underwent Brownian movement and vesicular fusion (Fig. 5C).

The superior photophysical features of semiconductor QDs (high fluorescence quantum yields and stability against photobleaching) are, however,

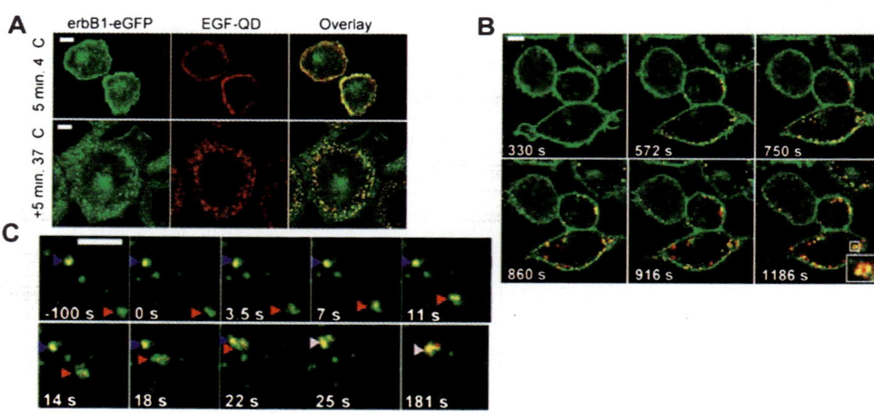

Fig. 5 Fluorescent imaging of: **A** Chinese hamster ovary (CHO) cell subjected to the expression of enbB1-eGFP acceptor treated with biotin-labeled EGF and imaged by steptavidin-functionalized QDs. (At 4 °C the ligand binds to the cell membrane, while at 37 °C intracellular penetration of the ligand is observed). **B** Time-dependent endocytosis of the ebrB1-eGFP/EGF-QDs. **C** Dynamics of endosomal fusion of the ebrB1-eGFP/EGF-QDs in the cells. (Reprinted with permission from [26], © 2004, Macmillan)

demonstrated in organic solvents, and their introduction into aqueous media is usually accompanied by a drastic decrease in the luminescence yields of the QDs. This is, presumably, due to the reaction of surface states with water, a process that yields the surface traps for the conduction-band electrons. As biorecognition events or biocatalytic transformations require aqueous environments as the reaction media, it is mandatory to preserve the luminescence properties of the QDs in aqueous systems. Different methods to stabilize the fluorescence properties of semiconductor QDs in aqueous media were reported, including their surface passivation with protective layers [27, 28], and the coating of the QDs with protecting silica films [29, 30] or polymers [31]. Although these methods preserve the photophysical properties of the QDs, the passivation layers limit the useful application of the QDs to follow biochemical processes. Quantum dots could be employed as optical labels that follow dynamic biological processes such as biocatalyzed transformations or structurally induced biomolecular changes, e.g., opening of hairpin nucleic acids by the hybridization of DNA, or changes in protein configurations, using fluorescence resonance energy transfer (FRET) or electron transfer quenching as photophysical probing mechanisms [32, 33]. The sensitivity of these photophysical processes to the distance separating the donor–acceptor or chromophore–quencher pairs prevents, however, the use of the fluorescent QDs as optical probes for dynamic bioprocesses due to the relatively thick capping layers that passivate the NPs. Hence, a very delicate nanostructuring of the capping layer is essential to allow the use of QDs as active components in energy/electron transfer reactions in aqueous media.

Several methods to stabilize semiconductor QDs with water-soluble monolayers that retain high fluorescence yields of the particles have been reported [6, 34, 35]. For example, CdSe QDs with a high fluorescence quantum yield of 25% and a narrow size distribution were synthesized in a single-step procedure in water using glutathione (GSH) as a stabilizing agent [36]. Similarly, the synthesis of glutathione-capped CdTe (GSH–CdTe) QDs with tunable fluorescence in the range of 500–650 nm and quantum yield as high as ca. 45% was described in aqueous media and employed for the staining of fixed-cell [37]. Another method for the preparation of water-soluble QDs relied on the exchange of native organic ligands, trioctylphosphine oxide (TOPO), on the surface with thiolated molecules. The bidentate surface ligands composed of derivatives of dihydrolipoic acid (DHLA) were used for the preparation of aqueous soluble CdSe/ZnS QDs [27, 38, 39]. The DHLA ligands provide stable interactions with the QD surfaces due to the bidentate chelate effect of the dithiol groups. Another method to stabilize semiconductor QDs in aqueous media is based on the exchange of organic ligands with mercaptoacetic acid (MAA) [28, 40, 41].

Different sensing schemes have been developed that use QDs as FRET donors. For example, CdSe/ZnS QDs conjugated to nucleic acids have been used to follow the biocatalyzed replication of DNA [42]. CdSe/ZnS NPs were functionalized with the DNA primer (**1**), which is complementary to a domain of M13mp18 DNA (**2**). Hybridization of the single-stranded M13mp18 DNA (**2**) with the (**1**)-nucleic acid-functionalized QDs, followed by the replication of the assembly in the presence of polymerase and the nucleotide (dNTP) mixture that included Texas Red-functionalized dUTP (**3**), resulted in the incorporation of the dye labels into the DNA replica (Fig. 6A). The Texas Red dye was selected as energy acceptor since its absorption spectrum overlaps the emission spectrum of the QDs. The FRET process from the semiconductor NPs to the incorporated dye units resulted in emission from the dye with concomitant quenching of the fluorescence of the QDs (Fig. 6B). This enabled not only the detection and amplification of the primary hybridization of the analyte DNA with the (**1**)-probe nucleic acid, but also monitoring of the dynamics of the time-dependent replication process.

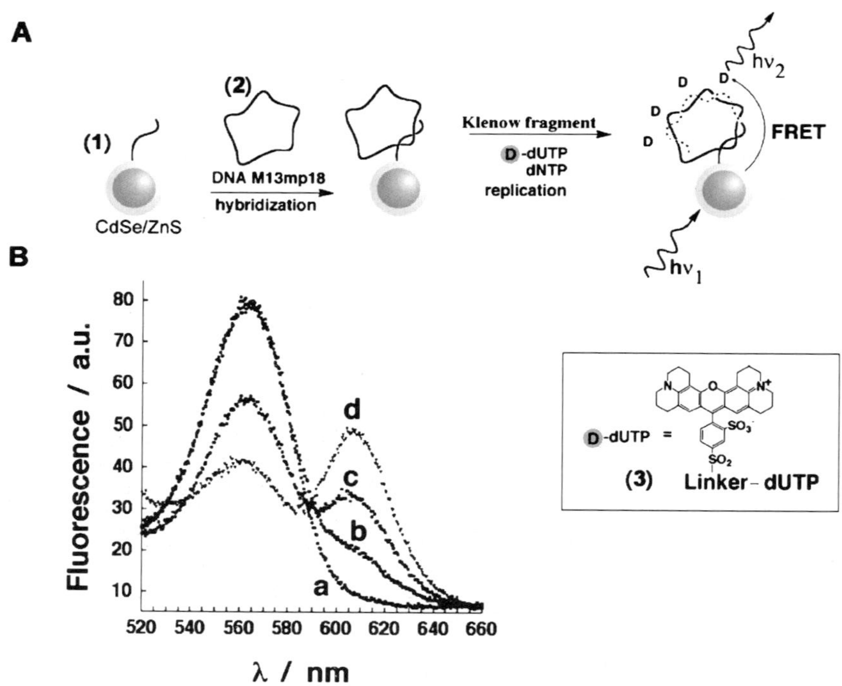

Fig. 6 **A** Optical detection of M13 phage DNA (**2**) by nucleic acid-functionalized CdSe/ZnS QDs. The replication of the analyte in the presence of the dNTP mixture that includes the Texas Red-labeled dUTP (**3**) results in the incorporation of the dye into the replica and stimulates a FRET process. **B** Time-dependent fluorescence changes upon incorporation of the dye, (**3**), into the DNA replica, and the analysis of the M13 phage DNA according to **A**: *a* 0, *b* 10, *c* 30, and *d* 60 min. (Reprinted with permission from [42], © 2003, American Chemical Society)

A similar approach was used to follow telomerase activity. The sensing of telomerase activity is important since telomerase is considered an important cancer marker. Nucleic acid residues consisting of the constant repeat units of the telomers protect the chromosomes. These units are constantly shortened during the cell's life cycles, and at a certain chain length of the telomers they provide an intracellular signal terminating cell proliferation [43, 44]. The accumulation of telomerase in certain cells results in the build up and elongation of the telomers chains parallel to their natural shortening and, as a result, immortal cells are formed. Indeed, in over 95% of different cancer cells, elevated amounts of telomerase were observed, and the enzyme is considered to be a versatile marker for malignant or cancerous cells [45, 46]. For analyzing telomerase activity, the CdSe/ZnS QDs were modified with a nucleic acid primer (**4**) that is recognized by telomerase (Fig. 7A). In the presence of telomerase and the nucleotide mixture dNTPs, which included

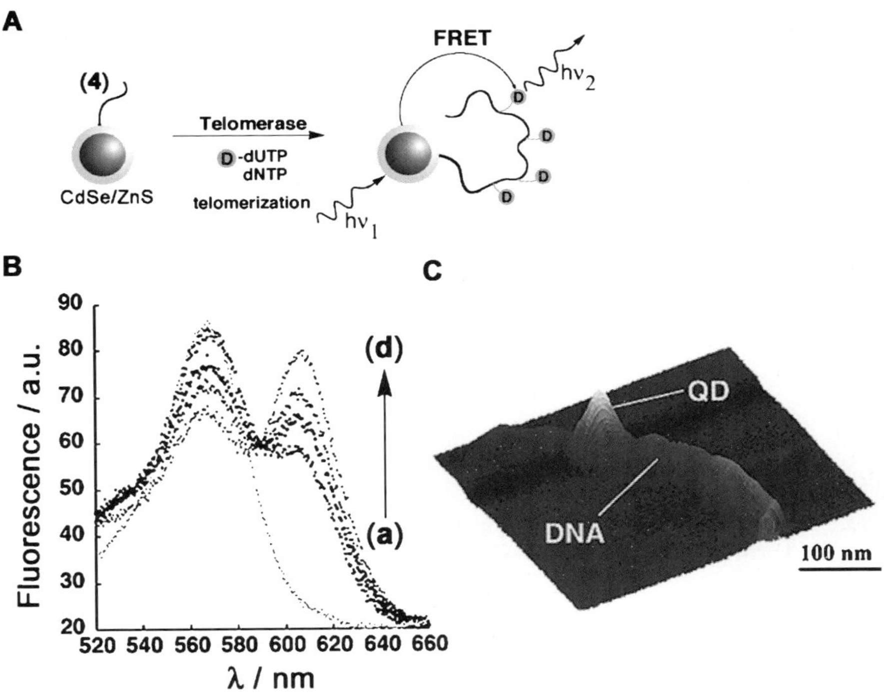

Fig. 7 **A** Optical analysis of telomerase activity by the incorporation of the Texas Red dUTP (**3**) into the telomers associated with CdSe/ZnS QDs. **B** Time-dependent fluorescence changes upon telomerization of the (**4**)-functionalized QDs in the presence of telomerase extracted from 10 000 HeLa cells, dNTPs, 0.5 mM, and Texas Red dUTP, 100 μM, at time intervals corresponding to: *a* 0, *b* 10, *c* 30, *d* 60 min. **C** AFM image of the telomers generated on the QDs. (Reprinted with permission from [42], © 2003, American Chemical Society)

Texas Red-functionalized dUTP (**3**), the telomerization of the nucleic acid associated with the QDs proceeded, while incorporating the Texas Red-labeled nucleotide into the telomers (Fig. 7A). The FRET process from the QDs to the dye units enabled the dynamics of the telomerization process to be followed (Fig. 7B) [42]. The formation of the telomers on the QDs was also imaged at the molecular level, using atomic force microscopy (AFM) (Fig. 7C).

Also, the association of maltose to the hybrid composed of the maltose-binding protein (MBP) was examined by the application of a CdSe/ZnS QD linked to the MBP [47]. CdSe/ZnS QDs were functionalized with MBP, and these were interacted with a β-cyclodextrin-QSY-9 dye conjugate (Fig. 8A). The β-cyclodextrin-QSY-9 dye conjugate resulted in the quenching of the luminescence of the QDs by the dye units. Addition of maltose displaced the quencher units, and this regenerated the luminescence of the QDs (Fig. 8B). This method enabled the development of a competitive QD-based sensor for maltose in solution. Similarly, a competitive QD-based assay for the detection of the explosive trinitrotoluene, TNT, was developed [48]. CdSe/ZnS QDs were functionalized with a single-chain antibody fragment that selectively binds TNT. The analog substrate trinitrobenzene (TNB) covalently linked to the quencher dye (BHQ10), (**5**), was bound to the QD/antibody conjugate, and the associated substrate quenched the fluorescence of the QDs (Fig. 9A). In the presence of the TNT analyte, the quencher TNB-BHQ10 conjugate was competitively displaced. This eliminated the FRET process between the QD and the dye, resulting in the switching on of the fluorescence of the QDs (Fig. 9B).

The hydrolytic functions of a series of proteolytic enzymes were analyzed by the application of QDs modified with peptides as reporter units for the biocatalytic transformations, and the FRET process as a readout mechan-

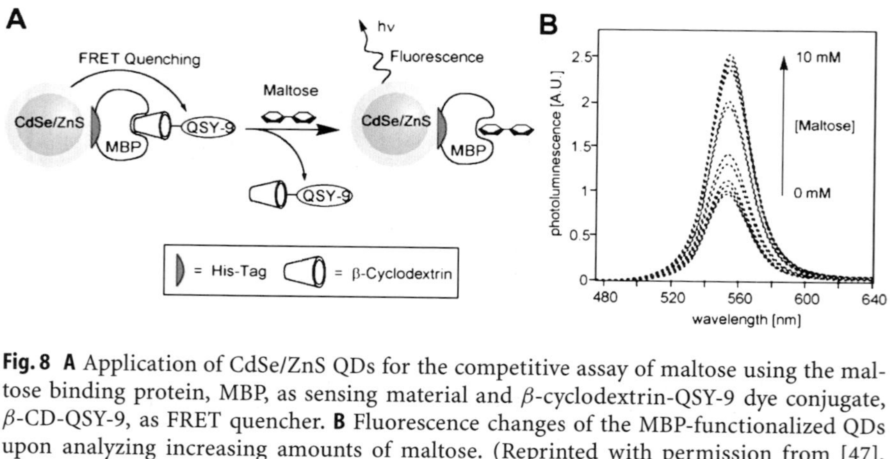

Fig. 8 **A** Application of CdSe/ZnS QDs for the competitive assay of maltose using the maltose binding protein, MBP, as sensing material and *β*-cyclodextrin-QSY-9 dye conjugate, *β*-CD-QSY-9, as FRET quencher. **B** Fluorescence changes of the MBP-functionalized QDs upon analyzing increasing amounts of maltose. (Reprinted with permission from [47], © 2003, Macmillan)

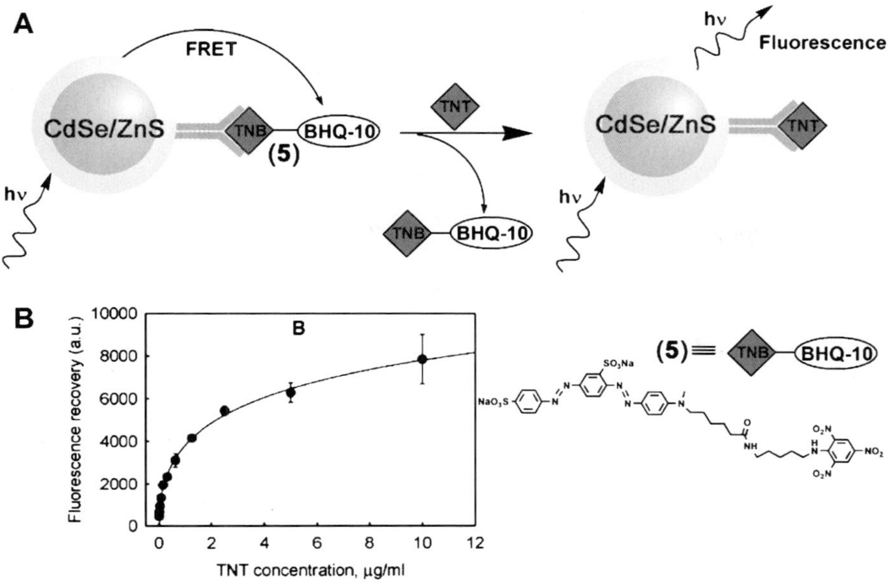

Fig. 9 **A** Competitive analysis of TNT by the anti-TNT Ab associated with CdSe/ZnS QDs and using BHQ-10 as quencher. **B** Derived calibration curve for the analysis of TNT. (Reprinted with permission from [48], © 2005, American Chemical Society)

ism [49, 50]. CdSe QDs were modified with different peptides that included specific cleavage sequences for different proteases, and quencher units were tethered to the peptide termini. The fluorescence of the QDs was quenched in the presence of the quencher–peptide capping layer. The selective hydrolytic cleavage of the peptides by the respective protease resulted in the removal of the quencher units, and this restored the fluorescence of the QDs. For example, collagenase was used to cleave the rhodamine Red-X dye-labeled peptide (**6**) linked to CdSe/ZnS QDs (Fig. 10A). While the tethered dye quenched the fluorescence of the QD, hydrolytic scission of the dye and its corresponding removal restored the fluorescence (Fig. 10B). In a related study, the activity of tyrosinase (TR) was analyzed by CdSe/ZnS QDs [51]. The QDs were capped with a protected L-DOPA monolayer. The tyrosinase-induced oxidation of the L-DOPA units to the respective dopaquinone units generated active quencher units that suppressed the fluorescence of the QDs (Fig. 11A). The depletion of the fluorescence of the QDs upon their interaction with different concentrations of tyrosinase (TR) was used to follow the tyrosinase activity and the time-dependent dynamics of oxidation of the L-DOPA residues (Fig. 11B). The analysis of TR activity by the QDs has practical utility, in addition to the basic demonstration that QDs follow biocatalytic processes. Elevated amounts of tyrosinase were found in melanoma cancer cells, and thus, the rapid optical detection of this biomarker by the QDs might

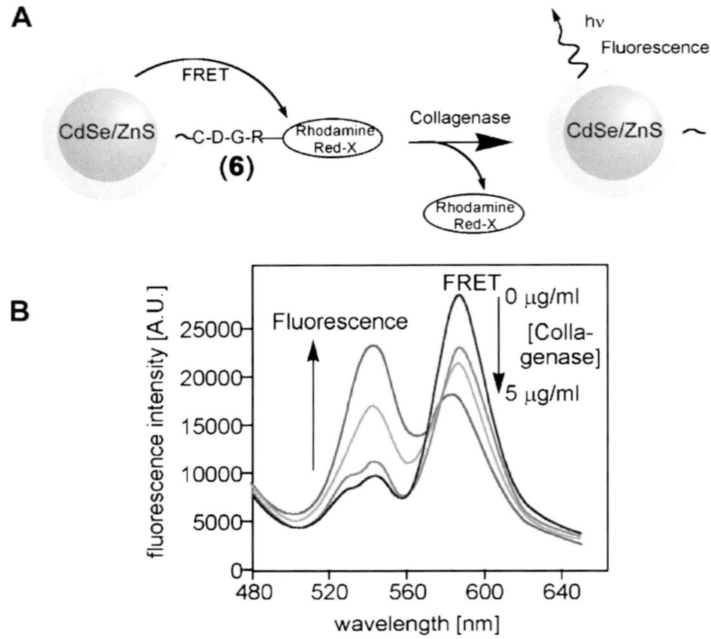

Fig. 10 **A** Application of CdSe/ZnS QDs for the optical analysis of the protease-mediated hydrolysis of the rhodamine Red-X-functionalized peptide (**6**). **B** Decrease in the fluorescence of the dye and the corresponding increase in the fluorescence of the QDs upon interaction with different concentrations of collagenase. (Reprinted with permission from [50], © 2006, American Chemical Society)

be envisaged. The tyrosinase-stimulated oxidation of phenol residues was further employed to use the QDs to monitor the activity of thrombin [51]. The CdSe/ZnS QDs were functionalized with the peptide (**7**) that includes the specific sequences for cleavage by thrombin and the tyrosine site. The tyrosinase-induced oxidation of tyrosine units yields the dopaquinone units that quenched the fluorescence of the QDs (Fig. 12A). The hydrolytic cleavage of the peptide by thrombin removed the quinone quencher units and restored the fluorescence of the QDs (Fig. 12B).

The FRET quenching of CdSe/ZnS QDs was used to follow the hybridization of DNA in QDs/molecular beacons conjugates [52, 53]. Molecular beacons (MBs) consisting of a QD and a quencher molecule attached to opposite ends of a single-stranded DNA oligonucleotide were used. In the absence of a target DNA the MBs existed in a hairpin structure in which the QDs and the quencher were in close proximity. In this configuration, the FRET quenching of the QDs occurred, and no fluorescence from the QDs was observed. The hybridization of the target DNA with the single-stranded loop of the beacon opened the beacon stem, and the QD and the quencher

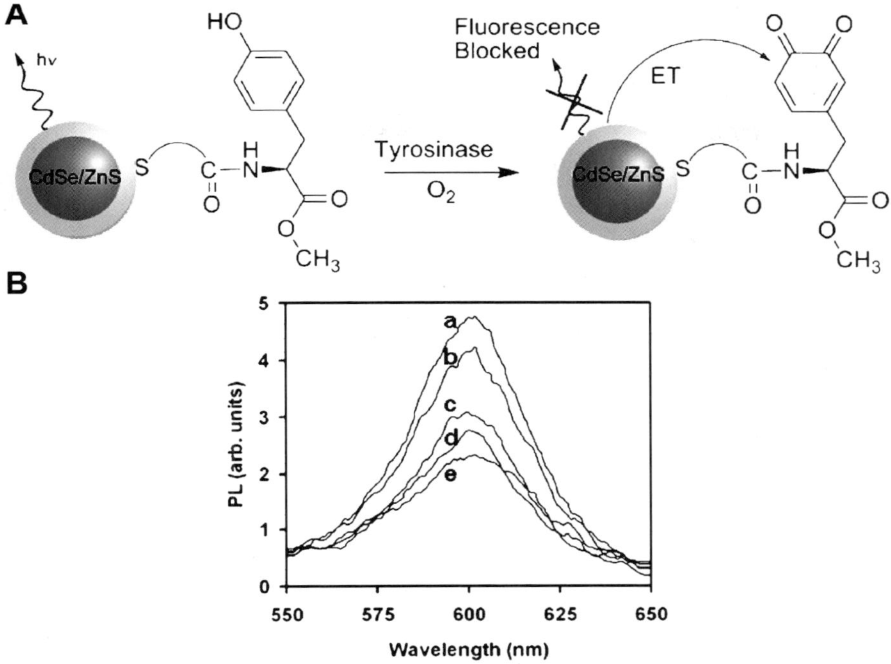

Fig. 11 **A** Analysis of tyrosinase activity by the biocatalytic oxidation of the methyl ester tyrosine-functionalized CdSe/ZnS QDs to the dopaquinone derivative that results in the electron transfer quenching of the QDs. **B** Time-dependent fluorescence quenching of the QDs upon the tyrosinase-induced oxidation of the tyrosine-functionalized QDs: *a* 0, *b* 0.5, *c* 2, *d* 5 and *e* 10 min. (Reprinted in part with permission from [51]. © 2006, American Chemical Society)

adapted an extended configuration. The spatial separation of the QD from the quencher restored the fluorescence of the QD (Fig. 13A). For example, CdSe/ZnS QDs conjugated to molecular beacons (MBs) attached to a 4-(4′-dimethylaminophenylazo)benzoic acid (DABCYL), quencher molecules, were used to detect single-base mismatches with long term optical stability [52]. Similarly, three different sized CdSe/ZnS QDs conjugated to MBs attached to the black-hole quencher-2^{TM} (BHQ2) were applied for the DNA detection [53]. The QDs-conjugated MBs with BHQ2 showed high target discrimination ability upon hybridization with the complementary target DNA, single-base mismatched target DNA (SMT), and the non-specific DNA (NST) (Fig. 13B). A 90% fluorescence intensity increase was observed in the presence of complementary target DNA as compared to the system that included a non-specific target DNA.

The FRET process occurring within a duplex DNA structure consisting of tethered CdSe/ZnS QDs and a dye was applied to probe DNA hybridization and the DNase I cleavage of the DNA [54]. Nucleic acid, (**8**),-

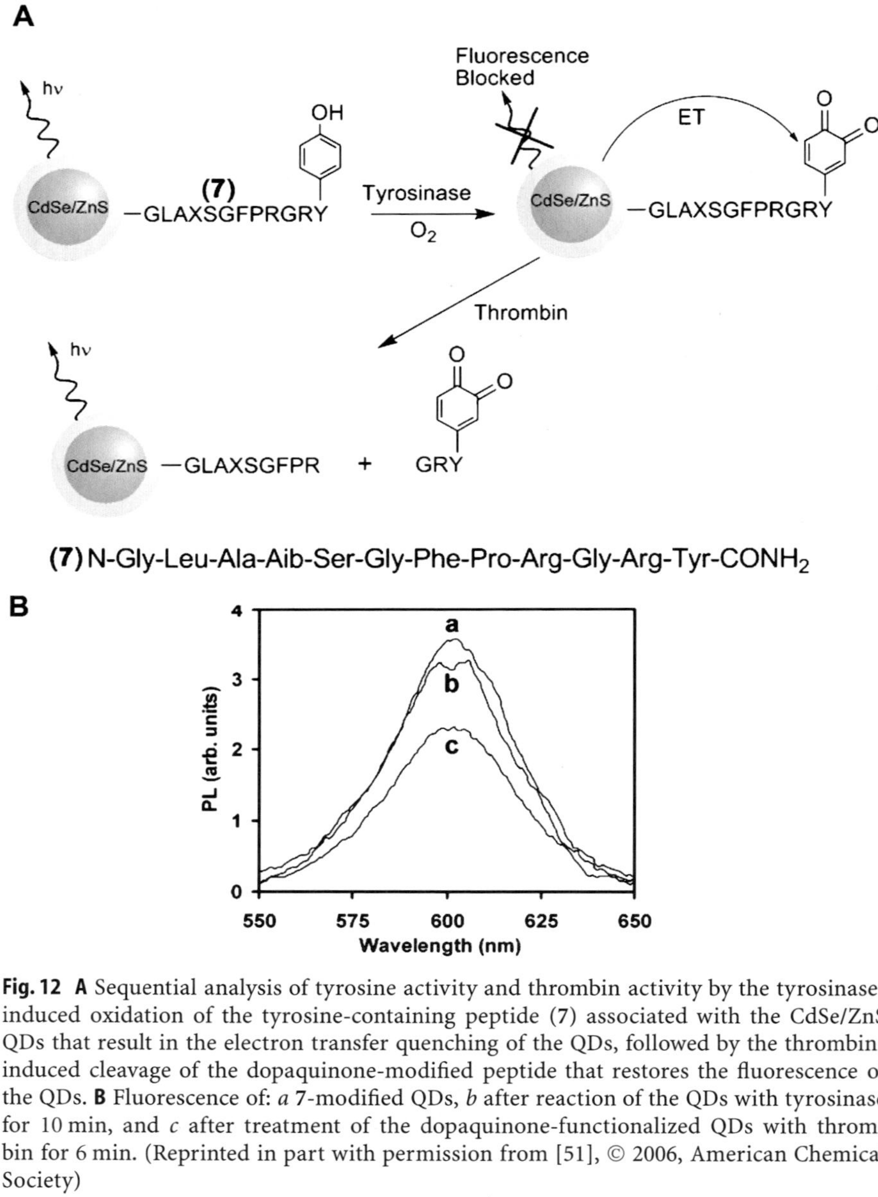

Fig. 12 **A** Sequential analysis of tyrosine activity and thrombin activity by the tyrosinase-induced oxidation of the tyrosine-containing peptide (7) associated with the CdSe/ZnS QDs that result in the electron transfer quenching of the QDs, followed by the thrombin-induced cleavage of the dopaquinone-modified peptide that restores the fluorescence of the QDs. **B** Fluorescence of: *a* 7-modified QDs, *b* after reaction of the QDs with tyrosinase for 10 min, and *c* after treatment of the dopaquinone-functionalized QDs with thrombin for 6 min. (Reprinted in part with permission from [51], © 2006, American Chemical Society)

functionalized CdSe/ZnS QDs were hybridized with the complementary Texas Red-functionalized nucleic acid (**9**) (Fig. 14A). The time-dependent resonance energy transfer from the QDs to the dye units was used to monitor the hybridization process. Treatment of the DNA duplex with DNase I resulted in the cleavage of the DNA and the recovery of the fluorescence proper-

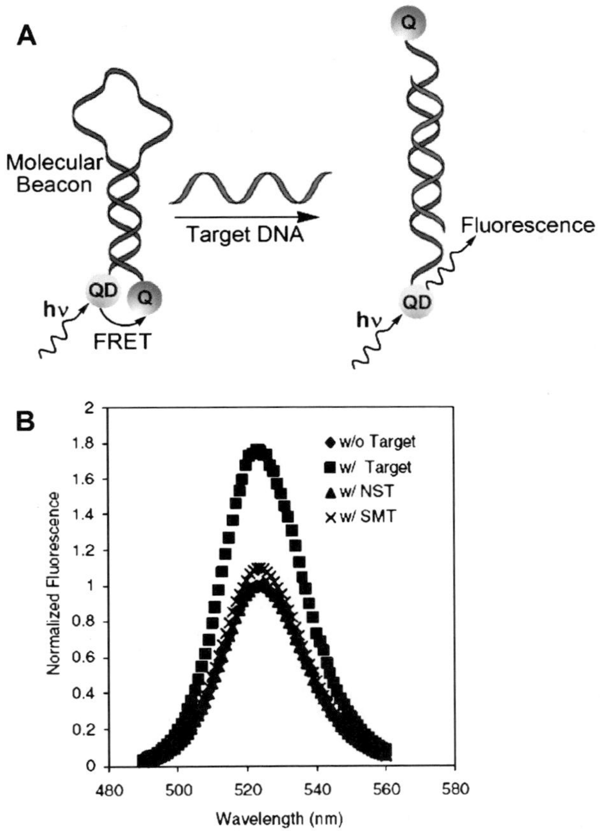

Fig. 13 **A** Optical detection of DNA by a hairpin nucleic acid functionalized with QD/quencher units. **B** Fluorescence intensities corresponding to: ♦ without target DNA, ■ with target DNA, 5′-AAACCCAAACCCAAA-3′, ▲ single-base mismatched target DNA, 5′-AAACCCGAACCCAAA-3′ (SMT), × non-complementary target DNA, 5′-AGG TATGCTCACCTT-3′ (NST). (Reprinted in part with permission from [53], © 2007, IOP)

ties of the CdSe/ZnS QDs. After cleavage of the double-stranded DNA with DNase I, the intensity of the FRET band of the dye decreased, and the fluorescence of CdSe/ZnS QDs increased (Fig. 14B). The luminescence properties of QDs were only partially recovered due to the non-specific adsorption of the dye onto QDs.

Aptamers are nucleic acid sequences that reveal specific binding properties towards proteins or low molecular weigh substrates. The aptamers are selected from a combinatorial library of 10^{15}–10^{16} DNAs using the systematic evolution of ligands by the exponential enrichment (SELEX) procedure. The recognition properties of aptamers were extensively used to develop electrochemical [55, 56] or optical [57, 58] sensor systems. QDs were also

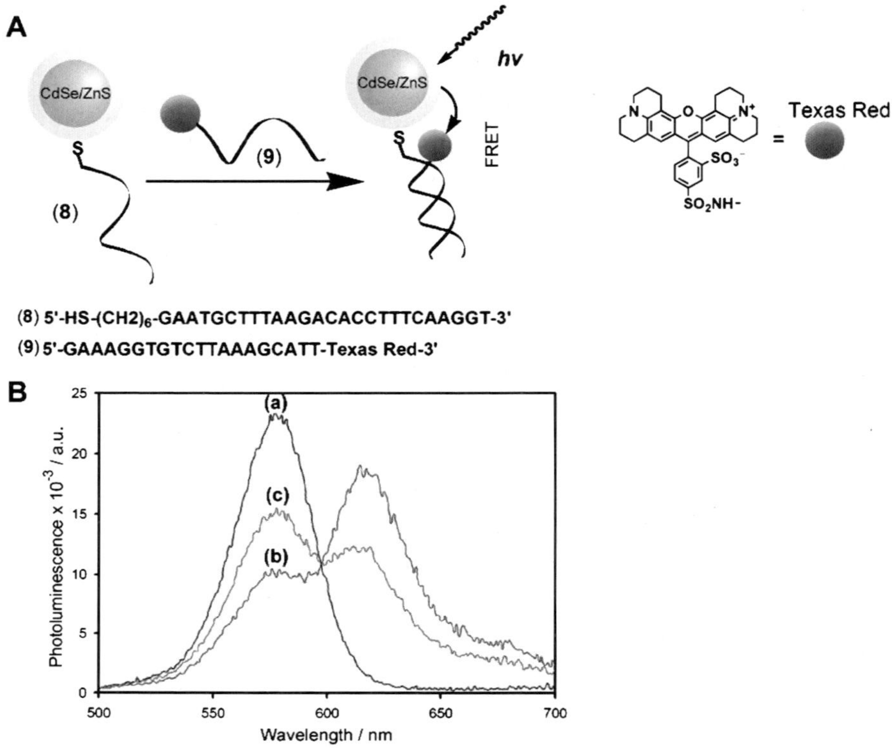

Fig. 14 **A** Assembly of the CdSe/ZnS and Texas Red-tethered duplex DNA. **B** Fluorescence spectra of: *a* the (**8**)-functionalized CdSe/ZnS QDs, *b* the (**8**)/(**9**) duplex DNA-tethered to the QDs and the Texas Red chromophore, and *c* after treatment of the duplex DNA tethered to CdSe/ZnS and the dye with DNase I. (Reprinted with permission from [54], © 2005, American Chemical Society)

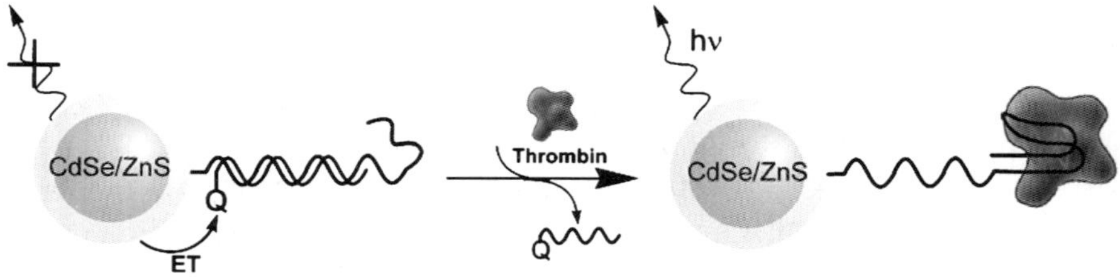

Fig. 15 Analysis of thrombin by the protein-induced separation of the anti-thrombin aptamer blocked by a quencher-functionalized nucleic acid that restores the fluorescence of the QDs

used to probe the formation of aptamer/protein complexes [59]. An anti-thrombin aptamer was coupled to QDs and the nucleic acid sequence was hybridized with a complementary oligonucleotide-quencher conjugate (Fig. 15).

The fluorescence of the QDs was quenched in the QD-aptamer/quencher-oligonucleotide duplex. In the presence of thrombin the duplex was separated and the aptamer underwent a conformational change to the quadruplex structure that binds thrombin. The displacement of the quencher units from the blocked aptamer activated the luminescence functions of the QDs and a ~19-fold increase in their fluorescence was observed. In a related study, thrombin was detected by the anti-thrombin aptamer conjugated to PbS QDs [60]. Upon binding of thrombin to aptamer-functionalized QDs, the selective fluorescence quenching was observed as a result of the charge transfer process from thrombin to QDs. This method enabled the sensing of thrombin with a detection limit corresponding to $\approx 1 \times 10^{-9}$ M, and it showed high selectivity in the presence of high background concentrations of interfering proteins.

3 Semiconductor Nanoparticles for Photoelectrochemical Bioanalytical Applications

The photoexcitation of semiconductor QDs does not only lead to luminescence properties, but the photogenerated electron-hole species might electrically communicate with electrode surfaces, (Fig. 1). The photoexcited QDs confined to electrode surfaces might directly communicate with the electrode surface by injection of the conduction-band electrons into the electrode, or by the sequential ejection of the conduction-band electrons to an electron acceptor solubilized in the electrolyte solution and the transfer of electrons to the valence-band holes. These two routes lead to the generation of an anodic photocurrent or a cathodic photocurrent by the photoexcited QDs, respectively. Different QD–DNA hybrid systems [61–63] or QD–protein conjugates [64, 65] were assembled on electrodes, and the control of the photoelectrochemical properties of the QDs by the biomolecules was demonstrated. A layer-by-layer deposition of nucleic acid-functionalized CdS QDs on electrodes was followed by the photoelectrochemical transduction of the assembly process [61]. The system demonstrated the amplified analysis of a DNA by photocurrent transduction. Semiconductor CdS NPs (2.6 ± 0.4 nm) were functionalized with one of the two thiolated nucleic acids **10** or **11** that are complementary to the 5′ and 3′ ends of a target DNA molecule (**12**). An array of CdS NP layers was then constructed on an Au electrode by a layer-by-layer hybridization process using the target DNA as crosslinker of CdS QD-functionalized with nucleic acids (**10** or **11**) complementary to the two ends of the DNA target (Fig. 16A). Illumination of the array in the presence of the sacrificial electron donor triethanolamine (TEOA) resulted in the generation of a photocurrent. The photocurrents increased with the number of layers of CdS NPs generated on the electrode (Fig. 16B). The photocurrent action spectra followed the absorbance features of the CdS NPs, implying that the photocurrents originated from the photoexcitation of the CdS nanoparticles. The ejection of the conduction-band electrons into the electrode occurred from the QDs that were in intimate contact with the electrode support. This was supported by the fact that $Ru(NH_3)_6{}^{3+}$ units ($E^0 = -0.16$ V vs. SCE), which were electrostatically bound to the DNA, enhanced the photocurrent from the DNA–CdS array. The $Ru(NH_3)_6{}^{3+}$ units

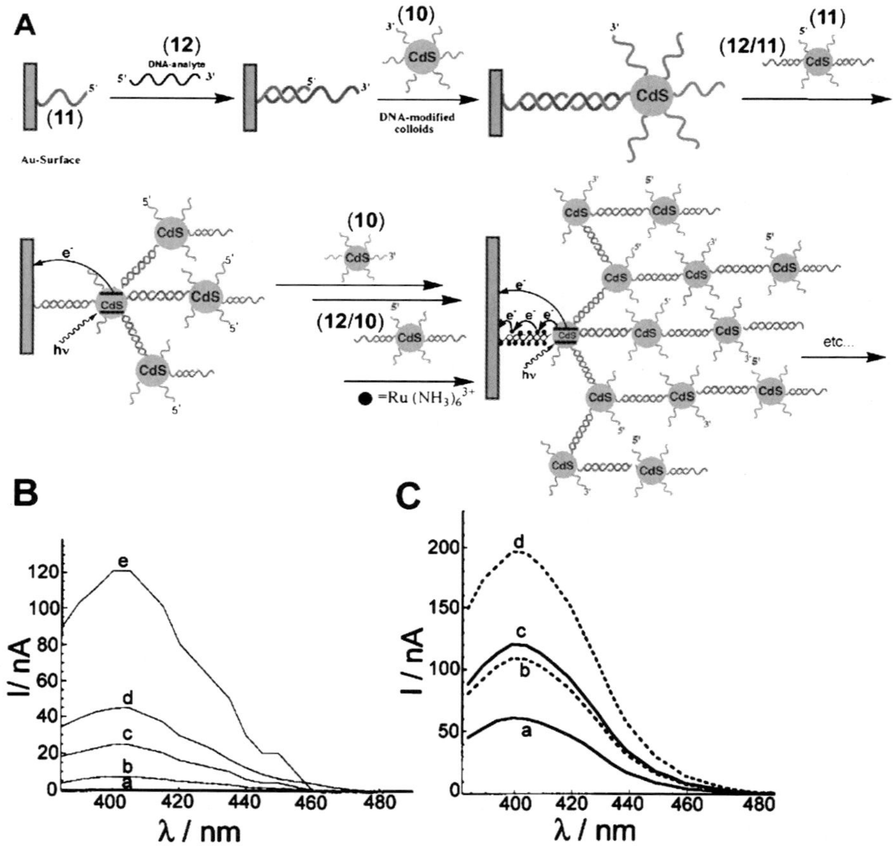

Fig. 16 **A** Layer-by-layer deposition of CdS NPs using (**10**)-and (**11**)-functionalized NPs and (**12**) as crosslinker. The association of $Ru(NH_3)_6{}^{3+}$ to the DNA array facilitates charge transport and enhances the resulting photocurrent. **B** Photocurrent action spectra generated by different numbers of CdS NP layers: *a* no CdS NPs, *b–e* one to four layers, repectively. Photocurrents were recorded in the presence of triethanolamine 2×10^{-2} M. **C** Effect of $Ru(NH_3)_6{}^{3+}$, 5×10^{-6} M, on the intensities of the photocurrents of layered CdS NP assemblies: *a* and *b* two layers of CdS NP in the absence and presence of $Ru(NH_3)_6{}^{3+}$, respectively; *c* and *d* four layers of CdS NP in the absence and presence of $Ru(NH_3)_6{}^{3+}$, respectively. Photocurrents were recorded in the presence of 2×10^{-2} M triethanolamine. (Reproduced with permission from [61], © 2001, Wiley-VCH)

acted as charge mediators that facilitated the hopping of conduction-band electrons from CdS particles, which lack contact with the electrode, due to their separation by the DNA tethers (Fig. 16C). The electrical contacting of the photoexcited QDs with the electrode surface through the DNA bridging units is of fundamental interest.

The charge transport through DNA is a subject of scientific controversy, and while several studies claimed that DNA acts as a conducting matrix [66], most of the results suggest that nucleic acids with a random distribution of bases behave as insulators [67]. The low photocurrents observed with semiconductor QDs separated by duplex nucleic acids from the electrode might support the insulating features of DNA and suggest that the photocurrents originate from QD–nucleic acid hybrid structures where the particles are in close contact with the electrode. The ability to intercalate redox-active components into a duplex DNA might provide, however, a means to electrically contact the photoexcited semiconductor QDs with the electrode by charge transport through the intercalated units. Indeed, redox-active intercalators incorporated into double-stranded DNAs that bridge semiconductor QDs to electrodes demonstrated not only enhanced photocurrent generation, but they enabled the potential-biased control of the photocurrent direction generated by the QDs [62]. The dithiol-tethered single-stranded ssDNA (**13**) was assembled on an Au electrode, and it was subsequently hybridized with a complementary dithiolated ssDNA (**14**) to yield a double-stranded DNA. The resulting surface was treated with CdS NPs to yield a semiconductor nanoparticle interface bridged to the electrode surface, by means of the nucleic acid duplex (Fig. 17). The irradiation of the dsDNA/CdS NPs-modified

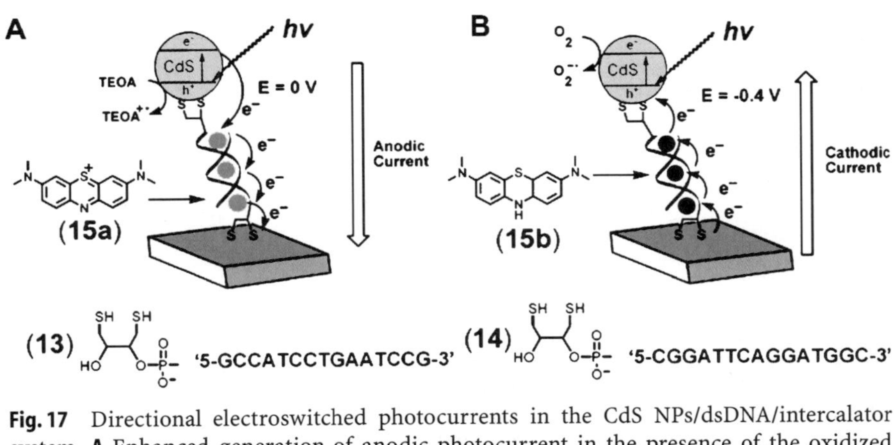

Fig. 17 Directional electroswitched photocurrents in the CdS NPs/dsDNA/intercalator system. **A** Enhanced generation of anodic photocurrent in the presence of the oxidized methylene blue intercalator (**15a**) (applied potential $E = 0$ V). **B** Enhanced generation of cathodic photocurrent in the presence of the reduced methylene blue intercalator (**15b**) (applied potential $E = -0.4$ V)

electrode, in the presence of triethanolamine (TEOA), as electron donor resulted in an anodic low intensity photocurrent. Subsequently, methylene blue, MB, (**15**) was intercalated into the (**13/14**)-dsDNA coupled to the CdS NPs (Fig. 17A,B). The cyclic voltammogram of the system implied that at potentials $E > -0.28$ V (vs. SCE) the intercalator exists in its oxidized form (**15a**), whereas at potentials $E < -0.28$ V (vs. SCE) the intercalator exists in its reduced leuco form (**15b**). Coulometric analysis of the methylene blue redox wave, $E° = -0.28$ V (vs. SCE), and knowing the surface coverage of the dsDNA, indicated that approximately two to three intercalator units were associated with the double-stranded DNA. An anodic photocurrent was generated in the system in the presence of TEOA as electron donor and methylene blue intercalated into the dsDNA, while applying a potential of 0 V (vs. SCE) on the electrode (Fig. 17A). At this potential MB existed in its oxidized state (**15a**), which acted as an electron acceptor. The resulting photocurrent was ca. fourfold higher than that recorded in the absence of the intercalator. The enhanced photocurrent was attributed to the trapping of conduction-band electrons by the intercalator units and their transfer to the electrode that was biased at 0 V, thus retaining the intercalator units in their oxidized form. The oxidation of TEOA by the valence-band holes led then to the formation of the steady-state anodic photocurrent. Biasing the electrode at a potential of −0.4 V (vs. SCE), a potential that retained the intercalator units in their reduced state (**15b**), led to the blocking of the photocurrent in the presence of TEOA and under an inert argon atmosphere. This experiment revealed that the oxidized intercalator moieties with the DNA matrix played a central role in the charge transport of the conduction-band electrons and the generation of the photocurrent.

Figure 18A, curve (b) shows the photocurrent generated by the (**13/14**)-dsDNA linked to the CdS NPs in the presence of the reduced intercalator

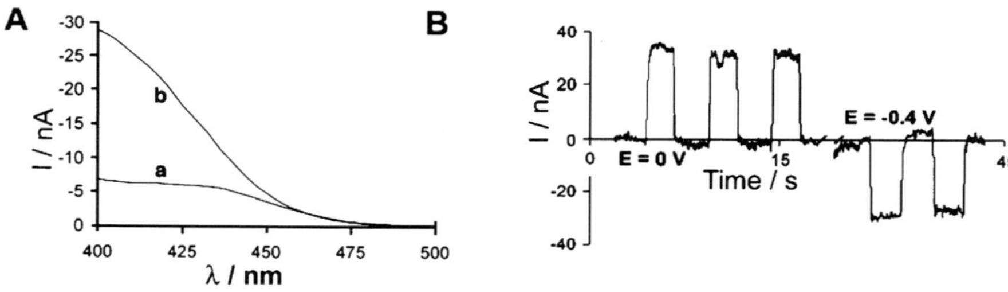

Fig. 18 **A** Cathodic photocurrents generated in the CdS NP/dsDNA system associated with the electrode: *a* in the absence and *b* in the presence of reduced methylene blue intercalator (**15b**). The data were obtained at the applied potential $E = -0.4$ V vs. SCE and in the presence of air. **B** Electrochemically switched anodic and cathodic photocurrents generated by the CdS NPs/dsDNA/**15a/15b** systems at 0 and −0.4 V, respectively, in the presence of TEOA, 20 mM and air. Photocurrents were generated upon irradiation at $\lambda = 420$ nm. (Reproduced with permission from [62]. © 2005, Wiley-VCH)

(**15b**) under conditions where the electrode was biased at –0.4 V (vs. SCE) and the system was exposed to air (oxygen). At the bias potential of –0.4 V (vs. SCE), the intercalator units exist in their reduced leuco form (**15b**) that exhibit electron-donating properties (Fig. 17B). Photoexcitation of the CdS NPs yields the electron-hole pairs in the conduction-band and valence-band, respectively. The transport of the conduction-band electrons to oxygen with the concomitant transport of electrons from the reduced intercalator units to the valence-band holes completed the cycle for the generation of the photocurrent. The fact that the electrode potential retained the intercalator units in their reduced state, and the infinite availability of the electron acceptor (O_2) yielded the steady-state cathodic photocurrent in the system. The introduction of TEOA and oxygen to the electrode modified with (**13/14**) DNA duplex and associated CdS NPs allowed the control of the photocurrent direction by switching the bias potential applied on the electrode. Figure 18B depicts the potential-induced switching of the photocurrent direction upon switching the electrode potential between –0.4 V (cathodic photocurrents) and 0 V (anodic photocurrents), respectively.

The self-assembly of CdS NPs on an electrode surface and the generation of a photocurrent were employed as a readout method for the operation of a DNA sensing machine [63] (Fig. 19A). The machine consisted of a DNA track (**16**) that included a recognition sequence (**I**), a nicking sequence generated upon the formation of a duplex structure (**II**), and a reporter sequence (**III**). Upon the hybridization of the analyte (**17**) and in the presence of polymerase/dNTPs and the nicking enzyme Nb.BbvcI, the DNA machine was activated. This was reflected by the replication of the track and the subsequent scission of the resulting duplex. The nicking process initiated the further replication of the track while displacing the nucleic acid (**18**). Accordingly, the sensing of (**17**) initiated the autonomous operation of the machine that generated (**18**) as a "waste product". The displaced product (**18**) was then interacted with an electrode modified with the nucleic acid (**19**), complementary to the 3′-end of the product (**18**), and CdS NPs functionalized with the nucleic acid (**20**), complementary to the 5′-end of the generated product (**18**). This resulted in the self-assembly of the CdS NPs on the electrode and the generation of a photocurrent (Fig. 19B).

The electrical contacting of semiconductor QDs with electrodes in the presence of proteins seems to be more problematic due to the insulating features of proteins [68]. The tailored organization of semiconductor QD–protein hybrid assemblies on electrodes may lead, however, to electrically contacted architectures that are controlled by the linked proteins. In the two configurations, the semiconductor QDs are directly linked to the electrode, and the proteins are associated with the QD surface, and the control of the photoelectrochemical functions of the semiconductor QDs may then be accomplished by two general paths: (i) Low molecular weight redox proteins might interact with the QDs, and the electrical contact between the

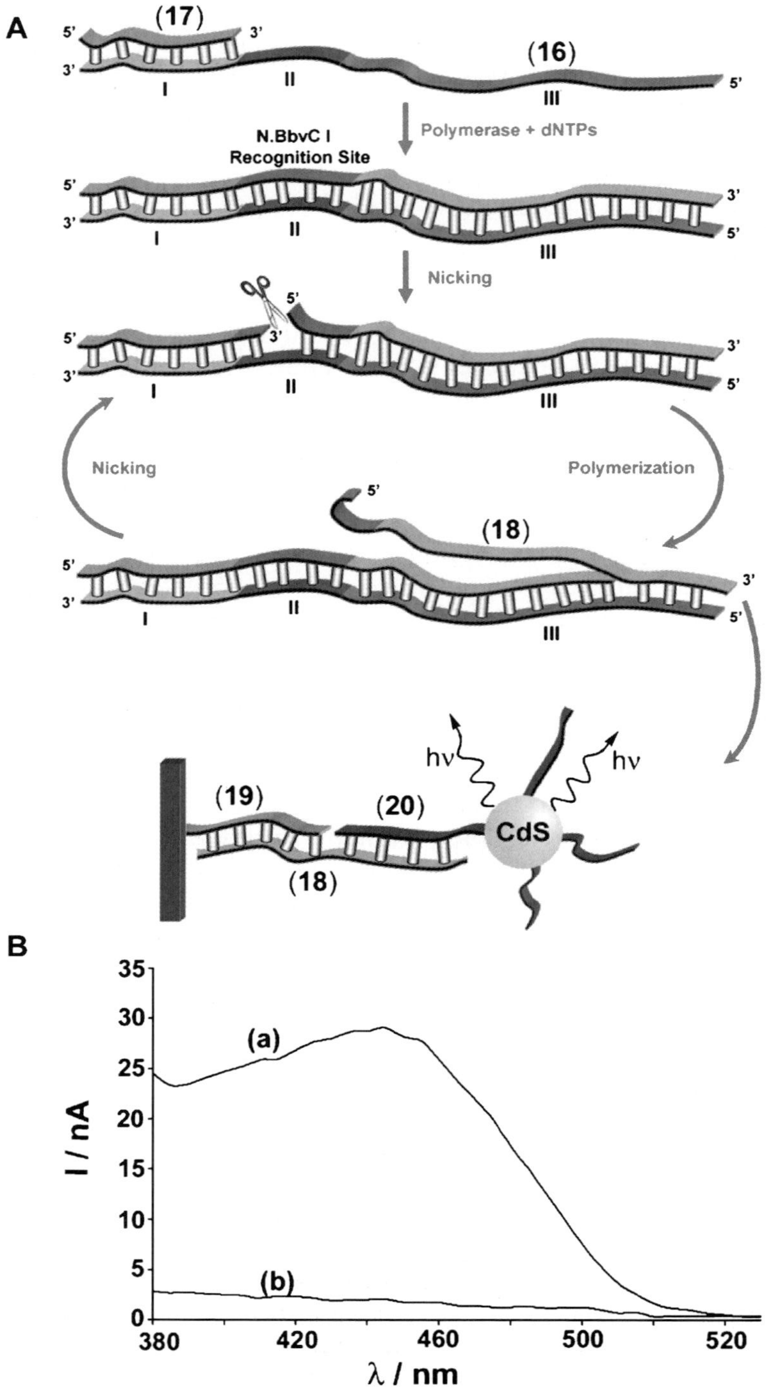
A
(17)
5'
3'
3'
(16)
5'
I
II
III
Polymerase + dNTPs
N.BbvC I
Recognition Site
Nicking
Polymerization
(18)
(19)
(20)
hν
hν
CdS
B
(a)
(b)
I / nA
λ / nm
0
5
10
15
20
25
30
35
380
420
460
500
520

◀ **Fig. 19** Photoelectrochemical readout of the sensing of a nucleic acid by a DNA machine: **A** Amplified analysis of the target DNA (**17**) by a DNA machine that generates the "waste" product (**18**). The association of the (**20**)-functionalized CdS NP on the electrode proceeds by the bridging of the NPs with the "waste" product (**18**). **B** Photocurrents observed: *a* in the absence of the target DNA, and *b* In the presence of the target DNA, 1×10^{-6} M. The DNA-machine was operated for 90 min. (Reproduced from [63] by permission of The Royal Society of Chemistry)

conduction-band electrons/valence-band holes with the redox centers of the proteins might affect the photocurrents of the hybrid system. (ii) Biocatalysts may be tethered to the QDs and their reaction products might act as electron acceptor or electron donor units. These may then activate the photoelectrochemical functions of the QDs.

Indeed, enzymes or redox proteins were linked to semiconductor QDs, and the resulting photocurrents were used to assay the enzyme activities, and to develop different biosensors. Cytochrome *c*-mediated biocatalytic transformations were coupled to CdS NPs, and the direction of the resulting photocurrent was controlled by the oxidation-state of the cytochrome *c* mediator [64]. The CdS NPs were immobilized on an Au electrode through a dithiol linker, and mercaptopyridine units, acting as promoter units that electrically communicate between the cytochrome *c* and the NPs were linked to the semiconductor NPs (Fig. 20). In the presence of reduced cytochrome *c*, the photoelectrocatalytic activation of the oxidation of lactate by lactate dehydrogenase (LDH) proceeded, while generating an anodic photocurrent (Fig. 20A). Photoexcitation of the NPs resulted in the ejection of the conduction-band electrons into the electrode and the concomitant oxidation of the reduced cytochrome *c* by the valence-band holes. The resulting oxidized cytochrome *c* subsequently mediated the LDH-biocatalyzed oxidation of lactate. Similarly, cytochrome *c* in its oxidized form was used to stimulate the bioelectrocatalytic reduction of NO_3^- to NO_2^- in the presence of nitrate reductase (NR), while generating a cathodic photocurrent (Fig. 20B). The transfer of conduction-band electrons to the oxidized, heme-containing cofactor generated reduced cytochrome *c*, while the transfer of electrons from the electrode to the valence-band holes of the NPs restored the ground state of the NPs. The cytochrome *c*-mediated biocatalyzed reduction of NO_3^- to nitrite enabled then the formation of the cathodic photocurrent, while biasing the electrode potential at 0 V vs. SCE. The photocurrents generated by the biocatalytic cascades at various concentrations of the different substrates are depicted in Fig. 20C. These results demonstrated that the photoelectrochemical functions of semiconductor NPs could be used to develop sensors for biocatalytic transformations. A related study has employed CdSe/ZnS QDs capped with mercaptosuccinic acid as protecting layer for the generation of photocurrents in the presence of cytochrome *c* [69].

The use of enzymes as catalysts that generate a product that stimulates the generation of a photocurrent by QDs was demonstrated with the de-

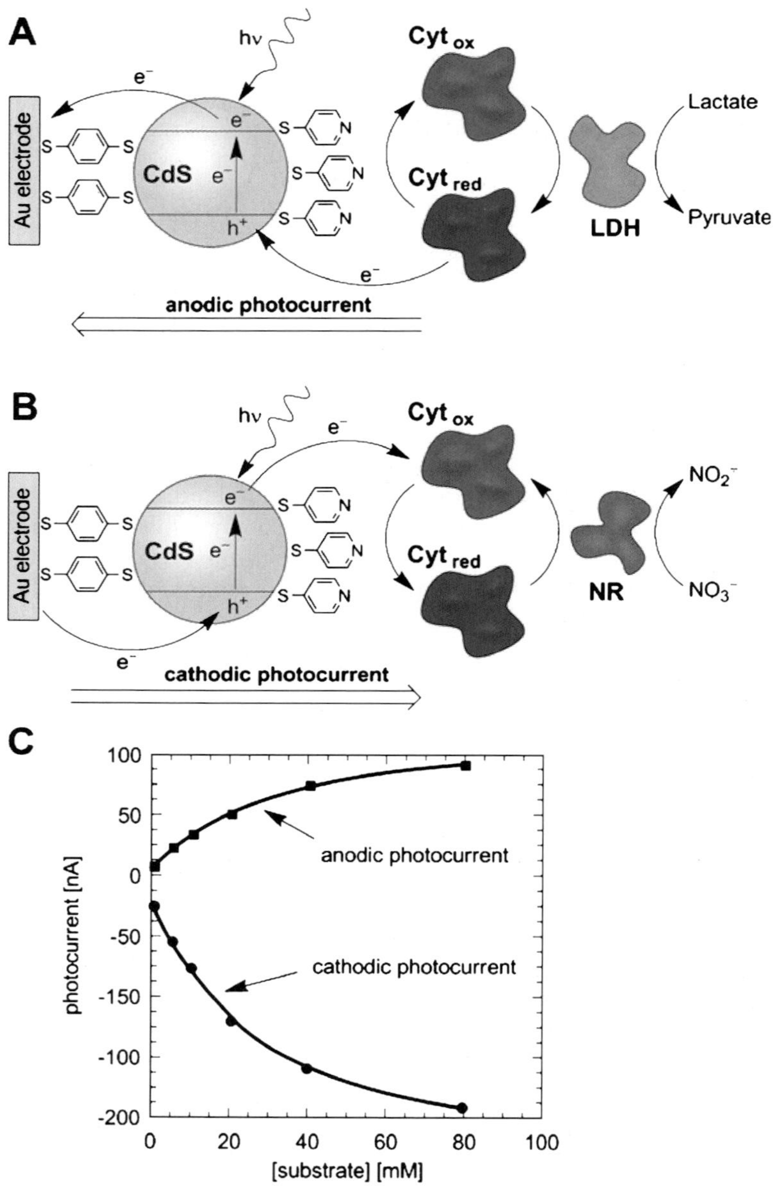

Fig. 20 Generation of photocurrents by the photochemically induced activation of enzyme cascades by CdS NPs. **A** The photochemical activation of the cytochrome c-mediated oxidation of lactate in the presence of LDH. **B** Photochemical activation of the cytochrome c-mediated reduction of nitrate (NO_3^-) by nitrate reductase (NR). **C** Photocurrents generated by the biocatalytic cascades in the presence of various concentrations of the substrates (lactate/nitrate). (Reproduced from [64] by permission of The Royal Society of Chemistry)

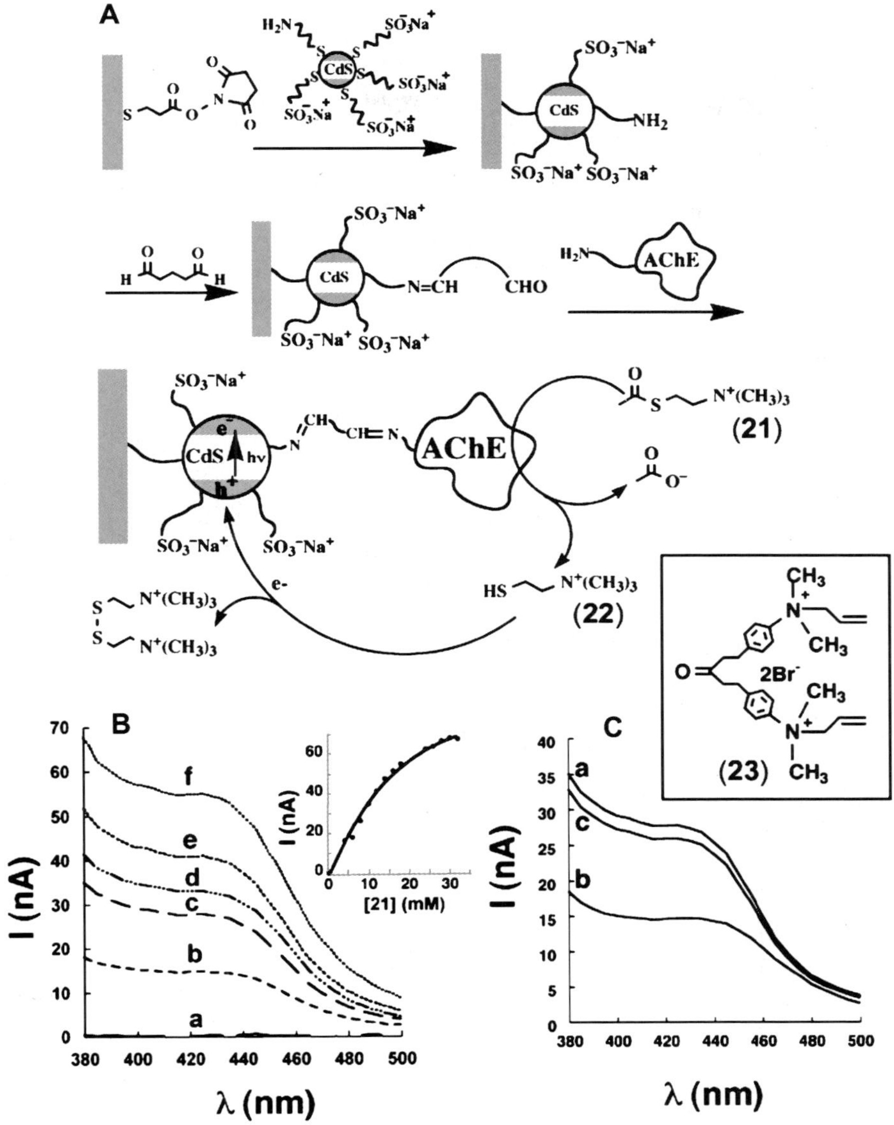

Fig. 21 **A** Assembly of the CdS NP/AChE hybrid system for the photoelectrochemical detection of enzyme activity (h = hole). **B** Photocurrent action spectra observed in the presence of acetylthiocholine (**21**): *a* 0 mM, *b* 6 mM, *c* 10 mM, *d* 12 mM, *e* 16 mM, and *f* 30 mM. *Inset*: Calibration curve of the photocurrent at $\lambda = 380$ nm at variable concentrations of **21**. **C** Photocurrent spectra for the CdS/AChE system in the presence of **21** (10 mM): *a* in the absence of inhibitor **23**, *b* upon addition of the inhibitor **23** (1×10^{-6} M), and *c* after rinsing the system, exclusion of the inhibitor, and addition of **23** (10 mM). (Reprinted with permission from [65]. © 2003, American Chemical Society)

velopment of an acetylcholine esterase-based biosensor of the enzyme inhibitors [65]. CdS NPs were assembled on an Au electrode, and the NPs were further modified with acetylcholine esterase, AChE, (Fig. 21A). The biocatalyzed hydrolysis of acetylthiocholine (**21**) by acetylcholine esterase generated thiocholine (**22**), which acted as an electron donor for the photogenerated holes in the valence-band of the CdS NPs. The resulting photocurrents were controlled by the concentration of the substrate (Fig. 21B), and the photocurrent intensities were depleted in the presence of inhibitors of acetylcholinesterase, such as 1,5-bis(4-allyldimethylammoniumphenyl)pentan-3-one dibromide (**23**) (Fig. 21C). The system was suggested as a potential sensor for chemical warfare agents that act as inhibitors of acetylcholinesterase.

4 Conclusions and Perspectives

The unique photophysical properties and functions of semiconductor QDs find growing interest in the area of nanobiotechnology. The high fluorescence yields of QDs, their stability against photobleaching, and the size-controlled luminescence features of QDs, paved ways to develop different biosensing platforms. Besides the advantages of using the QDs as effective luminescent labels, the multiplexed parallel analysis of targets by size-controlled luminescent QDs that act as coding labels is of great promise for future bioanalytical applications. The nanometer dimensions of QDs enabled their incorporation into cells. The in vitro imaging of the QDs in different intracellular domains has already been demonstrated. The incorporation of functionalized QDs into cells that permits the following of dynamic intracellular metabolic processes at the single cell level provides, however, exciting future opportunities. The charge-ejection functions of semiconductor QDs also hold great promise for future nanomedical applications of QDs. The ejection of conduction-band electrons of semiconductor NPs into oxygen, and the formation of cell destructive agents is well established [70], and such particles were suggested as cancer therapeutic materials. The successful incorporation of QDs into cells suggests that cell-targeted QDs could be synthesized and act as tumor cell killers. Experiments along these lines provide promising results.

The use of biomolecule–QD hybrid systems for photoelectrochemical application is still at its infancy. The use of such hybrid structures for biosensing has been demonstrated. Nonetheless, exciting future applications of these hybrid structures may be envisaged. One direction might involve the use of the hybrid structures as templates for the fabrication of devices. For example, DNA acts as a template for the synthesis of metallic nanowires by the reduction of metal ions linked to the DNA backbone. The QD-stimulated photoelectrochemical reduction of ions associated with DNA may then lead to metal–semiconductor–metal nanostructured devices. Furthermore, the activation of photoelectrochemical functions of QDs by enzymes hold great promises for the future development of light-to-electrical energy conversion systems, photo-biofuel cells. In view of the recent advances in the use of biomolecule–QD hybrids, the future perspectives of these systems are bright and promising.

Acknowledgements Our research on biomolecule–semiconductor quantum dots is supported by the Israel Ministry of Science and by the NanoSci-ERA consortium (NANO-LICHT Project).

References

1. Brus LE (1991) Appl Phys A 53:465
2. Alivisatos AP (1996) Science 271:933
3. Grieve K, Mulvaney P, Grieser F (2000) Curr Opin Colloid Interface Sci 5:168
4. Alivisatos AP (2004) Nat Biotechnol 22:47
5. Nirmal M, Brus LE (1999) Acc Chem Res 32:407
6. Medintz IL, Uyeda HT, Goldman ER, Mattoussi H (2005) Nat Mater 4:435
7. Chan WCW, Maxwell DJ, Gao X, Bailey RE, Han M, Nie S (2002) Curr Opin Biotechnol 13:40
8. Kamat PV (2007) J Phys Chem C 111:2834
9. Sapsford KE, Pons T, Medintz IL, Mattoussi H (2006) Sensors 6:925
10. Costa-Fernandez JM, Pereiro R, Sanz-Medel A (2006) Trends Anal Chem 25:207
11. Klostranec JM, Chan WCW (2006) Adv Mater 18:1953
12. Murphy CJ (2002) Anal Chem 74:520A
13. Goldman ER, Balighian ED, Mattoussi H, Kuno MK, Mauro JM, Tran PT, Anderson GP (2002) J Am Chem Soc 124:6378
14. Goldman ER, Anderson GP, Tran PT, Mattoussi H, Charles PT, Mauro JM (2002) Anal Chem 74:841
15. Goldman ER, Clapp AR, Anderson GP, Uyeda HT, Mauro JM, Medintz IL, Mattoussi H (2004) Anal Chem 76:684
16. Yang L, Li Y (2006) Analyst 131:394
17. Hoshino A, Fujioka K, Manabe N, Yamay S, Goto Y, Yasuhara M, Yamamoto K (2005) Microbiol Immunol 49:461
18. Gerion D, Chen F, Kannan B, Fu A, Parak WJ, Chen DJ, Majumdar A, Alivisatos AP (2003) Anal Chem 75:4766
19. Liang RQ, Li W, Li Y, Tan CY, Li JX, Jin YX, Ruan KC (2005) Nucl Acids Res 33:e17
20. Pathak S, Choi SK, Arnheim N, Thompson ME (2001) J Am Chem Soc 123:4103
21. Xiao Y, Barker PE (2004) Nucleic Acids Res 32:e28
22. Chan PM, Yuen T, Ruf F, Gonzalez-Maeso J, Sealfon SC (2005) Nucleic Acids Res 33:e161
23. Michalet X, Pinaud FF, Bentolila LA, Tsay JM, Doose S, Li JJ, Sundaresan G, Wu AM, Gambhir SS, Weiss S (2005) Science 307:538
24. Alivisatos AP, Gu W, Larabell C (2005) Annu Rev Biomed Eng 7:55
25. Gao X, Yang L, Petros JA, Marshall FF, Simons JW, Nie S (2005) Curr Opin Biotechnol 16:63
26. Lidke DS, Nagy P, Heintzmann R, Arndt-Jovin DJ, Post JN, Grecco HE, Jares-Erijman EA, Jovin TM (2004) Nat Biotechnol 22:198

27. Mattoussi H, Mauro JM, Goldman ER, Anderson GP, Sunder VC, Mikulec FV, Bawendi MG (2000) J Am Chem Soc 122:12142
28. Chan WCW, Nie S (1998) Science 281:2016
29. Gerion D, Pinaud F, Willimas SC, Parak WJ, Zanchet D, Weiss S, Alivisatos AP (2001) J Phys Chem B 105:8861
30. Bruchez M Jr, Moronne M, Gin P, Weiss S, Alivisatos AP (1998) Science 281:2013
31. Gao X, Cui Y, Levenson RM, Chung LWK, Nie S (2004) Nat Biotechnol 22:969
32. Clapp AR, Medintz IL, Mattoussi H (2006) ChemPhysChem 7:47
33. Sapsford KE, Berti L, Medintz IL (2006) Angew Chem Int Ed 45:4562
34. Smith AM, Duan H, Rhyner MN, Ruan G, Nie S (2006) Phys Chem Chem Phys 8:3895
35. Pellegrino T, Kudera S, Liedl T, Muñoz Javier A, Manna L, Parak WJ (2005) Small 1:48
36. Bäumle M, Stamou D, Segura JM, Hovius R, Vogel H (2004) Langmuir 20:3828
37. Zheng Y, Gao S, Ying JY (2007) Adv Mater 19:376
38. Medintz IL, Clapp AR, Mattoussi H, Goldman ER, Fisher B, Mauro JM (2003) Nat Mater 2:630
39. Uyeda HT, Medintz IL, Jaiswal JK, Simon SM, Mattoussi H (2005) J Am Chem Soc 127:3870
40. Willard DM, Carillo LL, Jung J, Orden AV (2001) Nano Lett 1:469
41. Wang S, Mamedova N, Kotov NA, Chen W (2002) Nano Lett 2:817
42. Patolsky F, Gill R, Weizmann Y, Mokari T, Banin U, Willner I (2003) J Am Chem Soc 125:13918
43. Moyzis RK, Buckingham JM, Cram LS, Dani M, Deaven LL, Jones MD, Meyne J, Ratcliff RL, Wu JR (1988) Proc Natl Acad Sci USA 85:6622
44. Bryan TM, Cech TR (1999) Curr Opin Cell Biol 11:318
45. Kim NW, Piatyszek MA, Prowse KR, Harley CB, West MD, Ho PL, Coviello GM, Wright WE, Weinrich SL, Shay JW (1994) Science 266:2011
46. Shay JW, Bacchetti S (1997) Eur J Cancer 33:787
47. Medintz IL, Clapp AR, Mattoussi H, Goldman ER, Fisher B, Mauro JM (2003) Nat Mater 2:630
48. Goldman ER, Medintz IL, Whitley JL, Hayhurst A, Clapp AR, Uyeda HT, Deschamps JR, Lassman ME, Mattoussi H (2005) J Am Chem Soc 127:6744
49. Medintz IL, Clapp AR, Brunel FM, Tiefenbrunn T, Uyeda HT, Chang EL, Deschamps JR, Dawson PE, Mattoussi H (2006) Nat Mater 5:581
50. Shi L, De Paoli V, Rosenzweig N, Rosenzweig Z (2006) J Am Chem Soc 128:10378
51. Gill R, Freeman R, Xu J, Willner I, Winograd S, Shweky I, Banin U (2006) J Am Chem Soc 128:15376
52. Kim JH, Morikis D, Ozkan M (2004) Sens Actuators B 102:315
53. Kim JH, Chaudhary S, Ozkan M (2007) Nanotechnology 18:1
54. Gill R, Willner I, Shweky I, Banin U (2005) J Phys Chem B 109:23715
55. Hansen JA, Wang J, Kawde AN, Xiang Y, Gothelf KV, Collins G (2006) J Am Chem Soc 128:2228
56. Polsky R, Gill R, Willner I (2006) Anal Chem 78:2268
57. Pavlov V, Xiao Y, Shlyahovsky B, Willner I (2004) J Am Chem Soc 126:11768
58. Liu J, Lu Y (2006) Angew Chem Int Ed 45:90
59. Levy M, Cater SF, Ellington AD (2005) ChemBioChem 6:2163
60. Choi JH, Chen KH, Strano MS (2006) J Am Chem Soc 128:15584
61. Willner I, Patolsky F, Wasserman J (2001) Angew Chem Int Ed 40:1861
62. Gill R, Patolsky F, Katz E, Willner I (2005) Angew Chem Int Ed 44:4554
63. Freeman R, Gill R, Beissenhirtz M, Willner I (2007) Photochem Photobiol Sci 6:416
64. Katz E, Zayats M, Willner I, Lisdat F (2006) Chem Comm, p 1395

65. Pardo-Yissar V, Katz E, Wasserman J, Willner I (2003) J Am Chem Soc 125:622
66. Porath D, Cuniberty G, Di Felice R (2004) Top Curr Chem 237:183
67. De Pablo PJ, Moreno-Herrero F, Colchero J (2000) Phys Rev Lett 85:4992
68. Willner I (2002) Science 298:2407
69. Stoll C, Kudera S, Parak WJ, Lisdat F (2006) Small 2:741
70. Bakalova R, Ohba H, Zhelev Z, Ishikawa M, Baba Y (2004) Nat Biotechnol 22:1360

Adv Biochem Engin/Biotechnol (2008) 109: 285–350
DOI 10.1007/10_2007_078

Published online: 13 November 2007

Microsystems Technology and Biosensing

Ramachandra Rao Sathuluri[1,2] · Shohei Yamamura[1] · Eiichi Tamiya[1,3] (✉)

[1]School of Materials Science, Japan Advanced Institute of Science and Technology, 1-1 Asahidai, Nomi City, Ishikawa, 923-1292, Japan

[2]Toyama New Industry Organization, 529 Takada, Toyama City, Toyama, 930-0866, Japan

[3]Department of Applied Physics, Graduate School of Engineering, Osaka University, 2-1 Yamadaoka, Suita City, Osaka, 565-0871, Japan
tamiya@ap.eng.osaka-u.ac.jp

Abstract This review addresses the recent developments in miniaturized microsystems or lab-on-a-chip devices for biosensing of different biomolecules: DNA, proteins, small molecules, and cells, especially at the single-molecule and single-cell level. In order to sense these biomolecules with sensitivity we have fabricated chip devices with respect to the biomolecule to be analyzed. The details of the fabrication are also dealt with in this review. We mainly developed microarray and microfluidic chip devices for DNA, protein, and cell analyses. In addition, we have introduced the porous anodic alumina layer chip with nanometer scale and gold nanoparticles for label-free sensing of DNA and protein interactions. We also describe the use of microarray and microfluidic chip devices for cell-based assays and single-cell analysis in drug discovery research.

1 Introduction

In recent years, miniaturized systems called lab-on-a-chip or the micro total analysis system (μ-TAS) have been focused on as new microsystems for biochemical analyses. These systems are expected to perform DNA, protein, and cell analyses for drug screening and development of novel therapies. Especially, microarray and microfluidic types of chip devices (so-called biochips) have been developed using micro- and nanotechnological techniques [1, 2]. These lab-on-a-chip systems can be used for high-throughput identification of large numbers of potential drug targets, for example DNA, protein, chemicals, etc. [3, 4]. Recent advances in the human genome project have prompted the use of the miniaturized chip devices for high-throughput analysis of the vast amount of information potentially available. In order to obtain as much information as possible in a short time with a minimal use of reagents, researchers require highly integrated and sophisticated devices, such as microarrays and microfluidics. Thus, biochips suitable for different biomaterial assay and detection technologies have been under intense investigation [1–4]. Microarrays have been mostly applied to the assessment of the presence of a specific base sequence, or which genes are expressed and at what level [5, 6]. They have also been employed in identification of peptides and proteins as pharmaceutical drugs [7, 8]. Research market study analysis (http://www.researchandmarket.com/reports/c21981) suggests the total biochip market size in 2004 was about US $2.0 billion and is forecast to grow to about $5.1 billion in 2009 with a compound annual growth rate (CAGR) of 20.2%. Although the number of human genes was reported to be approximately 30 000 from the human genome project, the functions and expression mechanism for most of the genes remain unknown, because the fate of the genes' functionality is determined after protein expression, but protein expression is controlled by cellular function. Therefore, it is necessary to develop microarray chip devices that can perform high-throughput screening and analysis of proteins and cells at the single-cell and single-molecule level [9–11, 203].

To achieve single-cell or single-molecule analysis, highly integrated microarray and microfluidic systems that can perform assays at pico- and nanoliter volume levels are highly desirable to realize post-genomic research, such as proteomics and cellomics. Single-cell analysis contributes to elucidating the functional mechanism of genes, proteins, and chemical responses, which leads to clinical diagnosis and drug discovery [12–14]. In recent work, some researchers reported cellular microarrays using biomaterial spotting techniques for investigating gene expression or differentiation of some kinds of cells [15, 16]. These microsystems can only screen and detect a group of cells but not single-cell-based assays from bulk cell suspension. Using these devices there is a great possibility of cross-contamination with neighboring cells due to the absence of a physical boundary. Therefore, it is necessary to construct a microarray or microfluidic system that can perform high-throughput analysis of single molecules or cells with quantitative detection.

However, there are no reports available to the best of our knowledge that describe high-throughput analysis of DNA, proteins, small chemical molecules, and cells at the single-molecule and single-cell level together in one place. Therefore, in this chapter we have addressed the analysis of DNA, proteins, and cells using a pico- or nanoliter chamber array system. The chapter is divided into three topics: DNA chip systems, protein chip systems, and cell chip systems. Furthermore, each topic is divided into different subsections of research, as shown in the contents section. The research results presented here are mainly from our own group at JAIST, Japan. The chapter ends with conclusions.

2
DNA Sensor Systems

2.1
Nanoliter Silicon Microchamber Array for Multiplexed Polymerase Chain Reaction from a Single Copy DNA

Since the discovery of PCR technology by Prof. Mullis, which earned him a Nobel Prize [17], it has opened new horizons for a limitless number of

DNA-based research possibilities. Qualitative PCR is a well-established and straightforward technology, but the quantification of specific target DNA sequences in a complex sample has been a difficult task. A number of variations, caused by the manipulation of nucleic acids, that may occur during sample preparation, storage, or the course of the reaction hampered accurate quantification. The exponential nature of the PCR amplification can significantly magnify even minor variations in reaction conditions. Normalizing the amount of PCR products of the specific template with respect to an internal reference template has been partly successful against these variations. Since the challenge of accurate DNA quantification stimulated many researchers, a great variety of protocols already exist for the utility of quantitative PCR [18–20]. However, these methods are nearly exclusively restricted to being applied for research purposes only because of two factors they have in common: they are difficult tasks and are costly to run.

To supply the demand for faster, more accurate, and more cost-effective PCR devices with a high-throughput capacity, three important properties have directed the development of the next generation of PCR systems: automation, standardization, and miniaturization. Recently, Yang et al. [21] reported a high-sensitivity PCR assay in polycarbonate plastic, disposable PCR microreactors. At a template concentration as low as ten *Escherichia coli* cells (equivalent to 50 fg of genomic DNA), 221-bp product was successfully amplified within 30 min. Lee et al. [22] have described a microfabricated PCR device for simultaneous DNA amplification and electrochemical detection on gold or indium tin oxide (ITO) electrodes patterned on a glass substrate. A miniaturized flow-through PCR with different template types in a silicon chip thermocycler was also reported to have minimum power consumption [23]. With the flow-through PCR device of Fukuba et al. [24], 580 and 1450 bp of DNA fragments were successfully amplified from *E. coli* genomic DNA and directly from untreated cells. For temperature control of their chip, six heaters made of ITO were placed on a glass substrate to act as three uniform temperature zones. Lee et al. [25] have recently reported a bulk-micromachined PCR chip. They validated that the proposed chip amplified the DNA related to the tumor suppressor gene BRCA 1 (127 bp at 11th exon) after 30 thermal cycles in a 200-nL-volume chamber. Although most of the recent assays are accurate and sensitive, they involve the definition of very stringent limits. The PCR products are usually separated by gel electrophoresis, and the band intensities are quantified by video imaging and densitometry. Additionally, Lagally et al. [26, 27] have shown that microfluidic systems are capable of multiplexed PCR reactions and robust on-chip detection.

Microchamber arrays etched on silicon or glass were also one of the most reported miniaturized devices for multiple simultaneous DNA amplification [28–30]. The minimum reported size for a microchamber for PCR was demonstrated by Leamon et al. in connection with PCR [31]. They reported a novel platform, namely PicoTiterPlate™, which enabled simultaneous amplification of 300 000 discrete PCR reactions in volumes as small as 39.5 pL. Following the PCR on the PicoTiterPlate™, the solution from each well was recovered, and then quantified by TaqMan assay. As for the easy integration with different applications, solid-phase amplification was also performed on PicoTiterPlate™ by immobilizing the PCR product to a DNA capture bead in each well. Thus, 370 000 beads bound with PCR product were obtained for parallel processing in numerous solid-phase applications. The volume of a microchamber for a successful PCR amplification was reduced down to 86 pL by Nagai et al. [28]. However, as the sample volume was decreased, the evaporation of sample solution, and an introduction method of quite a small amount of solution into the reaction microchamber, appeared as the major drawbacks.

In our research, for achieving simultaneous detection of several numbers of target DNA, the feasibility of our microchamber array was further improved by using TaqMan PCR [32, 33]. To the best of our knowledge, three different DNA sequences were amplified from three different DNA templates and detected in the same microchamber array simultaneously for the first time. In addition, quantification of the initial DNA concentration present in a microchamber was achieved from 0 to 12 copies per chamber, not only by monitoring the real-time fluorescence intensity but also by observing the end-point fluorescence signal. Therefore, this system proves to be a promising device for low-cost, high-throughput DNA amplification and detection for point-of-care clinical diagnosis, which can also be handled by nonspecialist users.

2.1.1 PCR Microchamber Array Chip System

Microchamber Array Chip Fabrication

The microchamber array chip for DNA amplification was fabricated using micromachining techniques, including photolithography and anisotropic wet etching on the optically polished side of a silicon (100) wafer; the details of the protocols are described in depth elsewhere [30]. Briefly, the chip substrate was designed to be 2.54×7.62 cm for compatibility with the dispensing system employed. A photo mask with 1248 chambers having 24×52 patterns was then printed on the surface that was coated with an OFPR-800 photoresist layer by a photolithographic process. After NMD-3 solution, HF/NH_4F, and tetramethylammonium hydroxide (TMAH) treatments, the resulting chamber array feature was observed by a Keyence digital microscope. Each chamber was a parallelepiped with dimensions of 650×650×200 μm, pitch of ~900 μm, and accommodated 50 nL (Fig. 1). The total number of chambers on each chip was 1248. To achieve precise introduction of sample mixture into the microchamber, only the inner wall surfaces of the microchamber were prepared as hydrophilic by leaving an oxidized layer on them with photolithographic techniques, as recently described by Felbel et al. [34].

A

B

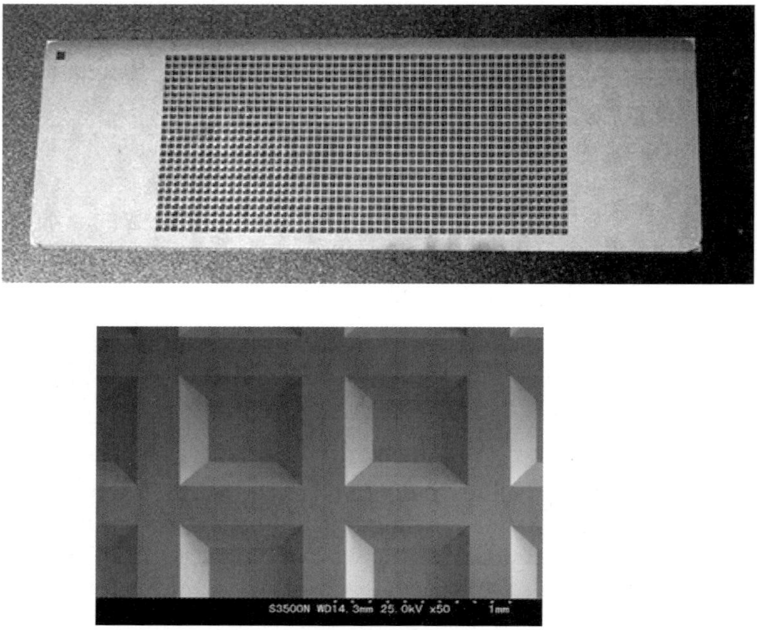

Fig. 1 **A** Photographic image of the microchamber array chip. The 1248 microchambers are integrated on the 1×3 in of a silicon chip. **B** Scanning electron micrograph (SEM) of a silicon microchamber array

Sample Loading with a Nanoliter Dispenser

The microchamber array chip was soaked in 1% (w/v) bovine serum albumin (BSA) solution overnight, then rinsed with deionized water and dried in order to prevent nonspecific adsorption by coating the chamber wall. The chip was placed onto the dispensing stage of a nanoliter dispenser from Cartesian Technologies. The precise dispensing of nanoliter volumes of solutions exactly at previously determined locations had become very simple by using their technology. The volume of dispensed solution in a single microchamber was 40 nL. Mineral oil as a cover layer was coated onto the template DNA-modified chip, and 40 nL of PCR mixture, which included target-specific primers and probe, was dispensed into all of the microchambers through the oil layer. After preparing this setup, the chip was placed onto a conventional thermal cycling system to achieve PCR reaction. Thermal cycling was initiated at 94 °C and held for 10 min, followed by 40 cycles of 94 °C for 10 s and 60 °C for 60 s. After the end of PCR amplification, the amplified DNA was observed using a charge-coupled device (CCD) camera (Hamamatsu Photonics, Japan), which was mounted on a fluorescence microscope (Leica, Heidelberg, Germany). The inner walls of the microchambers were rendered hydrophilic

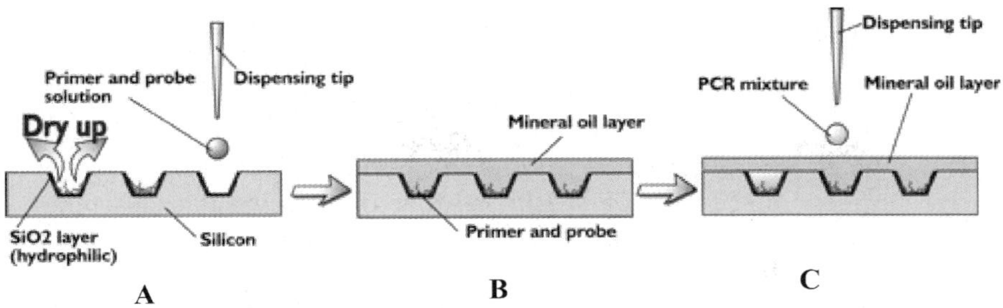

Fig. 2 Illustration of preparation steps for on-chip DNA amplification. **A** Different types of primers were dispensed into the microchamber and then dried. **B** Mineral oil was coated on the chip after the primer solution was dried. **C** PCR mixture (40 nL) with no primers was dispensed into each chamber through the oil layer. The solution sank to the bottom and then spread to the microchamber walls

by the formation of an oxidized layer on their surfaces. After coating the microchamber array with mineral oil, the remaining hydrophobic surface of the microchamber prevented the spread of the aqueous solution to the outside of the microchamber. After dispensing of the aqueous sample solution, it first formed a droplet, which in time was replaced with the oil in the microchamber, and settled inside the microchamber with the convection of the oil, as illustrated in Fig. 2. The thickness of the oil layer had a significant effect on the protection of nanoliter-scale solutions from evaporation. The oil layer was adjusted by controlling the volume of the oil drop. As the thickness of the oil layer increased, the dispensing of the sample solution became more erroneous (data not shown). The optimum thickness of the oil layer for introduction of the sample mixture was chosen to be $\sim$200 μm.

Multiplexed Detection of Different Target DNA on a Single Chip

The target DNA sequence was amplified specifically in a nanoliter-volume microchamber, and the microchambers, in which the fluorescence signal was released, were counted in consequence to TaqMan PCR. The cross-contamination between chambers was tested by using alternate dispensing of wells containing template and those without template. A high concentration of the template DNA was introduced on alternate dispensing into the microchambers. Fluorescence signals were obtained only from the chambers into which template DNA was introduced. No fluorescence was obtained from the remaining chambers into which no template DNA was dispensed. Thus, it was concluded that the selective distribution of the template DNA into the microchambers was achieved in our system.

There were several inhibition factors, such as nonspecific adhesion of biomaterials, variety in the distribution efficiency of sample dispensing, and the errors caused by PCR itself. Surface treatment of the microchamber was also an important factor affecting the efficiency of the PCR reaction. Several kinds of surface treatment methods were reported by Shoffner et al. [35] to be useful for avoiding the adsorption of biomaterials on the silicon surface. Erill et al. [36] reported a systematic analysis of material-related inhibition and adsorption phenomena in glass–silicon PCR chips. Their results suggested that the previously reported inhibition of PCR by silicon-related materials was caused mainly by the adsorption of Taq polymerase at the walls of the chip due to increased surface-to-volume ratios; thus, a straight chemical action of silicon-related materials on the PCR mixture was negligible. In contrast to Taq polymerase, DNA was not found to be adsorbed in significant amounts. The net effect of polymerase adsorption could be prevented by the addition of a titrated amount of a competing protein, BSA, and the ensuing reactions could be kinetically optimized to yield efficient PCR amplifications. In our system, we combined these advantageous points of previous reports. The surface of the microchamber walls was first modified by an oxidized layer [34, 35] and then coated with BSA [36]. Only a very low fluorescence signal could be observed when no BSA coating was employed. Thus, it was found necessary to coat the oxidized walls of the microchambers with BSA, in good agreement with the findings of Erill et al. [36]. To quantify DNA concentration, a certain number of microchambers were used as one region for only one concentration.

Figure 3A shows the fluorescence image of the chip after DNA amplification of three different target DNA sequences from three different DNA templates. Visual comparison of the positive fluorescence intensity signals with the negative ones greatly simplified the procedure to distinguish which chamber contained the target DNA. If the target DNA sequence was present in the dispensed sample, a high fluorescence signal was easily obtained in consequence to TaqMan PCR. Additionally, the background fluorescence intensity of the β-actin PCR system was found to be much lower than that of the other two probes by using both our chip and the SmartCycler real-time PCR system (Fig. 3B). The SmartCycler real-time PCR system results were in good agreement with the results of our chip. The difference in the background fluorescence intensity was caused by the bp distance between the FAM and TAMRA dyes of the TaqMan probes. In the TaqMan probe for the β-actin gene, FAM was only 6 bp away from TAMRA, but FAM and TAMRA were 26 and 31 bp away from each other in the probes for SRY and RhD genes, respectively. Such a short distance of 6 bp between the dyes caused the rapid quenching of the signal, and thus the β-actin system could release much lower fluorescence signals after amplification in comparison with the other systems. Although the PCR systems in this experiment had such different background fluorescence intensities, accurate detection of TaqMan amplification for all systems was achieved by using our chip. Since Rh(−) human female genomic DNA did not contain SRY and RhD genes, almost none of the microchambers showed a fluorescence signal. Both human male and female genomic DNA contained the β-actin gene; thus, a fluorescence signal

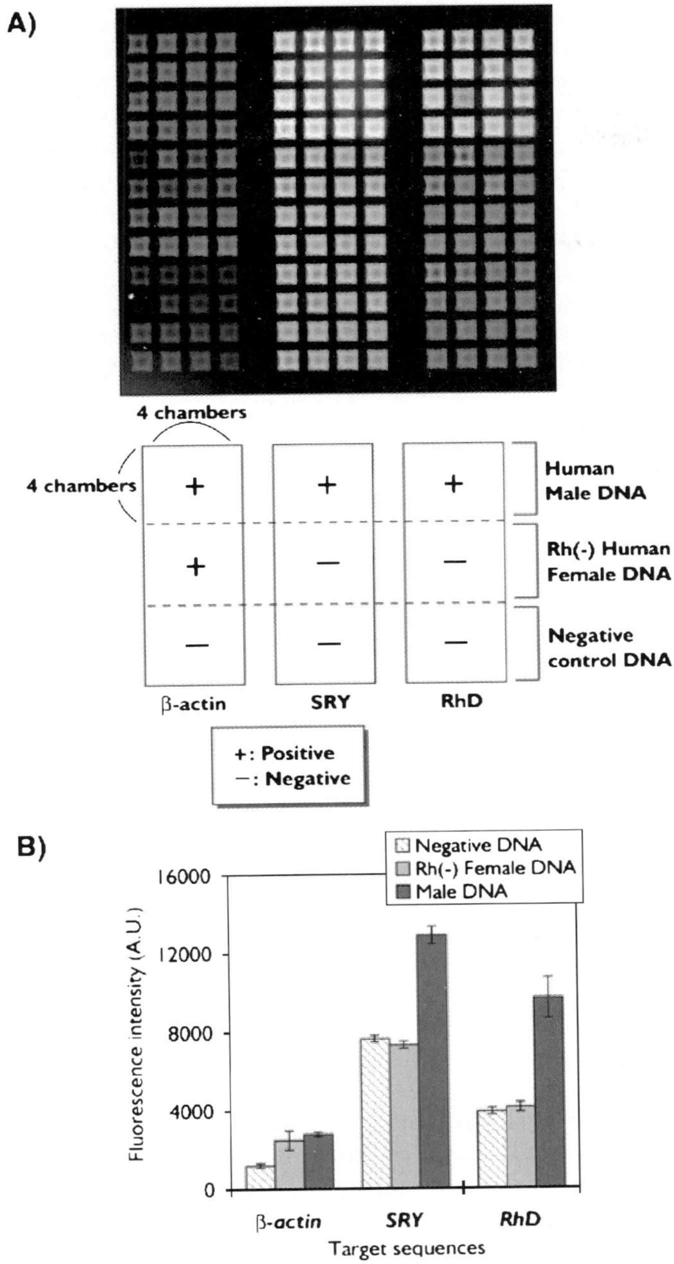

Fig. 3 Multiple PCR analysis of β-actin, SRY, and RhD genes using three different kinds of template DNA. **A** Photograph of fluorescence image of the microchamber array in consequence to TaqMan PCR. **B** Average fluorescence intensity values obtained from 16 microchambers

could be observed successfully in all microchambers of the related area on the chip. The fluorescence intensity of the microchambers was also scanned and evaluated using a DNA microarray scanner and its analysis system. The fluorescence intensity values were obtained from 16 chambers. A remarkable difference between the fluorescence intensities of the positive and negative controls was observed clearly for all three DNA templates. This result indicated that our system was capable of detecting different kinds of target DNA sequences from different DNA sources simultaneously. Since TaqMan PCR required the same thermal cycling protocol for the amplification of many kinds of target sequences, it was found to be the most suitable detection technique for the microchamber array PCR systems in this report. For example, the detection of genetic diseases, such as Down's syndrome [37], 22q11.2 deletion syndrome [38], and β-thalassemia [39], using TaqMan PCR has already been reported. The simultaneous detection of these clinically important diseases can also be performed by using our microchamber array-based PCR chip.

2.1.2
On-Chip Quantification of DNA

The quantification of the initial RhD gene concentration was also performed by using the microchamber array, as shown in Fig. 4A. The amplification of RhD gene was performed by dispensing different concentrations of target DNA into the microchambers. As target DNA was increased from 0 to 12 copies per chamber, the number of the microchambers with positive fluorescence signal also increased. PCR amplification in almost the whole block of the chip was achieved by using eight copies of the target DNA. When 0.4 copies of the target DNA were used, an average of two out of 60 chambers ($n = 3$) showed a signal above the threshold level. The average fluorescence intensity value of 1000 AU was determined as the threshold. The high fluorescence released in these two chambers could also be visually detected. Since 0.4 copies of the target DNA were enough to give a readable signal, this concentration was determined as our limit of detection. The chambers with positive fluorescence signals, which mean successful PCR amplification, showed easily distinguishable fluorescence intensity, as shown in Fig. 4A.

Figure 4B plots the number of chambers with positive fluorescence signal versus input template DNA copy number for the amplification of RhD (B) sequence. As target DNA increased from 0 to 12 copies/chamber, the number of microchambers with a positive fluorescence signal also increased. It was possible to fit the data from 0 to 8 copies/chamber into a straight line with the regression coefficients of 0.9879 for RhD sequence. The system reached a saturation plateau after a DNA concentration of 8 copies/chamber, indicating that a trace amount of target DNA was satisfactory for the detection process. PCR amplification in almost the whole block of the chip was achieved by using eight copies of the target DNA. Such a behavior indi-

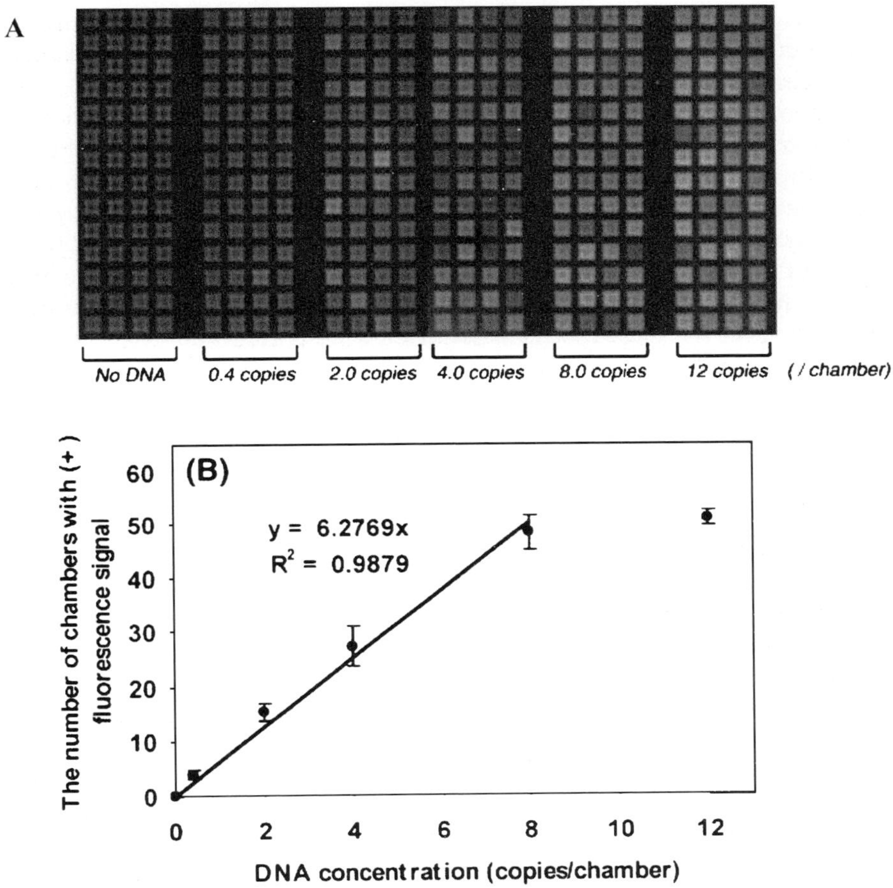

Fig. 4 **A** Photograph of fluorescence image for on-chip quantitative analysis of RhD gene. **B** Plot of the average number of chambers within a specific fluorescence signal range versus the number of target DNA copies related to the RhD gene after TaqMan PCR

cated the high detection capacity of our system, so that even a trace amount of DNA copies would be satisfactory for a precise quantification. The copy numbers above eight that would be amplified without any problems, however, could not be quantified. Even 0.4 copies of the target DNA was enough to give a readable signal, which was determined as our limit of detection. The average fluorescence intensity value of 1000 AU was determined as the threshold. When 0.4 copies were dispensed, only two microchambers reached within the 1000–1200 AU level. One would have expected to get four out of ten microchambers to be positive and six to be negative if we had an average of 0.4 copies/microchamber, and every copy was intact. The polymerase concentration, annealing temperature, $MgCl_2$ concentration, and the specific primers and TaqMan probes should have been kept under the optimum conditions in 40-nL volumes in the microchamber, which was a tedious task. The instrumental limitations also added to the inefficiency of PCR, when such a small volume containing such a trace amount of analyte was dispensed on a microchamber array. Thus, the combined negative effects of biochemical (resulting from the TaqMan PCR itself) and instrumental limitations caused the appearance of only two microchambers out of 60 with positive signals after TaqMan PCR on our system. Our method requires only counting the microchambers that show a positive fluorescence signal, in consequence to PCR amplification. No special equipment for the detection of real-time fluorescence intensity is required for determination of DNA copy numbers. A simple fluorescence microscope only, or a transilluminator used for gel electrophoresis, can be employed for accurate observation of the fluorescence released microchambers with positive signals.

The microchamber array chip presented in this report could be used to amplify multiple DNA targets in combination with a nanoliter dispenser. Theoretically, the chip could be used to amplify and detect ~1200 target DNA simultaneously. The size and the total number of the microchambers are determined by the dispensing system. The minimum solution volume that could be dispensed with reliability was optimized as 40 nL for our experiments. If the dispensing instrument could be improved to provide the dispensing of a lesser volume of solution, the size of microchamber would become smaller and the chip would become more integrated with a higher number of microchambers. Such further integration of our microarray PCR chips with a miniaturized thermal cycler unit is in progress in our laboratory. The microarray PCR chip reported here has a significant potential to be implemented for a wide range of applications. Overall, this system is a promising candidate for mass microfabrication due to its low-cost and high-throughput detection ability.

2.2 Quantitative Continuous-Flow PCR in Microfluidic Chip Systems for Genetically Modified Foods Detection

In this part on DNA sensor systems, we deal with the development of microfluidic chip systems for continuous-flow PCR and their application to detect genetically modified (GM) foods [204]. The development of a microfluidic device that would be suitable for high-temperature-based reactions becomes an important contribution toward the integration of micro total analysis systems (μTAS). These integrate several microdevices to perform diverse functions such as sample preparation [40, 41], mixing [42, 43], and detection [44, 45], and offer many advantages in the field of chemical and biochemical analysis studies [46]. These functions are controlled by a microvalve [47, 48], micropump [49, 50], or electrowetting switch [51], etc., which drive the liquid flow in the microchannels. These components provide the basis for the preparation of nano- or picoliter volumes of samples in the microchannels, which allow rapid detection at low experimental cost with less reagent consumption, less experimental waste generation, and reduction in the contamination risk together with less manual effort. Many kinds of reactions are being used in different chemical- or biochemical-mediated reactions, and some reactions require high temperatures, e.g., PCR. This reaction not only amplifies the specific gene sequence, but also quantifies the amplified DNA using an intercalator or a fluorescent probe, such as SYBR Green I, TaqMan probe, etc. [52]. Quantitative PCR is essential in several fundamental research fields, such as genomics and cell analysis. Furthermore, the technology is widely applied to analytical methods for gene diagnosis, and discrimination of biological identities, such as GM organisms. Improvement, however, is needed, since the current methods require time-consuming procedures and are expensive. Therefore, the development of a microfluidic device for high-temperature-based reactions would provide an important contribution in the integration of μ-TAS.

Kopp et al. [53] first demonstrated a continuous-flow PCR using a glass-based microfluidic device in 1998, and DNA amplification of 176 bp was performed in 1.5–18.7 min (20 cycles). A continuous-flow reverse transcription PCR that was integrated with a microfluidic device for reverse transcription was also reported [54]. Additionally, the same group performed the integration of a continuous-flow RT-PCR device with a laser-induced fluorescence (LIF) detection system, which allowed the detection of the amplified products without gel electrophoresis [55]. Although continuous-flow PCR has several advantages, such as a compact size, rapid DNA amplification, and easier integration with other microfluidic devices than the other PCR-based microdevices, for example microchamber-based PCR systems [28, 32], there are many obstacles that need to be overcome for their application in high-temperature-based reactions. One of the important problems lies in the thermal interference between the different temperature heaters, which are placed under the microfluidic device. Zhang et al. [56] reported temperature analysis of glass–glass-bonded and silicon–glass-bonded devices using finite element analysis (FEA). The temperature distribution in the microchannels was analyzed by introducing a temperature-sensitive dye, and the fluorescence spectrum was measured [57]. Another drawback in continuous-flow PCR is the generation of air bubbles in the microchannels, which decreases the solubility of gas in the liquid solution while increasing the temperature in the microchannels. This particularly occurs as a problem when the solutions flow to the zone maintained at high temperature: the air bubbles form just in front of the assay liquid, which makes the initial start-up flow rather unstable. The stabilization of the initial start-up flow is an important issue for microfluidic devices. To the best of our knowledge, previous reports did not discuss this aspect in detail. Therefore, we report the fabrication of a polydimethylsiloxane (PDMS)-based device and solve the problem of the initial start-up by introducing a highly viscous liquid with a high boiling point (fluorinated oil), just before the introduction of the sample solution, which in turn helped to increase the pressure of the sample solution in the microchannels. Additionally, the technique was adapted for continuous-flow PCR and applied to a practical case using a reference molecule for the quantification of GM maize [58], as a model system of quantitative PCR analysis on our chip.

2.2.1 Microfluidic Device Fabrication for Continuous-Flow PCR

Microfluidic devices with a microchannel geometry of 50×50 μm were made from PDMS (Dow Corning, Midland, Michigan) by standard soft-lithography techniques [59]. Access holes of 500-μm diameter were made for the inlet and the outlet to inject and dispose of the liquid solutions. A fluorinated ethylene propylene (FEP) tube (0.15 ± 0.05 mm i.d.) (BAS Inc., Tokyo, Japan) was inserted into these holes and sealed with small amounts of PDMS to prevent liquid leaks. The PDMS device was hermetically bonded to the glass substrate, which was coated with a thin film of PDMS using a spin-coater in connection with reactive ion etching.

2.2.2 Experimental Setup and Stabilization of the Initial Start-up Flow

The experimental setup consisted of a microsyringe pump (KD Scientific Inc., Holliston, MA, USA) and a 1-mL-volume syringe (Terumo Co., Tokyo, Japan). The heaters were composed of aluminum blocks at $10 \times 10 \times 70$ mm, which were inserted with a cartridge heater and temperature sensor (Kyushu-Nissho Co., Fukuoka, Japan). The temperature of the blocks was controlled within ± 1 °C by a thermocouple of type $\Phi 1.6$ K and a temperature controller (Kyushu-Nissho Co, Fukuoka, Japan). A fluorinated oil (Y04, Universal Co., Tokyo, Japan) with a viscosity of 6 mPa at 95 °C, measured by using a viscometer (model: VM-10A-L, CBC Materials Co. Ltd., Japan), was used in this experiment. The device was placed on a cartridge heater and the oil was introduced through the inlet by using a syringe pump. After introducing a moderate amount of the fluorinated oil, the flow was stopped and exchanged for a syringe with 20 μM Neutral Red (Wako, Osaka, Japan), which was dissolved in MilliQ to visualize the flow by using a digital microscope (VH-Z75, Keyence Co., Osaka, Japan). The generation of air bubbles before the introduction of the sample solutions under high-temperature conditions, which we call the initial start-up in this study, makes the flow rather unstable. To overcome this problem, moderate amounts of the fluorinated oil were added to the microchannels before the sample solution in order to increase the internal pressure of the latter solution (data not shown). The method suggested in this study would provide more flexibility and practical functionality for high-temperature reactions.

2.2.3
Quantitative Continuous-Flow PCR

The DNA plasmids pMu15 with 60, 800, and 10 000 copies/μL were purchased from Nippon Gene Co., Ltd. (Toyama, Japan) and used as the template DNA for all PCR analysis. Primer pair M810 2 (M810 2–5′ and M810 2–3′ with Mon-Taq) was used for the quantitation experiments using continuous-flow PCR [60]. The primers and TaqMan probes labeled with Cy5 and BHQ-2 at the 5′ and 3′ ends, respectively, were synthesized by Sigma Co., Ltd (Tokyo, Japan). The PCR mixture for continuous-flow PCR consisted of 0.04 μL AccuPrime Taq polymerase in 1 μL 10× AccuPrime PCR Buffer I (Invitrogen, Carlsbad, CA, USA), 0.2 μg/μL BSA to prevent the adsorption of PCR reagents, 0.5 μM each of the forward and reverse primers, 0.2 μM TaqMan probe, and template DNA. The microchannel geometry was designed to allow the PCR solutions to pass through two different temperature zones alternately for 50 cycles. First, the fluorinated oil was injected into the microchannels, and then the PCR assay mixture was introduced at a flow rate of 0.5 μL/min. Thermal cycle conditions consisted of denaturation at 95 °C for 20 s, and annealing and extension were set at 59 °C for 20 s. The PCR products were collected in a vial and further analyzed by gel electrophoresis (3%) with ethidium bromide staining.

2.2.4
Quantitative DNA Detection On-Device

First, a microfluidic device was fabricated for continuous-flow PCR, in which the structure of the microchannels was designed so that the PCR solution could flow alternately on different temperature zones, and such that it would allow one to perform PCR with 50 cycles (Fig. 5a). Additionally, the pressurizing channel that was laid on the unheated area was integrated into the outlet to prevent the generation of air bubbles at the outlet zone due to the low pressure. This concept was derived from understanding the pressure distribution in the case of laminar flow, for example, in the microchannels, as shown in Eq. 1.

$$-\Delta P = -(p_1 - p_2) \propto L \tag{1}$$

where p_1 and p_2 indicate the pressure in the upstream and downstream and L indicates the distance between the upstream and downstream. Air bubble generation in the steady state while the solution passes from the inlet to the outlet is another crucial drawback of working under high-temperature conditions. The solutions flowed to the microchannels in the following order: the fluorinated oil, 1 μL, followed by the PCR solution that contained 800 copies/μL of plasmid DNA at a flow rate of 0.5 μL/min. The plasmid used in this experiment was a standard molecule for which the quantification performance had already been validated [58]. The solutions were exhausted through the

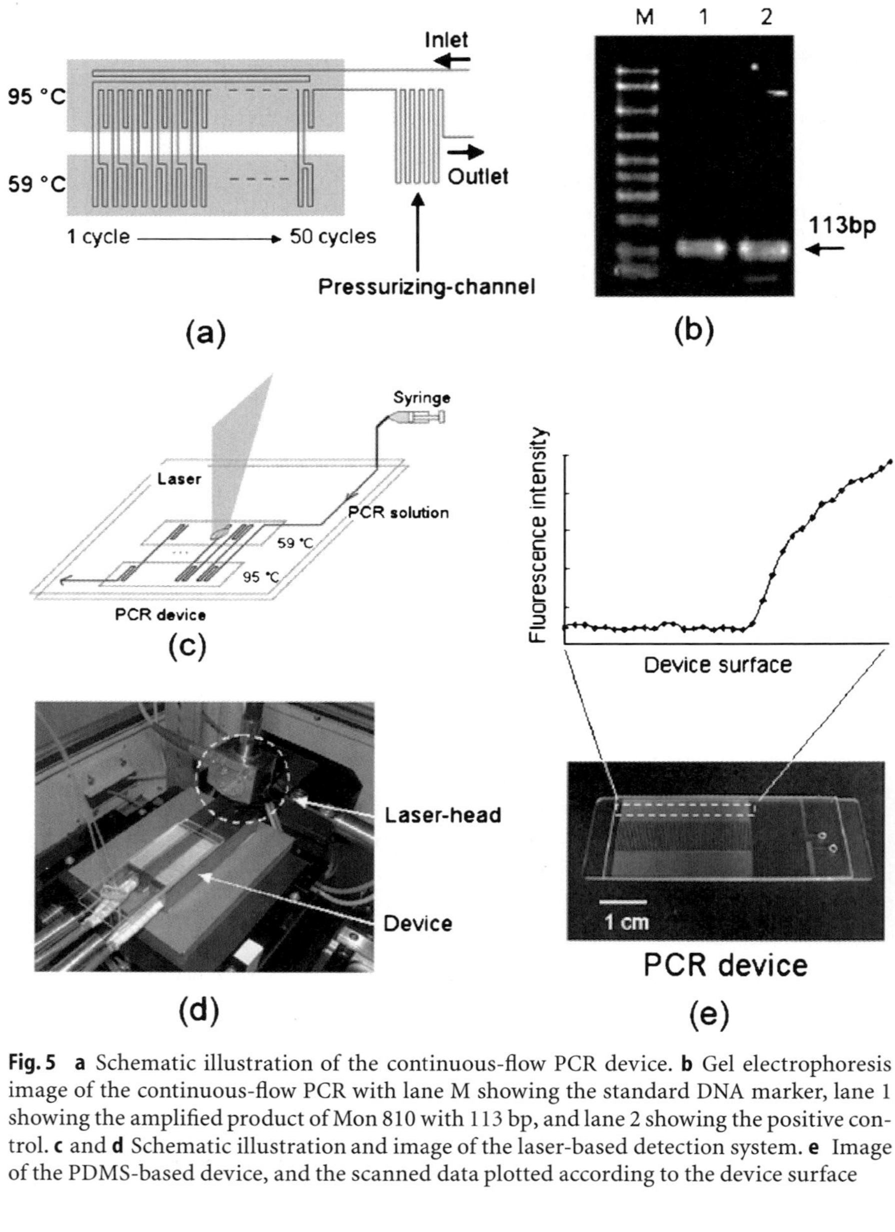

Fig. 5 **a** Schematic illustration of the continuous-flow PCR device. **b** Gel electrophoresis image of the continuous-flow PCR with lane M showing the standard DNA marker, lane 1 showing the amplified product of Mon 810 with 113 bp, and lane 2 showing the positive control. **c** and **d** Schematic illustration and image of the laser-based detection system. **e** Image of the PDMS-based device, and the scanned data plotted according to the device surface

outlet after about 40 min and collected in a vial. DNA amplification was then detected by running the samples on a gel electrophoresis system. The results shown in Fig. 5b demonstrated that specific DNA amplification was achieved. Additionally, DNA amplification was detected on the device by a laser-based detection system developed by the Industrial Research Institute of Ishikawa

(IRII), Ishikawa, Japan. The laser head, which could expose the excitation wavelength of Cy5 at 640 nm and focus the fluorescence signals from the target molecules by a confocal system, was applied at a speed of 4 mm/s with the simultaneous detection of the average fluorescence signals during intervals of once each millimeter. A schematic illustration and an image of the experimental setup are shown in Fig. 5c and d, respectively. The laser head scanned the surface of the device within the specified area as indicated in Fig. 5e. Quantitative continuous-flow PCR was performed using a wide range of DNA template

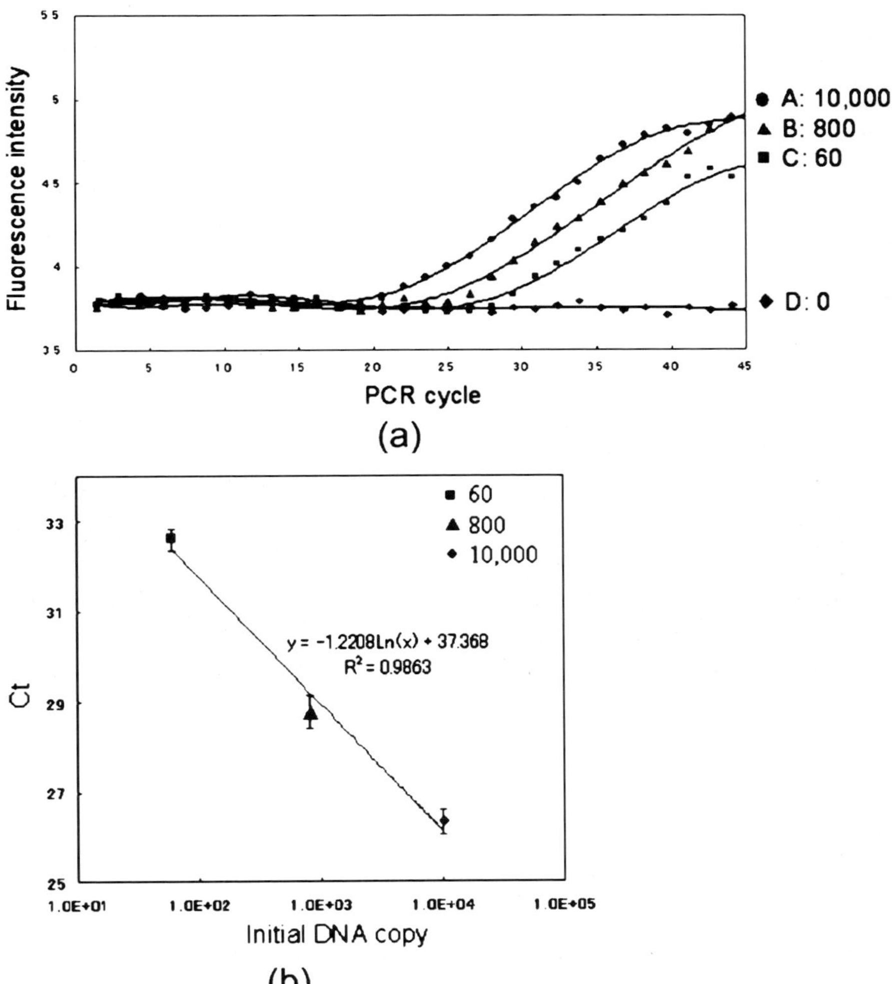

Fig. 6 **a** Plot showing the dependence of the fluorescence intensity on the initial DNA copies: *A* 10 000, *B* 800, *C* 6 copies/μL, *D* no DNA. **b** Relationship between the threshold cycle (Ct) and the initial DNA copy

concentrations, such as 60, 800, and 10 000 copies/μL. In this case, the solutions flowed to the microchannels in the following order: the fluorinated oil, 1 μL, followed by PCR solution at a flow rate of 0.5 μL/min, while the heaters were turned on. The increase in the fluorescence signal was delayed with decreasing DNA concentration, as we also observed in a typical quantitative PCR (Fig. 6a). Moreover, the relationship between threshold cycle (Ct) and the initial DNA concentration showed an almost linear dependence (Fig. 6b). These results showed that it would be possible to quantify the initial copies of the template DNA on our microfluidic device.

We have therefore performed a continuous-flow PCR which applied the novel liquid flow method for maintaining the stability of the initial start-up. Additionally, quantitative continuous-flow PCR was performed in connection with a laser detection system. Although Obeid et al. [55] reported that continuous-flow PCR was detected by LIF detection, their system could not allow quantitative detection. Quantitative DNA analysis is important for environmental inspection and food evaluation, etc., as well as biological studies; thus, our PCR system has promising potential for applications in a wide range of research.

2.3 Label-Free Optical Biosensor Based on Localized Surface Plasmon Resonance (LSPR) and Interferometry for Biomolecular Interactions

In life sciences, there is a continuously growing interest to find new methods and devices that would provide easy, highly reproducible, and sensitive sensing assays for biomolecular reactions. Surface plasmon resonance (SPR) has been the leading method in the label-free format. SPR relies on the changes in the refractive index (RI) induced by the biomolecular recognition events confined to the interface. The contributions to SPR have caused significant advances in label-free biosensor technology [61], and the portable SPR devices have become attractive for on-field applications [62–64]. Imaging SPR has allowed the implementation of an array format and significantly improved the utility of the SPR technique [65–67]; however, its instrumental setup is complicated for the development of hand-held devices. In efforts to overcome the problems associated with the conventional SPR-based systems, optical interferometric transducers provided a strong alternative [68–71]. A sensor based on inexpensive, optically flat, thin films of porous silicon was described by Sailor and coworkers [72, 73]. By illuminating the sensor from the top, the authors found an easy solution to an important drawback of SPR, the limited penetration depth, [74–76] and detected the biomolecules on the sensor surface. Recently, Pan and Rothberg [77] reported the optical detection of biomolecular reactions on nanoporous aluminum oxide templates.

The research toward miniaturization has led us to the unique optical characteristics of noble metals at nanoscale sizes, such as Au [78] and Ag [79] nanoparticles. The intense colors exhibited by colloidal solutions of noble metal nanoparticles are due to the colloidal SPR phenomenon. The properties of colloidal SPR are strongly dependent on the size, shape, and local environment of the nanoparticles, as described by Mie theory [80]. According to Mie theory, there is a restriction on the movement of electrons through the internal metal framework when the size of the metal particle is scaled down to the nanometer level (<100 nm) [81, 82]. The collective charge density oscillations of nanoparticles are defined as localized surface plasmon resonance (LSPR) [83, 84]. LSPR absorption bands are characteristic of the type of the nanomaterial, the diameter of the nanoparticles, and their distribution [85, 86]. LSPR can detect an immediate change in the interfacial RI of the surrounding medium [84–86], which is greatly affected by the attachment of biomolecules at the colloid/solution interface [87–89]. LSPR phenomena have been previously utilized to monitor biomolecular interactions by our group [90–92] and others [87, 89, 93].

In this work, the combination of LSPR with interferometry using a porous anodic alumina (PAA) layer chip for the detection of DNA hybridization is demonstrated [205]. Au-capped oxide nanostructures were prepared on the chip surface by gold deposition onto the chip to form a "caplike" layer on top of the oxide nanostructure in an orderly fashion. One of the major drawbacks of conventional Au nanoparticle-based LSPR chips is the complicated chemistry required to form a self-assembled monolayer (SAM) of Au nanoparticles. During the mass fabrication of numerous chips, the slightest defects in the uniformity of SAMs cause significant problems in the reproducibility and reliability of the results. When the Au-capped oxide nanostructures are illuminated with visible light from the top, the electromagnetic energy emitted from the nanostructures overlaps to form a linearly expanding region. Thus, a wide electromagnetic region provided high sensing capabilities for biomolecular binding events. It is well defined that the shift in the position of λ_{max} is small in LSPR observations [83–89], whereas there is a significant shift in the responses obtained from interferometric experiments [68–71, 90]. The combination of LSPR with interferometry on a PAA layer chip enables two important aspects of optical sensing systems: the shift in λ_{max} and the increment in the relative reflected intensity (RRI) in a new and highly sensitive format. The excitation of the optical characteristics and the detection were performed using only one optical fiber, which made our device user-friendly and suitable for the development of hand-held diagnostic devices.

2.3.1
Fabrication of Porous Anodic Alumina Layer Chip

For the preparation of the PAA layer on the aluminum (Al) substrate, we in our laboratory at the Japan Advanced Institute of Science and Technology developed a two-step anodizing method [94]. Al sheets (Aldrich 99.999%, 25 × 25 × 0.5 mm) were degreased in acetone and then annealed at 480 °C for 40 min to remove mechanical stress and recrystallize the metal. Subsequently, to smooth the surface morphology, the Al sheets were mechanically polished with 6-, 3-, 1-, and 0.25-μm diamond suspensions. After rinsing in acetone, ethanol, and distilled water, the Al sheets were electrochemically polished in $H_3PO_4 + H_2SO_4 + H_2O$ (8.5:1:0.5) + CrO_3 (35 g L^{-1}) for 10 min at 70 °C. The first anodizing step was conducted under the constant voltage of 40 V in a 0.3 M aqueous oxalic acid solution for 10–100 min (thickness of PAA layer: 0.5–5 μm). The anode temperature was kept constant at 10 °C, and the electrolyte was vigorously stirred during the process in order to maintain the temperature and the electrolyte concentration. The details of the mechanism of the ordered formation of holes in the textured Al are not clear at present; however, it is thought that each convex can induce the independent formation of a hole due to its geometrical effect. After the first anodizing step, the generated PAA layer was removed by immersing the sample in a solution containing a mixture of phosphoric (1.8%, w/v) and chromic acids (2%, w/v). After the removal of anodic Al, a textured pattern of concaves was obtained on the surface of the Al sheets. Then, the second anodizing step was carried out under the same conditions as described in the first anodizing step for 10–100 min. PAA layers with various thicknesses were obtained by changing the time for the anodizing steps. Finally, Cr and Au layers were deposited onto the PAA layer using a thermal evaporator (SVC-700TM/700-2, Sanyu Electron Co., Japan) to enhance the LSPR optical properties. The growth rate and pressure were monitored using a quartz crystal microbalance (QCM) (Model TM-200R, Maxtek Inc., Japan), and manually adjusted to 1.0 Å s^{-1} and 8×10^{-6} Torr, respectively. The PAA layer chip was plated with a 50-Å Cr layer and a 150-Å Au layer.

2.3.2 Optical System

The instruments that were used for evaluation of the optical properties of the PAA layer chip were based on an LSPR spectroscopy microscopy system (Ocean Optics Inc., USA). The optical system was equipped with a tungsten halogen light source (LS-1, wavelength range 360–2000 nm), spectrophotometer (USB2000 UV–visible, wavelength range 250–1100 nm), and an optical fiber probe bundle (R-400-7 UV–visible, fiber core diameter 200 μm, wavelength range 250–850 nm). The experimental setup for the optical properties evaluation is shown in Fig. 7. White light emerging from the optical fiber bundle was incident onto the PAA layer chip from the vertical direction. The reflected light was coupled into the detection fiber probe in the same bundle and analyzed using the UV–Vis spectrophotometer. All absorbance spectra were taken from 400 to 850 nm at room temperature.

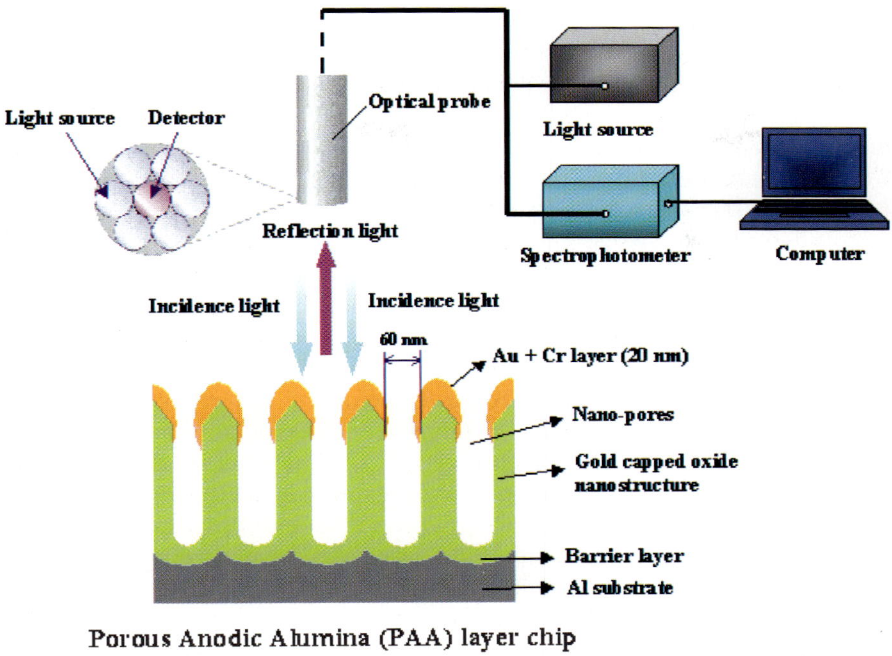

Fig. 7 Experimental setup and construction of LSPR and interferometry-based label-free optical biosensor with PAA layer chip

2.3.3 Label-Free Detection of DNA Hybridization

The synthetic oligonucleotides (23-mer) for the detection of apoE polymorphisms [95] were purchased from Fasmac Co. (Japan), and had the following sequences:

Probe DNA: 5′-thiol-CAG GCG GCC GCA CAC GTC CTC CA-3′ has a sequence that is complementary to a fragment of allele ε3 and surrounds codon 112.

Target DNA: 5′-TGG AGG ACG TG**T** GCG GCC GCC TG-3′ has a sequence that is complementary to a fragment of allele ε3 and surrounds codon 112.

Single-base mismatch target DNA: 5′-TGG AGG ACG TG**C** GCG GCC GCC TG-3′ has a sequence that is the same as a fragment of allele ε4 and surrounds codon 112. The single-base mismatch position is indicated with a boldface letter. Oligonucleotide primers were designed as described in Hixon and Vernier: 60 primer 1,5′-ACA GAA TTC GCC CCG GCC TGGTAC AC-3′, and primer 2,5′-TAA GCT TGG CAC GGC TGT CCAAGG A-3′.

An aliquot (20 μL) of 10 μM thiolated probe DNA solution containing 100 μM 6-mercaptohexanol (Dojindo Laboratories, Japan) was introduced to the Au-deposited PAA layer chip surface and incubated for 1 h. After the immobilization of probe DNA, the PAA layer chip surface was rinsed thoroughly with 20 mM phosphate-buffered saline (PBS, pH 7.4) and dried at room temperature. After the probe immobilization procedure, a desired concentration of target DNA solution in PBS (20 μL) was introduced to the probe immobilized PAA layer chip surface, and the hybridization reaction was allowed while incubating for 1 h at room temperature. Single-base mismatch DNA in PBS (20 μL) was also exposed to the probe DNA on the PAA layer chip surface as described above. After a stringent washing of the surface, the changes in the absorption spectrum caused by the hybridization reaction were monitored. PCR amplification of DNA that was extracted from peripheral blood of consenting adults of our lab was performed.

Figure 8A and 8B show the cross-sectional AFM analysis, corresponding to the diagonal lines, of a PAA layer chip that was established by the two-step anodizing process. The diameters of the nanopore and the interpore were ∼60 (1.2×10^{10} pores/cm^2) and ∼120 nm, respectively, before Au deposition on the PAA layer chip. The dimension of the nanopore became ∼47 nm after the deposition of the Au layer. The overlaid images showing the effect of Au deposition on the PAA layer are shown in Fig. 8C. An Au-capped oxide nanostructure ∼45 nm in height was established on all sides of the nanopore uniformly. The line profile was formed by a set of cones instead of the expected cylinder-shaped internal pore. This deviation in the expected shape resulted from the inability of the AFM probe tip to enter in-

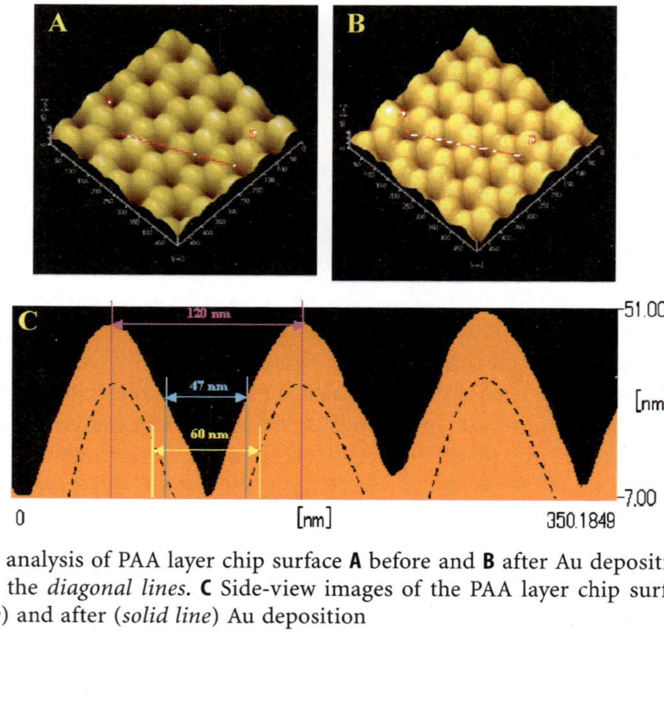

Fig. 8 AFM analysis of PAA layer chip surface **A** before and **B** after Au deposition corresponding to the *diagonal lines*. **C** Side-view images of the PAA layer chip surface before (*dotted line*) and after (*solid line*) Au deposition

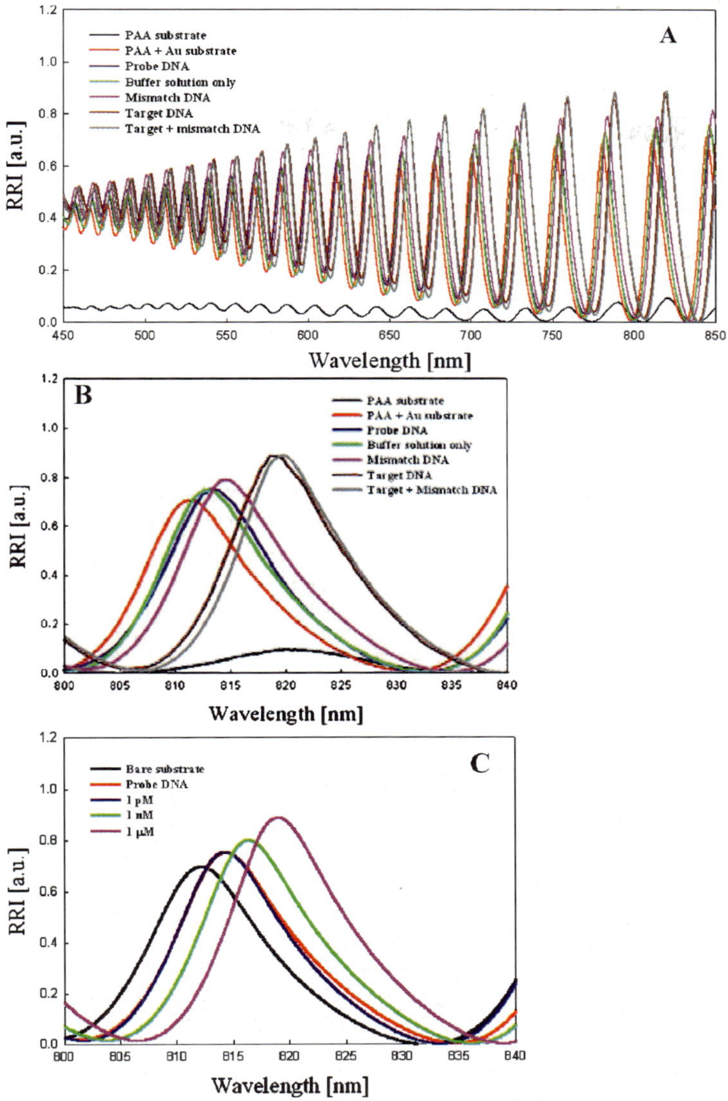

Fig. 9 The detection of hybridization between DNA sequences related to apoE gene polymorphisms using an Au-deposited PAA layer chip **A** in wide range and **B** from 800 to 840 nm, by monitoring the optical responses of the bare PAA layer chip (*black line*), Au-deposited PAA layer chip (*red line*), after 10 μM probe DNA immobilization on Au-deposited PAA layer chip (*blue line*), after incubation with the buffer solution only (*green line*), after hybridization with 10 μM single-base mismatch DNA (*magenta line*), after hybridization with 10 μM target DNA (*brown line*), and after hybridization with a mixture of target with single-base mismatch DNA in 1 : 1 ratio (*gray line*). **C** Optical characteristics obtained after hybridization of 10 μM probe DNA with serial dilutions of target DNA from 800 to 840 nm with the calibration plots for the dependence of the RRI

side the nanopores due to its conical geometry with a relatively low aspect ratio. Following Fig. 8C, the limited penetration of the tips into the nanopores was evident when the expected depth of the nanopores was several micrometers. However, the fabrication of the PAA layer chips resulted in similarly prepared nanopores with exactly the same dimensions and optical characteristics. Binding of probe DNA immobilized on the PAA layer substrate to its corresponding target DNA strand resulted in a change in the RI of the layer medium, and was detected as a corresponding shift and an increase in the RRI. Figure 9A shows the superimposed absorbance spectrum profiles obtained from PAA layer chips, and Fig. 9B shows the spectrum from 800 to 840 nm. When the bare PAA layer was excited, we could observe only the interferometric pattern (black line). After Au deposition, the absorbance intensity increased significantly as a result of the LSPR characteristics (red line). When the probe DNA was immobilized on an Au-deposited PAA layer surface, we could observe both the red shift and an increase in the absorbance intensity (green line). After a stringent washing of the surface, there was no significant change in the probe DNA responses. Under similar conditions, but in the presence of 10 μM single-base mismatch DNA sequences, a slight red shift in wavelength of the interference fringe and LSPR patterns were detected with minor amplitude fluctuations (blue line). However, the hybridization reaction between 10 μM target and probe DNA caused a significant red shift in wavelength and an enhancement in the RRI (magenta line). In the presence of target DNA with concentrations ranging from 1 pM to 1 μM, pronounced wavelength shifts and RRI occurred (Fig. 9C).

The large dynamic range and high sensitivity of the PAA layer chip with respect to analyte concentration enables the quantification of biomolecules in small volumes in a rapid and simple format. Our biosensor system represents a unique approach to performing interferometry and LSPR that utilizes a very simple and cost-effective optical setup with disposable chips. The PAA layer chip appears to be scalable down to a single nanopore without loss of sensitivity, and is amenable to integration with microfluidics for high-throughput determinations.

3 Protein Sensor Systems

3.1 On-Chip Cell-Free Protein Synthesis Using a Picoliter Chamber Array

The progress in analyzing the human genome has shifted the focus of research from genes to proteins [96–100]. Although the number of human genes is reported to be 28 000–38 000 [101], the functions of most of them remain unknown. A rapid and easy method for synthesizing gene products has yet to be developed [206]. Thus, in this section we report an in vitro protein synthesis system that was designed and constructed on the microarray to make a protein library chip. The chip has proteins arranged in an array, and can detect target molecules. Gene cloning and expression is widely used in the preparation of proteins. However, some kinds of proteins often cannot be expressed well in host cells. Our cell-free protein synthesis system could be suitable for expressing such proteins. This protein synthesis system has other advantages as well, such as labeling proteins with isotopes for detection by NMR spectroscopy [102], easy purification of the synthesized protein, and short protein synthesis time.

A highly integrated protein chip is a powerful tool for accelerating post-genomic research. Our aim is to develop protein chips directly from a DNA library using the in vitro protein synthesis system. Recently, a cell-free protein synthesis system from *E. coli*, rabbit reticulocytes, and wheat germ was commercialized [103]. In this research, a rapid translation system from *E. coli* was used for protein expression. Previously, we reported the development of large-scale integrated picoliter microchamber arrays for PCR [28], the introduction of a novel nanoliter dispensing system suitable for DNA amplification on a microchamber array chip [30], the development of a simultaneous multianalyte immunoassay method for detecting human immunoglobulins based on a protein chip and imaging detection [104], and the development of a new approach for manufacturing encoded microstructures used as versatile building blocks for miniaturized multiplex bioassays [105]. Others have reported the construction of protein chips [106, 107] or cell-free protein synthesis in small chambers [108]. Kukar et al. detected eight samples simultaneously on one chip [106], and Kojima et al. [107] constructed an electrochemical immunochip including an assembly of 36 electrodes. There have been reports of high-throughput screening of a mutated anti-human serum albumin single-chain antibody (anti-HSA-scFv) using an in vitro protein synthesis system [109, 110]. In these reports, two amino acids were mutated randomly, and over 600 mutations were screened on 96-well plates. Our newly developed chip could also be a powerful tool in similar applications.

High-throughput screening is required for the rapid elucidation of protein functions. Microscale reactions have the advantages of short reaction time and the use of small amounts of samples and reagents. Especially in a high-throughput screening system, numerous samples must be analyzed simultaneously and, if possible, economically. Thus, we made a highly integrated protein microchamber array chip by using microfabrication techniques and PDMS. PDMS micromolding techniques have been used to fabricate microfluidic systems [111, 112]. Unlike traditional microfabrication materials, such as silicon and glass, PDMS can be bonded and manufactured easily and efficiently [113]. In addition, PDMS has some properties that are advantageous for biochemical applications, such as high transparency in the 230–700 nm wavelength range and high permeability to gases.

3.1.1 Cell-Free Protein Synthesis Chip Fabrication

The photolithography process [113] has been used to fabricate thin microarray sheets using PDMS. A master pattern was formed on a silicon wafer using SU-8 photoresist. The PDMS prepolymer (Sylgard-184: Dow Corning, USA) mixture was poured onto the master and covered with a transparency film (overhead projector sheet). A multilayer stack of aluminum plates, the master pattern, PDMS, a transparency film, a glass wafer, and rubber sheets were clamped tightly and the PDMS prepolymer was baked and cross-linked at 80 °C for 2 h. The resulting thin PDMS sheet that has over 200 000 microchambers (microholes) was put on a slide glass in acetone and treated with oxygen plasma to bind the sheet to the glass. A reactive ion etching (RIE) system was used for the oxygen plasma treatment. The PDMS sheet has microholes and a hydrophobic surface. Thus, only the bottoms of the microchambers were hydrophilic and the solution easily remained in the chambers. Three different types of chips were designed and fabricated (Fig. 10).

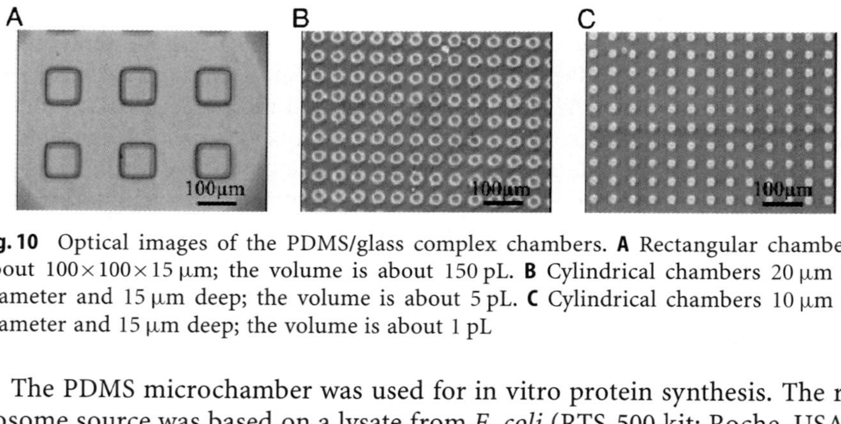

Fig. 10 Optical images of the PDMS/glass complex chambers. **A** Rectangular chambers about 100×100×15 μm; the volume is about 150 pL. **B** Cylindrical chambers 20 μm in diameter and 15 μm deep; the volume is about 5 pL. **C** Cylindrical chambers 10 μm in diameter and 15 μm deep; the volume is about 1 pL

The PDMS microchamber was used for in vitro protein synthesis. The ribosome source was based on a lysate from *E. coli* (RTS-500 kit: Roche, USA). The wild-type GFP gene contained in the kit was used as a reporter gene and expressed on the chips. Cell-free protein synthesis reagents were prepared according to the supplier's directions. The reaction solution was composed of a mixture of 0.25 mL of *E. coli* lysate solution, 0.75 mL of the reconstituted reaction mixture, 50 μL of the enzyme mixture, and the GFP vector at a final concentration of 10 μg/mL. First, we dripped the in vitro protein synthesis solution on the chip and removed the surplus. Next, the microchamber chip was covered with a gap cover glass and sealed to prevent evaporation. There was a 20-μm gap between the chip surface and the cover glass. This gap prevented capillary action among the chambers. The chip was held at 30 °C, and GFP expression was detected by an optical fluorescence microscope with an FITC filter (excitation: 450–490 nm; emission: 515–565 nm).

3.1.2
Cell-Free Protein Synthesis Using a Microchamber Array

RIE treatment (oxygen plasma treatment) was used to bind PDMS to a glass slide. Three different types of PDMS chips were designed and fabricated (Fig. 10). Since the PDMS and glass construction gives the chamber structure a hydrophobic surface and a hydrophilic bottom substrate, an aqueous solution poured onto the chip enters through the holes of the array and remains only in these microchambers. This phenomenon prevents cross-contamination between the microchambers. This chip is also suitable for optical observations because of its transparency over a wide wavelength range. Protein synthesis was carried out on the microchamber chips with the GFP gene used as a reporter gene. The expression of GFP was detected by fluorescence using an optical microscope. Thus, cell-free protein synthesis on the chip resulted (data not shown). The fluorescence intensity was detected within 1 h of incubation, and remained constant. In a batch system, protein synthesis is said to be inhibited by a lack of substrate or accumulated waste within 2 h [114]. In our system, cell-free protein synthesis stopped within 2 h. This result is in agreement with the report of Spirin et al. [114]. However, the formation of the GFP fluorescent group is known to take 1–2 h [114]; thus, it may be considered that protein synthesis stopped before the GFP fluorescence became constant. However, a similar shift in GFP fluorescence was shown in chambers with 10 (Fig. 11) or 20 μm i.d.

The lowest concentration of DNA template necessary for the detection of the GFP signal was determined to be only ten molecules of DNA per chamber. A type of microchamber array chip with cylindrical chambers 10 μm wide and 15 μm deep was used in this experiment. The volume of this chamber is about 1 pL. Thus, the concentration of a solution containing ten molecules of DNA is about 4×10^{-5} mg/mL. This concentration is about 1/100 to 1/500 of the DNA concentration utilized in conventional cell-free protein synthesis protocols. The use of a small-volume chamber increases the possibility of contact between DNA and reagents, making it possible to express a protein using a trace amount of DNA.

The distribution of over 10 000 samples using a DNA spotter would take a very long time; therefore, self-layout of samples containing the DNA library was used in this study. DNA-immobilized beads were used as DNA carriers. The amount of DNA immobilized on one bead was about 200 molecules on Dynabeads M-270 Carboxylic Acid, and 10 000 molecules on Dynabeads M-280 Streptavidin (Dynal, USA). The microchamber array chip with a 10-μm-i.d. chamber was used for bead arrangement. About 60% of the chambers contained one bead (Fig. 11A); however, some chambers had multiple beads because the diameter of a bead, which is 2.8 μm, is much smaller than that of the chamber. The design and fabrication of a new chip with

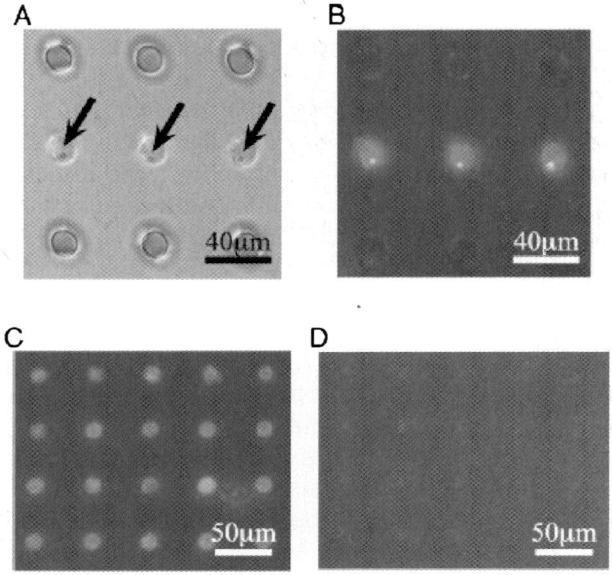

Fig. 11 Optical images of GFPuv expression from DNA-immobilized beads. **A** Optical image under white light; the arrows indicate the presence of beads. **B** Fluorescence image of the chamber; ex.: 400–440 nm, em.: 475 nm. **C** Fluorescence image of a positive control containing cell-free protein synthesis reagents with 10 fg/pL pGGFPH vector. **D** Fluorescence image of a negative control containing only cell-free protein synthesis reagents (no template DNA). The cylindrical chambers shown in the figure were 10 μm in diameter

smaller chambers to allow the entry of only one bead into each chamber are currently under way in our laboratory.

With the aid of lipid, it was possible to disperse the beads into the chambers. Rhodamine-modified lipid was used instead of phosphatidylcholine, and the chip covered with lipid solution was observed both in air and in water. Interestingly, the lipid moved into the chamber when the chip was soaked in water. A chip with beads containing the GFP gene was used for in vitro protein synthesis. As shown in Fig. 11, fluorescence was observed after 1 h of incubation only in chambers that contained DNA beads. As a positive control, pGGFPH vector solution was added to the in vitro protein synthesis reagents at 10 fg/pL; no template DNA was added to the in vitro protein synthesis reagents as a negative control. The results of these control experiments are shown in Fig. 11C and d. A comparison between DNA-immobilized beads and DNA in solution suggests that the amount of protein per DNA molecule in solution is greater than that on DNA-immobilized beads. Nevertheless, easy and fast manipulation of DNA-immobilized beads prompted us to use this method. In the experiments shown in Fig. 11, DNA-immobilized beads with biotin–streptavidin conjugate were used. Similar results were observed when primer-immobilized beads were used. The concentration of GFP solution was about 10 μg/mL, estimated from the intensity of the fluorescence. The amount of GFP protein per chamber was about 10 fg. The GFP solution did not diffuse from the chambers, indicating that the solution in each chamber is physically separated from that in other chambers.

Tabuchi et al. have made a microchamber chip with a chamber volume per chip of 10 μL [108]. In our case, the chamber number is much larger and the chamber volume is much smaller, which is advantageous for high-throughput applications. However, the analysis of over 10^4 chambers takes a long time. Therefore, a scanner-type analyzer for our system is being developed to enable automatic screening in the near future.

A new method for making a highly integrated protein chip from a DNA library using in vitro protein synthesis on a microchamber array is demonstrated. The chambers are of three types based on their volume capacity: 1, 5, and 150 pL, and the total number of chambers per chip is 10 000 (150 pL) and 250 000 (both 1 and 5 pL). The array has a hydrophobic surface of PDMS and a hydrophilic glass bottom. These structural properties provide the advantage of preventing cross-contamination among the chambers. In vitro protein synthesis using these chambers was achieved. The fluorescence of GFP expressed on the microchamber was rapidly detected. GFP expression was also achieved using immobilized DNA molecules on polymer beads, which allows easy handling of the DNA molecules. Brenner et al. [115] described a method for cloning nucleic acid molecules onto the surfaces of 5-μm microbeads rather than in biological hosts. A unique tag sequence was attached to each cDNA molecule, and the tagged library was amplified. A unique tag was also attached to each bead, and the tagged library was conjugated with the tagged beads.

This method allows the immobilization of one kind of DNA on a bead. Because such clones are segregated on microbeads, they can be manipulated simultaneously and then assayed separately. If this method can be applied to the chip described in this report, it will be possible to analyze easily a whole DNA library on a chip in a short time. In the future, this system will be used for the exhaustive expression of proteins included in target cells, the functional analysis of proteins expressed from unknown genes, and the screening of artificially mutated proteins.

In this study, we have reported a highly integrated protein microarray chip using in vitro protein synthesis from DNA-conjugated microbeads. The protein microarray system made it possible to perform high-throughput screening and analysis for multiplexed gene expression on single beads in each picoliter chamber. In the future, this system will be applied to the expression analysis of proteins in multiplexed single cells and the functional analysis of proteins expressed from unknown genes or cells.

3.2
Multiple Label-Free Detection of Antigen–Antibody Interactions Using LSPR-Based Optical Biosensor

The post-genome era of life sciences is rapidly moving beyond functional genomics to proteomics. As an emerging field in life sciences, proteomics is not only based on, but also being developed beyond, genomics [116]. However, a large-scale, rapid, and ultrasensitive assay system is highly desired. The large-scale and partly high-throughput characterization of the human proteome has become possible with the sophisticated biochemical techniques. Although enzyme-linked immunosorbent assay and two-dimensional gel electrophoresis are currently the most widely used bioanalysis tools for monitoring protein–protein interactions, they bring along disadvantages with regard to throughput, reproducibility, and sensitivity [117]. SPR is a strong alternative to the existing technologies for monitoring biomolecular interactions. Unfortunately, conventional SPR reflectometry requires sophisticated optical instrumentation associated with the detection system. This latter limitation is significant, because biochips are urgently in demand for high-throughput and cost-effective monitoring. To overcome these disadvantages of conventional detection systems, biochip assay systems employing micro- and nanoelectromechanical systems (MEMS and NEMS) have been developed [118]. A highly developing trend in the application of MEMS and NEMS technologies in life sciences and biotechnology is directed toward miniaturization and multidetection. These chip technologies have several advantages over the conventional bioanalysis systems. First, the chip-based assays enable rapid analysis of a large number of samples in a single experiment. Second, the amount of material required is significantly small. Reaction volumes are lower than the amount that is typically used in conventional microtiter plates. Third, the signal-to-noise ratio exhibited by micro- or nanofabricated biochips is much better than that observed for conventional microtiter plate assay systems [119, 120]. In addition to the advantages, such as reduced reagent consumption and laboratory space conservation, lab-on-a-chip technology offers new prospects for laboratory innovation and automation. Biochips in proteomics have evolved into powerful tools for quantifying proteins and qualifying their state of activation in complex biological samples. Until now, several kinds of biochips for proteomics using MEMS and NEMS technologies have been developed, such as an array-based chip [32, 33, 104] and a microfluidic biochip [13, 87, 90, 121, 122]. However, these biochips require the labeling of the proteins with different types of reagents, such as fluorescent dyes or enzymes. The labeling procedure with these reagents is a difficult and time-consuming task, and may cause an inhibition of the biofunctions of the native protein. To overcome these disadvantages, we developed a novel LSPR-based nanochip for label-free monitoring of biorecognition events.

LSPR has been the leading method in the label-free format for measuring the biomolecular interactions. LSPR-based monitoring methods can detect an immediate increase in thickness of a biomolecular layer caused by the reaction between the solution component under study and the receptor layer immobilized on the surface [123, 124]. The excitation of LSPR by visible light, at an incident wavelength where resonance would occur, results in the appearance of an intense absorption band. The intensity and position of the SP absorption band are characteristic of the type of the material, the diameter of the nanoparticles, and their distribution. The methods can detect immediate change in the interfacial RI of the surrounding medium [87], which is greatly affected by the attachment of biomolecules at the colloid/solution interface. Biomolecule binding induces a change in the color of the sensor substrate, thus providing an easily detectable optical signal. Mie theory predicts a red shift in the position of the absorbance peak (λ_{max}) and an increase in its intensity. Previously, our group achieved the label-free detection of antigen–antibody reactions [90] and peptide nucleic acid (PNA)–DNA and DNA–DNA hybridization using our LSPR-based nanochip. On the basis of the characteristics of our previous chips, we fabricated a multiarray LSPR-based nanochip for the multiple detection of six different proteins in this research.

3.2.1 Preparation of a Multiarray LSPR-Based Nanochip

An LSPR-based nanochip was constructed with a core–shell structured nanoparticle layer (Fig. 12A). The surface-modified silica nanoparticle was used as the "core", and the "shell" was applied as the top and bottom gold layers that were deposited using thermal deposition. For the preparation of the surface-modified silica nanoparticles, silica nanoparticles (particle diameter: 100 nm) were dried for 24 h at 55 °C and reacted with 1% (v/v) γ-APTES solution in ultrapure water for 24 h at room temperature (RT) by stirring continuously. After the surface modification, the γ-APTES solution was removed in the centrifugal operation for 1 h at 3500 rpm, and the recovered nanoparticles were washed with ultrapure water. Both the washing and centrifugal operations were repeated three times, then surface-modified nanoparticles were dried for 5 min at 120 °C. Silica nanoparticles modified with amino groups were thus obtained. The surface-modified nanoparticles were stored in a desiccator until use, then a colloidal solution of the surface-modified silica nanoparticles was simply prepared by dispersing the desired amount in ultrapure water; a thermal evaporator was used at a base pressure of 8×10^{-6} Torr. The growth rate was monitored using a QCM and manually adjusted to 1.0 Å/s. Gold and chromium with 99.99% purity were obtained from Furuya Metal (Tokyo, Japan). A chromium layer of 5 nm and a bottom gold layer of 40 nm were deposited onto the glass substrates. Additionally, after the

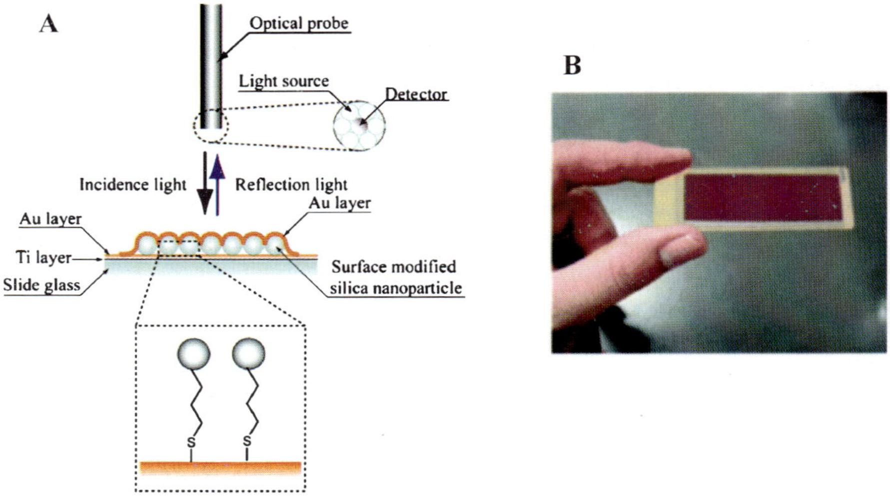

Fig. 12 **A** Construction of the multiarray LSPR-based nanochip. The surface-modified silica nanoparticles were aligned on the gold-deposited glass substrate surface. Subsequently, the gold layer was deposited on the silica nanoparticle layer. **B** Photograph of the multiarray gold-capped nanoparticle layer substrate. The antibodies were immobilized on the multiarray gold-capped nanoparticle layer substrate surface using a nanoliter dispensing system

formation of the silica nanoparticle monolayer, 30 nm of another gold layer was evaporated on the top.

For the fabrication of an LSPR-based nanochip using surface-modified silica nanoparticles, 1 mM of DDA solution was introduced to the gold-deposited glass slide substrate surface, and the SAM was formed in 1 h. SAM functionalization was carried out with 400 mM 1-ethyl-3-(3-dimethylamino-propyl)carbodiimide (EDC) for 1 h. EDC activated the carboxyl groups of the DDA, and therefore the amino groups of silica nanoparticles could form esters with the activated carboxyl groups. The surface-modified silica nanoparticles that were modified with amino groups (1% w/v) by silane coupling reagent in ultrapure water were exposed to the activated SAM-modified gold substrate surface for 1 h. The nanoparticle layer modified substrates were rinsed thoroughly with ultrapure water to remove the excess surface-modified nanoparticles and dried at RT. During the preparation of the nanoparticle layer, a silicon chamber was attached to the gold layer-deposited glass slide surface to control the spread of the sample solutions. Finally, a top gold layer (30 nm) was deposited onto the nanoparticle layer modified substrates using a thermal evaporator. Thus, the LSPR band was introduced in the visible range. A multiarray (20×60 mm) of the core–shell structured monoparticle layer substrate could be obtained (Fig. 12B).

3.2.2
Immobilization of Antibodies onto Multiarray LSPR-Based Nanochip and Label-Free Detection of Proteins Using the LSPR-Based Nanochip

Antibody immobilization onto the multiarray LSPR-based nanochip surface was carried out in a similar fashion, with the formation of a nanoparticle monolayer (Fig. 13A). DDA at 1 mM was introduced to the LSPR-based biochip surface, and a SAM was formed in 1 h. SAM functionalization was carried out with 400 mM EDC for 1 h, and then 100 mM NHS solution was added to the

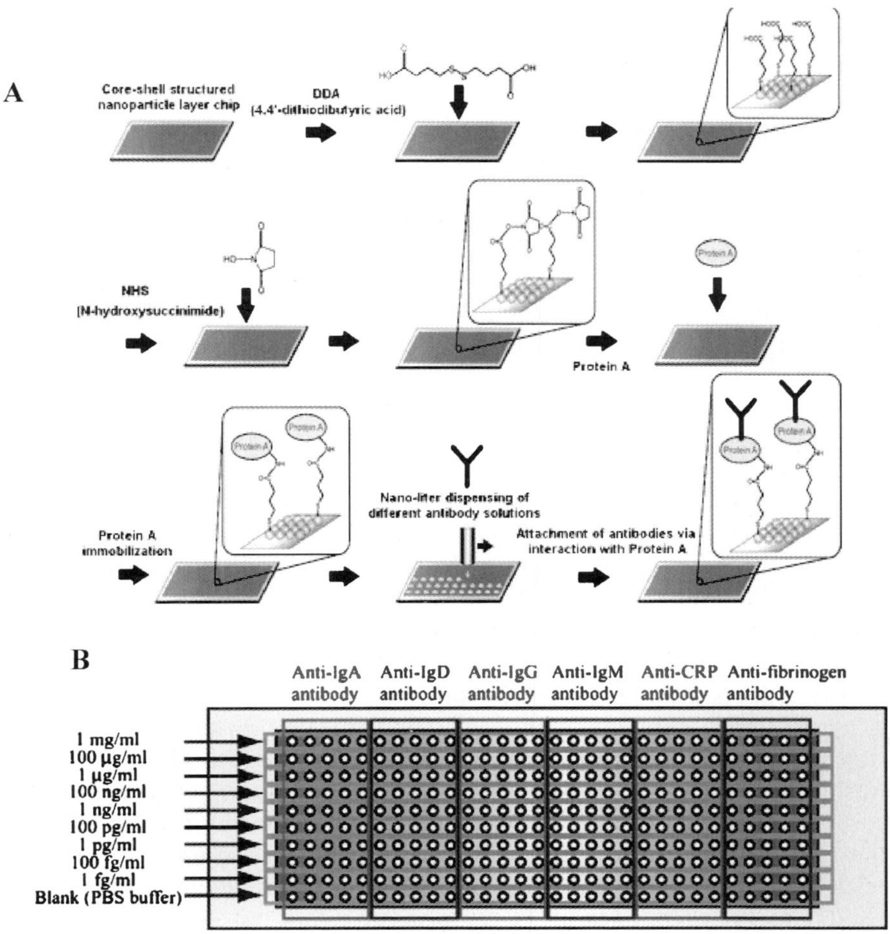

Fig. 13 Experimental conditions of the multiarray LSPR-based nanochip. **A** Immobilization of antibodies on the surface of the multiarray LSPR-based protein array biochip. **B** Construction of the antibody immobilized spots and antigen concentrations. Six kinds of antibodies and antigens were spotted onto the multiarray LSPR-based nanochip surface

SAM-functionalized surface for 1 h. Protein A at 100 μg/mL was immobilized on the surface for 1 h, which reacted with the Fc region of the immunoglobulin G (IgG) antibodies. Antibodies against IgA, IgD, IgG, IgM, C-reactive protein (CRP), and fibrinogen at 100 μg/mL were spotted onto the protein A-modified surfaces using the nanoliter dispensing system at a volume of 100 nL (Fig. 13B) and incubated for 1 h. Finally, the antibody-immobilized surface was rinsed thoroughly with 20 mM PBS (pH 7.4) and dried at RT. In total, 300 antibody-immobilized spots were formed on the chip surface.

After the immobilization of the antibodies on the multiarray chip, different concentrations of antigen solutions (∼0 to 100 μg/mL) were introduced onto the 300 spots using the nanoliter dispensing system and incubated for 30 min. Especially, the total sample volume was significantly reduced. Each spot was reacted with 100 nL of sample solution containing the antigens at varying concentrations. After an incubation period of 30 min, a stringent washing procedure with 1% (v/v) Tween-20 was applied to suppress the non-specific adsorption. Subsequently, evaluation of the optical characteristics of the chip was carried out. All absorbance spectra were taken in the range of 400–800 nm on the UV–Vis spectrometer at RT. White light emerging from the optical fiber bundle was incident onto the nanochip from the vertical direction. The reflected light was coupled into the detection fiber probe of the optical fiber bundle and analyzed by the UV–Vis spectrometer. The evaluation of the results obtained from the nanochip surface revealed that the specific absorbance strength change was directly related to the applied antigen concentration on the spot.

3.2.3 Multiple Detection of Proteins using a Microarray LSPR-Based Nanochip

The antibody solutions were spotted on the multiarray LSPR-based nanochip surface using a nanoliter dispensing system. In this research, a total of 300 antibody immobilized spots were prepared on the nanochip surface by dispensing six different antibodies. Each antibody was dispensed onto 50 predetermined spots. Additionally, each spot was separated with a pitch of 1 mm to prevent cross-contamination. LSPR spectra were measured after incubation with the different concentrations of antigens for 30 min, and then the washing procedure was carried out as described in the experiment. The absorbance strength change was recorded depending on the concentration of the antigens IgA, IgD, IgG, IgM, CRP, and fibrinogen, as shown in Fig. 14A–F, respectively. The multiarray LSPR-based nanochip provided a limit of detection of 100 pg/mL for all of the proteins (data not shown). The wide detection range of our nanochips was linear up to 1 μg/mL. Our method using multiarray LSPR-based nanochips is a promising candidate for the low-cost and highly sensitive detection of multiple analytes in a simple and rapid format.

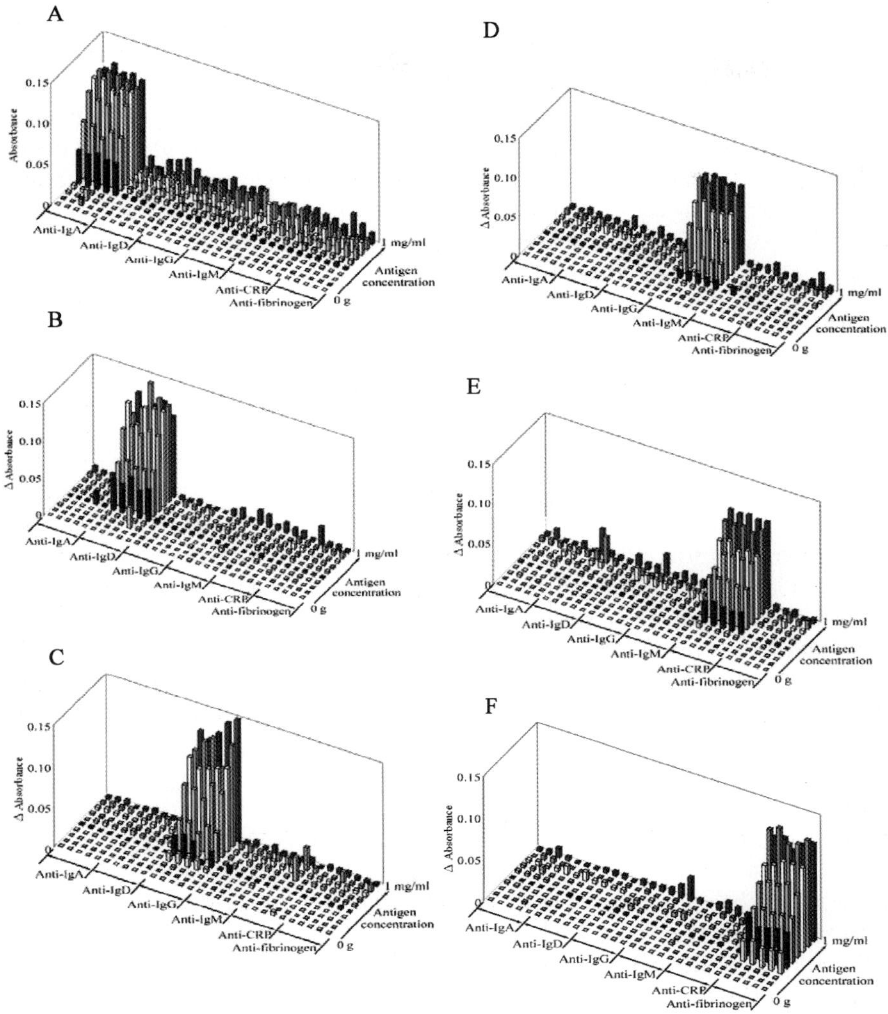

Fig. 14 Absorbance measurements at each spot using the multiarray LSPR-based nanochip. After the introduction of antigens at different concentrations using the nanoliter dispensing system, the LSPR spectra were recorded for **A** immunoglobulin A (IgA), **B** IgD, **C** IgG, **D** IgM, **E** C-reactive protein (CRP), and **F** fibrinogen

Future trends in diagnostics will continue in miniaturization of biochip technology toward the nanoscale. Our multiarray LSPR-based nanochip provides a convenient, low-cost, and label-free method for specific and highly sensitive detection of the biomolecular interactions in a parallel format. We anticipate that this technology will extend the limits of current molecular diagnostics and enable point-of-care diagnosis as well as promote the development of personalized medicine. The multiparallel possibilities of biosensing applications have the potential to allow the optimization of biomarker research, cancer diagnosis, and also the detection of infectious microorganisms for biodefense. Especially, the cost for the fabrication of our multiarray LSPR-based nanochip, including the optical apparatus, is significantly lower than that of a conventional SPR system. This LSPR method is "easy to operate" even by "nonspecialists". The LSPR-based multiarray nanochip presents a highly versatile method that is readily applicable to the other kinds of bioassays, such as metabolomics and cellomics.

4 Cell Chip Device Systems

4.1 Neuronal Patterning in Biochips

Long-term potentiation (LTP) of synaptic transmission in the hippocampus is the primary experimental model for investigating the synaptic basis of learning and memory in vertebrates [125]. LTP is believed to be linked to the glutamate cycle [126] and activation of amino acid receptors, such as the *N*-methyl-D-aspartate receptor complex [127]. Studies in this area have involved electrophysiological study with patch clamping techniques and high-resolution fluorescence imaging with confocal microscopy. This work aimed to improve this method with the introduction of two techniques: patterning of neurons to the level of individual synapses and super high-resolution near-field optical microscopy [128–130]. The patterned growth of in vitro neurons is a key feature in the development of both fundamental and applied research in the neuroscience field. Moreover, patterning of neurons will allow for better studies of neuron connectivity, synapse formation, and studies into LTP. Neuron guidance can be carried out through physical, chemical, or electrical cues [131, 132]. A number of studies on the patterning of cell lines, such as PC12 and BCE cell lines, have already been done [133]. However, primary cell lines have proven to be more difficult to pattern. Some microfluidics have been shown to be capable of patterning primary culture cells [134, 135]. The patterning method consists of chemical guidance with an element of physical confinement, and allows for ultrafine patterning of neural growth on transparent glass substrates. The substrates consist of microfabricated perfluoropolymer barrier structures on glass. Poly-L-lysine was selectively deposited using a silicone-based microfluidic stencil aligned to the perfluoropolymer/glass substrate. Topographical patterns are formed by etching of amorphous perfluoropolymer on top of the glass substrates. It is believed that hydrophobic perfluoropolymer absorbs albumin proteins from the growth medium, which is repulsive to cells [136]. Chemical patterning of a poly-L-lysine growth matrix on the glass substrates has been carried out with microfluidic stencils. Poly-L-lysine growth matrix contains many amide groups, which promote cell adhesion. Thus, this method creates structures, which alternate between chemically attractive and repulsive regions. Use of a microstencil demonstrates a way to pattern molecules on surfaces where the topographical patterns are already formed. Using this method it is possible to create 10-μm paths and achieve guidance of individual neurites. It is also desirable to integrate the patterning system with microelectrode arrays and to apply this technique to commercial chips, such as the Med64 system from Panasonic [137].

Primary culture neurons were extracted from 8-day-old chicks, patterned successfully, and grown for 3 days to form good networks at a concentration of 8.5×10^5 cells/mL, while three-dimensional patterning was possible at high concentrations ($> 1 \times 106$ cells/mL). Cell viability was evaluated with fluorescein diacetate (Fig. 15A). Tests were done to see the effect of poly-L-lysine on the Cytop/glass patterns without using the microfluidic stencil. For substrates dipped for long periods of time, the growth was nonspecific. The samples that were dipped in the poly-L-lysine for 2–3 min showed preferential growth in the glass channels with patterning down to the level of individual neurites (Fig. 15B). Fluorescence imaging was carried out on both the cell viability during growth and immune-tagged microtubule-associated proteins on the neurites. Successful growth of primary neurons was also achieved in the microfluidic patterns, where growth of cell lines were found to be good when compared with primary culture (Fig. 15C) [138].

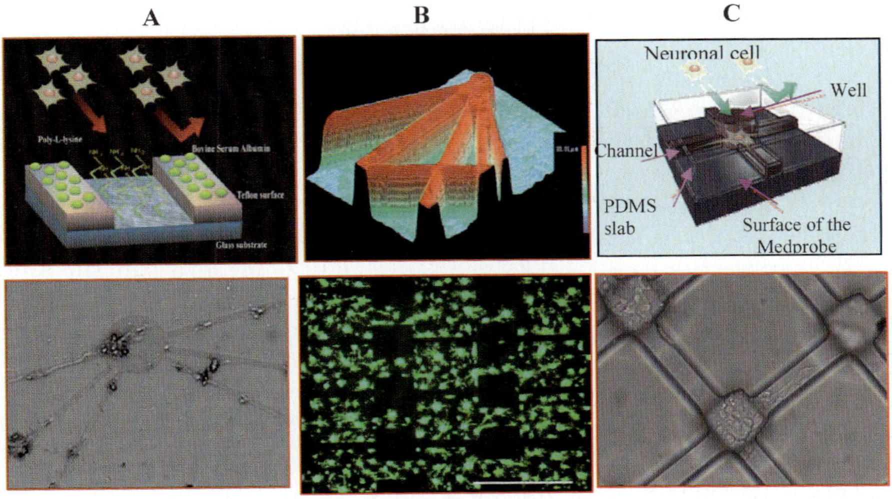

Fig. 15 Neuronal patterning methods: **A** chemical patterning, **B** physical patterning, **C** microfluidic patterning. Primary cultures are grown in Cytop-coated patterns, poly-L-lysine-coated deep Si structures, and microfluidic chips, respectively

In vitro culture of small neuronal networks with predefined topological features is particularly desirable when the electrical activity of such assemblies can be monitored for long periods of time. Indeed, it is hoped that such networks, with predetermined connectivity, will provide unique insights into the structure/function relationship of biological neural networks and their properties of self-organization. However, the experimental techniques that have been developed so far for that purpose have either failed to provide very long-term pattern definition and retention, or they have not shown potential for integration into more complex microfluidic devices. To address this problem, three-dimensional microfluidic systems in PDMS were fabricated and used in conjunction with both custom-made and commercially available planar microelectrode arrays (pMEAs). Various types of primary neuronal cell cultures (cortical cells from chick embryos and mouse embryos) were established inside these systems (Fig. 16A–C). Extracellular

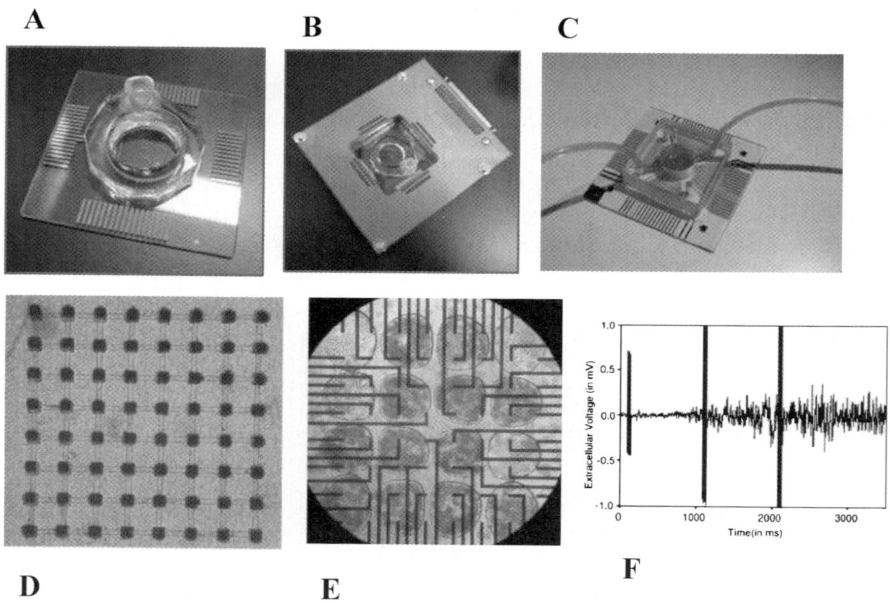

Fig. 16 **A** Microfluidic chip assembled on the Panasonic Medprobe dish. **B** Electrode interface. **C** After assembling total microfluidic chip for electrical stimulation and recordings. **D** Commercial pMEA fitted with patterns with 50-μm^2 chambers and 25-μm-wide channels (day 8 in vitro). Electrodes are 50-μm^2 squares and give the scale of the picture. **E** Custom pMEA fitted with larger patterns, 5 h after seeding. Chambers are 600 μm in diameter and channels are 30 μm wide. Wiring tracks are 50 μm wide and give the scale of the picture. **F** Typical signals obtained in response to stimulation from the cells inside the PDMS microsystem shown in **d** and **e**. The three artifacts due to stimulation are clearly visible. The resulting composite signal is difficult to analyze due to the impossibility of extracting single spikes

electrical signals were successfully recorded from all types of cells placed inside the patterns, and this bioelectrical activity was present for several weeks (Fig. 16D–F). We have shown that microfluidic systems made up of wells and channels could accommodate neuronal cell cultures for several weeks and guide their growth on top of microelectrode arrays. Our results prove that confinement inside PDMS microstructures with characteristic dimensions as small as 20 μm does not impair the ability of the cells to generate electrical activity. In contrast to most existing patterning methods for neural networks, which emphasize the fineness of cell patterns, the approach presented here seems especially suitable for isolating small groups of neuronal cells inside chambers with characteristic dimensions of a few hundred micrometers and constraining the topology of the synaptic connections between these groups [139]. Such a setup could be used to further constrain the connection topology between cell clusters by applying differential chemical and/or electrical stimulation to certain chambers. This technique will enable exploration of cellular self-organization mechanisms at a new scale, although further development of microfluidic technology is first required. The advantage of this approach is that it can be further integrated with microfluidic devices and pMEAs to yield, for example, complex neuron-based biosensors or chips for pharmacological screening.

4.2 Cell-Based Assays for High-Throughput Anticancer Drugs and Neuropeptide Molecule Screenings Using Microchamber Array Platforms

4.2.1 On-Chip Multiplexed Screening of Anticancer Drugs

Over the past decade, a variety of scientific advances including the growing number of potential therapeutic targets emerging from the field of functional genomics, the rapid development of large compound libraries derived from parallel and combinatorial chemical synthesis techniques, and the ever-increasing pressure to reduce developmental costs have driven the need for improved drug discovery screening technology [140, 141]. Despite the huge increase in research and development spending in pharmaceuticals, the number of approved drugs as new chemical entities has only slightly increased. The high productivity and low cost of new technologies were greatly needed in every step of the drug discovery process, for example, a rapid, reliable, and sensitive high-throughput screening (HTS) technique with extremely low cost. To achieve this goal, assay miniaturization with precision and automation in analyte delivery becomes a prerequisite [142]. The recent development of microarray and microfluidics technologies has facilitated the development of high-density, low-volume assays. HTS assays play a pivotal role in the search for novel drugs and potential therapeutics. Traditionally, most of the

HTS assays performed using multiwell microtiter plates as platforms in identification of potential bioactive compounds, enzyme inhibitors, and small-molecule targets involve in vitro biochemical assays [143–147]. In the last few years, cell-based assays have become an increasingly attractive alternative to in vitro biochemical assays for HTS. The requirements for such in vivo assays are the ability to examine a specific cellular process triggered by a defined target and a means to readily measure its output in a HTS system [142, 148, 149]. Cell-based assays have notable advantages over in vitro assays. Firstly, these in vivo assays do not require purification of the target protein and therefore eliminate investment of resources to gain the necessary knowledge for obtaining a biochemical target. Secondly, the conformation and the activity of the target protein, as well as the readout to monitor the effect of compounds, are examined in a cellular context that most likely represents the natural physiological state more closely than in vitro assays. Thirdly, cell-based assays can immediately select against compounds that are generally toxic. Thus, hit and lead compounds that are identified through cell-based assays have passed important validation steps.

The availability of this information provides a head start compared to many in vitro assays, and can save valuable time and costs in the development of the drug. There have been growing number of publications reported in the literature on high-throughput cell-based assays for various applications: unraveling novel intracellular pathways [150], discovery of anticancer drugs [151], yeast cell-based assays for functional genomics and drug discovery [152], the discovery of inositol phosphatase inhibitors [153], and evaluation of plant extracts as anticancer drugs [154], etc. All these reports mainly used 96- or 384-well plates as formats, which require a minimum of 100 μL assay mixture, which is bulky and suggests further miniaturized assay formats in order to reduce the total cost of drug screening. Testing very small samples is very attractive because it requires only minute amounts of both the component being assayed and the components of the assay, both of which may be quite expensive or available in very limited amounts. There are no reports, so far, on using miniaturized microchamber array chip formats coupled with an automatic nanoliter sample dispensing system for screening of bioactive compounds of choice with a cell proliferation assay. Therefore, we have constructed different sizes of miniaturized microchamber arrays with silicon and polymer as chip substrates in our laboratory by using micromachining and soft lithography techniques, and employed them in various applications.

In this research, we report applications of this silicon microchamber array platform for HTS of carotenoids astaxanthin, canthaxanthin, β-carotene, and lutein, and bacteria-derived carotenoids lutein and zeaxanthin, as cytotoxic anticancer drugs against human cervical cancer cell line HeLa 229 using a fluorescence-based cell proliferation assay, and quantitation of their cytotoxic effects [207]. Further, we have compared the on-chip cell proliferation assays by performing the same assays with conventional 96-well plates.

Figure 17A illustrates a flow diagram of nanoliter silicon microchamber array platforms for fluorescent cell-based high-throughout screening of cytotoxic anticancer drugs and quantitative determination of cytotoxicity of screened drugs. First, the chip was placed onto the dispensing stage of a nanoliter dispenser, and 20 nL of HeLa 229 cell suspension in MEM consisting of 250 cells per chamber was distributed precisely into 192 chambers (24×8) of a microarray chip, as a block for each concentration, by using a Cartesian nanoliter dispensing system. The samples were incubated for 2 h in a humidified chamber at 37 °C with 5% CO_2 atmosphere to allow the cells to adhere to the chamber wall surfaces. We intentionally left two lanes of microchambers (24×2) empty between each chemical treatment for easy distinction. It was followed by introduction of 20 nL of authentic carotenoids, such as astaxanthin, canthaxanthin, lutein, and β-carotene, as well as carotenoidal extracts from *Stenotrophomonas* containing lutein as chief constituent and zeaxanthin from *Flavobacterium* carotenoidal extracts in ethanol at different concentrations, depending on the chemical availability. Materials were distributed into the microchambers ($550 \times 550 \times 250\,\mu m^3$) of the chip array by a dispensing system while maintaining the humidity over 90% inside the system, which allows cells to have native activity inside the chambers and also avoids dryness [155, 156]. The effect of each chemical concentration was quantified from the 192 microchambers. The chip containing the cells and chemical target to be screened was then transferred into an incubator maintained at 37 °C with 5% CO_2 atmosphere for 18 h. At the end of incubation, the chip was processed with fluorescent dye Calcein-AM (Dojindo, Japan), to detect fluorescence release from the live cells, which was determined as given in the Dojindo Technical Manual [157]. The chip was loaded with 1 μM Calcein-AM that was dissolved in 50 μM PBS buffer (pH 7.4) and allowed to stand for 30 min at room temperature while protecting the chip from the effects of light by covering with aluminum foil. The cells were then washed three times with PBS buffer and treated with 1 : 1 volumes of acetone and methanol (Wako Pure Chemicals, Japan) for 1 min to dry the cells. The chip was then scanned with a CRBIO IIe FITC microarray scanner (Hitachi Software, Japan) and the fluorescence readout was recorded with a filter having excitation and emission of 473 and 532 nm, respectively. The cell viability was measured from the intensity of the fluorescent calcein released by the action of membrane esterase present in the living cells, and the cell death was quantified using DNA SIS (R) Array software, version 2.1.

Figure 17B shows a fluorescence image of the microarray scanned chip after simultaneous screening of different carotenoids including astaxanthin, canthaxanthin, lutein, and β-carotene, and bacteria-derived lutein (represented in D-1 extract) and zeaxanthin, as well as a control (no chemical) against HeLa cell proliferation. To achieve this, we constructed a miniaturized microchamber array chip with 1248 chambers with a 24×52 pattern on a 1×3 in. silicon wafer by using soft lithography and anisotropic wet etching

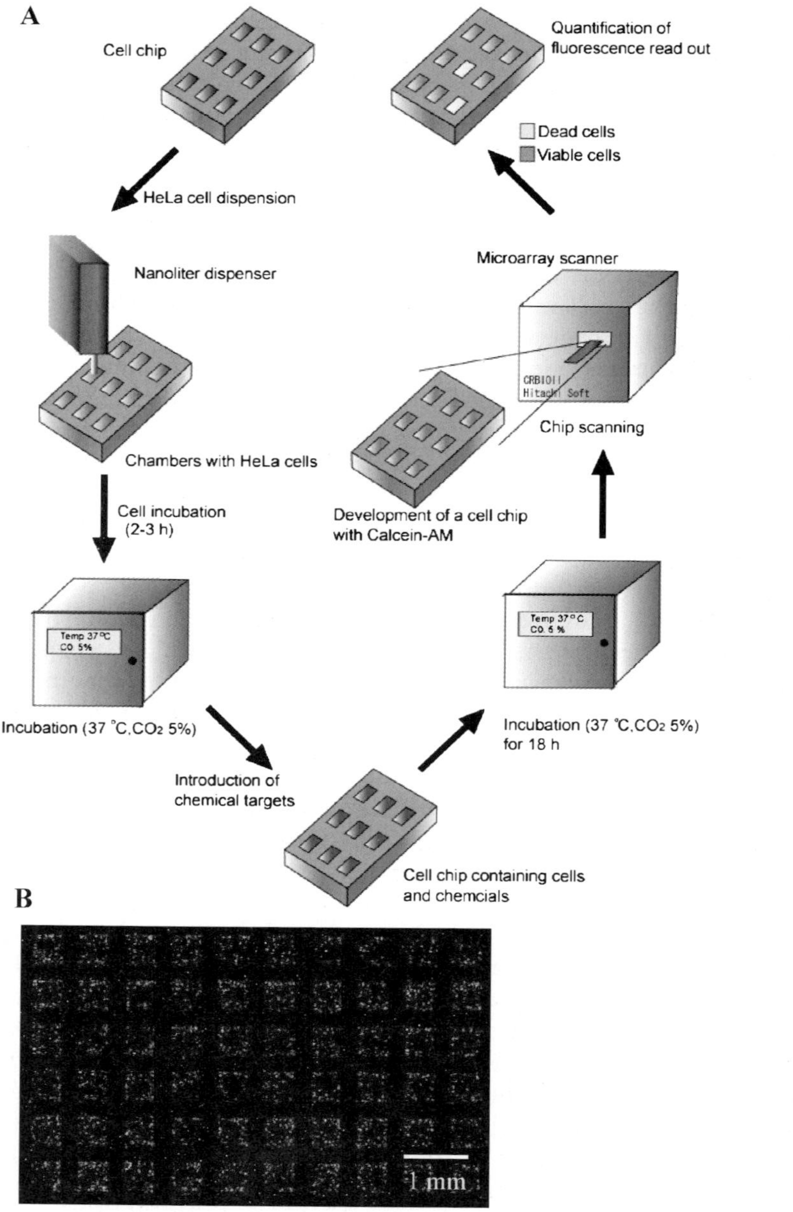

Fig. 17 **A** Schematic illustration of a novel on-chip cell proliferation assay in nanoliter volumes using highly integrated silicon microchamber array chip platform for high-throughput cytotoxic anticancer drug screenings and the quantitative determination of cytotoxicity of carotenoids. **B** Fluorescence image of microarray-scanned chip releasing fluorescence from the microchambers introduced with different carotenoids after cell proliferation assay

techniques. This microchamber array platform enabled us to carry out reactions as small as 40-nL volumes in each microchamber, which was combined with an automatic nanoliter sample delivery system, for precise introduction of the HeLa cell suspension and chemicals into microchambers, and a microarray scanner for fluorescent cell-based HTS assays to screen compounds derived from natural sources and/or combinatorial chemistry for determining cytotoxic anticancer effects. The rendering of the inner wall surfaces of the microchamber to be hydrophilic by an oxidized layer allows the cells to adhere to their surfaces, which are amenable for cell cultivation in the nanoliter chambers to measure the cytotoxic effects of compounds based on the cell proliferation assay. This also minimizes cross-contamination to a certain extent between chambers, due to the hydrophobic surfaces of the chip which prevented the spread of the aqueous solution to the outside of the microchamber, as shown in Fig. 17B. There is a need to control evaporation in open systems such as this, as having high surface area to volume ratios becomes increasingly important and volumes are reduced. We overcame this by keeping the microchamber array chip on a petri dish containing distilled water, which avoided the chip becoming dry during incubation. We have reported recently the optimization of this screening system including a nanoliter delivery system for precise introduction of samples into microchambers and the number of cells required per chamber for the cell proliferation assay [155]. We have delivered 20 nL of HeLa 229 suspension consisting of 250 cells per chamber precisely by using a nanoliter dispensing system. After a brief incubation, the chip was distributed with 20 nL of five different standard carotenoids and two bacteria-derived carotenoidal extracts at different doses, along with a control that received no chemical, into 192 chambers of each chip to determine their cytotoxic anticancer activities against HeLa 229 cells. Since each microchamber array chip consists of 1248 chambers and accommodates a few nanoliters as assay volume, it enables simultaneous screening of large numbers of sparsely available compounds against a specific target (cell line) or vice versa, which is otherwise difficult using conventional 96- or 384-well plate formats, to identify compounds with cytotoxic anticancer effects. The miniaturized silicon microchamber array platform is highly customizable and reduces the consumption of cells 200-fold, while the consumption of compounds decreases by >2500-fold, compared to 96-well plate assays.

A significant inhibition in the HeLa cell proliferation was found as well as cell death inside nanoliter microchambers delivered with standard carotenoids astaxanthin, canthaxanthin, lutein, and β-carotene and bacteria-derived carotenoidal extracts, suggesting their cytotoxic anticancer activities against human cervical cancer cells, HeLa 229. It was evident from the data that the cytotoxicity is proportional to the increasing concentrations of carotenoids and carotenoidal extracts studied (data not shown). The cytotoxic anticancer effects of carotenoids and carotenoidal extracts were visualized from the microarray scanned chip image, where various levels of fluorescence intensity are seen depending on the effect of the carotenoid or carotenoidal extract (Fig. 17B). Further, this integrated system not only allows screening and detection of positive compounds that show cytotoxic effects against a cell line, but also quantifies the cytotoxicity of the carotenoids, which is measured by subtracting the fluorescence intensity of all viable cells in each chemical treatment from the intensity of control chambers that are not treated with any of the chemicals. The difference in the intensity was a measure of the cytotoxic anticancer activity of these chemicals against HeLa 229 cells. The cytotoxic effect of carotenoids was found to be in the order of astaxanthin > canthaxanthin > β-carotene > lutein >bacteria-derived carotenoidal extracts. Of all the carotenoids, astaxanthin showed the highest inhibition in HeLa cell proliferation with 73% at 10 μM (146% for 20 μM on a calculated basis), followed by canthaxanthin with 94%, and β-carotene with 78%, which was followed by standard lutein, lutein-rich carotenoidal extract (from D-1 cultures), and zeaxanthin-rich extract (from *Flavobacterium*) with 35, 55, and 39%, respectively (Fig. 18A). The increase in the cytotoxicity of bacteria-derived carotenoidal extracts is expected because they contain mixtures of carotenoids, which probably acted synergistically and contributed to increased inhibitions of cell proliferation over standard lutein cultures. We compared zeaxanthin-rich carotenoid extracts by taking lutien as standard, due to the nonavailability of standard pure zeaxanthin. Though carotenoids have long been known for their beneficial effects as antioxidants, free radical scavengers, and cytotoxic agents for some cells [32–34], there are no reports so far available on their cytotoxic anticancer effects, especially against human cervical cancer cells (HeLa 229). Thus, this study highlights the development of a novel on-chip cell proliferation assay in nanoliter microchamber arrays for high-throughput screening of anticancer compounds and quantification of their cytotoxic effects, and also addresses the fact that these carotenoids and bacteria-derived carotenoidal extracts containing lutein and zeaxanthin were potent against in vitro inhibition of human cervical cancer cells.

Figure 18B represents the effects of authentic carotenoids and bacteria-derived carotenoidal extracts on HeLa cell growth inhibition carried out in 100-μL reaction volumes using conventional 96-well plates by MTS assay. This assay was carried out to compare the nanoliter microchamber-based cell proliferation assays with multiwell plate assays for HTS studies. It was observed that the inhibition in the HeLa cell growth is proportional to the increasing concentration of carotenoids and carotenoidal extracts tried. The profile of carotenoid and carotenoidal extract induced HeLa cell death obtained in the nanoliter microchamber array-based assays was almost comparable to the profile of 96-well plate formats with bulk reaction volumes, suggesting that on-chip cell proliferation assays in nanoliter volumes with microchamber array platforms are reliable and sensitive enough to employ in drug discovery screenings as well as in quantitative determination of their cytotoxic effects.

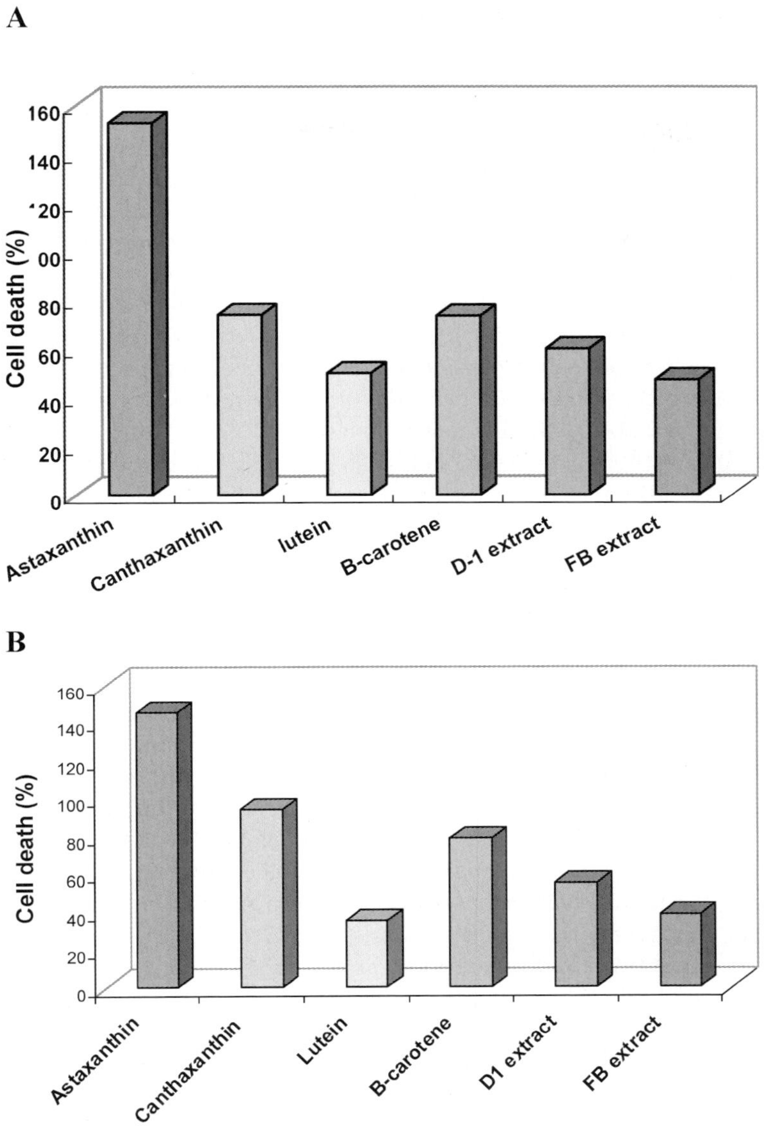

Fig. 18 **A** Comparison of cytotoxic anticancer effects of standard carotenoids and bacteria-derived carotenoidal extracts on HeLa 229 cell death using microchamber array chips and **B** 96-well plate formats. Cell death is calculated and represented for 20 μM in the figures for better comparison

In addition, microchamber array-based assays provide multiplexed screening of large numbers of potential compounds at tiny assay volumes and facilitate HTS assays of very short durations economically.

4.2.2 Microchamber Array-Based Cell Chips to Screen Combinatorial Peptide Libraries to Identify Neurite Outgrowth Inducing Peptides

Nerve growth factor (NGF) is a polypeptide and a prototype of the neurotrophins family that plays a pivotal role in the survival and differentiation of many types of neuronal cells during development [158]. Apart from this, it also plays an active role in the repair, regeneration, and protection of neurons and thereby serves as a therapeutic agent in the treatment of neurodegenerative diseases, such as Alzheimer's disease [159, 160], which is the most common cause of dementia in the elderly population [161]. Current estimates report that 50% of people worldwide over age 85 are afflicted with Alzheimer's disease, and this number is believed to be 16 million by the year 2050 (see http://www.alz.org/Media/newsreleases/2002/072202_researchE.asp). The total cost of Alzheimer's disease to the US economy has been estimated to be US $183–207 billion in 2001 [160]. The treatment of Alzheimer's disease using NGF has been mainly practiced by a direct infusion into the brain, which is an expensive option. NGF, being a high molecular weight protein, demonstrates its inability to cross the blood–brain barrier (BBB) [162]. This can be a major impediment in the treatment of diseases related to the central nervous system, since many drugs including NGF are unable to reach this organ at therapeutic concentrations [163]. The demands for searching and screening for novel small molecules that can cross the BBB to treat neurodegenerative Alzheimer's disease are constantly increasing. Combinatorial approaches offer hundreds of thousands of peptides with randomly defined substitutions that serve as a source of peptides with special biological functions [164, 165], as potential therapeutics [166], anticancer peptides [167–169], and bioactive molecules [170], and also employed as a source of molecules with specially designed functions [171]. It is also known from previous reports that small peptides containing hydrophobic amino acids cross the BBB without any hurdles, suggesting that hydrophobic amino acids facilitate the peptides in crossing the BBB [172–174]. For example, transport of amino acids, especially phenylalanine, across the BBB and the implications for treatment of maternal phenylketonuria were reported [175]. The drug penetration through the BBB was believed to depend on lipid solubility and molecular weight [176]. Peptide lipophilicity was also reported to be of major importance in determining penetration through the BBB [177]. There were several reports on peptide and small-molecule microarrays for detection of peptides, ligands, small molecules, or enzymes, and cell adhesion etc. [178–181]. However, the lack of screening assay systems that enable screening of combinatorial peptide libraries together with cells to identify peptides with neurite outgrowth inducing effects has been a great limitation. There were, however, reports in the literature on screening of synthetic peptide libraries using indirect bulk assays to identify a neuritogenic ligand C3 peptide [182] and NCAM binding peptide [183] when the target was a well-known protein–NCAM, a membrane associated glycoprotein, which plays a key role in morphogenesis of the nervous system and synaptic plasticity of neuronal connections [184], based on streptavidin affinity binding [179].

We therefore report a microfabricated microchamber array as a platform to screen large numbers of synthetic peptide libraries, to identify peptide molecule(s) that are capable of modulating a special functional response in the nervous system without prior knowledge of a molecular target. This approach may be of interest to researchers in the field of neurosciences where the functional end point, such as neuritogenesis, neuronal differentiation, or neuronal cell survival, can be monitored in cultures, and where the development of unknown peptide molecules promoting these functions is of importance. We have synthesized nearly one million pentapeptide-conjugated beads consisting of 3125 varieties using the hydrophobic amino acids Ile, Leu, Met, Phe, and Val by solid-phase split-pool synthesis. Each bead expresses only one kind of peptide with approximately 10^{13} copies conjugated on to a 100-μm bead [185]. Screening of this large number of peptide library mixtures individually in the presence of cells using multiple-well plate or petri dish assay formats makes it almost impossible to identify neurite outgrowth inducing peptides. Therefore, we constructed a miniaturized silicon microchamber array by employing micromachining and soft lithography techniques, which serve as a desirable platform and allow the physical confinement of peptide-conjugated beads inside the microchambers [208].

Microchamber Array-Based Cell Chips for Combinatorial Peptide Library Screening

A schematic illustration of microchamber array platforms for cell-based screening of combinatorial peptide-bead libraries, to identify peptides that induce neurite outgrowths of PC12 cells, is shown in Fig. 19A. When a combinatorial peptide-bead library of approximately 5000 peptides was dispensed onto the microchamber array chip (Fig. 1B), it was observed that over 70% of the total microchambers (1248) confined an average of three to five peptide-conjugated beads (Fig. 19B) by physical adsorption. PC12 cell suspension on a collagen-coated glass slide was simultaneously and aseptically prepared. Then a silicon microchamber array that contained peptide beads was placed over the glass slide upside down, thus facing the peptide-conjugated beads toward the PC12 cell suspension, and its four sides were sealed with a parafilm to prevent the cells from drying. The total setup was transferred into a humidified chamber maintained at 37 °C and with a 5% CO_2 atmosphere. Likewise, screening of a 100 000 peptide library was achieved by employing over 20 silicon microchamber arrays. It was observed under a microscope that only two peptide beads in two separate microchambers demonstrated neurite outgrowths of PC12 cells (Fig. 19C) at the end of incubation, which were then retrieved by a micromanipulator for sequence determination. The beads

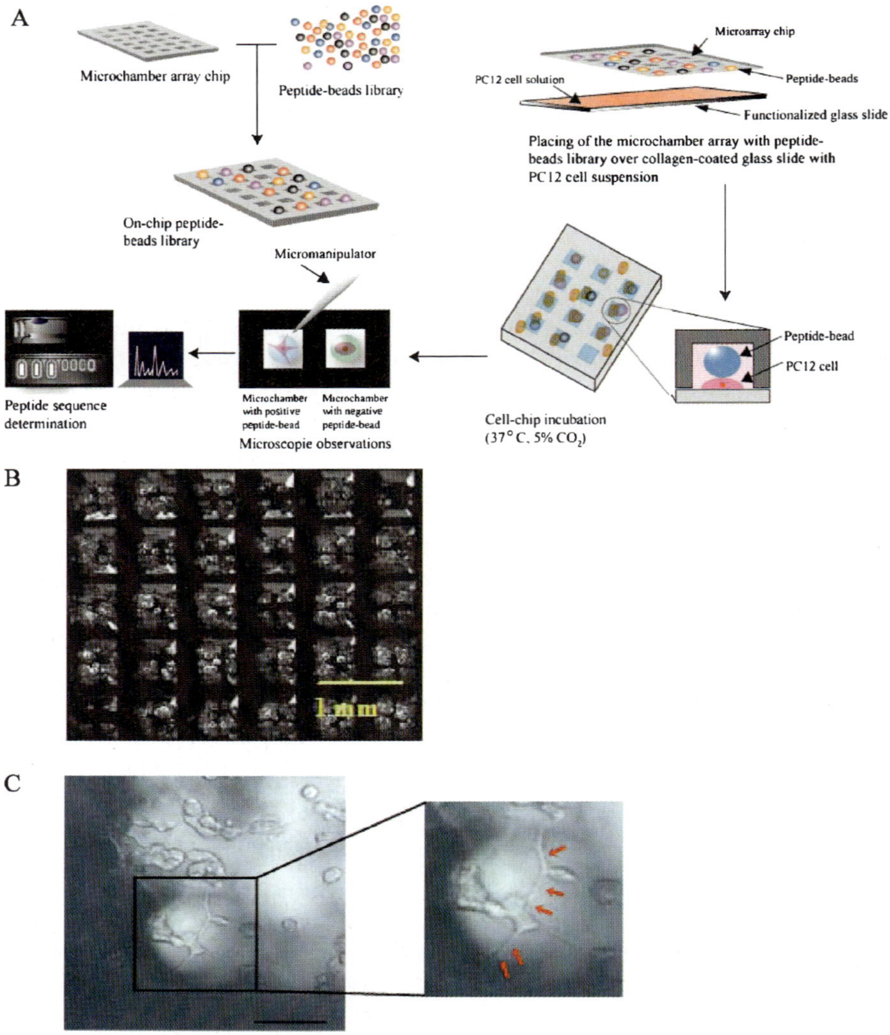

Fig. 19 Schematic illustration of the microchamber array-based cell chips for screening of combinatorial peptide libraries. **A** Microchamber array platform for high-throughput cell-based screening of combinatorial peptide libraries to identify peptides that induce PC12 cell neurite outgrowth extensions. A microchamber array chip that contained the peptide bead library was then placed upside down, thus facing the peptides downward on the collagen-coated glass slide with PC12 cell suspension, and incubation resulted in the emergence of neurite outgrowths in chambers that had affinity peptide–cell interactions. This can be seen in only a few chambers; the corresponding peptide beads were retrieved by a micromanipulator and peptide sequences were determined by microsequencing. **B** A real picture of a microfabricated chamber accommodated with beads. **C** Microscopic photograph showing PC12 cell neurite outgrowth on beads conjugated with peptides in a microchamber

were microsequenced with Edman chemistry, using an automatic protein sequencer. The data analysis revealed only part of the sequences (three amino acids) of two positive peptide beads, which were determined to be M-L-M-X-X and M-M-V-X-X, respectively. The reason for not being able to determine the sequences of the remaining two amino acids in the above positive peptides may be that they would have been degraded by the proteases that are present in the serum of the medium during manipulation.

To determine the missing two amino acids of the above two pentapeptides, we synthesized all the possible combinations of amino acids of these peptides ($5^2 \times 2 = 50$) by using a manually operable parallel peptide synthesizer (SRM96, Hipep laboratories, Japan) (data not shown). This resulted in 50 different pentapeptides, which were tested individually for their neurite outgrowth inducing effects in 96-well plates for 3 days under the above culture conditions. Four peptide inducing PC12 cell neurite outgrowths were identified, and their sequences were determined to be M-M-V-I-F, M-M-V-M-M, M-L-M-V-F, and M-L-M-V-L. Of these, M-M-V-I-F stimulated prolific neurite outgrowth inductions of PC12 cells. Therefore, M-M-V-I-F was selected to study further its quantitative neurite outgrowth effects with respect to time and dose, evaluation of functionality of neurons derived by measuring neurotransmitter release, and elucidation of its neurosignaling. These findings demonstrate that microchamber arrays allow cell-based screening of large numbers of combinatorial peptides in parallel to identify peptides with neurite outgrowth inducing effects and their label-free detection, facilitating easy retrieval of peptide-conjugated beads under the microscope by direct visualization.

4.3 High-Throughput Single-Cell Analysis System Using Microarray and Microfluidics

Each B-cell clone expresses antigen receptors, antibodies, with a unique antigen specificity: an antigen-specific monoclonal antibody derived from a single B-cell clone finds applications in antibody medicine and clinical diagnosis. Though each B cell has 10^7 to 10^8 varieties of monoclonal antibody on these surfaces, only a small percentage of B cells respond and produce a specific monoclonal antibody. It is reported that only one or two cells in a total of 10 000 B cells become active and produce antigen-specific antibodies after stimulation with a hepatitis B virus (HBV) surface antigen (HbsAg) [186]. It would be extremely difficult to make a HBsAg-specific antibody taking this ratio of positive B cells into account using currently available technologies. Flow cytometry allows us to monitor individual cells that flow through a sheath fluid, but the signals of the cells become background noise, which consists of 0.1 to sometimes 1% of total cells; thus, it is quite difficult to monitor the signal of a minor population of cells whose signals are buried in the noise from a flow cytometer. Further, we cannot compare the states of each cell before and after stimulation by using flow cytometry. On the other hand, fluorescence microscopy allows us to observe the states of cells before and after stimulation. However, it is difficult to observe signals of a large number of cells under a microscope. Accordingly, it is difficult to monitor Ca^{2+} mobilization of a minor population of cells.

Therefore, it is necessary to construct a microarray platform that can confine a large number of single cells and detect antigen-specific single B cells before and after stimulation with an antigen from a bulk cell suspension. For high-throughput single-cell separation and analysis, Thorsen et al. reported high-density microfluidic chips that contain plumbing networks with thousands of micromechanical valves and hundreds of individually addressable chambers, and showed the separation of single *E. coli* cells in each chamber [187]. To achieve single-cell separation, they diluted cells to create a median distribution of 0.2 cells per compartment, so that reliable capturing of cells in each chamber was difficult. In another recent report, Anderson et al. tested biomaterial microarrays for their effects on human embryonic stem cell growth and differentiation using populations of human embryonic stem cells [15]. However, single-cell-based assay seemed to be impossible using this microarray format. Also, we recently reported a microchamber array and microfluidic chip for the high-throughput measurement of cellular fluorescence [155, 188, 189]. Here, we report an improved microchamber array to monitor Ca^{2+} mobilization of over 25 000 cells simultaneously at the single-cell level [190]. We have developed a novel high-throughput screening and analysis system for antigen-specific single B cells using a microarray, which was carried out by detecting antigen-specific single B cells against an antigen of interest and their retrieval by a micromanipulator for antibody DNA analysis [190]. The single-cell microarray system developed in this study does not need to use myeloma, as in the case of the conventional hybridoma technique, and can screen the antigen-specific single B cells directly from cell suspension and analyze antigen-specific antibody DNA at a single-cell level. This system is simple and easy in its operation, and quick enough for making monoclonal antibodies when compared to conventional techniques. Moreover, this system can perform high-throughput single-cell analysis using chip devices.

4.3.1 Picoliter Microarray for Single-Cell Studies

The single-cell microarray chip that was constructed by using the LIGA process has over 200 000 cylindrical microchambers with a flat bottom (10 μm diameter, 12 μm depth), 1.4 cm^2 in area on a plastic wafer (2×8 cm^2) (Fig. 20A–C). The characterization of the microarray chip for single-cell studies was performed using mouse splenic B lymphocytes. The microchamber

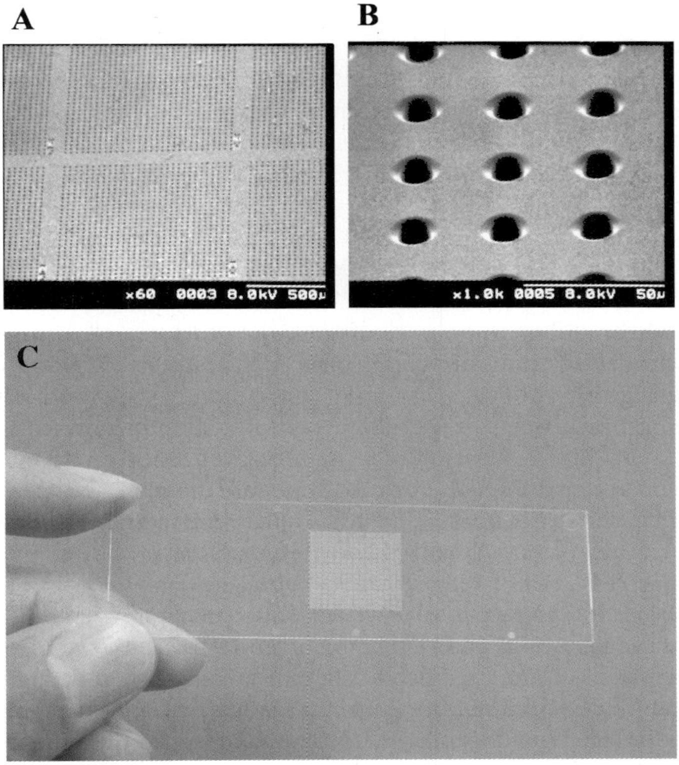

Fig. 20 Construction of the single-cell microarray chip. **A** and **B** SEM images and **C** a real picture of the microarray chip device. The microarray chip is made from polystyrene with over 200 000 microchambers (10 μm width, 12 μm depth, and 30 μm pitch)

design was made such that each chamber allowed entry of a single B cell (approximately 8 μm diameter) and hence there is no possibility of two cells in a single microchamber. For achieving single B-cell confinement in each microchamber as well as cell retrieval from microchambers, the hydrophilicity of the microarray chip surface was controlled by adopting different timings of RIE exposure. As a result of chip surface treatment using RIE, lymphocytes derived from mouse spleen or human blood were spread on the microarray, and over 80% of the microchambers achieved single-cell status. In addition, this novel microarray system demonstrated easy retrieval of positive single B cells from microchambers by a micromanipulator under a microscope.

In this study, we report the data from approximately 30 000 microchambers for detecting antigen-specific single B cells, because the assay we have adopted, i.e., measurement of the increase in the intracellular calcium after antigenic stimulus using Fluo-4, lasts for only a few minutes. During this short period of time a microarray scanner could scan the analyzable area of

30 000–40 000 microchambers. This number of microchambers is sufficient to screen and identify as little as 0.1% of antigen-specific single cells in a total B-cell population, which is the limitation of the flow cytometer as described above. The ultimate potential of this single-cell microarray can be realized by improving the assay system and the scanning speed of the microarray scanner for detecting the total number of microchambers on the microarray chip.

4.3.2
Single-Cell Microarray System for Analysis of Antigen-Specific Single B Cells

To evaluate the utility of a novel single-cell microarray system for detecting intracellular calcium, the $(Ca^{2+})_i$ level of individual cells, we employed anti-mouse IgM antibody as a stimulant for mouse splenic B cells, which delivers the signals through B-cell antigen receptors and induces a transient increase in the $(Ca^{2+})_i$ levels. Mouse lymphocytes loaded with Ca^{2+} indicator, Fluo-4, were applied to the chip and their fluorescence was detected with a microarray scanner before and after stimulation with anti-mouse IgM antibody (Fig. 21). Fluo-4 is a standard fluorescent calcium indicator [191], which enters into the cells and generates fluorescence after binding to $(Ca^{2+})_i$. It is known that the concentration of $(Ca^{2+})_i$ increases after B cells respond to antigenic stimulation [192]. In this experiment, it was observed that the majority of single B cells on the microarray showed increases in the $(Ca^{2+})_i$ after stimulation (Fig. 21A). The fluorescence intensity reached its maximum level after 1 min stimulation, maintained a high magnitude for 2 min, and then decreased gradually (data not shown). Thus, we used the 2 min stable duration for analysis. Scattered plot analysis of an individual cell's fluorescence intensity (Fig. 21B) enabled us to discriminate B cells that were activated with anti-mouse IgM antibody. The data from an area of over 30 000 microchambers in the microarray chip revealed that 68% (18 130 cells) of total splenic lymphocytes (26 650 cells) showed more than two times higher fluorescence after stimulation with anti-mouse IgM (Fig. 11B). Cells with over five times increase in fluorescence (3883 cells) existed as 14% of the total splenic lymphocytes (Fig. 21B(I)), whereas most of the lymphocytes incubated with the control antibody (anti-human IgM) showed less than five times increase in fluorescence intensity (Fig. 21B(II)).

Each B cell expresses membrane-bound antibodies with unique antigen specificity as antigen receptors. When B cells are stimulated with an antigen instead of anti-IgM antibody, only a minor population of total B cells is stimulated and their $(Ca^{2+})_i$ increases. Previous studies reported diverse frequencies in the number of antigen-specific B cells, ranging from 1 in 10^2 for rabies virus [193] to 7 in 10^5 for myelin basic protein [194]. However, it is quite difficult to screen and identify minor cell populations of antigen-specific single B cells (< 0.1%) with a flow cytometer, because fluorescent signals of minor populations of cells are buried in the noise signals of nonspecific cells.

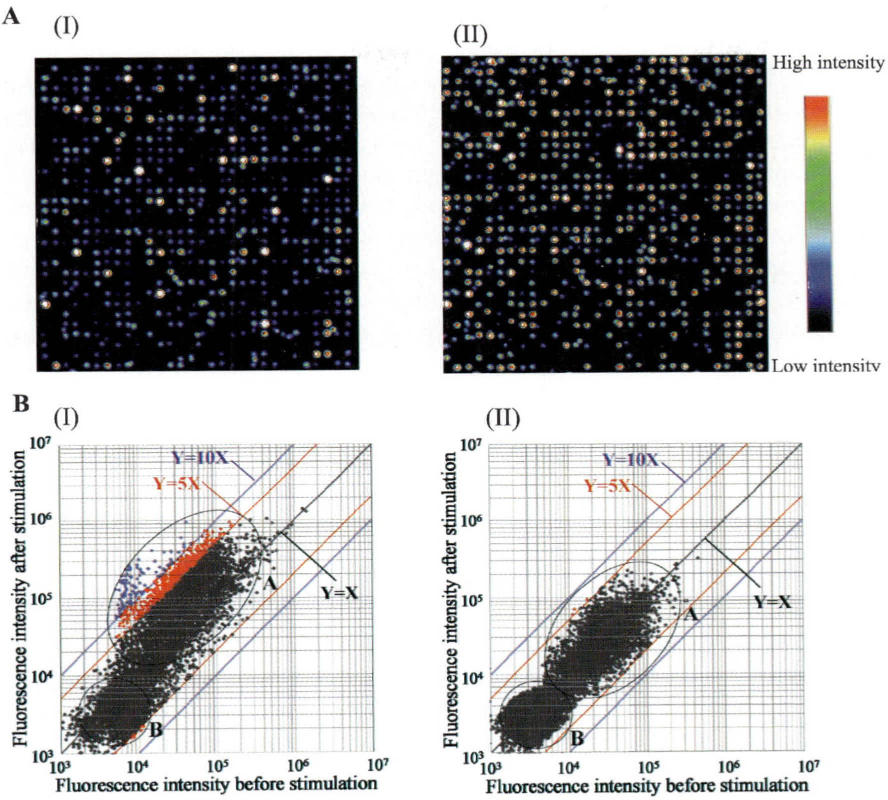

Fig. 21 Detection of activated single B cells using Fluo-4 upon stimulation with anti-mouse IgM antibody. **A** Scanned images of single-cell microarray in a single cluster area of 900 (30×30) microchambers on a microarray (I) before and (II) after stimulation. Color scale represents the intensity of fluorescence emission. **B** Scattered plot analysis of single B-cell response in 32 400 microchambers after stimulation with (I) anti-mouse IgM antibody and (II) negative control antibody. *Circle A*, microchambers with lymphocytes; *Circle B*, empty microchambers; *X*, fluorescence intensity before stimulation; *Y*, fluorescence intensity after stimulation; *blue line* ($Y = 10\times$), ten times higher fluorescence intensity; *red line* ($Y = 5\times$), five times higher fluorescence intensity; *green line* ($Y = 2.5\times$), 2.5 times higher fluorescence intensity

The single-cell microarray platform developed in this study could successfully screen and detect a low frequency of antigen-specific single B cells using a single chip in one run. In addition, it also analyzes the same single cells before and after stimulation with antigen, which flow cytometry could not.

Further, we could retrieve the highest active antigen-specific cells from the respective microchambers, and antibody cDNA is recovered to develop antigen-specific monoclonal antibodies. To this end, we have successfully performed retrieval of a single B cell from the microarray by a micromanipulator system under a microscope, and the following single-cell RT-PCR amplification to determine its antibody cDNA (data not shown). Monoclonal antibodies are usually produced using the hybridoma techniques developed by Milstein and Koeller [195]. Using conventional hybridoma methods, one hybridoma is routinely produced from 10^5 splenocytes [186] and therefore not all B lymphocytes are screened for antigen specificity. However, the single-cell microarray system developed in this study could screen directly all the B-cell population to detect antigen-specific single B cells and analyze the antigen-specific antibody DNA at a single-cell level. Our chip system is simple and easy in its operation, and can perform high-throughput single-cell analysis. This explorative study throws more light on a novel high-throughput single-cell analysis system that has tremendous potential to analyze antibody DNA of an antigen-specific single B cell for developing highly specific monoclonal antibodies as antibody medicines. It might also be applicable for detection of antigen-specific T cells, which could lead to immune therapy or gene therapy in the future.

4.3.3 Compartmental Microfluidic Chip Systems for Single-Cell Analysis

Most of the current clinical diagnosis is based on observation and analysis of a large number of cells, which provides averaged or integrated information about the disease state. Many diseases, however, start from a small number of mutated or unhealthy cells while most other cells remain normal. In order to detect these precursors as early as possible, a high-throughput single-cell-based detection method is required for the analysis of large numbers of cells from blood samples. Sorting and analysis of individual cells from bulk suspensions plays an important role in the prognosis and early diagnosis of diseases, thus facilitating timely and correct treatment. Although conventional fluorescence-activated cell sorters (FACS) are widely used for their efficient cell sorting, these devices have certain drawbacks [196]. First, the devices are rather expensive (US \$250 000), mechanically complex, and require trained personnel for operation and maintenance. Flow cytometry has a great limitation in detecting rare cells which are present in 0.1% of a total population, due to cell signals being buried in the background noise. Further, we cannot compare the states of each cell before and after stimulation by using flow cytometry, and in addition it is not feasible for assay purposes. Microfluidics has been used to perform a variety of biological assays, with advantages that include rapid efficient reaction, minimal reagent consumption, short reaction time, and less waste generation etc. The use of microfluidics as a versatile and powerful research tool is becoming a ubiquitous trend within increasingly diverse technological disciplines. Because of their size, microfluidic devices are able to exploit unique transport properties and provide the capability for significant parallelization and high throughput. Microfluidic approaches have been developed to allow more precise control of cell positioning and reagent introduction in analyzing single cells [197–200]. Implementations of these methods for large-scale single-cell analysis are yet to be realized. A complementary technique to form a high-throughput regular array of single cells and their analysis was reported by our own group recently [188, 190]. To perform high-throughput assays in a multiplexed manner, we have constructed a novel and inexpensive microfluidic device coupled with two-phase liquid systems, enabling high-throughput single-cell sorting and analysis from bulk lymphocyte suspensions in the picoliter aqueous compartments before and after stimulation with an antigen of choice, as shown in Fig. 22 [190, 201, 202]. The detection of a positive single B cell is based on the increased intrinsic Ca^{2+} levels after stimulation with antigen.

In this research, we have developed a novel microfluidic device where individual cells are encapsulated in picoliter aqueous compartments separated by nonaqueous (oil) compartments in a fast-flowing microchannel. We have characterized these devices for high-throughput compartmentalization, and single-cell sorting, screening, and analysis from bulk cell suspensions.

The PDMS microchannel system ($50 \times 50\ \mu m^2$ wide and deep) shown in Fig. 23A was constructed by casting PDMS (Sylgard 184 silicon elastomer, Dow Corning) over an SU-8 50 photoresist (Microchem) mold on a silicon wafer by soft photolithography as previously described [190]. These devices were evaluated for establishment of compartmentalization of two-phase liquids (eosin liquid, mineral oil) under microsyringe pump pressure-driven force. The effect of the flow rates of liquids, oil phase viscosity, and sample channel geometry (20×50, 30×50, $50 \times 50\ \mu m^2$) on the size of aqueous compartments and cell sorting was investigated.

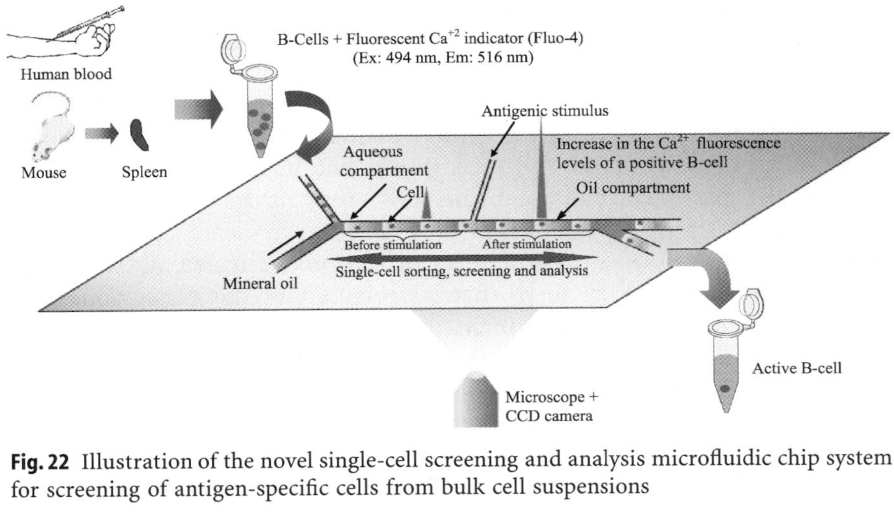

Fig. 22 Illustration of the novel single-cell screening and analysis microfluidic chip system for screening of antigen-specific cells from bulk cell suspensions

System Setup and Compartmentalization of Two-Phase Liquids in the Microchannels

A microfluidic device was mounted on a fluorescence stereomicroscope (Leica MZ FLIII, Switzerland) coupled with a Fluo III filter system and APO 1.0× objective lens. Lighting was provided by Intralux 5000-1A (Volpi, Switzerland). A 100-W mercury 106Z lamp (Leica, Switzerland) provided epifluorescence excitation and the fluorescence image was collected by a GFP 3 filter. The real-time imaging of compartments of two-phase liquids was captured by a color CCD camera (30 fps) (Toshiba Teli, Tokyo). Aqueous and nonaqueous mineral oil solutions were pumped using 100-μL Hamilton gas-tight syringes (1710 series, TLL) and 1.0-mL disposable syringes (Terumo, Japan). The syringes were attached to microfluidic devices by means of Hamilton Teflon needles (22 gauge, standard hub). Microsyringe pumps from EiCom (ESP-64, Japan) and KD Scientific (USA) were used to introduce the aqueous liquid and nonaqueous mineral oil solutions.

Figure 23 shows microfabricated PDMS–glass microfluidic chip devices with a Y-shaped microchannel (200 mm length, $50 \times 50\,\mu m^2$ width and depth). The visualization of compartmentalization of the aqueous Neutral Red or eosin (Wako, Japan) liquids and the immiscible phase (mineral oil, Sigma) was achieved when they were introduced under microsyringe pump pressure-driven force from respective inlet channels; flows occurred all along the microchannel and intersected with each other at the Y-junction. Continuous and alternate compartments of aqueous and oil phases were generated in the main microchannel and transported throughout the 200-mm-long microchannel in a stable manner in all the 50×50, 30×50, and $20 \times 50\,\mu m^2$ areas at flow rates of $0.1{-}1.0\,\mu L/min$, due to the viscosity and interfacial tension of the two-phase liquids (Fig. 23C). High-throughput picoliter compartmentalization of both the liquids in the microfluidic device is shown in Fig. 23B and D; they were transported throughout the microchannel with stability to the outlet in the $50 \times 50\,\mu m^2$ chip devices tested. Characterization of compartmental microfluidics revealed that this system generates compartments at a very high rate (30 comp./s) in $30 \times 50\,\mu m^2$ chip devices, with a size of $300{-}150$ pL based on the flow ratio of the two-phase liquids. It was also noticed that increasing flow rates and flow ratios of the two-phase liquids favored reduction in the size of the aqueous compartments in $50 \times 50\,\mu m^2$ devices (data not shown). Measurement of randomly selected aqueous compartments throughout the microchannel reveals that the liquid volume is equally distributed around the microchannel at different flow ratios of liquids, suggesting the stability of these devices (data not shown). Further manipulation of the flow ratios of the two-phase liquids to give different ratios ranging from 0.001 to $0.1\,\mu L/min$ between the aqueous and oil phase resulted in the smallest size compartment containing 70-pL volumes (data not shown).

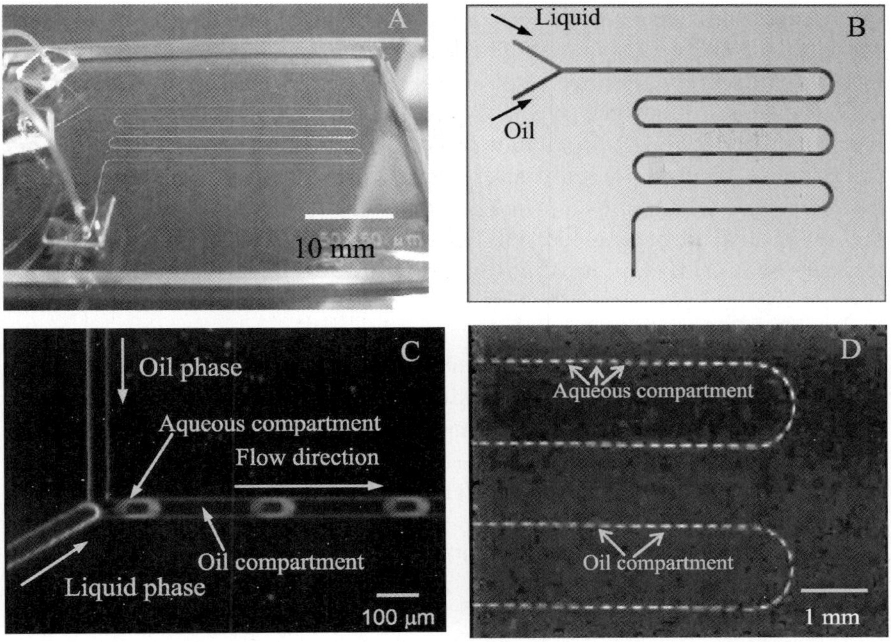

Fig. 23 **A** Photograph of a microfluidic device. **B** Illustration showing the compartmentalization of two-phase liquids in a microchannel. **C** Bright-field microscopic image displaying the formation of compartments of mineral oil and aqueous phase. **D** Microphotograph showing the high-throughput picoliter compartmentalization of eosin liquid and mineral oil in the microchannels ($50\times50\,\mu m^2$)

Multiplexed Single B Cell Sorting and Detection

Before proceeding with cell sorting, PDMS–glass microchannels have to be modified with PC modifier polymer (Ai. Biochips, Japan) to prevent nonspecific adhesion of cells to the microchannel surfaces, which otherwise disturbs the compartmental phenomenon. In this study, 20×50 and $30\times50\,\mu m^2$ chip devices were employed. Mouse B-cell suspensions (2×10^5 cells/mL) were loaded with 0.5-µM Cell Tracker Orange CMTMR (Molecular Probes) (excitation 541 nm, emission 565 nm) and incubated for 30 min at RT before introduction into the chip device. Cell suspension and mineral oils served as the two-phase liquids introduced into the chip device (0.1 µL/min each), and observation of the entrapping of compartments with cells was recorded by real-time imaging using an Olympus IX70 fluorescence microscope coupled with an Axiocam HSm CCD camera (60 fps). The percentage of single-cell sorting was quantified to determine sample channel width effects. When mouse spleen derived B lymphocytes loaded with Cell Tracker Orange and mineral oil were introduced as aqueous and immiscible phases into the chip device, over 42% of the total 47% aqueous compartments achieved single-cell status (Fig. 24B and D) in devices with 20-μm sample channel width, while in a 30-μm sample channel width, 32% of the total 40% B-cell sorting was recorded (Fig. 24B and D) at a flow rate of 0.1 μL/min. The fluorescence intensity from a single B cell was measured using Image J software to detect the positive B cells before and after stimulation (Fig. 24C). It is thought that by decreasing the sample channel height, there is a possibility of increasing the encapsulation of single B cells in aqueous compartments. Furthermore, we have built cell assay chip devices where analysis of the same B cell before and after stimulation was achieved, and identified the B cells that are specific to mouse anti-IgM stimulation.

The newly developed compartmental microfluidic system demonstrated generation of high-throughput picoliter volumes of compartments of two-phase liquids without using any complicated chip geometry and only using a pressure-driven microsyringe. The size of the compartments can be controlled by maintaining the stability. This system allows multiplexed analyses in a single microchannel without any cross-contamination between aqueous compartments. Further, we demonstrated that this system can sort single

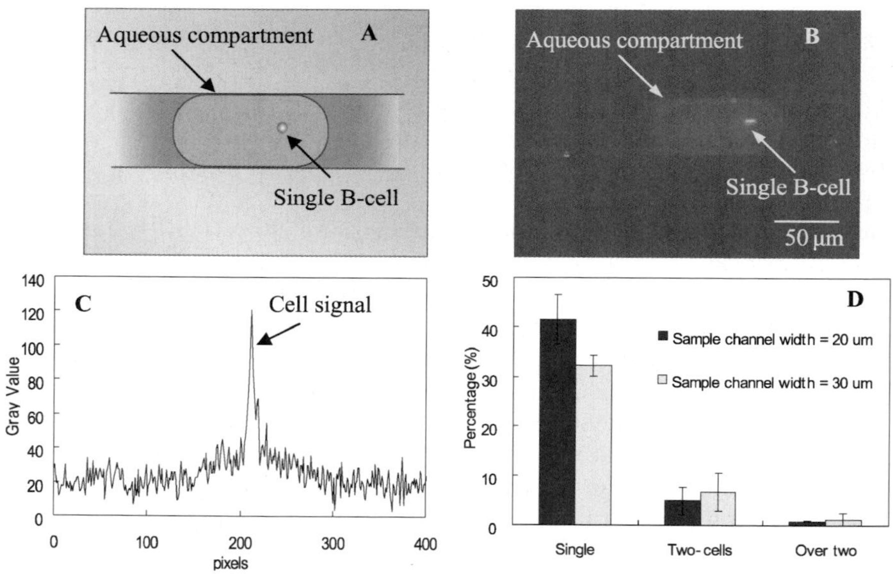

Fig. 24 **A** Illustration depicting the encasing of a single B cell in the aqueous compartment. **B** Fluorescence microscopic image showing the entrapment of a single B cell in the aqueous compartment. **C** Measurement of the signal from a single B cell encapsulated in the aqueous compartment. **D** Percentage of B-cell sorting in devices with 20×50 and 30×50 μm^2 wide and deep channels at 0.1 μL/min flow rates of both the two-phase liquids. Scale bar indicates 50 μm

cells in a high-throughput manner from bulk suspensions, thereby allowing screening and analysis of single cells as a tool in disease diagnosis applications. In addition, it also serves as a novel bioanalytical tool to perform rapid and sensitive assays to screen novel drugs at picoliter volumes on a single chip platform.

5 Conclusions

The pico- and nanoliter microsystems developed in this research showed a great potential to detect and analyze DNA, proteins, and cells at the single-molecule or single-cell level.

In the first part, we demonstrated use of the microchamber (40 nL) array PCR chip to amplify and detect quantitatively multiple DNA targets in combination with a nanoliter dispenser. This demonstrated the possibility of using the microarray PCR chip to amplify and detect ~1200 target DNAs simultaneously; in addition, target gene analysis at the single-copy level was achieved. Thus, the microarray PCR chip system has significant potential to be implemented for a wide range of purposes, such as detection of pathogenic microorganisms and clinical diagnosis. We have also developed continuous-flow microfluidic chip devices for real-time PCR-based detection, for detecting a sample with as little as 60 copies of genetically modified DNA, which shows its potential for the rapid and sensitive detection of genetically modified food. Further, we also introduced label-free PAA chips as optical biosensors for measuring DNA–DNA hybridization based on LSPR and interferometry principles.

The highly integrated protein microarray chip presented here successfully performed in vitro protein synthesis using a picoliter chamber array. It was possible to perform screening and analysis for multiplexed gene expression on single beads in each chamber. In future, this system will be applied to high-throughput screening and functional analysis of proteins expressed from unknown genes or cells. We have also developed a LSPR-based nanochip for multiple antigen–antibody interactions. We further employed this platform for quantitative detection of metabolites from cells for a specific stimulus, and it served as a hand-held device for measuring biomolecular interactions.

In the cell chip part of the chapter, we introduced our neuron chips for selective guided growth of neuronal cells on different functionalized chip substrates, which shed light on the interactions between neuronal cells; furthermore, we measured the response of neurons upon electric potentiation. We also employed our nanoliter microchamber array chips for high-throughput screening of chemicals against groups of HeLa 229 cells at nanoliter volumes based on cell proliferation. Further, we screened neurite outgrowth peptide from millions of peptides synthesized by combinatorial chemistry. This methodology we have developed is a simple and direct way of screening small molecules with a cell-based assay.

We also introduced our single-cell microarray and microfluidic devices for screening and analysis of rare cells to a specific stimulus. The single-cell microarray system that can accommodate single cells in picoliter chambers made it possible to perform high-throughput screening and analysis of antigen-specific single B cells and analyze multiplexed single-cell status before and after antigenic stimulation on a single platform. Compartmental microfluidics can encase single cells in each aqueous compartment, and a rapid and highly sensitive screening assay can be achieved due to vortex mixing at the picoliter volume level, which avoids integration of mixers in the chip device. Further, both the microarray and microfluidic systems facilitated retrieval of positive antigen-specific single B cells from the microarray and channel, thus demonstrating their novelty and ease of use over conventional methods for making monoclonal antibodies. These single-cell analysis systems have tremendous potential to analyze antibody DNA of an antigen-specific single B cell for developing highly specific monoclonal antibodies as antibody medicines. They might also be applicable to cell therapy for personal diagnosis and development of personalized medicine in the future.

Thus, the pico- and nanoliter chamber array microsystems developed in this study could become potential tools for high-throughput analysis of DNA, proteins, and cells at the single-molecule or single-cell level, which could lead to efficient clinical diagnosis and drug discovery.

Acknowledgements The research of in vitro protein synthesis on a picoliter chamber array was supported by grants from the New Energy and Industrial Technology Development Organization (NEDO) and the Japan Society for the Promotion of Science (JSPS). The research on porous anodic alumina chips for biomolecular interactions was supported by a Grant-in-Aid for Scientific Research on Priority Areas "Lifesurveyor" from the Monbukagausho, Japan. The work on the single-cell microarray chip and research on the compartmental microfluidic chip was carried out under the Toyama Medical Biocluster project, which was sponsored by the Ministry of Education, Culture, Sports, and Science, Japan.

References

1. Dittrich S, Rachikawa K, Manz A (2006) Anal Chem 78:3887
2. Yi C, Li C-W, Ji S, Yang M (2006) Anal Chim Acta 560:1
3. Oosterbroek E, Van den Berg A (2003) Lab-on-a-chip: miniaturized systems for biochemical analysis and synthesis. Elsevier, Amsterdam
4. Figeys D, Pinto D (2000) Anal Chem 72:330
5. Heller MJ (2002) Annu Rev Biomed Eng 4:129
6. Eisen MB, Brown PO (1999) Methods Enzymol 303:179
7. Panicker RC, Huang X, Yao SQ (2004) Comb Chem High Throughput Screen 7:547
8. Figeys D (2002) Proteomics 2:373
9. Smith JC, Lambert J-P, Elisma F, Figeys D (2007) Anal Chem 79:4325

10. Craighead H (2006) Nature 442:387
11. El-Ali J, Sorger PK, Jensen KF (2006) Nature 442:403
12. Cottingham K (2004) Anal Chem 76:235
13. Di Carlo D, Lee L (2006) Anal Chem 78:7918
14. Price AK, Culbertson CT (2007) Anal Chem 79:2614
15. Anderson DG, Levenberg S, Langer R (2004) Nat Biotechnol 22:863
16. Ziauddin J, Sabatini DM (2001) Nature 411:107
17. Mullis KB, Faloona FA (1987) Methods Enzymol 155:335
18. Irino T, Kitoh T, Koami K, Kashima T, Mukai K, Takeuchi E, Hongo T, Nakahata T, Schuster SM, Osaka M (2004) J Mol Diagn 6:217
19. Hirt C, Schuler F, Dolken L, Schmidt CA, Dolken G (2004) Blood 104:904
20. Nurmi J, Wikman T, Karp M, Lovgren T (2002) Anal Chem 74:3525
21. Yang J, Liu Y, Rauch CB, Stevens RL, Liu RH, Lenigk R, Grodzinski P (2002) Lab Chip 2:179
22. Lee TMH, Carles MC, Hsing IM (2003) Lab Chip 3:100
23. Schneegass I, Brautigam R, Kohler M (2001) Lab Chip 1:42
24. Fukuba T, Yamamoto T, Naganuma T, Fujii T (2004) Chem Eng J 101:151
25. Lee DS, Park SH, Yang H, Chung KH, Yoon TH, Kim SJ, Kim K, Kim YT (2004) Lab Chip 4:401
26. Lagally ET, Emrich CA, Mathies R (2001) Lab Chip 1:102
27. Lagally ET, Medintz I, Mathies RA (2001) Anal Chem 73:565
28. Nagai H, Murakami Y, Yokoyama K, Tamiya E (2001) Biosens Bioelectron 16:1015
29. Nagai H, Murakami Y, Morita Y, Yokoyama K, Tamiya E (2001) Anal Chem 73:1043
30. Matsubara Y, Kobayashi M, Morita Y, Tamiya E (2002) Arch Histol Cytol 65:481
31. Leamon JH, Lee WL, Tartaro KR, Lanza JR, Sarkis GJ, deWinter AD, Berka J, Lohman KL (2003) Electrophoresis 24:3769
32. Matsubara Y, Kerman K, Kobayashi M, Yamamura S, Morita Y, Takamura Y, Tamiya E (2004) Anal Chem 76:6434
33. Matsubara Y, Kerman K, Kobayashi M, Yamamura S, Morita Y, Tamiya E (2005) Biosens Bioelectron 20:1482
34. Felbel J, Bieber I, Pipper J, Kohler JM (2004) Chem Eng J 101:333
35. Shoffner MA, Cheng J, Hvichia GE, Kricka LJ, Wilding P (1996) Nucleic Acids Res 24:375
36. Erill I, Campoy S, Erill N, Barbe J, Aguilo J (2003) Sens Actuators B 96:685
37. Zimmermann B, Holzgreve W, Wenzel F, Hahn S (2002) Clin Chem 48:362
38. Kariyazono H, Ohno T, Ihara K, Igarashi H, Joh-o K, Ishikawa S, Hara T (2001) Mol Cell Probes 15:71
39. Chiu RW, Lau TK, Leung TN, Chow KC, Chui DH, Lo YM (2002) Lancet 360:998
40. Breadmore MC, Wolfe KA, Arcibal IG, Leung WK, Dickson D, Giordano BC, Power ME, Ferrance JP, Feldman SH, Norris PM, Landers JP (2003) Anal Chem 75:1880
41. Holden MA, Kumar S, Castellana ET, Beskok A, Cremer PS (2003) Sens Actuators B 92:199
42. He B, Burke BJ, Zhang X, Zhang R, Regnier FE (2001) Anal Chem 73:1942
43. Lin Y, Gerfen GJ, Rousseau DL, Yeh S-R (2003) Anal Chem 75:5381
44. Lim T-K, Ohta H, Matsunaga T (2003) Anal Chem 75:3316
45. Nakamura H, Karube I (2003) Anal Bioanal Chem 377:446
46. Vilkner T, Janasek D, Manz A (2004) Anal Chem 76:3373
47. Yu C, Mutlu S, Selvaganapathy P, Mastrangelo CH, Svec F, Frechet JMJ (2003) Anal Chem 75:1958
48. Melin J, Roxhed N, Gimenez G, Griss P, van der Wijngaart W, Stemme G (2004) Sens Actuators B 100:463

49. Takamura Y, Onoda H, Inokuchi H, Adachi S, Oki A, Horiike Y (2003) Electrophoresis 24:185
50. Lim K, Kim S, Hahn JH (2003) Sens Actuators B 92:208
51. Huh D, Tkaczk AH, Bahng JH, Chang Y, Wei H-H, Grotberg JB, Kim C-J, Kurabayashi K, Takayama S (2003) J Am Chem Soc 125:14678
52. Valasek MA, Repa JJ (2005) Adv Physiol Educ 29:151
53. Kopp MU, de Mello AJ, Manz A (1998) Science 280:1046
54. Obeid PJ, Christopoulos TK, Crabtree HJ, Backhouse CJ (2003) Anal Chem 75:288
55. Obeid PT, Christopoulos TK (2003) Anal Chim Acta 494:1
56. Zhang Q, Wang W, Zhang H, Wang Y (2002) Sens Actuators B 82:75
57. Sun K, Yamaguchi A, Ishida Y, Matsuo S, Misawa H (2002) Sens Actuators B 84:283
58. Shindo Y, Kuribara H, Matsuoka T, Futo S, Sawada C, Shono J, Akiyama H, Doda Y, Toyoda M, Hino A (2002) J AOAC Int 85:1119
59. Duffy DC, McDonald JC, Schueller OJA, Whitesides GM (1998) Anal Chem 70:4974
60. Kuribara H, Shindo Y, Matsuoka T, Takubo K, Futo S, Aoki N, Hirao T, Akiyama H, Doda Y, Toyoda M, Hino A (2002) J AOAC Int 85:1077
61. Pattnaik P (2005) Appl Biochem Biotechnol 26:79
62. Lahiri J, Isaacs L, Tien J, Whitesides GM (1999) Anal Chem 71:777
63. Nakamura R, Muguruma H, Ikebukuro K, Sasaki S, Nagata R, Karube I, Pedersen H (1997) Anal Chem 69:4649
64. Kai E, Sawata S, Ikebukuro K, Iida T, Honda T, Karube I (1999) Anal Chem 71:796
65. Jordan CE, Frutos AG, Thiel AG, Corn RM (1997) Anal Chem 69:4939
66. Goodrich TT, Lee HJ, Corn RM (2004) J Am Chem Soc 126:4086
67. Wark AW, Lee HJ, Corn RM (2005) Anal Chem 77:3904
68. Woodruff SD, Yeung ES (1982) Anal Chem 54:1174
69. Wang Z, Bornhop DJ (2005) Anal Chem 77:7872
70. Markov DA, Swinney K, Bornhop DJ (2004) J Am Chem Soc 126:16659
71. Dancil K-PS, Greiner DP, Sailor MJ (1999) J Am Chem Soc 121:7925
72. Lin VS-Y, Motesharei K, Dancil K-PS, Sailor MJ, Ghadiri MR (1997) Science 278:840
73. Doan VV, Sailor MJ (1992) Science 256:1791
74. Boussaad S, Pean J, Tao NJ (2000) Anal Chem 72:222
75. Sota H, Hasegawa Y, Iwakura M (1998) Anal Chem 70:2019
76. Lu J, Strohsahl CM, Miller BL, Rothberg LJ (2004) Anal Chem 76:4416
77. Pan S, Rothberg LJ (2003) Nano Lett 3:811
78. Eustis S, el-Sayed MA (2006) Chem Soc Rev 35:209
79. McFarland AD, Van Duyne RP (2003) Nano Lett 3:1057
80. Mie G (1908) Ann Phys 25:377
81. Barnes WL, Dereux A, Ebbesen TW (2003) Nature 424:824
82. Kreibig U, Vollmer M (1995) Optical properties of metal clusters. Springer, Heidelberg
83. Prasad NP (2004) Nanophotonics. Wiley, Hoboken, p 129
84. Stuart DA, Haes AJ, Yonzon CR, Hicks EM, Van Duyne RP (2005) IEE Proc Nanobiotechnol 152:13
85. Freeman RG, Grabar KC, Allison KJ, Bright RM, Davis JA, Guthrie AP, Hommer MB, Jackson MA, Smith PC, Walter DG, Natan MJ (1995) Science 267:1629
86. Nath N, Chilkoti A (2004) Anal Chem 76:5370
87. Nath N, Chilkoti A (2002) Anal Chem 74:504
88. Mock JJ, Smith DR, Schultz S (2003) Nano Lett 3:485
89. Raschke G, Kowarik S, Franzl T, Sönnichsen C, Klar TA, Feldmann J, Nichtl A, Kürzinger K (2003) Nano Lett 3:935

90. Endo T, Kerman K, Nagatani N, Takamura Y, Tamiya E (2005) Anal Chem 77:6976
91. Endo T, Yamamura S, Nagatani N, Morita Y, Takamura Y, Tamiya E (2005) Sci Technol Adv Mater 6:491
92. Endo T, Kerman K, Nagatani N, Hiep HM, Kim D-K, Yonezawa Y, Nakano K, Tamiya E (2006) Anal Chem 78:6465
93. Mock JJ, Smith DR, Schultz S (2003) Nano Lett 3:485
94. Masuda H, Satoh M (1996) Jpn J Appl Phys 35:126
95. Corder EH, Saunders AM, Strittmatter WJ, Schmechel DE, Gaskell PC, Small GW, Roses AD, Haines JL, Pericak-Vance MA (1993) Science 261:921
96. Rudert F (2000) Curr Opin Mol Ther 2:633
97. Srinivas PR, Srivastava S, Hanash S, Wright GL Jr (2201) Clin Chem 47:1901
98. Figeys D, Pinto D (2001) Electrophoresis 22:208
99. Reid G, Gan BS, She YM, Ens W, Weinberger S, Howard JC (2002) Appl Environ Microbiol 68:977
100. Tabuchi M, Baba Y (2002) Electrophoresis 23:1138
101. Venter JC et al Science 291:1304
102. Pavlov MY, Freistroffer DV, Ehrenberg M (1997) Biochimie 79:415
103. http://www.roche-applied-science.com/sis/proteinexpression/
104. Zhi ZL, Murakami Y, Morita Y, Hasan Q, Tamiya E (2003) Anal Biochem 318:236
105. Zhi ZL, Morita Y, Hasan Q, Tamiya E (2003) Anal Chem 75:4125
106. Kukar T, Eckenrode S, Gu Y, Lian W, Megginson M, She JX, Wu D (2002) Anal Biochem 306:50
107. Kojima K, Hiratsuka A, Suzuki H, Yano K, Ikebukuro K, Karube I (2003) Anal Chem 75:1116
108. Tabuchi M, Hino M, Shinohara Y, Baba Y (2002) Proteomics 2:430
109. Rungpragayphan S, Kawarasaki Y, Imaeda T, Kohda K, Nakano H, Yamane T (2002) J Mol Biol 318:395
110. Nakano H, Okumura R, Goto C, Yamane T (2002) Biotechnol Bioprocess Eng 7:311
111. McDonald JC, Whitesides GM (2002) Acc Chem Res 35:491
112. Ng JM, Gitlin I, Stroock AD, Whitesides GM (2002) Electrophoresis 23:3461
113. Jo BH, Van Lerberghe LM, Motsegood KM, Beebe DJ (2000) J Microelectromech Syst 9:76
114. Spirin AS, Baranov VI, Ryabova LA, Ovodov SY, Alakhov YB (1988) Science 242:1162
115. Brenner S, Williams SR, Vermaas EH, Storck T, Moon K, McCollum C, Mao JI, Luo S, Kirchner JJ, Eletr S, DuBridge RB, Burcham T, Albrecht G (2000) Proc Natl Acad Sci USA 97:1665
116. Drewes G, Bouwmeester T (2003) Curr Opin Cell Biol 15:199
117. Petach H, Gold L (2002) Curr Opin Biotechnol 13:309
118. Talapatra A, Rouse R, Hardiman G (2002) Pharmacogenomics 3:527
119. Workman J Jr, Koch M, Veltkamp D (2005) Anal Chem 77:3789
120. Gardeniers H, Van Den Berg A (2004) Int J Environ Anal Chem 84:809
121. Murakami Y, Endo T, Yamamura S, Nagatani N, Takamura Y, Tamiya E (2004) Anal Biochem 334:111
122. Shaikh KA, Ryu KS, Goluch ED, Nam JM, Liu J, Thaxton CS, Chiesl TN, Barron AE, Lu Y, Mirkin CA, Liu C (2005) Proc Natl Acad Sci USA 102:9745
123. Haes AJ, Chang L, Klein WL, Van Duyne RP (2005) J Am Chem Soc 127:2264
124. Himmelhaus M, Takei H (2000) Sens Actuators B 63:24
125. Bliss TVP, Collinbridge GL (1993) Nature 361:31
126. Ehlers MD, Mammen AL, Lau L, Hunganir L (1996) Curr Opin Cell Biol 8:484
127. Petrlia R, Yokotani N, Wenthold R (1994) J Neurosci 14:667

128. Tamiya E, Nagatani N, Iwabuchi S, Murakami Y, Sakaguchi T, Yokoyama K (1997) Anal Chem 69:3697
129. Iwabuchi S, Muramatsu H, Yamamoto N, Murakami Y, Yokoyama Y, Tamiya E (1996) Opt Rev 3:470
130. Degenaar P, Murakami Y, Yokoyama K, Tamiya E (2000) Proc NFO 6:271
131. Chiu DT, Jeon NL, Huang S, Kane RS, Wargo CL, Cho IS, Ingber DE, Whitesides GM (2000) Proc Natl Acad Sci USA 97:2408
132. Matsuda T, Sugawara T, Inoue K (1992) ASAIO J 243
133. Mrksich M, Dike EL, Tien J, Ingber DE, Whitesides GM (1997) Exp Cell Res 235:305
134. Matinoia S, Bove M, Tedesco M, Margesin B, Grattarola M (1999) J Neurosci Methods 87:35
135. Degenaar P, Griscom L, Le Pfioufle B, Murakami Y, Yokoyama K, Fujita H, Tamiya E (2001) Proc Biochips 2001
136. Givangrandi L (1999) PhD thesis, École polytechnique Fédérale de Lausanne. Lausanne, Switzerland
137. Shirakawa T, Honma S, Katsuno Y, Oguchi H, Honma K (2000) Eur J Neurosci 12:2833
138. Degenaar P, Le Pioufle B, Griscom L, Tixier A, Akagi Y, Morita Y, Murakami Y, Yokoyama K, Fujita H, Tamiya E (2001) J Biochem 130:367
139. Morin F, Nishimura N, Griscom L, Le Pioufle B, Fujita H, Takamura Y, Tamiya E (2006) Biosens Bioelectron 21:1093
140. Wolcke J, Ullmann D (2001) Drug Discov Today 6:637
141. Hertzberg RP, Pope AJ (2000) Curr Opin Biotechnol 4:445
142. Sundberg SA (2000) Curr Opin Biotechnol 11:47
143. Houseman BT, Huh JH, Kron SJ, Mrksich M (2002) Nat Biotechnol 20:270
144. Lavery P, Brown MJB, Pope AJ (2001) J Biomol Screen 6:3
145. Mere L, Bennett T, Coassin P, England P, Hamman B, Rink T, Zimmerman S, Negulescu P (1999) Drug Discov Today 4:363
146. Kariv I, Cao H, Marvil PD, Bobkova EV, Bukhitiyarov YE, Yan YP, Patel U, Coudurier L, Chung TDY, Oldenburg KR (2001) J Biomol Screen 6:233
147. Turconi S, Shea K, Ashman S, Fantom K, Earnshaw DL, Bingham RP, Haupts UM, Brown MJB, Pope AJ (2001) J Biomol Screen 6:275
148. Sittampalam GS, Kahn SD, Janzen WP (1997) Curr Opin Chem Biol 1:384
149. Bhandariraju K, Chen CS (2002) Drug Discov Today 7:612
150. Giese K, Kaufmann J, Pronk GJ, Klippel A (2002) Drug Discov Today 7:179
151. Gonzalez-Nicolini V, Fux C, Fussenegger M (2004) Invest New Drugs 22:253
152. Tucker CL (2002) Drug Discov Today 7:S125
153. Zheng W, Brandish PE, Kolodin DG, Scolnick EM, Strulovici B (2004) J Biomol Screen 9:132
154. Lindholm P, Gullbo J, Claeson P, Goransson U, Johansson S, Backlund A, Larsson R, Bohlin L (2002) J Biomol Screen 7:333
155. Akagi Y, Sathuluri RR, Morita Y, Tamiya E (2004) Sci Technol Adv Mater 5:343
156. Oldenburg KR, Zhang JH, Chen T, Maffia A III, Blom KF, Combs AP, Chung TDY (1998) J Biomol Screen 1:55
157. Anonymous (2003) Cell counting kit-F. Technical manual of DOJINDO, p 260
158. Greene LA, Tischeler AS (1976) Proc Natl Acad Sci USA 73:2424
159. Lad SP, Neet KE, Mufson EJ (2003) Curr Drug Targets CNS Neurol Disord 2:315
160. Salehi A, Delcroix JD, Swabb DF (2004) J Neural Transm 111:323
161. Stevens T, Livingston G, Kitchen G, Manela M, Walker Z, Katona C (2002) Br J Psychiatry 180:270

162. Poduslo JF, Curran GL (1996) Brain Res Mol Brain Res 36:280
163. Friden PM, Walus LR, Watson P, Doctrow SR, Kozarich JW, Backman C, Bergman H, Hoffer B, Bloom F, Granholm AC (1993) Science 259:373
164. Falciani C, Lozzi L, Pini A, Bracci L (2005) Chem Biol 12:417
165. Blondelle SE, Perez-Paya E, Houghten RA (1996) Antimicrob Agents Chemother 40:1067
166. Dooley CT, Ny P, Bidlack JM, Houghten RA (1998) J Biol Chem 273:18848
167. Hernandez J, Schoeder K, Blondelle SE, Pons FG, Lone YC, Simora A, Linglade-Demoyen P, Wilson DB, Zanette M (2004) Eur J Immunol 34:2331
168. Oyama T, Sykes KF, Samli KN, Minna JD, Johnston SA, Brown KC (2003) Cancer Lett 202:219
169. Takahashi S, Mok H, Parrott B, Marini FC III, Andreeff M, Brenner MK, Barry MA (2003) Cancer Res 63:5213
170. Bae YS, Park EY, Kim Y, He R, Ye RD, Kwak JY, Suh PG, Ryu SH (2003) Biochem Pharmacol 66:1841
171. Kuruvilla HW, Shamji AF, Sternson SM, Hergenrother PJ, Schreiber SL (2002) Nature 416:653
172. Kai J, Nakamura K, Masuda T, Ueda I, Fujiwara H (1996) J Med Chem 29:2621
173. Pardridge WM (2005) NeuroRx 2:3
174. Pardridge WM (2003) Mol Interv 3:90
175. Gardiner RM (1990) J Inherit Metab Dis 13:627
176. Oldendorf WH (1974) Proc Soc Exp Bio Med 147:813
177. Banks WA, Kastin AJ (1985) Brain Res Bull 15:287
178. Macbeath G, Koehler AN, Schreiber SL (1999) J Am Chem Soc 121:7967
179. Liu R, Marik J, Lam KS (2002) J Am Chem Soc 124:767
180. Chen GYJ, Uttamchandani M, Ziu Q, Wang G, Yao SQ (2003) ChemBioChem 4:336
181. Kodadek T, Bachhawat-Sikder K (2006) Mol Biosyst 2:25
182. Ronn LCB, Olsen M, Ostergaard S, Kiselyov V, Berezin V, Mortensen MT, Lerche MH, Jensen PH, Soroka V, Saffells JL, Doherty P, Poulsen FM, Bock E, Holm A (1999) Nat Biotechnol 17:1000
183. Ronn LCB, Olsen M, Soroka V, Ostergaard S, Dissing S, Poulsen FM, Holm A, Berezin V, Bock E (2002) Eur J Neurosci 16:1720
184. Luthi A, Laurent JP, Figurov A, Luller D, Schachner M (1994) Nature 372:777
185. Lam KS, Lebl M, Krchnak V (1997) Chem Rev 97:411
186. Shokrgozar MA, Shokri F (2001) Immunology 101:75
187. Thorsen T, Maerkl SJ, Quake SR (2002) Science 298:580
188. Yamamura S, Sathuluri RR, Omori M, Tokimitsu Y, Kondo S, Kishi H, Muraguchi A, Takamura Y, Tamiya E (2004) Proc Micro Total Anal Systems, vol 1. Royal Society of Chemistry, Cambridge, p 78
189. Sathuluri RR, Yamamura S, Takamura Y, Tamiya E (2004) Proc Micro Total Anal Systems, vol 1. Royal Society of Chemistry, Cambridge, p 61
190. Yamamura S, Kishi H, Tokimitsu Y, Kondo S, Honda R, Sathuluri RR, Omori M, Tamiya E, Muraguchi A (2005) Anal Chem 77:8050
191. Haugland RP (2002) In: Handbook of fluorescent probes and research products, 9th edn. Molecular Probes, Eugene
192. Nishida M, Sugimoto K, Hara Y, Mori E, Mori T, Kurosaki T, Mori Y (2003) EMBO J 22:4677
193. Ueki Y, Goldfarb IS, Harindranath N, Gore M, Koprowski H, Notkins AL, Casali P (1990) J Exp Med 171:19

194. Jingwu Z, Henderikx P, Ying C, Medaer R, Raus JCM (1989) J Immunol Methods 123:153
195. Dessain SK, Adekar SP, Stevens JB, Carpenter KA, Skorski ML, Barnoski BL, Goldsby RA, Weinberg RA (2004) J Immunol Methods 291:109
196. Martin JC, Swartzendruber DE (1980) Science 207:199
197. Wheeler AR, Thronsdset WR, Whelan RJ, Leach AM, Zare RN, Liaor YH, Farrell K, Manger ID, Daridon A (2003) Anal Chem 75:3581
198. Wu HK, Wheeler A, Zare RN (2004) Proc Natl Acad Sci USA 101:12809
199. Peng XY, Li PCH (2004) Anal Chem 76:5273
200. Yang MS, Li CW, Yang J (2002) Anal Chem 74:3991
201. Sathuluri RR, Kitamura M, Yamamura S, Tamiya E (2006) Proc Micro Total Anal Systems, vol 1. Royal Society of Chemistry, Cambridge, p 957
202. Sathuluri RR, Kitamura M, Yamamura S, Tamiya E (2006) Proc IEEE Sensors, p 642
203. Yamamura S, Sathuluri RR, Tamiya E (2007) Pico/nano-liter chamber array chips for single-cell, DNA and protein analyses. In: Kumar C (ed) Nanomaterials for Biosensors, vol 8. Wiley-VCH, Weinheim, p 368
204. Nakayama T, Kurosawa Y, Furui S, Kerman K, Kobayashi M, Sathuluri RR, Yonezawa Y, Nakano K, Hino A, Yamamura S, Takamura Y, Tamiya E (2006) Anal Bioanal Chem 386:1327
205. Kim DK, Kerman K, Saito M, Sathuluri RR, Endo T, Yamamura S, Kwon YS, Tamiya E (2007) Anal Chem 79:1855
206. Kinapara T, Mizuno R, Murakami Y, Kobayashi M, Yamamura S, Hasan Q, Morita Y, Nakano H, Yamane T, Tamiya E (2004) J Biochem (Tokyo) 136:149
207. Sathuluri RR, Akagi Y, Morita Y, Tamiya E (2002) Proc Micro Total Anal Systems, vol 2. Kluwer Academic Publishers, Netherlands, p 862
208. Akagi Y, Sathuluri RR, Morita Y, Takamura Y, Tamiya E (2004) Proc Micro Total Anal Systems, vol 2. Royal Socitey of Chemistry, Cambridge, p 309

Adv Biochem Engin/Biotechnol (2008) 109: 351–394
DOI 10.1007/10_2007_085

Published online: 15 November 2007

Development of Microbial Sensors and Their Application

Hideaki Nakamura · Mifumi Shimomura-Shimizu · Isao Karube (✉)

School of Bionics, Tokyo University of Technology, 1404-1 Katakura, Hachioji, 192-0982 Tokyo, Japan
karube@bs.teu.ac.jp

Abstract Many types of microbial sensors have been developed as analytical tools since the first microbial sensor was studied by Karube et al. in 1977. The microbial sensor consists of a transducer and microbe as a sensing element. The characteristics of the microbial sensors are a complete contrast to those of enzyme sensors or immunosensors, which are highly specific for the substrates of interest, although the specificity of the microbial sensor has been improved by genetic modification of the microbe used as the sensing element. Microbial sensors have the advantages of tolerance to measuring conditions, a long lifetime, and cost performance, and also have the disadvantage of a long response time. In this review, the long history of microbial sensor development is summarized.

Abbreviations

ATP	Adenosine triphosphate
BL	Bioluminescence
BMP	Bacterial magnetic particle
BOD	Biological (or biochemical) oxygen demand
DCIP	2,6-Dichlorophenolindophenol
DM	Double mediator
DO	Dissolved oxygen
FIA	Flow injection analysis
GGA	Glucose–glutamic acid
HCF	Hexacyanoferrate
HPLC	High-performance liquid chromatography
ISFET	Ion-sensitive field-effect transistor
LAS	Linear alkylbenzene sulfonate
LOD	Lactate oxidase
MFC	Microbial fuel cell
OECD	Organization for Economic Cooperation and Development
OSS	OECD synthetic sewage
PD	Photodiode
PMT	Photomultiplier
PNA	Peptide nucleic acid
RCI	Redox color indicator
RSD	Relative standard deviation
SPR	Surface plasmon resonance
SPV	Surface photovoltage
TAS	Total assimilable sugar
TCE	Trichloroethylene

Microbes

Acetobacter aceti, Achromobacter sp. *Acinetobacter calcoaceticus, Arthrobacter nicotianae, Arxula adeninivorans, Aspergillus niger, Aspergillus ustus, Bacillus licheniformis, Bacillus subtilis, Bacterium cadavers, Botrytis cinerea, Brevibacterium lactofermentum, Candida* sp., *Chlorella* sp., *Citrobacter freundii, Clostridium butyricum, Escherichia coli, Gluconobacter oxydans, Hansenula anomala, Issatchenkia orientalis, Klebsiella oxytoca, Kluyveromyces marxianus, Lactobacillus fermenti, Leuconostoc mesenteroides, Methylomonas flagellata, Moraxella* sp., *Nitrobacter* sp., *Nitrosomonas europaea, Penicillium* sp., *Photobacterium phosphoreum, Pichia methanolica, Pseudomonas aeruginosa, Pseudomonas fluorescens, Pseudomonas putida, Pseudomonas stutzeri, Rhodococcus erythropolis, Saccharomyces cerevisiae, Salmonella typhimurium, Sarcina faecium, Sarcina flava, Scenedesmus* sp., *Serratia marcescens, Spirulina* sp., *Staphylococcus* sp., *Sulfolobus* sp., *Thiobacillus ferrooxidans, Thiobacillus thioparus, Treponema* sp., *Torulopsis candida, Trichosporon brassicae, Trichosporon cutaneum, Vibrio* sp., *Yarrowia lipolytica*

1 Introduction

Since the first biosensor was studied by Updike and Hicks in 1967 [1], several kinds of biosensors have been developed [2] and after 10 years, the first microbial sensor was produced [3]. The general developments in microbial biosensors were introduced by many literature reports [4–14]. In microbial sensor development, many kinds of microorganisms have been employed by applying their characteristics to adapt to the analyte. For example, a microbial consortium or single kind of microorganism was used to estimate organic pollution in industrial wastewater. Photobacteria as a photoindicator, thermophilic bacteria as a stable biosensing element, and genetically modified microorganisms were also used in the development of microbial biosensors.

Microbial sensors are suitable for online control of biochemical processes and for environmental monitoring. In this review, the microbes used and the principles and applications of the microbial sensors are described. In the section on microbial sensor development, the fields of food, clinical, and environmental analysis are introduced.

2 Microbes

The microorganisms are classified into the three kingdoms of eukarya, eubacteria, and archaea based on the sequence homology of 16S (or 18S) RNA. The eukaryote has a nuclear envelope, is not limited to unicellular organisms, and involves unicellular protozoa, yeasts (*Candida* sp., *Saccharomyces* sp., *Trichosporon* sp., etc.), algae, and some species of molds (*Aspergillus* sp., *Penicillium* sp., etc.). The application of yeasts and molds to the development of microbial biosensors was reviewed recently [15].

Eubacteria and the archaea, categorized as prokaryote, are defined as organisms having no nuclear envelope and are limited to unicellular organisms. Eubacteria involve all bacteria (*Escherichia coli*, *Staphylococcus* sp., *Treponema* sp., *Vibrio* sp., etc.) and cyanobacteria, except for the archaea. The archaea are considered as an ancestor of eukaryotes, and involve methanogenic bacteria, halobacteria, thermophilic bacteria (such as extreme thermophilic bacteria which can survive at over 75 °C and hyperthermophilic bacteria which can survive at over 90 °C), and acidophilic bacteria (which include sulfur-metabolizing bacteria such as *Sulfolobus* sp. and *Thiobacillus* sp.).

In addition, microorganisms are also classified into three types: aerobe, facultative anaerobe, and anaerobe. The aerobe requires oxygen for growth. By utilizing aerobic energy metabolism, the aerobe can efficiently obtain chemical energy from compounds of saccharides, amino acids, and lipids by assimilation into the cell and save the energy as adenosine triphosphate (ATP).

The anaerobe survives under anaerobic conditions for growth. *Clostridium* sp. (anaerobic fermentative microbe), sulfate-reducing bacteria, and methanogenic bacteria are known as strict anaerobes. Anaerobic energy metabolism does not consume oxygen and is not efficient compared with that in aerobes.

Facultative anaerobes can survive in both aerobic and anaerobic conditions. Under aerobic conditions, the microbe survives by aerobic respiration. In contrast, under anaerobic conditions, the microbe survives by anaerobic fermentation or respiration. Yeasts, *E. coli*, denitrifying bacteria, and *Pseudomonas aeruginosa* are known as facultative anaerobes.

Furthermore, genetically modified microorganisms were recently studied and applied to the development of microbial biosensors. For example, genetically modified bioluminescent microbes were successfully employed as a highly specific biosensing element for the analyte, and have mainly been applied in the fields of toxicity assay [16]. In addition, by utilizing the production mechanism of green fluorescent protein (GFP), genetically engineered microbes were also applied to many microbial sensor developments for highly specific biosensing [17].

3 Principles

In principle, the microbial sensors mainly involve changes in respiratory activity, or the production of electrochemically active metabolites. The former can be categorized into two groups: activation of microbial respiration by assimilation of organic compounds and inactivation of the respiration by inhibitory substances. These changes can be monitored by using a dissolved oxygen (DO) electrode. Further, a mediator type of microbial sensor has recently been developed by substitution of DO indication. The latter can be monitored directly by an electrochemical device.

3.1 Dissolved Oxygen Measuring Systems

A DO electrode is the most general transducer for the microbial sensor. The membrane type electrode (Clark type) is widely used [18]. The Clark electrodes are classified as either galvanic electrodes or polarographic electrodes. The galvanic electrode has a lead anode and silver (or platinum) cathode, and gives rise to a potential difference. Therefore, it is a self-driven electrode and does not require an externally supplied voltage. This type of electrode is very simple; however, it has disadvantages since it shows a slower response and a shorter stability than a polarographic electrode. The polarographic electrodes consist of a platinum cathode and a silver (or silver/silver) chloride anode, both immersed in the same solution of saturated potassium chloride. A suitable polarization voltage between the anode and cathode selectively reduces oxygen at the cathode. The results of these chemical reactions are shown as a current which is proportional to the DO concentration.

In addition, for portable and disposable use of the microbial sensor, miniaturized disposable type DO electrodes have been developed. In the first development of a micro-oxygen electrode, there was a problem due to leakage of electrolyte caused by the poor adhesivity of the gas-permeable membrane to the substrate of the electrode. By improving the problem, a micro-oxygen electrode was developed based on semiconductor fabrication technology [19]. The improvements were carried out by using agarose gel as a porous material to support the electrolyte solution and a hydrophobic polymer (negative photoresist) as the gas-permeable membrane, and directly casting it over the porous material. In addition, a paper-based O_2 electrode was also studied by Yang et al. [20].

Changes due to activation of the aerobe's or facultative anaerobe's respiration caused by assimilation of chemical compounds (organic and/or inorganic compounds) are detected by a DO electrode. These changes can be estimated as substrate concentrations. The principle in this sensor is shown in Fig. 1a. Aerobic microorganisms are used in these sensors. When the microbial sensor is dipped into sample solution saturated with DO, the respiratory

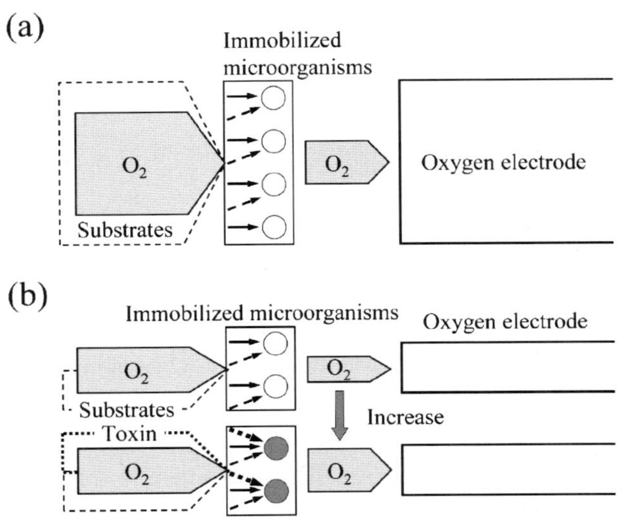

Fig. 1 Principles of the microbial biosensor (oxygen electrode). **a** Respiration activity measurement type for assimilable compounds; **b** respiration activity measurement type for toxic compounds

activity of the microorganisms is increased, which causes a decrease in DO concentration near the membrane. Using a DO electrode, substrate concentration can be measured from the oxygen decrease.

Changes due to inactivation of the microorganism's respiration caused by toxic compounds are also detected by the DO electrode. These changes can be estimated as concentrations of the toxic compound. The principle in this sensor type is shown in Fig. 1b. Aerobic microorganisms are used in these sensors. When the microbial sensor is dipped into sample solution saturated with DO, the respiratory activity of the microorganisms is decreased, which causes an increase in DO concentration near the membrane. Using a DO electrode, the concentration of a toxic compound can be estimated from the oxygen increase.

In addition, the first optical sensing device for biochemical oxygen demand (BOD) using an oxygen-sensitive Ru complex was reported by Preininger et al. in 1994 [21]. The sensor was based on the oxygen quenching of luminescence. Since the work was reported, several BOD biosensors were developed using different kinds of microbes [22–26].

3.2 Electron Transfer Measuring Systems

On the other hand, assimilation of organic compounds by microorganisms can also measure analytes using redox-active substances which can serve as electron shuttles between microorganism and electrode. Electron transfers such as "mediator" or "redox color indicator (RCI)" have been applied to the construction of microbial sensors. The principle of this sensor type is shown in Fig. 2.

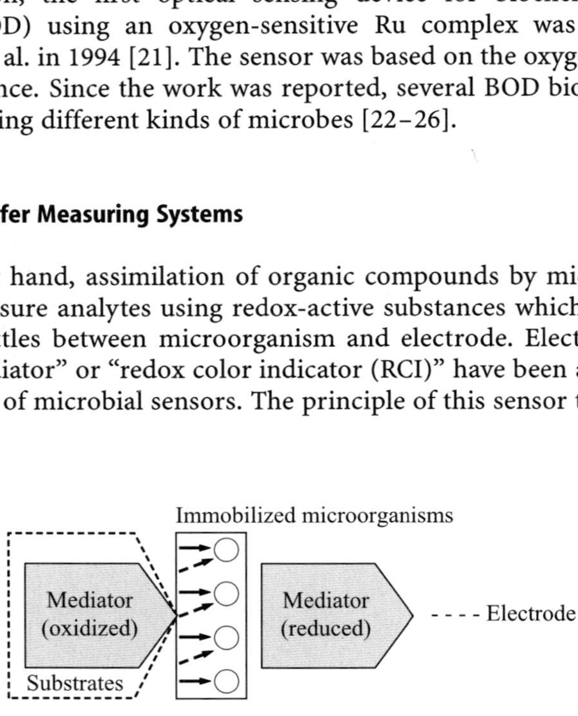

Fig. 2 Principles of mediator measurement type microbial biosensor

As a mediator, potassium hexacyanoferrate(III) [HCF(III)] was used for microbial sensor development (see Fig. 3). Generally, organic substances are oxidized by microorganisms during aerobic respiration. However, when HCF(III) is present in the reaction medium, it acts as an electron acceptor and is preferentially reduced to HCF(II) during the metabolic oxidation of organic substances. The reduced HCF(III) is then reoxidized at a working electrode

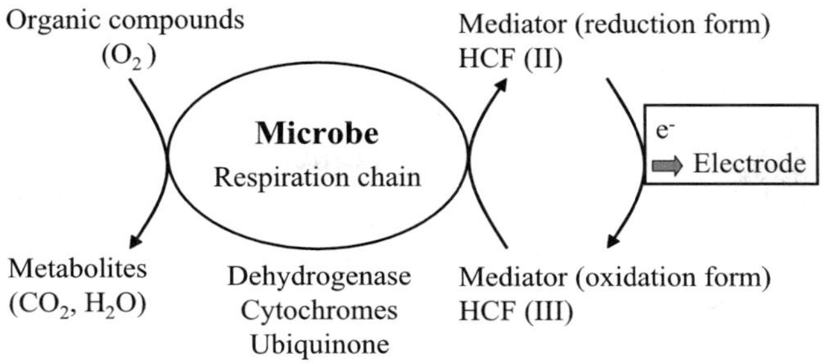

Fig. 3 Principles of the amperometry-mediated biosensor

(anode) which is held at a sufficiently high electric potential. Consequently, a current is generated and detected using the electrode system.

The mediator type of sensor enables measurement of a certain amount of target substance without the influence of DO concentration in the analyte sample. The sensor system utilizing electron transfer has many advantages. The solubility of mediator or color indicator is much higher than that of DO. The sensor system does not require an aeration system and can be greatly simplified to a mobile type device. The detectable potential of the mediator is low compared with that of DO. Therefore, the measurement is not influenced by reducing compounds and can be sufficiently performed with a normal battery, due to the fact that the electric power can be kept at the low detection potential.

3.3 Metabolite Measuring Systems

Electrically active metabolites, such as H_2, CO_2, NH_3, and organic acids, which are secreted from microorganisms, can also be detected as the microbial sensors. This type of sensor mostly uses a gas-permeable membrane to detect gaseous compounds in aqueous or gaseous samples. The principle of this sensor type is shown in Fig. 4. Microbes employing this kind of sensor are not limited to aerobes, and anaerobes can also be employed.

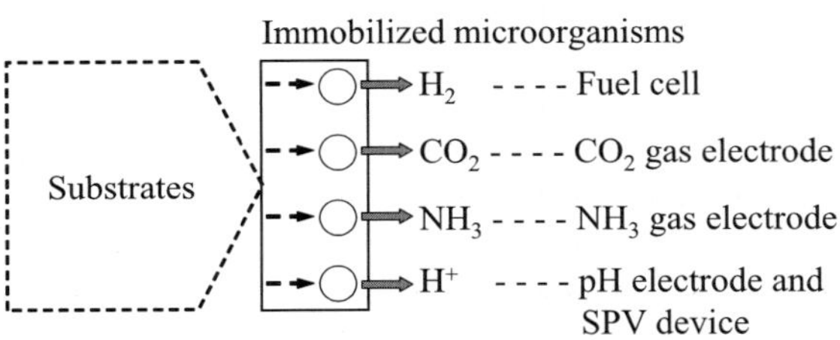

Fig. 4 Principles of metabolite measurement type microbial biosensor

As the transducers, fuel cell type electrodes (H_2 detection), CO_2 electrodes, NH_3 electrodes, or pH electrodes (including ion-sensitive field-effect transistors (ISFETs)) have been used in the microbial sensors. As most of these electrodes, except for the fuel cell type, are based on potentiometry, although they have a wide measurable range they respond to other contaminants and have the limitation of lower detection limits. In addition, several microbial sensors using ISFETs have been developed as microbiosensors.

3.4
Other Measuring Systems

Several other devices can also be applied to microbial sensors. To measure the metabolic heat evolved by the immobilized microorganisms, microbial sensors can be constructed by placing them in proximity to a thermistor.

A combination of photobacteria and a photodetector (e.g., photomultiplier (PMT) or photodiode (PD)) could be constructed as a highly sensitive microbial sensor. The luminescence intensity of photobacteria (luminobacteria) is dependent on metabolic activity. Therefore, nutrients of the microorganisms (e.g., glucose, amino acids) and inhibitors (e.g., toxicants, heavy metals) could be detected using this type of device. In general, luminescence intensity is a more sensitive parameter for metabolic activity than respiration activity or heat generation. Obviously only photobacteria can be used for this purpose.

A combination of microbe and the RCI as an electron transfer system can be constructed for the microbial sensor. As well as the mediator type, several advantages are expected as the features of the microbial sensors. Simultaneous measurements are also applied by using the RCI, for example, simultaneous spectroscopic measurement using 96 or 364 microwells [27]. Field monitoring systems applying a portable spectroscopic self-monitoring blood glucose (SMBG) device can also be applied using a light-emitting diode (LED) and a PD [28].

The surface photovoltage (SPV) technique can be applied to a microbial sensor [29]. The SPV device as a transducer is sensitive to the surface pH, ionic strength, and physical adsorption. A silicon-based SPV device or a light-addressable potentiometric sensor (LAPS) measures the surface potential of the device, especially the pH of the solution near the surface. The device can be easily fabricated by using a silicon chip. SPV devices have been employed in some applications such as chemical sensors for quantification of enzyme-linked immunoassays, taste sensors, hydrogen sensors, and monitoring sensors for metabolism in mammalian cells.

Surface plasmon resonance (SPR) spectroscopy is a detector of refractive index and thickness of a sample close to a metal layer (generally a few hundred nanometers). In that sense, the SPR has been used mainly to measure chemical and biochemical compounds, and several applications were studied in the field of microbial biosensor development.

4
Microbial Sensor Developments

Microbial biosensors can be categorized by three applications, i.e., food analysis (see Table 1), clinical analysis (see Table 2), and environmental analysis (see Tables 3–7).

4.1
Food Analysis

Microbial biosensors for food analysis and fermentation processes have been developed and the works were reviewed in several literature reports [30–36]. For fermentation process control in brewing or fuel production, monitoring of alcohol concentration is important. For monitoring of gaseous methanol in a liquid sample, *Trichosporon brassicae* CBS 6382 [37] or unidentified bacterium AJ 3993 [38] was used as sensing element in each biosensor system which employed an acetylcellulose membrane and the oxygen electrode (see Table 1Ia and Ib). Gaseous ethanol was also determined by *T. brassicae* CBS 6382 [37–39] (see Table 1Ic) or *Gluconobacter oxydans* with ferricyanide mediation [40]. *Acetobacter aceti* IAM 1802 was used for gaseous alcohol measurements with the ISFET element [41] (see Table 1Id). In addition, several microbial biosensors were recently developed for ethanol determination using yeast [42–44]. For koji quality control in sake brewing as the fermentation process, *Saccharomyces cerevisiae* K701 and K9 were employed in the SPV device [45]. The pH change due to the production of organic acids in sake brewing was determined by the SPV device. Yeast activity in alcoholic fermentation was measured by the mediator system combining HCF(III) and menadione [46].

For organic acid measurements, a specific microbial biosensor for formic acid was developed [47] (see Table 1IIa). Formic acid is found in culture media, urine, blood, and gastric content as a product of many chemical reactions. It is a commonly occurring intermediate of cellular metabolism and is attracting attention as an intermediate of biomass conversion which is easily converted to hydrogen. For gaseous formic acid determination, the anaerobe *Clostridium butyricum* IFO 3847 was immobilized in agar gel on an acetylcellulose membrane, and a fuel cell system for hydrogen detection was used. This system showed highly specific performance to formic acid, although hydrogen in the sample influenced the formic acid measurements. An acetic acid biosensor is required in fermentation processes and was developed using *T. brassicae* [48] (see Table 1IIb). For determination of tannic acid, the fungus *Aspergillus ustus* immobilized in poly(vinyl alcohol) was used in a batch system [49]. For amino acid determination, *E. coli* ATCC 8739 was used for L-glutamic acid [50] (see Table 1IIc), *Sarcina faecium* was used for L-arginine [51], *Bacterium cadavers* was used for L-aspartic acid [52], and

Table 1 Characteristics of microbial biosensors developed by our group for food analysis and fermentation processes

	Target	Microbe	Indicator	Transducer	System	Calibration range (mg L^{-1})	Measuring time (min)	Stability (assays) (days)	References
Alcohol									
Ia	Methanol (gaseous)	*Trichosporon brassicae* CBS 6382	DO	Electrode	Batch-flow & membrane	2–22.5	< 10	> 21 (1000)	[37]
Ib	Methanol (gaseous)	Unidentified bacterium AJ 3993	DO	Electrode	Batch-flow & membrane	5.5–22.3	–	–	[38]
Ic	Ethanol (gaseous)	*T. brassicae* CBS 6382	DO	Electrode	Batch-flow & membrane	2–22.3	< 10	> 21 (2100)	[38] ([37, 39])
Id	Alcohol (gaseous)	*Acetobacter aceti* IAM 1802	pH	ISFET	Batch & chip	0.1–70 mM	15	15 h	[41]
Organic acid									
IIa	Formic acid (gaseous)	*Clostridium butyricum* IFO 3847	H_2	Fuel cell	Batch & membrane	10–1000	20	> 20	[47]
IIb	Acetic acid (gaseous)	*T. brassicae* CBS 6382	DO	Electrode	Batch-flow & membrane	5–54	8	> 21 (1500)	[48] ([37, 39])
IIc	Glutamic acid	*Escherichia coli* ATCC 8739	CO_2	Glass electrode	Flow & membrane	100–800	5	> 21 (1500)	[50]
Sugar									
IIIa	Glucose	*Pseudomonas fluorescens* IFO 3081	DO	Electrode	Batch & membrane	2.0–20	10	> 14 (150)	[56]
IIIb	Total assimilable sugars	*Brevibacterium lactofermentum* AJ 1511	DO	Electrode	Flow & membrane	< 1 mM	1	> 10 (960)	[59]

Table 1 (continued)

	Target	Microbe	Indicator	Transducer	System	Calibration range ($mg\,L^{-1}$)	Measuring time (min)	Stability (assays) (days)	References
Ammonia									
IVa	Ammonia (gaseous)	Nitrifying bacteria	DO	Electrode	Batch & membrane	3.5–42	4	> 10 (200)	[37] ([39])
IVb	Ammonia (gaseous)	Nitrifying bacteria	DO	Electrode	Batch-flow & membrane	0.05–1.3	8	> 14 (1400)	[64]
IVc	Ammonia (gaseous)	Nitrifying bacteria	DO	Electrode	Batch & membrane	0.1–42	< 4	> 10 (200)	[65]
IVd	Ammonia (gaseous)	Nitrifying bacteria	DO	Electrode	Batch & membrane	0.45–10 μM	< 7	> 10 (200)	[66]
Vitamin									
Va	Vitamin B_1	*Lactobacillus fermenti* ATCC 9338	NAD[a] or FAD[b]	Fuel cell	Batch & suspension	0.001–0.05	15	–	[68]
Vb	Vitamin B_{12}	*E. coli* 215	DO	Electrode	Batch & membrane	0.005–0.025	120	25 stored at −25 °C	[69] ([70])

[a] Nicotinamide adenine dinucleotide
[b] Flavin adenine dinucleotide

Sarcina flava was used for L-glutamine [53]. The biosensor for L-glutamic acid determination utilized glutamate decarboxylase (GD) which was contained in *E. coli* cells and used a CO_2 gas sensor [50]. This sensor influenced oxygen and glutamine and did not influence the other amino acid. Lactic acid was determined by a mediator type amperometric biosensor based on carbon paste electrodes modified with *S. cerevisiae* [54].

For sugar measurements, glucose was determined by *Aspergillus niger* with mediator [55] and *Pseudomonas fluorescens* IFO 3081 [56] (see Table 1IIIa). For xylose detection, *G. oxydans* cells were used in the field-effect transistor (FET) device [57] and in an amperometric device [58]. For glucose sensing, *P. fluorescens* was immobilized in a collagen gel on a gas-permeable membrane [56]. Determination of total assimilable sugars (TASs) is required for online measurements of substrate in culture broths in the fermentation industry. TASs (glucose, fructose, and sucrose) were determined using *Brevibacterium lactofermentum* AJ 1511 which was immobilized in a dialysis membrane and a Teflon™ membrane [59] (see Table 1IIIb).

Microbial biosensors for simultaneous determination were required for food analysis and fermentation control. Simultaneous determination of glucose, sucrose, and lactose was performed by *G. oxydans, S. cerevisiae*, or *Kluyveromyces marxianus* [60]. Simultaneous determination of glucose and ethanol was performed by a nonselective microbial sensor for both glucose and ethanol using *G. oxydans* and a glucose electrode with glucose oxidase [61] and by *G. oxydans* or *Pichia methanolica*. The bacterial cells of *G. oxydans* were sensitive to both substrates, while the yeast cells of *P. methanolica* oxidized only ethanol [62]. For the simultaneous determination of mono- and disaccharides, a microbial biosensor array with transport mutants of *E. coli* K12 was developed [63].

Ammonium ion or ammonia gas monitoring is not only required for food analysis and fermentation processes, but also for environmental and clinical analysis. Thus, several types of microbial biosensors for them were studied. For the detection of gaseous ammonia in a liquid sample, nitrifying bacteria (unidentified) were used [37, 39, 64–66] (see Table 1IV). The *Nitrobacter* sp. was immobilized on a gas-permeable Teflon™ membrane in a batch system and the biosensor measured 3.5 to 42 $mg\,L^{-1}$ NH_3 [37]. Other microbial immobilization methods and measuring systems were also studied for the determination of gaseous targets [64–66]. *Trichosporon cutaneum* was immobilized in a membrane for long-term stability and was used for continuous monitoring of ammonium ion in sewage [67].

Vitamin sensors were also studied [68–72] (see Table 1V). Vitamin B_1 (thiamine) in culture broth was measured by using *Lactobacillus fermenti* ATCC 9338 with a fuel cell system [68] or *S. cerevisiae* with a DO meter [72]. In this study, a possible mechanism of current generation is discussed. A vitamin B_{12} determination system using *E. coli* 215 and an oxygen electrode was constructed [69, 70].

4.2 Clinical Analysis

Several types of microbial biosensors were studied for clinical application. Hybrid biosensor systems using both materials of microbe and enzyme as sensing elements were mainly developed for clinical applications [73, 74]. Phenylalanine as an indicator of phenylketonuria was determined by incubation with *Leuconostoc mesenteroides* ATCC 8042 (6 h) followed by use of a lactate electrode (immobilized lactate oxidase (LOD) and an O_2 electrode) to measure the lactate produced [75] (see Table 2Ia). A linear relationship is obtained between the current decrease and the phenylalanine concentration between 75 and 600 $\mu g\,L^{-1}$. Next, the microbes were immobilized to agar on an acetylcellulose membrane and packed into a column as a reactor to improve measuring time [76] (see Table 2Ib). As a result, the measuring time was shortened to 90 min. A hybrid urea biosensor was developed using nitrifying bacteria from activated sludge and urease [77] (see Table 2Ic). This sensor measured ammonia (NH_3), which was produced from urea by urease reaction, by an electrode, immobilized nitrifying bacteria, and made possible the determination of 2–200 mM within 7 min. The determination of creatinine in serum and urine is a diagnostically important test for external dialysis. Then, a hybrid creatinine sensor was developed using creatininase and nitrifying bacteria, and showed linear responses from 5 to 100 $mg\,L^{-1}$ creatinine [78] (see Table 2Id). A hybrid biosensor for L-tyrosine determination, consisting of immobilized L-tyrosine decarboxylase and CO_2 utilizing chemoautotrophic bacterium, was developed and performed 10 to 50 μM detection for L-tyrosine [79] (see Table 2Ie).

Mutagen sensing for clinical analysis was performed by several microbial biosensors. Mutagen for the marker of cancer was detected using two types of *Bacillus subtilis* [80] (see Table 2IIa). By a similar way, mutagens were detected by different immobilization of microbes to O_2 electrodes [81, 82] (see Table 2IIb). Other biosensors for mutagen detection were also developed using *Salmonella typhimurium* TA 100 [83] (see Table 2IIc) and a combination of *E. coli* GY5027 (lysogenic strain) and *E. coli* GY5026 [84] (see Table 2IId). The latter type of biosensor was improved to a bioluminescence (BL) detection system using luminescent bacteria [85] (see Table 2IIe).

A cephalosporin (antibiotics) measuring system was developed using the hydrogen producing anaerobe of *Citrobacter freundii* B-0652 and a glass electrode (H_2) in a flow and reactor system [86] (see Table 2IIIa). The biosensor enabled measurement of cephalosporins between 62.5 and 300 $mg\,L^{-1}$ for 7-phenylacetylamidodesacetoxysporanic acid (phenylacetyl-7 ADCA), cephaloridine, and cephalothin, and between 62.5 and 125 $mg\,L^{-1}$ for cephalosporin c. A nystatin biosensor was also developed as an antibiotic sensor using *S. cerevisiae* and had a linear range between 0.5 and 80 $U\,mL^{-1}$ [87]

Table 2 Characteristics of microbial biosensors developed by our group for clinical analysis

	Target	Microbe	Indicator	Transducer	System	Calibration range ($mg\,L^{-1}$)	Measuring time (min)	Stability (assays) (days)	Refs.
Hybrid biosensors for constituent in body fluid									
Ia	Phenylalanine	*Leuconostoc mesenteroides* ATCC 8042	DO (lactate)	Electrode immob. LOD	Flow & reactor (hybrid)	0.075–0.6	6 h	–	[75]
Ib	Phenylalanine	*L. mesenteroides* ATCC 8042	DO (lactate)	Electrode immob. LOD	Flow & reactor (hybrid)	0.1–50	90	–	[76]
Ic	Urea	Nitrifying bacteria (and urease)	DO (NH_4^+)	Electrode	Batch & membrane (hybrid)	2–200 mM	< 7	10 (150)	[77]
Id	Creatinine	Nitrifying bacteria (and creatininase)	DO (NH_4^+)	Electrode	Batch & membrane (hybrid)	5–100 $mg\,dL^{-1}$	< 3	> 21 (300)	[78]
Ie	L-Tyrosine	Chemoautotrophic bacterium	DO (CO_2)	Electrode	Batch & membrane (hybrid)	10–50 μM	–	5	[79]
Mutagens									
IIa	Mutagens	*Bacillus subtilis* M45 Rec^- & *B. subtilis* H17 Rec^+	DO	Electrode	Batch & membrane	1.6 AF-2[a]	< 1 h	6 months at – 20 °C	[80]
IIb	Mutagens	*B. subtilis* M45 Rec^- & *B. subtilis* H17 Rec^+	DO	Electrode	Batch & membrane	1.6–2.8 AF-2	–	–	[81] ([82])
IIc	Mutagens	*Salmonella typhimurium* TA 100	DO	Electrode	Batch & membrane	0.001–0.006 AF-2	< 10 h	–	[83] ([82])
IId	Mutagens	*Escherichia coli* GY5027 (lysogenic strain), GY5026	DO	Electrode	Batch & membrane	0.01–0.2 AF-2	2.5–4.0 h	–	[84]

Table 2 (continued)

	Target	Microbe	Indicator	Transducer	System	Calibration range (mg L^{-1})	Measuring time (min)	Stability (assays) (days)	Refs.
IIe	Mutagens	*E. coli* GY5026 (lysogenic strain)	BL	PMT	Batch & suspension	10^2–10^4 ng	< 1 h	–	[85]
Antibiotics									
IIIa	Cephalosporins (antibiotics)	*Citrobacter freundii* B-0652	H_2	Glass electrode	Flow & reactor	62.5–300	10	7 (1 month at 5 °C)	[86]
IIIb	Nystatin (antibiotics)	*Saccharomyces cerevisiae*	DO	Electrode	Batch & membrane	0.5–80 U mL^{-1}	< 1 h	–	[87]
Microbial populations									
IV	Microbial populations	Five kinds of microbes were used	DO	Electrode	Batch & membrane	> 10^6 bact. cells, > 10^5 yeast cells	10	–	[95] ([96, 97])

[a] 2-(2-Furyl)-3-(5-nitro-2-furyl)acrylamide

(see Table 2IIIb). As another application, *E. Coli* DPD2749 for mitomycin C was studied [88]. Then, a collagen membrane was used instead of a nylon net. For diagnosis of leukemia, an enzyme activity measuring system in single cells has been developed using a microcell with a positionable dual electrode. Peroxidase activities in single neutrophils and single acute promyelocytic leukemia cells were measured by this method [89].

To prevent contamination especially that caused by pathogenic bacteria in diagnostic and environmental fields and food processing, microbial detection techniques are required. Thus, several biosensing techniques have been developed [90, 91]. Zhou et al. developed an enhanced fluorescent fiber-optic biosensor system for *S. typhimurium* detection [92]. Ertl and Mikkelsen developed an electrochemical biosensing array employing lectin-lipopolysaccharide recognition for the identification of microorganisms, and six microbial species including Gram-negative and Gram-positive microbes were examined for identification using this method [93]. As a microbial pathogen sensor, bacterial elicitor flagellin was detected with a bioelectronic portable system employing plant living cells [94]. Determination of microbial populations is also required in many fields; thus, several biosensors have been developed [95–98] (see Table 2IV).

4.3 Environmental Analysis

For environmental analysis and monitoring, microbial sensors have been most extensively studied and reviewed in the literature [17, 99–102]. In recent years, BOD sensors and toxicity sensors have been actively developed. In this review article, the BOD sensors were categorized into several types, i.e., BOD_{DO} as DO type, BOD_{BL} as bioluminescence type, BOD_{SPV} as SPV type, BOD_{MD} as mediator type, BOD_{DM} as double mediator type, BOD_{RCI} as RCI type, and BOD_{MFC} as microbial fuel cell type. Table 3 shows a comparison of the biodegradation characteristics of organic compounds with these BOD sensors.

4.3.1 BOD_{DO} Sensors

In environmental analysis, the most important application is BOD determination in polluted water or effluent. The first microbial BOD sensor applying the principle of microbial fuel cells was developed by our group [3] (see Table 4a). Anaerobe *C. butyricum* was isolated from soil according to the (JIS) method. The microorganisms were immobilized by polyacrylamide gel and fixed to a glass cell. This sensor system realized a short measuring time around 30 min. The measuring time was dramatically shortened by the invention of this BOD sensor compared with the conventional 5-day BOD (BOD_5)

Table 3 Comparison of BOD values of various organic samples [a]

Substrate [b]	BOD_5 [107]	BOD_{DO} [106]	BOD_{BL} [152]	BOD_{SPV} [29]	BOD_{MD} [167]
Glucose	0.50–0.78	0.72	0.62	0.66	1.54
Fructose	0.71	0.54	0.57	0.73	0.35
Sucrose	0.49–0.76	0.36	0.50	0.45	0.07
Lactose	0.45–0.72	0.06	0.31	0.04	0.02
Soluble starch	0.22–0.71	0.07	0.02	0.07	–
Asparagine	0.58	–	0.48	–	0.29
Alanine	0.55	–	–	–	0.73
Glycine	0.52–0.55	0.45	0.50	0.36	–
Glutamic acid	0.64	0.70	0.73	0.40	0.59
Histamine	0.55	0.35	–	0.34	0.27
Acetic acid	0.34–0.88	1.77	0.27	0.39	0.32
Citric acid	0.63–0.88	0.72	–	0.18	–
Lactic acid	0.40	0.17	0.32	0.14	0.66
Ethanol	0.93–1.67	2.90	0.25	0.49	0.30
Propanol	0.47–1.50	0.28	0.28	–	0.29
Glycerol	0.62–0.83	0.51	0.53	0.44	0.05

[a] Values are expressed in mg O_2 mg^{-1} substrate
[b] Concentrations of each pure organic substance were 100 mg L^{-1} for glucose, 500 mg L^{-1} for sucrose, lactose, and glycerol, and 200 mg L^{-1} for others

method (see Fig. 5). Thus, the possibility of BOD monitoring of effluent was suggested by this sensor development.

Several microbial BOD sensors employing the batch and microbial membrane system have subsequently been developed by utilizing different immobilization methods with different aerobes [103, 104] (see Table 4b–d). Microorganisms from soil were immobilized in collagen gel membrane, and anaerobes *C. butyricum* IFO 3847 were immobilized in a polyacrylamide gel membrane [103]. This system differs from typical batch systems with the existence of an air supply system. *C. butyricum* was grown and immobilized under anaerobic conditions, and the sensor system using *C. butyricum* was modified as a microbial fuel cell. In this case, collagen gel was degraded by the soil microbes. Thus, the sensor system using *C. butyricum* subsequently improved to simplify the operation [104]. However, anaerobe is not suitable for BOD estimation. Therefore, other approaches for microbial BOD sensors were required.

A flow system utilizing indication of respiration activity by aerobe was subsequently developed for BOD_{DO} estimation [105] (see Table 4e). Soil microorganisms were used and immobilized in collagen gel membrane as in the previous study [103]. Good reproducible results of 3% relative standard deviation (RSD) were obtained by the glucose–glutamic acid (GGA) standard

Table 4 Characteristics of BOD sensors developed early by our group

		Microbe	Indicator	Transducer	System	Calibration range ($mg\ O_2\ L^{-1}$)	Measuring time (min)	Stability (d)	Refs.
a	BOD_{DO}	Soil consortium	DO	Electrode	Batch & membrane	200–600	About 30	10	[3]
b	BOD_{DO}	Soil consortium	DO	Electrode	Batch & membrane	5.0–22	< 15	At least 10	[103]
c	BOD_{MFC}	*Clostridium butyricum* IFO 3847[a]	DO	Fuel cell	Batch & membrane	30–300	40	> 30	[103]
d	BOD_{MFC}	*C. butyricum* IFO 3847	DO	Fuel cell	Batch & membrane	6–400	30–40	30	[104]
e	BOD_{DO}	Soil consortium	DO	Electrode	Flow & membrane	44–132	30	10	[105]
f	BOD_{DO}	*Trichosporon cutaneum* IFO 10466, AJ 4816	DO	Electrode	Flow & membrane	10–40	< 18	17 (400 tests)	[106]
g	BOD_{DO}	*T. cutaneum*	DO	Electrode	Flow & membrane	2.0–20	1 h[b]	30–90	[109]

[a] Anaerobe
[b] The measuring time includes calibration time

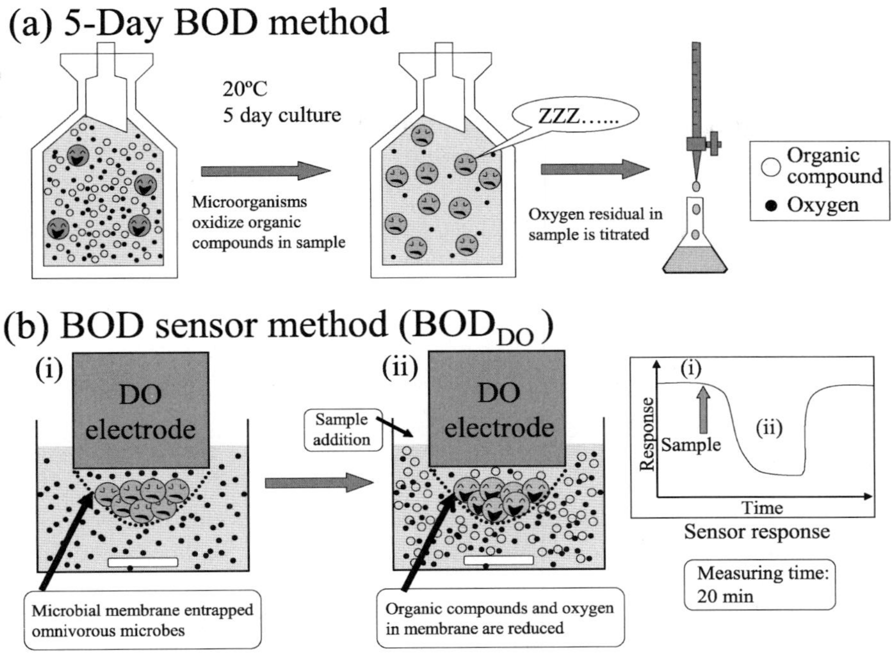

Fig. 5 Conceptual diagram of the principles of the conventional BOD_5 method (**a**) and a batch type BOD sensor (**b**)

solution, although the stability of the gel membrane was not enough and the same as the previous study. Then, the problem was raised as to which microorganisms of the soil consortium or single strain should be used for the BOD_{DO} sensor. For example, the biodegradability of the consortium to many kinds of organic compounds is greater than that of a single strain. However, reproducible results cannot be obtained by the use of a consortium, because to obtain reproducible culture conditions in a consortium is mostly impossible.

Thus, we next examined the use of a single strain and fabrication of a continuous flow system for automatic BOD_{DO} estimation [106] (see Table 4f). *T. cutaneum* AJ 4816 (IFO 10466) was employed as a single strain for a BOD sensor (see Fig. 6). Comparison of biodegradation characteristics of organic compounds between the conventional BOD_5 method [107] and the BOD_{DO} sensor method using *T. cutaneum* AJ 4816 [106] was performed. The sensor showed low BOD values compared with the BOD_5 method when lactose and soluble starch were employed for experiments, probably due to the slow decomposition rate of these compounds by the immobilized yeasts. On the other hand, the sensor showed high BOD values compared with the BOD_5 method when acetic acid and ethyl alcohol were employed. These results suggested the oxidation rate of acetate and ethyl alcohol to be faster than that of

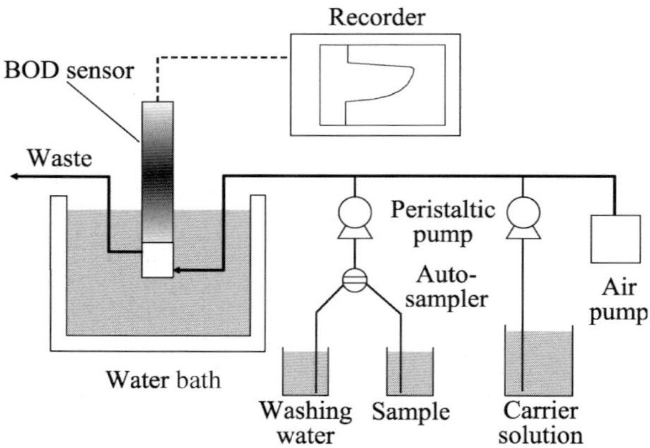

Fig. 6 Construction of a flow and porous membrane type BOD sensor

some standard substrates such as glucose and glutamic acid. Therefore, the GGA solution was defined as a standard solution for the BOD estimation of effluents. Before development of the BOD sensor, such primary effluent of sewage or wastewater from the food or pulp industry was hard to control by BOD values obtained by the BOD_5 method. As a new method for the BOD determination of primary effluent, a sensor method was established which was defined as a JIS (JIS K 3602) in 1990 [108]. Finally, an automatic flow system was developed for practical use [106], and about 800 instruments have been sold since 1983 [109] (see Table 4g). Figure 7 shows two BOD sensors of a deferred type (a) and a desktop type (b). These are available from Central Kagaku Co.

Yang et al. developed disposable DO electrode chips that can be applied to the BOD estimation for field monitoring. At first, the single DO electrode ($15 \times 2 \times 0.4$ mm) was constructed on silicon substrates using micromachining techniques [110] (see Table 5IIa). This electrode is of the Clark type and *T. cutaneum* was directly immobilized on the electrode surface using an ultraviolet cross-linking resin (ENT-3400). This DO electrode chip enabled measurements between 1.0 and 18 mg O_2 L^{-1} BOD. For BOD estimation using the DO electrode chip, a dynamic transient measuring method was adopted and compared with the steady-state measuring method as a conventional method [111]. In the study, the measuring time was dramatically reduced.

Subsequently, they fabricated an array type DO electrode (Clark type) using thin film technology [112] (see Table 5IIb). However, these BOD_{DO} sensor chips developed for field monitoring have several remaining problems. The biggest problem is that the concentration of DO is limited to about 8 mg O_2 L^{-1}. Therefore, measurement of samples with high a BOD value influenced the DO concentration.

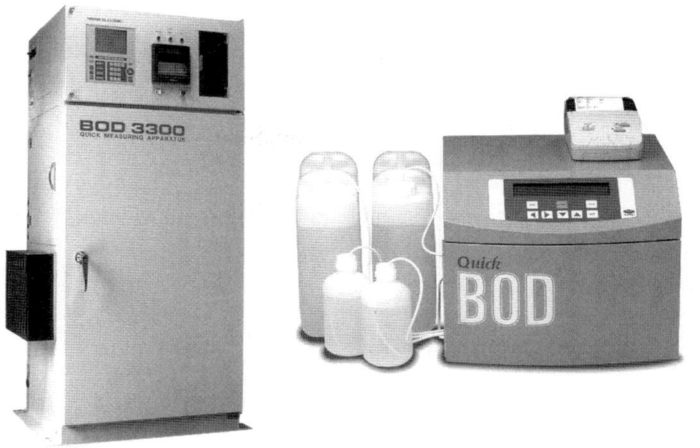

(a) Deferred type (b) Desktop type

Fig. 7 Photographs of BOD sensors. **a** Deferred type and **b** desktop type. These photographs are from a catalog of Central Kagaku Co.

Quality monitoring of secondary effluents (less than 10 mg O_2 L^{-1} BOD) is required for public organizations such as sewage plants and industries. In addition, the sensitive BOD_{DO} sensor was also required for monitoring low BOD values in river waters. Chee et al. developed highly sensitive BOD_{DO} sensors and pretreatment methods for the measured samples [23, 113–117] (see Table 5III and IV).

Secondary effluents and the upper stream of river water generally indicate low BOD values, and they mostly contain refractory organic compounds such as humic acid, lignin, tannic acid, gum arabic, and surfactants [118, 119]. Thus, for the precise estimation of such samples, several microbes from sewage plants were screened and *Pseudomonas putida* SG10 was isolated to apply to highly sensitive BOD_{DO} sensors.

In the first step of the development of highly sensitive BOD_{DO} sensors using DO electrodes, a basic batch system was studied and characterized to obtain the optimum conditions for low BOD estimation [113]. The calibration curve was obtained by the GGA solution from 0.5 to 10 mg O_2 L^{-1} BOD and the detection limit was 0.5 mg O_2 L^{-1} BOD ($n = 5$). Next, in the BOD determination of various river waters, the sensor's results generally indicated somewhat lower values than those obtained by the BOD_5 method. The reason could be that the compounds were not easily assimilable to the sensor in such a short time.

In the second step, a highly sensitive BOD_{DO} estimation was studied using an optical fiber probe as an oxygen sensor, based on fluorescence quenching by oxygen [23]. Then, sufficient reproducibility of this optical fiber biosensor

Table 5 Characteristics of BOD sensors developed recently by our group

		Microbe	Indicator	Transducer	System	Calibration range ($mg\,O_2\,L^{-1}$)	Measuring time (min)	Stability (d)	Refs.
I	BOD_{DO}	Thermophilic bacteria (not identified)	DO	Electrode	Batch & membrane	1.0–10	< 7	> 40	[135]
IIa	BOD_{DO}	*Trichosporon cutaneum* IFO 10466	DO	Chip electrode	Chip & batch	1.0–18 (det. lim. 0.2)	7–20	3	[110]
IIb	BOD_{DO}	*T. cutaneum* IFO 10466	DO	Chip electrode (five electrode array)	Chip (disposable)	8.0–32	–	–	[112]
IIIa	BOD_{DO}	*Pseudomonas putida* SG10	DO	Electrode	Batch & membrane	0.2–10	5	–	[23]
IIIb	BOD_{DO}	*P. putida* SG10	DO	Electrode	Batch & membrane	0.5–10	2–15	> 10	[113]
IIIc	BOD_{DO}	*P. putida* SG10	DO	Electrode	Batch & membrane	1.0–10	–	–	[114]
IV	BOD_{DO}	*P. putida* SG10	DO	Optical fiber probe (fluorescence)	Batch & membrane	1.0–10	15	–	[115]
V	BOD_{BL}	*Photobacterium phosphoreum* IFO 13896	Lumines-cence	Photodiode	Batch	20–160	< 15	–	[152]
VI	BOD_{SPV}	*T. cutaneum* IFO 10466 (AJ 4816)	pH	SPV device	Flow & membrane	10–100	25	> 98	[29]
VIIa	BOD_{MD}	*P. fluorescens* biovar V	Mediator	Chip electrode	Chip & batch	15–200	15	–	[167]
VIIb	BOD_{MD}	*P. fluorescens* biovar V	Mediator	Chip electrode	Chip (disposable)	15–260	15	> 35	[169]

Table 5 (continued)

	Microbe	Indicator	Transducer	System	Calibration range (mg O_2 L^{-1})	Measuring time (min)	Stability (d)	Refs.
VIIc BOD_{MD}	*P. fluorescens* biovar V	Mediator	Chip electrode (ten electrode system)	Chip (disposal)	10–120 for SES[a], 25–250 for OECD	–	–	[172]
VIII BOD_{RCI}	*P. fluorescens* biovar V	Color indicator	Photodiode	Chip (disposable)	50–3000	10	Low	[28] ([27])

[a] Sludge extract solutions

was not obtained. Therefore, the DO electrode was used again for subsequent studies.

In the third step, ozonation as a pretreatment method for low BOD samples was examined [114]. Ozonation is known as a treatment method to decompose organic compounds, especially for refractory organics in industrial wastewater, municipal effluent, and river water. In this study, ozonation of river water samples was successfully established as a new pretreatment method for low BOD_{DO} estimation combined with the sensitive sensor method. However, this sensor method requires the ozonation equipment, which is large and expensive, and ozone gas requires careful handling. Therefore, a photocatalytic pretreatment method was next applied for this purpose.

Refractory organics containing samples can also be decomposed to biodegradable substances by photocatalytic preoxidation using illuminated titanium dioxide (TiO_2). Accordingly, the photocatalytic pretreatment method was applied to the sensitive BOD sensor [115]. The sensor responses to various river water samples, which were pretreated by photocatalysis, were successfully improved. The results obtained by this sensor method and the conventional BOD_5 method corresponded ($r = 0.983$). As the next steps, both a flow type sensor system and stopped-flow system based on the above mentioned studies have been developed [116, 117]. In another highly sensitive BOD sensor, a microbial fuel cell was applied using an improved cathode reaction [118].

Other BOD_{DO} sensors were also developed using different kinds of microbes, for example, *Arxula adeninivorans* (salt-tolerant yeast) [120–123], *B. subtilis* [124], *Hansenula anomala* [125], *Klebsiella oxytoca* [126], *Klebsiella* sp. [127], *Serratia marcescens* LSY 4 [128], *Torulopsis candida* [129], combinations of two microbes such as *B. subtilis* and *B. licheniformis* [130–132], *B. subtilis* and *T. cutaneum* [133], *Rhodococcus erythropolis* and *Issatchenkia orientalis* [134], thermophilic bacteria [135] (see Table 5I), microbial fuel cell [136, 137], a mixture of microorganisms (consortium) [138–140], activated sludge [141, 142], slime mold [143, 144], and dead microbial cells [145–149]. In addition, a disposable BOD sensor for measuring nitrification (N-BOD) and inhibition of nitrification in wastewater was developed using nitrifying bacteria [150]. As the other approaches, a respirographic BOD sensor was developed utilizing indication of CO_2 concentration (BOD_{CO_2}) [151]

4.3.2
BOD_{BL} Sensors

Using luminous bacteria (*Photobacterium phosphoreum* IFO 13896), a bioluminescence BOD sensor has been developed by Hyun et al. [152] (see Table 5V). The luminous bacteria, which are isolated from marine sources, emit light at a constant rate for fairly long periods of time as a result of normal metabolic processes. In this process, the substrate produced by catabolic degradation or organic compounds in the microbial cell are shunted to the bioluminescence reaction which is coupled to the electron transport pathway [153]. In this study, *P. phosphoreum* was put into a transparent glass dish, which was located on a PD in a light-protected box. Signals from the PD were amplified and recorded. The BOD responses of the bacterial reagents could be observed between 20 and 160 mg O_2 L^{-1} BOD within 15 min with 7% RSD. By the efforts of the research (Tamiya et al.), the BOD_{BL} sensor was produced by Ishikawa Seisakusho Ltd. In addition, the sensor system was applied to an array chip for on-site BOD sensing [154].

4.3.3 BOD_{SPV} Sensors

According to Murakami et al., the SPV technique was applied to a BOD sensor using *T. cutaneum* as a biosensing element [29] (see Table 5VI). The biodegradation profiles of this biosensor were examined. The results in the BOD estimation of wastewaters obtained by both SPV-based sensor and conventional BOD_5 methods were well correlated between 0 and about 180 mg O_2 L^{-1} BOD.

4.3.4 BOD_{MD} Sensors

The mediator has been applied to the fabrication of microbial fuel cells [155, 156] and to microbial detection [157, 158]. It has been suggested that reduction of the redox mediator, rather than DO, is due to metabolic reactions of microorganisms [159]. Thus, instead of DO, HCF(III) has been used as an electroactive compound for the development of amperometric biosensors using microorganisms [160, 161].

Thus, the first BOD_{MD} sensor was developed using HCF(III) and *E. coli* by Pasco et al. in 2000 [162]. Since the study was reported, many BOD_{MD} biosensors have been intensively developed in recent years [163–166]. We have also developed several BOD_{MD} sensors. As the first, in our BOD_{MD} sensor which was reported in the same year as the report by Pasco et al., we employed a batch system using HCF(III) and *P. fluorescens* biovar V which was isolated from a municipal sewage treatment plant [167] (see Table 5VIIa). A combination chip consisting of a working electrode, on which *P. fluorescens* was immobilized, and a counter electrode was constructed. Under optimized conditions, synthetic sewage determined using the Organization for Economic Cooperation and Development (OECD) method was used as a standard solution approximating real wastewater to determine the BOD_{MD} using the sensor [168]. With the OECD synthetic sewage (OSS), evidence that is was possible to measure BOD without the influence of DO was shown. By this work, the possibility of a new estimation method for BOD_{MD} determination was shown.

As the next step, a mobile type BOD sensor was studied [169] (see Table 5VIIb). In this work, storing conditions for the practical use of the BOD sensor chips were investigated and a handy type amperometer was fabricated for field monitoring. Under the conditions of limited nutrients, the microorganisms survive by metabolizing the endogenous substrates in the cell [170]. It has also been reported that bacteria in the "starved" condition can concurrently metabolize various exogenous substrates [171], which is advantageous for BOD measurements. In fact, the linearity of the calibration curves was improved up to 68 h of aeration. Finally, the sensor response decreased to approximately half the original after at least 35 days in storage.

The BOD_{MD} sensor was improved to a ten-channel system for compost monitoring [172] (see Table 5VIIc). Screen-printed disposable sensor chips for single use were fabricated by incorporating *P. fluorescens* and HCF(III) immobilized in sodium alginate gel. In conclusion, the relative change in BOD sensor values determined using our system corresponded well with the BOD_5 values obtained using the standard BOD_5 method during 58 days of composting.

4.3.5 BOD_{DM} Sensors

A double mediator (DM) system combining HCF(III) and menadione was studied using the eukaryote *S. cerevisiae* [173, 174]. Baronian et al. revealed that menadione (vitamin K_3; lipophilic mediator) can penetrate the outer cell membrane. Roustan et al. applied the DM system for measuring yeast activity in alcoholic fermentation [46]. Heiskanen et al. compared menadione and menadione bisulfite using yeast, and they revealed that hydrophobic menadione was superior to its water-soluble bisulfite derivative for probing living cells [175]. Yeasts are easily handled, omnivorous to many kinds of organic substances, and stable even in saline solutions. Thus, we have applied the DM system to BOD estimation employing baker's yeast *S. cerevisiae* [176].

4.3.6 BOD_{RCI} Sensors

A simple, multiple simultaneous spectrophotometric method for new BOD determination using 2,6-dichlorophenolindophenol (DCIP) as the RCI was realized [27] (see Table 5VIII). The absorbance of DCIP decreases due to the metabolism of organic substances in aqueous samples by *P. fluorescens* (see Fig. 8). As a new technical tool, a microplate reader for a plate having 96 wells was used for multiple simultaneous spectrophotometric BOD_{RCI}

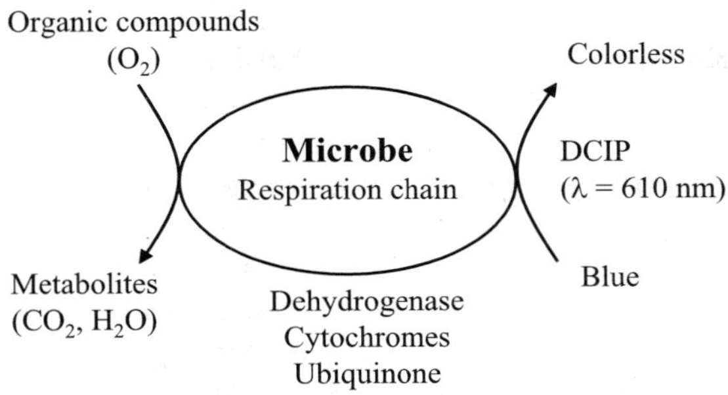

Fig. 8 Principle of BOD determination using DCIP by the spectrophotometric method

estimation. Thus, real samples, blanks, standards, and replicates could be determined concurrently, ensuring the accuracy of this method. The process required only 20 min per determination. This method gives a linear response ($r = 0.971$) to the OSS samples from 50 to 430 mg O_2 L^{-1}. Next, a compact optical device and disposable sensing strips for the simultaneous determination of the BOD of multiple samples were developed [27]. Using this strip, real samples from a wastewater treatment plant and a kitchen in our factory were examined, and good correlations were observed between the BOD_{RCI} values derived using this system and those determined by the conventional BOD_5 method.

Recently, we have developed a highly sensitive and reproducible BOD_{RCI} sensor using baker's yeast and a temperature-controlling system providing a three-consecutive-stir unit [177]. A calibration curve for GGA concentration was obtained between 1.1 and 22 mg O_2 L^{-1} ($r = 0.988$, six points, $n = 3$) when the incubation mixture was incubated for only 10 min at 30 °C. The reproducibility of the optical responses in the calibration curve was 1.77% (average of RSDs). This method was superior to the available BOD sensor (Central Kagaku Co.) in the detection limit (available BOD sensor's value, 2 mg O_2 L^{-1}), dynamic range (2–20 mg O_2 L^{-1}), reproducibility (5%), and measuring interval (30 min).

4.3.7 BOD_{MFC} Sensors

The first BOD biosensor, which was developed by Karube et al. in 1977, was based on a microbial fuel cell (MFC) using the hydrogen produced by anaerobe *C. butyricum* immobilized on the electrode [3]. Recently, applications of the mediatorless MFC to BOD sensor development have been performed using an electrochemically active metal-reducing bacterium [178–181]. The BOD_{MFC} sensors can be used as continuous monitoring systems and have long-term stability.

4.3.8 Surfactant Sensors

Linear alkylbenzene sulfonates (LASs) are most commonly used for the production of detergents in synthetic anionic surfactants. LAS is more easily biodegradable than nonlinear alkylbenzene sulfonate (ABS), and the biodegradation by microorganisms requires several days. However, LAS has toxicity in itself and also contributes to increase the toxicity of other pollutants in the aquatic environment. A large amount of LASs contained in domestic wastewater are flowed into streams and rivers. For the simple and rapid determination of LAS concentration, detergent biosensors indicating LAS were developed for river water monitoring [182, 183].

The first LAS biosensor was constructed using LAS degrading bacteria (strain A), which were isolated from activated sludge of a sewage treatment plant in Tokyo [182]. The LAS biosensor developed in this study is shown in Fig. 9. This is a reactor type sensor system consisting of immobilized LAS degrading bacteria and an oxygen electrode. The bacteria were immobilized in calcium alginate beads and the beads were packed into two columns (reactors). The two-column system was employed to raise the effects of LAS degradation. A sample solution was circulated in this flow type sensor system several times. Then, the LAS and DO concentrations in the sample were decreased by the LAS degrading bacteria packed in the two columns. By monitoring DO consumption in the sample, the LAS concentration was indirectly measured as well as showing the principle of the early BOD sensor. A calibration curve was obtained between 0 and 4 mg L^{-1} LAS. Finally, the LAS degradability of strain A in this sensor system was confirmed by using the high-performance liquid chromatography (HPLC) method.

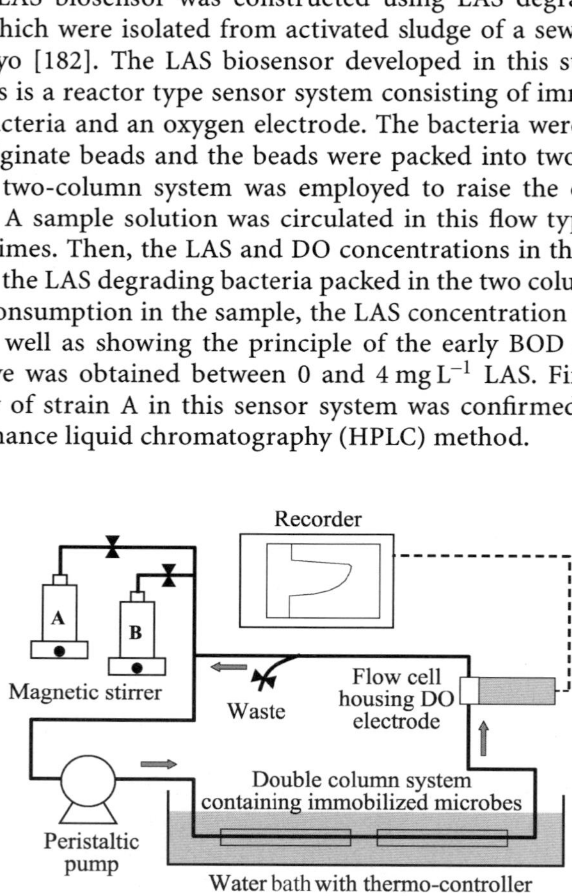

Fig. 9 Schematic diagram of the LAS measuring system using LAS degrading bacteria (strain A). A, sample; B, buffer or $CaCl_2$ solution. Measurement was performed within 15 to 30 min at 25 to 45 °C with a flow rate of 1.0 mL min^{-1}

Subsequently, a LAS biosensor for river water monitoring was developed using the strain A and *T. cutaneum* [183]. In this study, to prevent the influence of coexisting substances dissolved in a real sample, a dual sensing system was constructed. Changes of LAS concentration in river water were monitored and mostly corresponded with the human life cycle in a day.

As other surfactant sensors, a fungal conidial cell (*Botrytis cinerea*) was used for determining cationic surfactants [184] and *Pseudomonas* sp. and *Achromobacter* sp. were used for determining anionic surfactants [185].

4.3.9
Chlorinated Hydrocarbon Sensors

Trichloroethylene (TCE) is an extensively used industrial solvent, and due to its broad application and high volatility, it has become a major contaminant of ground water. In Japan, it was found by the Environmental Agency in 1982 that many ground waters were contaminated by TCE with a higher concentration than the environmental standard (0.03 mg L^{-1}). To realize simple and rapid determination of TCE, whole cell biosensors have been developed using specific bacteria to degrade TCE [186, 187]. The bacteria *P. aeruginosa* JI104 were isolated from soil near a gasworks [186]. This sensor system showed a linear concentration range for TCE from 0.1 to 4 mg L^{-1}. The response time was less than 10 min. This sensor characteristic makes it suitable for the detection of TCE in industrial wastewater. For improvement of this sensor system, a flow injection analysis (FIA) system was employed [187]. Finally, the sensor signals were linearly proportional to the concentration of TCE in the range from 0.03 to 2 mg L^{-1}. The sensor was applied to real samples and showed a good response for ground water.

As other chlorinated hydrocarbon sensors, a *Rhodococus* sp. was used for determining chlorinated and brominated hydrocarbons [188] and transformed *E. coli* was used for determining halogenated organic acids [189].

4.3.10
Toxicity Sensors

Toxicity assays are required for the prevention of environmental destruction and obstacles to the health of the human body. The toxicities are categorized into general toxicity and specific toxicity. The general toxicity includes the acute toxicity, i.e., 50% lethal dose (LD_{50}), subacute toxicity, and chronic toxicity. Toxicity assays have been widely required not only in the environmental field, but also in the food and diagnosis fields. Toxic substances are generally detected by using physical or chemical analytical systems, for example, HPLC and ion-selective electrodes (ISEs). However, the toxicity of such chemical substances cannot be measured. Only biological methods using living biomaterials enable measurement of the toxicity. As the most convenient way to detect or determine the toxicity, organisms such as bird, fish, and microorganisms have been used, especially for detecting acute toxicity. In particular, several microorganisms were used for biosensing of toxicants such as cyanides, pesticides, antibiotics, and other acute toxicants.

To detect the chemical toxicity which affects organisms by their acute toxicities, several biosensors were developed. In this field, the development of toxicity sensors can be classified into three types, i.e., indicating metabolic activity, respiration activity, or biodegradability of toxic chemicals. The former kind of biosensor was developed for total toxicity sensing. The latter two kinds of biosensors were mainly developed for cyanide detection, and these biosensors were reviewed in the literature [190].

In metabolic activity indicating methods, the "Microtox™" method using luminous bacteria was developed for the assessment of toxicity in aquatic samples and marketed as a bioassay system [191, 192]. In our group, several toxicity sensors were also developed using in vivo luminescence occurring in bacterial cells for the online monitoring or continuous measurement of toxic compounds.

The first bioluminescence-based toxicity sensor was studied in 1991 [193, 194] (see Table 6Ia). The toxicity was measured based on the decrease of in vivo luminescence intensity emitted by recombinant *E. coli* HB101 (pRSV), which was affected by cell metabolic inactivator. Toxicants such as sodium azide and fluoroacetic acid, which are components of ATP-inhibiting pesticides, and antibiotics were detected at or below the $\mu g\ L^{-1}$ level by this system. Using a similar measuring system and *P. phosphoreum* [195] (see Table 6Ib) or recombinant *E. coli* [196] (see Table 6Ic), pesticides and antibiotics were also detected [195, 196] (see Table 6Ib and Ic). Finally, the toxin sensor was improved to a batch-flow system using recombinant *E. coli* immobilized on magnetic beads [197] (see Table 6Id). In this system, illustrated in Fig. 10, an electromagnet is placed below the photomultiplier. The microbe-immobilized magnetic beads were incubated with toxicant solution for 30 min at 30 °C. After washing with buffer solution, the beads flowed inside the tube and were held above the electromagnet. The antibiotic chloramphenicol was detected in this system.

Cyanide is a deadly poison that inhibits respiration [198]. Nevertheless, it is widely utilized in industrial applications, especially for electroplating. Occasional accidents have taken place when industrial plants discharge cyanide into environmental water. The lethal dose is in the range of 0.5–3.5 $mg\ kg^{-1}$ body weight. Therefore, to regulate the discharge of cyanide into the environment, the Water Pollution Control Law in Japan stipulates 1 $mg\ L^{-1}$ of cyanide (38.5 μM) as the maximum concentration of cyanide allowed in wastewater. Several cyanide biosensors have been developed following this principle [199–201] (see Table 6II).

Table 6 Characteristics of toxicant sensors

	Target	Microbe	Indicator	Transducer	System	Calibration range ($mg\,L^{-1}$)	Measuring time (min)	Stability (d)	Refs.
Ia	Pesticide[a] and antibiotics[b]	*Recombinant Escherichia coli* HB101 (pRSV)	BL	PMT	Batch	0.05–10 (nonlinear)	30	–	[193]
Ib	BC[c], SDS[d], Cr(VI)	*Photobacterium phosphoreum* MT10204	BL	PMT	Flow & membrane	28.2–110 nM BC	< 2	–	[195]
Ic	Antibiotics[e] and herbicides[f]	*Recombinant E. coli* 207.T4	BL	PMT	Batch	0.5–10 CP, 0.1–100 Met	< 2	–	[196]
Id (Ib)	Antibiotics[e]	*Recombinant E. coli*	BL	PMT	Batch-flow	1.0–10	30	–	[197]
IIa	Cyanide ion (CN^-)	*Saccharomyces cerevisiae*IFO 0337	DO	Electrode	Batch & membrane	0.008–4	–	–	[199]
IIb	CN^-	*S. cerevisiae*	DO	Electrode	Flow & reactor	0.004–0.4	3	16	[200]
IIc	CN^-	*S. cerevisiae*	Two DO	Electrodes	Flow & reactor	0–0.4	7	9	[201]
IIIa	CN^-	*Pseudomonas fluorescens* NCIMB 11764	DO	Electrode	Batch & membrane	0.004–0.4	2	14	[202]
IIIb	CN^-	*P. fluorescens* NCIMB 11764	DO	Electrode	Batch & membrane	0.04–0.4	–	30	[203]
IIIc	CN^-	*P. fluorescens* NCIMB 11764	DO	Electrode	Flow & reactor	0.02–0.4	5	30	[204]

[a] Sodium azide, fluoroacetic acid [b] Chloramphenicol (CP), neomycin sulfate [c] Benzalkonium chloride [d] Sodium dodecyl sulfate [e] CP
[f] Metoxuron (Met), isoproturon, ioxynil, and propanil

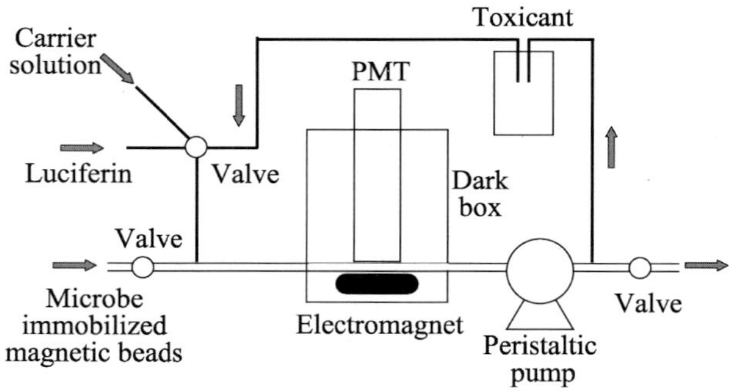

Fig. 10 Schematic diagram of the batch-flow type toxicity detecting system

Cyanide Sensor Indicating Respiration Activity of Aerobe

The first cyanide sensor was made by employing an oxygen electrode entrapped in yeast with an oxygen-permeable membrane, which was confirmed as a cyanide sensor [199]. The yeast (*S. cerevisiae* IFO 0337) was selected as a sensitive microorganism to cyanide. The yeast was entrapped between two porous cellulose nitrate membranes. The yeast in the microbial membrane can take up oxygen, glucose, and other nutritious substances through the porous membrane, being exposed to cyanide as well when it exists in the solution. This sensor was able to determine the cyanide ion (CN^-) with a linear range between 8.0 and 4000 $\mu g\,L^{-1}$ (0.3–150 μM), which indicates that this cyanide biosensor employing a flow system and a reactor has possible applications for the monitoring of cyanide.

The second sensor employed a flow system. [200]. Yeast was immobilized on glass beads acting as support; then, the beads were packed into a column acting as a reactor. The system employed two electrodes, and the reactor was placed between them. This system was able to detect the cyanide ion with a linear range between 0 and 400 $\mu g\,L^{-1}$ (0–15 μM), under the conditions of a flow rate of 4.5 $mL\,min^{-1}$ and 25 °C. These results indicate the possible use of this sensor in the construction of a flow sensor system that can be applied to continuous monitoring for preventing the discharge of cyanide from wastewater.

The third sensor was developed for determining cyanide in river water using an improved previous sensor system. The sensor employed a double electrode system and the reactor was set up between the two electrodes. As a result, the sensor was sensitive enough to detect cyanide contamination from industrial plants in river water.

Biosensors employing inhibition of microbial respiration are also affected by other toxic compounds, such as pesticides and herbicides. Therefore, this kind of sensor can be used for estimating total toxicity around a polluted water area. However, it is also important to improve the selectivity of the sensor as a cyanide detection system. Thus, subsequently, biosensors using cyanide-degrading bacteria for the selective determination of cyanide were developed.

Cyanide Sensor Using Cyanide-Degrading Microbe

P. fluorescens NCIMB 11764 having cyanide oxidase and cyanase aerobically biodegrades cyanide as a sole nitrogen source. Cyanide oxidase produces cyanate from cyanide, consuming oxygen. Then, the cyanate is hydrolyzed by cyanase, and ammonia and carbon dioxide are produced. Using this microbial degrading mechanism of cyanide, several biosensors were developed [202–204] (see Table 6III).

The first microbial sensor using this bacterial degradation of cyanide was employed in a batch system. Cyanide can be measured using an oxygen electrode to determine the decrease of DO by degrading cyanide of *P. fluorescens* NCIMB 11764. *P. fluorescens* was immobilized between two membranes. After characterization, the multiple effects of numerous substances on the sensor response were investigated using water from the Watarase River in Japan. The sensor responses to CN^- concentrations between 80 and 400 $\mu g\,L^{-1}$ (3.0–15 μM) were determined. However, the membrane type sensor was affected by the concentrations of nutrients, such as glucose and glutamate [202].

To improve selectivity, a cyanide-selective sensor using a gas-permeable membrane (PTFE; polytetrafluoroethylene) with a batch system was developed. Cyanide in water forms molecular acid hydrogen cyanide (HCN) and/or free cyanide ion (CN^-). Over pH 4, cyanides are easily converted into HCN. Liquid HCN can be volatile from water in the gas phase. By using the gas-permeable membrane, *P. fluorescens* immobilized in the microbial membrane was exposed to only volatilized gases such as oxygen and HCN. In this system, the nutrients did not affect the sensor response. A comparison of the JIS method (pyridine method) with the gas-phase biosensor was carried out using several river waters, adding CN^- with a concentration between 40 and 400 $\mu g\,L^{-1}$ (1.5–15 μM), with a good correlation ($r = 0.995$, $n = 4$). In conclusion, the sensitivity, selectivity, and stability of the gas-phase biosensor were considered adequate for practical use, suggesting the usefulness of this sensor. Therefore, cyanide biosensors employing a reactor with a flow system have been implemented for the continuous monitoring of cyanide in river water.

A flow and reactor type cyanide sensor using an immobilized *P. fluorescens* column was developed for environmental monitoring. *P. fluorescens* was entrapped in calcium alginate gel beads. Cyanide dissolved in the sample solution was degraded by *P. fluorescens*, and then consumed oxygen was detected by the DO electrode. After optimization, a calibration curve for CN^- was obtained with a linear response between 20 and 400 $\mu g\,L^{-1}$ (0.75–15 μM) and a response time of 5 min. The result obtained had a wider range than that of the previous sensors. The sensor response was obtained after at least 30 days. This sensor demonstrated its possible application for continuous detection of cyanide and online monitoring of river water, in the same manner as the flow and reactor type yeast sensor. However, this sensor's selectivity needs to be improved. Therefore, the development of a gas-phase biosensor employing a flow system is required to avoid the influences of nutrients in river water.

Other Toxicity Sensors

As other developments, luminescence-based genetically engineered microbial biosensors for water toxicity monitoring were developed using *S. cerevisiae* [205], freshwater bacterium [206], *E. coli* [207], or *E. coli* with an array system [208]. Luminescent bacteria were used for bioavailable toxic metals and metalloids from natural water samples [209]. Stress-responsive luminous bacteria were used for toxicants [210]. For gas toxicity monitoring, *E. coli* GC1 [211] or GC2 [212] was employed. For heavy metal toxicity sensing, a cardiac cell-based biosensor was developed [213]. Early detection of wastewater toxicity was performed by using a microbial sensing system [214]. A respirometric biosensor system was developed using *Nitrosomonas europaea* for the detection of inhibitors of ammonia oxidation in wastewater [215]. Green fluorescent protein-based genetically engineered microbial biosensors were developed using *S. cerevisiae* for genotoxicity monitoring [216]. For sediment quality control, application of microbial toxicity tests was reviewed [217]. As a new approach, biological toxicity detection was performed using a SPR system [218].

4.3.11 Agricultural Agent Sensors

As a novel approach to detect atrazine, SPR determination of P450 mRNA levels in *S. cerevisiae* was carried out [219]. A linear relationship was obtained between the SPR response and atrazine concentration in the range of 1 ng–1 mg L^{-1} ($r = 0.993$). This method was fast (15 min), highly sensitive, simple to use, and gave higher precision (< 2%) than the conventional enzyme-linked immunosorbent assay (ELISA) method. In addition, a highly sensitive toxicity measurement system was also developed utilizing quantification of induced P450 mRNAs by toxic chemicals and a FIA system based on a SPR method [220]. The DNA and peptide nucleic acid (PNA) probes containing a complementary sequence to a part of P450 mRNA hybridized to the probes were quantified. In this method, 10 ng L^{-1} (10 ppt) of atrazine was detected by using both DNA and PNA probes, and highly sensitive detection was achieved by amplifying the target P450 mRNA based on nucleic acid sequence-based amplification (NASBA).

As other agricultural agent sensors, organophosphorus aromatic nitro insecticides and *p*-nitrophenol were determined by the microbial-cell respiratory activity of *P. putida* C-11, *P. putida* BA-11, and *Acinetobacter calcoaceticum* A-122 [221], and organophosphate pesticides (nerve agents; paraoxon, parathion, and methyl parathion) were determined by using recombinant microorganism with surface expressed organophosphorus hydrolase [222]. Chloroform was detected by anaerobic microbial consortia [223].

4.3.12 Acid Rain Sensors

Acid rain is caused mainly by release into the atmosphere of oxides of sulfur (SO_x) and nitrogen (NO_x). To detect nitrogen dioxide (NO_2) gas, a high-performance biosensor was developed using *Nitrobacter* sp. which immobilized onto $CaCO_3$ particles [224, 225] (see Table 7Ia). The NO_2 gas sensor showed linear responses between 0.51 and 255 mg L^{-1} NO_2 with no interfering substances. For liquid samples, a sodium nitrite ($NaNO_2$) biosensor was also developed using *Nitrobacter* sp. which immobilized to an acetylcellulose membrane [226] (see Table 7Ib). The $NaNO_2$ biosensor showed linear responses between 10 and 590 μM $NaNO_2$ with no interfering substances.

For sulfite ion (SO_3^{2-}) sensing, *Thiobacillus thioparus* TK-m was used in a batch system [227] (see Table 7Ic). The SO_3^{2-} biosensor showed linear responses between 4 and 280 μM SO_3^{2-}. In this sensor system, it was found that sodium thiosulfate and sodium sulfate were interfering substances, although dimethyl sulfide did not influence the sensor response. To measure sulfate ion (SO_4^{2-}), a microbial sensor was also developed using *Thiobacillus ferrooxidans* strain 15 [228] (see Table 7Id). The SO_4^{2-} biosensor showed linear responses between 4 and 200 μM SO_4^{2-}. In this sensor system, it was found that sodium nitrate was an interfering substance, although NaCl did not influence the sensor response. In addition, these two biosensors had poor stability.

4.3.13 Carbon Dioxide Sensors

Carbon dioxide (CO_2) is known as a greenhouse effect gas and measurements of CO_2 are also important to clinical analysis and the fermentation process. Using *Pseudomonas* sp. S-17 as a CO_2 utilizing atrophic bacterium, the first biosensor for CO_2 gas was studied for a liquid sample [229] (see Table 7IIa). After optimization, the biosensor showed linear responses between 5 and 200 mg L^{-1} CO_2. Acetic acid was found to be an interfering substance; however, formic acid, ethanol, and butanol did not affect the sensor response. Using the same microbe, a chip type biosensor was studied by developing a miniature O_2 electrode using microfabrication techniques for semiconductors [230] (see Table 7IIb). This biosensor chip could measure from 0.5 to

Table 7 Characteristics of microbial biosensors developed by our group for environmental analysis

	Target	Microbe	Indicator	Transducer	System	Calibration range ($mg\,L^{-1}$)	Measuring time (min)	Stability (d)	Refs.
Acid rain									
Ia	NO_2 (gaseous)	*Nitrobacter* sp.	DO	Electrode	Semi gas flow & membrane	0.51–255	< 3	> 24 (400)	[224] ([225])
Ib	$NaNO_2$	*Nitrobacter* sp.	DO	Electrode	Flow & membrane	0.01–0.59 mM	< 10	> 21 (400)	[226]
Ic	SO_3^{2-}	*Thiobacillus thioparus* TK-m	DO	Electrode	Batch & membrane	4–280 µM	–	Some days	[227]
Id	SO_4^{2-}	*T. ferroxidans* strain 15	DO	Electrode	Batch & membrane	4–200 µM		1	[228]
Greenhouse effect gas									
IIa	CO_2 (gaseous)	*Pseudomonas* sp. S-17	DO	Electrode	Batch & membrane	5–200	3	30	[229]
IIb	CO_2 (gaseous)	*Pseudomonas* sp. S-17	DO	Electrode	Batch & chip	0.5–3.5 mM $NaHCO_3$	2 to 3	10 times	[230]
IIc	CO_2 (gaseous)	Thermophilic bacteria	DO	Electrode	Batch & membrane	1–8 mM	5 to 10	> 30	[231]
IId	CO_2 (gaseous)	Thermophilic bacteria	DO	Electrode	Batch & membrane	3–12%	–	> 30	[231]
IIe	Methane (gaseous)	*Methylomonas flagellata* AJ 3670	DO	Electrode	Gas-flow & reactor	13.1 µM–6.6 mM	< 1	> 20 (500)	[232]
IIf	Methane (gaseous)	*M. flagellata* AJ 3670	DO	Electrode	Gas-flow & reactor	3 µM–6.6 mM	< 1	> 10 (250)	[233]

3.5 mM $NaHCO_3$ within 3 min using ten measurements. In addition, both types of CO_2 gas biosensors for liquid and gaseous samples were fabricated using thermophilic bacteria [231] (see Table 7IIc and IId). Each biosensor measures between 1 and 8 mM CO_2 in liquid samples and between 3 and 12% CO_2 in gas samples.

Methane (CH_4) is known as a greenhouse effect gas and one of the indicators for the gasification process. *Methylomonas flagellata* AJ 3670 was used for CH_4 gas measurements and two gas-flow and reactor systems were constructed [232, 233] (see Table 7IIe and IIf). The first developed biosensor utilizing immobilized *M. flagellata* could measure 13.1 μM to 6.6 mM CH_4 gas, and the next developed sensor using a cell suspension improved the linear range of CH_4 to 3 μM.

4.3.14
Other Environmental Sensors

Microbial sensors for other pollutants have also been developed. The photosynthetic activities of microalgae of *Scenedesmus* sp., *Chlorella* sp., and *Spirulina* sp. were measured by a system consisting of an O_2 electrode integrated with optical fibers [234]. Phenolic compounds were detected by using *R. erythropolis* [235], *Moraxella* sp. [236], marine bioluminescent bacteria [237], and *P. putida* [238]. Naphthalene was detected by *P. putida* BS238 carrying the naphthalene degradable plasmid pBS2 [239]. Phosphate ion, which causes eutrophication, was detected using a luminescent cyanobacterial reporter strain [240].

5
Outlook

In this review of microbial biosensors, many kinds of developments were introduced. In most microbial biosensors, the selectivity is a shortcoming. Thus, genetically transformed microbes have recently been engineered as molecular recognition elements mainly for selective measurements [17]. For example, BOD sensing yeast, *T. cutaneum* IFO 10466, was genetically transformed using a plasmid, pAN 7-1, for luminous BOD sensing [241]. *Pseudomonas stutzeri* AK61 having the enzyme cyanidase was isolated from wastewater at a metal-plating plant by Watanabe et al. [242]. A gene encoding cyanidase was taken into *E. coli* for cloning and expression [243]. Further, the cyanidase was improved to increase the *Km* value for cyanide by site-directed mutagenesis [244]. Magnetic bacteria can be applied for the development of a new type of microbial sensor utilizing magnetic particles (bacterial magnetic particles; BMPs) [245], because the BMPs have been widely applied to the development of biochips and other biosensing devices [246].

References

1. Updike SJ, Hicks GP (1967) Nature 214:986
2. Nakamura H, Karube I (2003) Anal Bioanal Chem 377:446
3. Karube I, Mitsuda S, Matsunaga T, Suzuki S (1977) J Ferment Technol 55:243
4. Wilson K, Goulding KH (1986) A biologist's guide to principles and techniques of practical biochemistry, 3rd edn. Edward Arnold, London
5. Karube I (1987) Micro-organisms based sensor. In: Turner APF, Karube I, Wilson GS (eds) Biosensors: fundamentals and application. Oxford University Press, New York, p 13
6. Karube I, Suzuki M (1990) Microbial biosensors. In: Cass AEG (ed) Biosensors: practical approach. IRL, Oxford, p 155
7. Kubo I, Sode K, Karube I (1991) Whole-organism based biosensors and micro-biosensors, vol 1. In: Turner APF (ed) Advances in biosensors. JAI, London, p 1
8. Karube I, Chang SM (1991) Microbial biosensors. In: Blum LJ, Coulet PR (eds) Handbook of biosensors. Marcel Dekker, New York, p 267
9. Nakanishi K, Karube I (1994) Curr Opin Biotechnol 5:54
10. Karube I, Nomura Y (1996) Biosensors using immobilized living cells. In: Willaert RG, Baron GV, De Backer L (eds) Immobilized living cell systems. Wiley, New York, p 345
11. Bousse L (1996) Sens Actuators A 34:270
12. Nomura Y, Karube I (2000) Biosensors, vol 1. In: Lederberg J, Alexander M, Bloom BR, Hopwood D, Hull R, Iglewski BH, Laskin AI, Oliver SG, Schaechter M, Summers WC (eds) Encyclopedia of microbiology, Vol. 1, 2nd edn. Academic, Tokyo, p 611
13. D'Souza SF (2001) Biosens Bioelectron 16:337
14. Lei Y, Chen W, Mulchandani A (2006) Anal Chim Acta 568:200
15. Baronian KHR (2004) Biosens Bioelectron 19:953
16. Steinberg SM, Poziomek EJ, Engelmann WH, Rogers KR (1995) Chemosphere 30:2155
17. Rogers KR (2006) Anal Chim Acta 568:222
18. Clark LC Jr, Wolf R, Granger D, Taylor Z (1953) J Appl Physiol 6:189
19. Suzuki H, Tamiya E, Karbe I (1988) Anal Chem 60:1078
20. Yang Z, Suzuki H, Sasaki S, Karube I (1997) Anal Lett 30:1797
21. Preininger C, Klimant I, Wolfbels OS (1994) Anal Chem 66:1841
22. Li XM, Ruan FC, Ng WY, Wong KY (1994) Sens Actuators B 21:143
23. Chee GJ, Nomura Y, Ikebukuro K, Karube I (2000) Biosens Bioelectron 15:371
24. Kwok NY, Dong S, Lo W, Wong KY (2005) Sens Actuators B 110:289
25. Lin L, Xiao LL, Huang S, Zhao L, Cui JS, Wang XH, Chen X (2006) Biosens Bioelectron 21:1703
26. Jiang Y, Xiao LL, Zhao L, Chen X, Wang X, Wong KY (2006) Talanta 70:97
27. Yoshida N, McNiven SJ, Morita T, Nakamura H, Karube I (2002) Anal Lett 35:1541
28. Yoshida N, McNiven SJ, Yoshida A, Morita T, Nakamura H, Karube I (2001) Field Anal Chem Technol 5:222
29. Murakami Y, Kikuchi T, Yamaura A, Sakaguchi T, Yokoyama K, Tamiya E (1998) Sens Actuators B 53:163
30. Karube I, Suzuki S (1983) Application of biosensor in fermentation processes. In: Tsao GT (ed) Annual reports on fermentation processes, vol 6. Academic, New York, p 203
31. Karube I (1985) Biosensors in fermentation and environmental control. In: Arbor BA (ed) Handbook of biotechnology. Science Publishing, New Jersey, p 135
32. Karube I, Tamiya E (1987) Food Biotechnol 1:147

33. Karube I, Sode K (1989) Biosensors for fermentation process control. In: Ghose TK (ed) Bioprocess engineering: the first generation. Chapman Hall, London, p 207
34. Karube I, Sode K (1991) Microbial sensors for process and environmental control. In: Wise DL (ed) Bioinstrumentation and biosensors. Marcel Dekker, New York, p 1
35. Ohashi E, Karube I (1993) Food Control 4:183
36. Karube I (1994) Application of biosensors in measurement of food. In: Karube I (ed) On-line sensors for food processing. Gordon and Breach, Yverdon, p 1
37. Karube I, Suzuki S, Okada T, Hikuma M (1980) Biochimie 62:567
38. Hikuma M, Kubo T, Yasuda T, Karube I (1979) Biotechnol Bioeng 21:1845
39. Karube I, Suzuki S (1980) Enzyme Eng 5:263
40. Tkac J, Vostiar I, Gemeiner P, Sturdik E (2002) Bioelectrochemistry 56:127
41. Kitagawa Y, Tamiya E, Karube I (1987) Anal Lett 20:81
42. Tkac J, Vostiar I, Gorton L, Gemeiner P, Sturdik E (2003) Biosens Bioelectron 18:1125
43. Rotariu L, Bala C, Magearu V (2004) Anal Chim Acta 513:119
44. Liao MH, Guo JC, Chen WC (2006) J Magn Magn Mater 304:e421
45. Chiyo T, Matsui K, Murakami Y, Yokoyama K, Tamiya E (2001) Biosens Bioelectron 16:1021
46. Roustan JL, Sablayrolles JM (2003) J Biosci Bioeng 96:434
47. Matsunaga T, Karube I, Suzuki S (1980) Eur J Appl Microbiol Biotechnol 10:253
48. Hikuma M, Kubo T, Yasuda T, Karube I, Suzuki S (1979) Anal Chim Acta 109:33
49. Zhao YB, Wen ML, Liu SQ, Liu ZH, Zhang WD, Yao Y, Wang CY (1998) Microchem J 60:201
50. Hikuma M, Obana H, Yasuda T, Karube I, Suzuki S (1980) Anal Chim Acta 116:61
51. Rechnitz GA, Kobos RK, Riechel SJ, Gebauer R (1977) Anal Chim Acta 94:357
52. Kobos RK, Riechel SJ (1977) Anal Lett 10:751
53. Rechnitz GA, Riechel SJ, Kobos RK, Meyerhoff ME (1978) Science 199:440
54. Garjonytea R, Melvydasb V, Malinauskas A (2006) Bioelectrochemistry 68:191
55. Katrlk J, Brandsteter R, Svorc J, Rosenberg M, Miertus S (1997) Anal Chem 356:217
56. Karube I, Mitsuda S, Suzuki S (1979) Eur J Appl Microbiol Biotechnol 7:343
57. Reshetilov AN, Donova MV, Dovbnya DV, Boronin AM, Leathers TD, Greene RV (1996) Biosens Bioelectron 11:401
58. Reshetilov AN, Iliasov PV, Donova MV, Dovbnya DV, Boronin AM, Leathers TD, Greene RV (1997) Biosens Bioelectron 12:241
59. Hikuma M, Obana H, Yasuda T, Karube I, Suzuki S (1980) Enzyme Microb Technol 2:234
60. Svitel J, Curilla O, Tkac J (1998) Biotechnol Appl Biochem 27:153
61. Reshetilov AN, Lobanov AV, Morozova NO, Gordon SH, Greene RV, Leathers TD (1998) Biosens Bioelectron 13:787
62. Lobanov AV, Borisov IA, Gordon SH, Greene RV, Leathers TD, Reshetilov AN (2001) Biosens Bioelectron 16:1001
63. Held M, Schuhmann W, Jahreis K, Schmidt HL (2002) Biosens Bioelectron 17:1089
64. Hikuma M, Suzuki H, Yasuda T, Karube I, Suzuki S (1980) Anal Chem 52:1020
65. Karube I, Okada T, Suzuki S (1981) Anal Chem 53:1852
66. Okada T, Karube I, Suzuki S (1982) Anal Chim Acta 135:159
67. Ikeda M, Hachiya H, Ito S, Asano Y, Imato T (1998) Biosens Bioelectron 13:531
68. Matsunaga T, Karube I, Suzuki S (1978) Anal Chim Acta 98:25
69. Karube I, Wang Y, Tamiya E, Kawarai M (1987) Anal Chim Acta 199:93
70. Wang E, Karube I, Suzuki S (1986) Chin J Biotechnol 2:51
71. Kalman A, Caelen I, Svorc J (2006) J AOAC Int 89:819
72. Akyilmaz E, Yasa I, Dinckaya E (2006) Anal Biochem 354:78

73. Karube I, Kubo I, Suzuki S (1984) Ann NY Acad Sci 424:508
74. Karube I (1988) Methods Enzymol 137:131
75. Karube I, Matsunaga T, Teraoka N, Suzuki S (1980) Anal Chim Acta 119:271
76. Matsunaga T, Karube I, Teraoka N, Suzuki S (1981) Anal Chim Acta 127:245
77. Okada T, Karube I, Suzuki S (1982) Eur J Appl Microbiol Biotechnol 14:149
78. Kubo I, Karube I, Suzuki S (1983) Anal Chim Acta 151:371
79. Suzuki H, Tamiya E, Karube I (1989) Anal Lett 22:15
80. Karube I, Matsunaga T, Nakahara T, Suzuki S (1981) Anal Chem 53:1024
81. Karube I, Suzuki S (1982) Enzyme Eng 6:417
82. Karube I (1984) Trends Anal Chem 3:40
83. Karube I, Nakahara T, Matsunaga T, Suzuki S (1982) Anal Chem 54:1725
84. Karube I, Sode K, Suzuki M, Nakahara T (1989) Anal Chem 61:2388
85. Lee SM, Suzuki M, Kumagai M, Ikeda H, Tamiya E, Karube I (1992) Anal Chem 64:1755
86. Matsumoto K, Seijo H, Watanabe T, Karube I (1979) Anal Chim Acta 105:429
87. Karube I, Matsunaga T, Suzuki S (1979) Anal Chim Acta 109:39
88. Gu MB, Gil GC, Kim JH (1999) Biosens Bioelectron 14:355
89. Gao N, Zhao M, Zhang X, Jin W (2006) Anal Chem 78:231
90. Hobsen NS, Tothill IE, Turner APF (1996) Biosens Bioelectron 11:455
91. Ivinski D, Abdell-Hamid I, Atanasov P, Wilkins E (1999) Biosens Bioelectron 14:599
92. Zhou C, Pivarnik P, Rand G, Lecher SV (1998) Biosens Bioelectron 13:495
93. Ertl P, Mikkelsen SR (2001) Anal Chem 73:4241
94. Oczkowski T, Zwierkowska E, Bartkowiak S (2007) Bioelectrochemistry 70:192
95. Matsunaga T, Karube I, Nakahara T, Suzuki S (1981) Eur J Appl Microbiol Biotechnol 12:97
96. Ishimori Y, Karube I, Suzuki S (1981) Appl Environ Microbiol 42:632
97. Matsunaga T, Karube I, Suzuki S (1979) Appl Environ Microbiol 37:117
98. Ramsayl G, Turner APF (1988) Anal Chim Acta 215:61
99. Karube I, Tamiya E (1987) Pure Appl Chem 59:545
100. Karube I, Nomura Y, Arikawa Y (1995) Trends Anal Chem 14:295
101. Yano K, Sasaki S, Nomura Y, Ikebukuro K, Karube I (1998) Enzyme Eng 14:23
102. Belkin S (2003) Curr Opin Microbiol 6:206
103. Karube I, Matsunaga T, Suzuki S (1977) J Solid-Phase Biochem 2:97
104. Karube I, Matsunaga T, Mitsuda S, Suzuki S (1977) Biotechnol Bioeng 19:1535
105. Suzuki S, Karube I (1978) Enzyme Eng 4:329
106. Hikuma M, Suzuki H, Yasuda T, Karube I, Suzuki S (1979) Eur J Appl Microbiol Biotechnol 8:289
107. Bond RG, Straub CP (eds) (1973) Handbook of environmental control. CRC, Cleveland
108. Japanese Industrial Standard Committee Testing Methods for Industrial Waste Water, edited JIS, Tokyo (1993), JIS K 3602
109. Central Kagaku Co. http://www.aqua-ckc.jp/ (Nov. 2007)
110. Yang Z, Suzuki H, Sasaki S, Karube I (1996) Appl Microbiol Biotechnol 46:10
111. Yang Z, Suzuki H, Sasaki S, McNiven S, Karube I (1997) Sens Actuators B 45:217
112. Yang Z, Suzuki H, Sasaki S, Karube I (1997) Anal Chim Acta 357:41
113. Chee GJ, Nomura Y, Karube I (1999) Anal Chim Acta 379:185
114. Chee GJ, Nomura Y, Karube I (1999) Anal Chim Acta 394:65
115. Chee GJ, Nomura Y, Ikebukuro K, Karube I (2001) Sens Actuators B 80:15
116. Chee GJ, Nomura Y, Ikebukuro K, Karube I (2005) Biosens Bioelectron 21:67
117. Chee GJ, Nomura Y, Ikebukuro K, Karube I (2007) Biosens Bioelectron 22:3092

118. Kang KH, Jang JK, Pham TH, Moon H, Chang IS, Kim BH (2003) Biotechnol Lett 25:1357
119. Murakami K, Hasegawa K, Watanabe H, Komori K (1978) Public Works Research Institute Technical Memorandum, no 1421
120. Tanaka H, Nakamura E, Minamiyama Y, Toyoda T (1994) Water Sci Technol 30:215
121. Lehmann M, Chan C, Lo A, Lung M, Tag K, Kunze G, Riedel K, Gruendig B, Renneberg R (1999) Biosens Bioelectron 14:295
122. Chan C, Lehmann M, Tag K, Lung M, Gotthard K, Riedel K, Gruendig B, Renneberg R (1999) Biosens Bioelectron 14:131
123. Chan C, Lehmann M, Chan K, Chan P, Chan C, Gruendig B, Kunze G, Renneberg R (2000) Biosens Bioelectron 15:343
124. Riedel K, Renneberg R, Kuhn M, Scheller F (1988) Appl Microbiol Biotechnol 28:316
125. Kulys J, Kadziauskiene K (1980) Biotechnol Bioeng 22:221
126. Ohki A, Shinohara K, Ito O, Naka K, Maeda S, Sato T, Akano H, Kato N, Kawamura Y (1994) Int J Environ Anal Chem 56:261
127. Kim MN, Park KH (2001) Sens Actuators B 80:9
128. Kim MN, Kwon HS (1999) Biosens Bioelectron 14:1
129. Sangeetha S, Sugandhi G, Murugesan M, Madhav VM, Berchmans S, Rajasekar R, Jeyakumar D, Rao GP (1996) Electroanalysis 8:698
130. Tan TC, Neoh KG, Li F (1992) Sens Actuators B 8:167
131. Li F, Tan TC, Lee YK (1994) Biosens Bioelectron 9:197
132. Li F, Tan TC (1994) Biosens Bioelectron 9:315
133. Jia J, Tang M, Chen X, Qi L, Dong S (2003) Biosens Bioelectron 18:1023
134. Heim S, Schnieder I, Binz D, Vogel A, Bilitewski U (1999) Biosens Bioelectron 14:187
135. Karube I, Yokoyama K, Sode K, Tamiya E (1989) Anal Lett 22:791
136. Kim HB, Chang IS, Cheol GG, Park HS, Kim HJ (2003) Biotechnol Lett 25:541
137. Chang IS, Jang JK, Gil GC, Kim M, Kim HJ, Cho BW, Kim BH (2004) Biosens Bioelectron 19:607
138. Stand SE, Carlson DA (1984) J Water Pollut Control Fed 56:464
139. Rastogi S, Kumar A, Mehra NK, Makhijani SD, Manoharan A, Gangal V, Kumar R (2003) Biosens Bioelectron 18:23
140. Rastogi S, Rathee P, Saxena TK, Mehra NK, Kumar R (2003) Curr Appl Phys 3:191
141. Sakai Y, Abe N, Takeuchi S, Takahashi F (1995) J Ferment Bioeng 80:300
142. Liu J, Bjornsson L, Mattiasson B (2000) Biosens Bioelectron 14:883
143. Suzuki M, Takahashi S, Ishibashi M, Natsume K (1995) Sens Mater 7:159
144. Suzuki M, Ogata K, Tanaka M, Natsume K (1995) Denki Kagaku 63:1128
145. Tan TC, Qian Z (1997) Sens Actuators B 40:65
146. Qian Z, Tan TC (1998) Chem Eng Sci 53:3281
147. Qian Z, Tan TC (1998) Water Res 32:801
148. Qian Z, Tan TC (1998) Water Res 33:2923
149. Tan TC, Wu C (1999) Sens Actuators B 54:252
150. König A, Bachmann TT, Metzger JW, Schmid RD (1999) Appl Microbiol Biotechnol 51:112
151. Viaopoulou E, Melidis P, Kampragou E, Aivasidis A (2005) Biosens Bioelectron 21:365
152. Hyun CK, Inoue N, Tamiya E, Takeuchi T, Karube I (1993) Biotechnol Bioeng 41:1107
153. Hastings JW, Potrikus CJ, Cupta SC, Kurfurst M, Makemson JC (1985) Adv Microb Physiol 26:235
154. Sakaguchi T, Morita Y, Yamasaki M, Iwanaga J, Beppu K, Maeda H, Morita Y, Tamiya E (2007) Biosens Bioelectron 22:1345

155. Bennetto HP, Stiring JL, Tanaka K, Vega CA (1983) Biotechnol Bioeng 25:559
156. Roller SD, Bennetto HP, Delaney GM, Mason JR, Stirling JL, Thurston CF (1984) J Chem Technol Biotechnol 34B:3
157. Nishizawa S, Sakai S, Karube I, Matsunaga T, Suzuki S (1982) Appl Environ Microbiol 43:814
158. Learoyd SA, Kroll RG, Thurston CF (1992) J Appl Biotechnol 72:479
159. Kalab T, Skladal P (1994) Electroanalysis 6:1004
160. Kubiak WW, Wang J (1989) Anal Chim Acta 221:43
161. Katrlik J, Brandsteter R, Svorc J, Rosenberg M, Miertus S (1997) Anal Chim Acta 356:217
162. Pasco N, Baronian K, Jeffries C, Hay J (2000) Appl Microbiol Biotechnol 53:613
163. Morris K, Catterall K, Zhao H, Pasco N, John R (2001) Anal Chim Acta 442:129
164. Catterall K, Morris K, Gladman C, Zhao H, Pasco N, Richard John R (2001) Talanta 55:1187
165. Pasco N, Hay J, Webber J (2001) Biomarkers 6:83
166. Catterall K, Zhao H, Pasco N, John R (2003) Anal Chem 75:2584
167. Yoshida N, Yano K, Morita T, McNiven SJ, Nakamura H, Karube I (2000) Analyst 12:2280
168. OECD Guideline for Testing of Chemicals (1993) 209:1
169. Yoshida N, Hoashi J, Morita T, McNiven SJ, Nakamura H, Karube I (2001) J Biotechnol 88:269
170. Cabral JPS (1994) Microbios 78:47
171. Leefeldt RH, Martin A (1980) J Bacteriol 142:645
172. Yoshida N, Hoashi J, Morita T, McNiven SJ, Yano K, Yoshida A, Nakamura H, Karube I (2001) Analyst 126:1751
173. Yashiki Y, Yamashoji S (1996) J Ferment Bioeng 82:319
174. Baronian KHR, Downard AJ, Lowen PK, Pasco N (2002) Appl Microbiol Biotechnol 60:108
175. Heiskanen A, Yakovleva J, Spegel C, Taboryski R, Koudelka-Hep M, Emneus J, Ruzgas T (2004) Electrochem Commun 6:219
176. Nakamura H, Suzuki K, Ishikuro H, Kinoshita S, Koizumi R, Okuma S, Gotoh M, Karube I (2007) Talanta 72:210
177. Nakamura H, Kobayashi S, Hirata Y, Suzuki K, Mogi Y, Karube I, Anal Biochem 369:168
178. Gil GC, Chang IS, Kim BH, Kim M, Jang JK, Park HS, Kim HJ (2003) Biosens Bioelectron 18:327
179. Moon H, Chang IS, Kang KH, Jang JK, Kim BH (2004) Biotechnol Lett 26:1717
180. Chang IS, Moon H, Jang JK, Kim BH (2005) Biosens Bioelectron 20:1856
181. Kumlanghan A, Liu J, Thavarungkul P, Kanatharana P, Mattiasson B (2007) Biosens Bioelectron 22:2939
182. Nomura Y, Ikebukuro K, Yokoyama K, Takeuchi T, Arikawa Y, Ohno S, Karube I (1994) Anal Lett 27:3095
183. Nomura Y, Ikebukuro K, Yokoyama K, Takeuchi T, Arikawa Y, Ohno S, Karube I (1998) Biosens Bioelectron 13:1047
184. Cabral SMJCS, Vasconcelos MTSD, Cabral JPS (1995) J Microbiol Methods 24:11
185. Taranova L, Semenchuk I, Manolov T, Iliasov P, Reshetilov A (2002) Biosens Bioelectron 17:635
186. Han TS, Kim YC, Sasaki S, Yano K, Ikebukuro K, Kitahara A, Nagamine T, Karube I (2001) Anal Chim Acta 431:225

187. Han TS, Sasaki S, Yano K, Ikebukuro K, Kitayama A, Nagamune T, Karube I (2002) Talanta 57:271
188. Peter J, Hutter W, Stollnberger W, Hampel W (1996) Biosens Bioelectron 11:1215
189. Taube M, Rosen R, Belkin S (2001) Talanta 55:959
190. Arikawa Y, Ikebukuro K, Karube I (1998) Microbial biosensors based on respiratory inhibition. In: Mulchandani A, Rogers KR (eds) Methods in biotechnology, vol 6. Humana, New Jersey, p 225
191. Bulich AA, Isenberg DL (1981) ISA Trans 20:29
192. Kamlet MJ, Doherty RM, Veith GD, Taft RW (1986) Environ Sci Technol 20:690
193. Lee SM, Suzuki M, Tamiya E, Karube I (1991) Anal Chim Acta 244:201
194. Lee SM, Suzuki M, Tamiya E, Karube I (1991) Chem Express 6:415
195. Lee SM, Sode K, Nakanishi K, Marty JL, Tamiya E, Karube I (1992) Biosens Bioelectron 7:273
196. Lee SM, Suzuki M, Tamiya E, Karube I (1992) Anal Chim Acta 257:183
197. Karube I, Asakura J, Arikawa Y, Takei T, Sasaki S, Suzuki M, Tamiya E (1995) Denki Kagaku 63:1134
198. Ikebukuro K, Nakamura H, Karube I (2000) Cyanide. In: Nollet LML (ed) Handbook of water analysis. Marcel Dekker, New York, p 367
199. Nakanishi K, Ikebukuro K, Karube I (1996) Appl Biochem Biotechnol 60:97
200. Ikebukuro K, Honda M, Nakanishi K, Nomura Y, Masuda Y, Yokoyama K, Yamauchi Y, Karube I (1996) Electroanalysis 8:876
201. Ikebukuro K, Miyata A, Cho SJ, Nomura Y, Chang SM, Yamauchi Y, Hasebe Y, Uchiyama S, Karube I (1996) J Biotechnol 48:73
202. Lee JI, Karube I (1995) Anal Chim Acta 313:69
203. Lee JI, Karube I (1996) Biosens Bioelectron 11:1147
204. Lee JI, Karube I (1996) Electroanalysis 8:1117
205. Hollis RP, Killham K, Glover LA (2000) Appl Environ Microbiol 66:1676
206. Cho JC, Park KJ, Ihm HS, Park JE, Kim SY, Kang I, Lee KH, Jahng D, Lee DH, Kim SJ (2004) Biosens Bioelectron 20:338
207. Lee JH, Gu MB (2005) Biosens Bioelectron 20:1744
208. Lee JH, Mitchell RJ, Kim BC, Cullen DC, Gu MB (2005) Biosens Bioelectron 21:500
209. Tauriainen SM, Virta MPJ, Karp MT (2000) Water Res 34:2661
210. Belkin S, Smulski DR, Dadon S, Vollmer AC, Van Dyk TK, Larossa RA (1997) Water Res 31:3009
211. Gil GC, Mitchell RJ, Chang ST, Gu MB (2000) Biosens Bioelectron 15:23
212. Gil GC, Kim YJ, Gu MB (2002) Biosens Bioelectron 17:427
213. Liu Q, Cai H, Xu Y, Xiao L, Yang M, Wang P (2007) Biosens Bioelectron 22:3224
214. Liao JD, Wang SH, Hsu DJ (2001) Sens Actuators B 72:167
215. Cui R, Chung WJ, Jahng D (2005) Biosens Bioelectron 20:1788
216. Billinton N, Baker MG, Michel CE, Knight AW, Heyer WD, Goddard NJ, Fielden PR, Walmsley RM (1998) Biosens Bioelectron 13:831
217. Beelen P (2003) Chemosphere 53:795
218. Choi JW, Park KW, Lee DB, Lee W, Lee WH (2005) Biosens Bioelectron 20:2300
219. Lym TK, Oyama M, Ikebukuro K, Karube I (2000) Anal Chem 72:2856
220. Oyama M, Ikeda T, Lim TK, Ikebukuro K, Masuda Y, Karube I (2001) Biotechnol Bioeng 71:217
221. Guliy OI, Ignatov OV, Makarov OE, Ignatov VV (2003) Biosens Bioelectron 18:1005
222. Mulchandani P, Chen W, Mulchandani A, Wang J, Chen L (2001) Biosens Bioelectron 16:433

223. Pollice A, Rozzi A, Concetta Tomei M, Di Pinto AC, Laera G (2001) Water Res 35:1179
224. Okada T, Karube I, Suzuki S (1983) Biotechnol Bioeng 25:1641
225. Suzuki S, Karube I (1982) Enzyme Eng 6:387
226. Karube I, Okada T, Suzuki S (1982) Eur J Appl Microbiol Biotechnol 15:127
227. Marty JL, Sode K, Karube I (1992) Electroanalysis 4:249
228. Sasaki S, Hayashi C, Arikawa Y, Numata M, Yokoyama K, Tamiya E, Karube I (1997) Anal Chim Acta 347:275
229. Suzuki H, Tamiya E, Karube I (1987) Anal Chim Acta 199:85
230. Suzuki H, Kojima N, Sugama A, Takei F, Ikegami K, Tamiya E, Karube I (1989) Electroanalysis 1:305
231. Suzuki H, Tamiya E, Karube I, Oshima T (1988) Anal Lett 21:1323
232. Okada T, Karube I, Suzuki S (1981) Eur J Appl Microbiol Biotechnol 12:102
233. Karube I, Okada T, Suzuki S (1982) Anal Chim Acta 135:61
234. Takeuchi T, Yokoyama K, Imai O, Masuda Y, Utsunomiya K, Kobayashi K, Suzuki M, Tamiya E, Karube I (1993) Anal Chim Acta 276:65
235. Emelyanova EV, Reshetilov AN (2002) Process Biochem 37:683
236. Mulchandani P, Lei Y, Chen W, Wang J, Mulchandani A (2002) Anal Chim Acta 470:79
237. Philp JC, Balmand S, Hajto E, Bailey MJ, Wiles S, Whiteley AS, Lilley AK, Hajto J, Dunbar SA (2003) Anal Chim Acta 487:61
238. Timur S, Pazarlioglu N, Pilloton R, Telefoncu A (2003) Talanta 61:87
239. Reshetilov AN, Iliasov PV, Filonov AE, Gayazov RR, Kosheleva IA, Boronin AM (1997) Process Biochem 32:487
240. Schreiter PPY, Gillor O, Post A, Belkin S, Schmid RD, Bachmann TT (2001) Biosens Bioelectron 16:811
241. Sakaguchi T, Amari S, Nagashio N, Murakami Y, Yokoyama K, Tamiya E (1998) Biotechnol Lett 20:851
242. Watanabe A, Yano K, Ikebukuro K, Karube I (1998) Microbiology 144:1677
243. Watanabe A, Yano K, Ikebukuro K, Karube I (1998) Appl Microb Biotechnol 50:93
244. Watanabe A, Yano K, Ikebukuro K, Karube I (1998) Biochim Biophys Acta 1382:1
245. Sakaguchi T, Burgess JG, Matsunaga T (1993) Nature 365:47
246. Matsunaga T, Sakaguti T (2000) J Biosci Bioeng 90:1

Adv Biochem Engin/Biotechnol (2008) 109: 395–432
DOI 10.1007/10_2007_076

Published online: 13 November 2007

Strategies for Label-Free Optical Detection

Guenter Gauglitz (✉) · Guenther Proll

Institute of Physical and Theoretical Chemistry, Eberhard Karls University,
Auf der Morgenstelle 8, 72070 Tuebingen, Germany
guenter.gauglitz@ipc.uni-tuebingen.de

Abstract A large number of methods using direct detection with label-free systems are known. They compete with the well-introduced fluorescence-based methods. However, recent applications take advantage of label-free detection in protein–protein interactions, high-throughput screening, and high-content screening. These new applications require new strategies for biosensors. It becomes more and more obvious that neither the transduction principle nor the recognition elements for the biomolecular interaction process alone determine the quality of the biosensor. Accordingly, the biosensor system has to be considered as a whole.

This chapter focuses on strategies to optimize the detection platform and the biomolecular recognition layer. It concentrates on direct detection methods, with special focus on optical transduction. Since even this restriction still leaves a large number of methods, only microrefractometric and microreflectometric methods using planar transducers have been selected for a detailed description and a listing of applications. However, since many review articles on the physical principles exist, the description is kept short. Other methods are just mentioned in brief and for comparison. The outlook and the applications demonstrate the future perspectives of direct optical detection in bioanalytics.

Keywords Direct optical detection · Reflectometric · Refractometric · Optical transduction · Surface modification

Abbreviations

Ab	Antibody
AFM	Atomic force microscopy
AMD	Amino dextran
ATR	Attenuated total reflection
BIA	Biomolecular interaction analysis
BRE	Biological recognition elements
CCD	Charge coupled device
CMD	Carboxy dextran
DNA	Deoxyribonucleic acid
ELISA	Enzyme-linked immunosorbent assay
IR	Infrared spectroscopy
IR-ATR	Infrared attenuated total absorption
ITC	Isothermal titration calorimetry
ITO	Indium tin oxide
LNA	Locked nucleic acid
MALDI	Matrix-assisted laser desorption ionization
MIP	Molecular imprinted polymer
ODTR	Optical time domain reflectometry
OEG	Oligoethylene glycol
PDMS	Polydimethylsiloxane
PEG	Polyethylene glycol
PNA	Peptide backbone nucleic acid
POCT	Point-of-care testing
RIfS	Reflectometric interference spectroscopy
RNA	Ribonucleic acid
SAM	Self-assembled monolayer
SAW	Surface acoustic wave
SEIRA	Surface enhanced infrared absorption
SELEX	Systematic evolution of ligands by exponential enrichment
SERS	Surface-enhanced Raman spectroscopy
SPR	Surface plasmon resonance
THz	Terahertz spectroscopy
TE	Transversal electric
TM	Transversal magnetic
TOF	Time-of-flight
UV	Ultra-violet spectroscopy

1 Introduction

Whereas fluorescence-based assays [1] are standards in bioanalytics, label-free detection has gained some interest only since BiaCore entered the market some years ago [2]. Accordingly, the number of reviews on fluorescence-based methods is rather high; therefore only some of the more recent reviews will be mentioned here [3–12]. Many of these methods have been commercialized, as have microarrays (e.g., in gene expression analysis [13]) and chip technology (see chapter by Bier in this volume). However, moving from genomics to proteomics and also glycomics (–omics), in many cases direct optical detection can offer advantages. Fluorescence-based and enzyme-linked assays have a long tradition. They offer the possibility of high-throughput screening, are well-tested, highly suitable for routine analysis down to very low limits of detection, and even highly parallel, not considering just the Affymetrix chip [14]. State-of-the-art of microarray fluorescence analysis has been reviewed recently [15].

Nevertheless, there are problem with labeling of interaction partners (e.g., in the case of proteins) with respect to cost, loss of bioactivity, and toxicity of quantum dots (even though they are sometimes preferred because of their low photodegradation, high fluorescence enrichment factor, and other advantageous optical properties such as variation of fluorescence wave length even using a single excitation wave length). This has focused the interest in academia and in industry on substitution of the fluorescence assays in some applications by direct optical detection methods. It turns out that these optical approaches have opened new applications, allowed various test formats, and can be used in cases where fluorescence assays are not applicable.

It is true that fluorescence-based assays are rather simple, are well-introduced (especially in diagnostics), can be used in heterogeneous as well as in homogeneous assays, allow very low limits of detection, and provide high multianalyte detection even at small spots. Dual lifetime referencing has overcome problems with drift, turbidity, coloration and even photodegradation [16]. Matrix effects can be overcome more easily as fluorescence assays fight less against non-specific interaction. Dissemination, state of the art long-term development, and success in genomics have brought these assays into the position of being the leading assays they are at present.

However, upcoming research and development in optics, the variety of detection principles [15, 17, 18], and modern surface chemistry indicate that in the near future direct optical detection can take over many of the fields of application where fluorescence assays are still currently used.

Any type of biomolecular interaction process monitoring, which is the basis of all the –omics, screening, and diagnostic approaches, rely neither on the optical transducer nor on the quality of the interacting partners (binding constants, cross-reactivities) alone, but also on the whole system (including electronics, data acquisition, data handling). Thus, both aspects have to be discussed in this chapter with respect to recent developments. Trends will be outlined and perspectives will be given to allow an understanding of the optimization strategies for biosensors relying on direct optical detection.

In biosensing, the influence of the sensitive layer is considerable, therefore the wide field of chemosensors (being less specific because of the poor selectivity of the chemical sensitive layer) will not be discussed in detail, with the exception of a few applications using biomimetics in the section on applications (Sect. 3). This chapter concentrates on direct optical detection methods, namely quartz microbalance, electrochemical methods (which are reviewed in more detail in the chapter by Schuhmann in this volume), Terahertz spectroscopy, and calorimetric methods.

An essential part of the biosensor is determined by the sensitive layer, which is a low molecular weight ligand, a recognition sequence, or even a protein. However, to stabilize these recognition elements, sophisticated surface chemistry and surface modification have to be included in any biosensor development. Accordingly, a large part of this chapter will be dedicated to this approach, referring to the state of the art and giving some perspectives on using new biological recognition elements (BRE).

The section on applications (Sect. 3) will demonstrate developments in the past, show trends, and identify new aspects such as miniaturization and parallelization. Since topics such as evaluation of kinetics and thermodynamic data have been reviewed elsewhere, these aspects will be covered only in brief [19, 20]. However, miniaturization and parallelization will be an essential. The same applies to recently discussed imaging techniques under the aspect whether they just allow answers like "yes" or "no", or whether they offer the chance to obtain binding curves at a large number of detection sites on small chips, with validated reproducibility and calibration.

2 Biosensor Components: Transduction and Reception

2.1 Transduction: General Survey

In biosensing, a large variety of transduction principles have been introduced, many of which rely on direct detection principles avoiding labels. Accordingly, principles and applications have been published for electrochemical, electronic, calorimetric, and optical methods, which makes it rather difficult to get an overview and a good comparison between the large number of realizations [21]. Nevertheless, there are reviews comparing some of the direct optical detection methods [17, 22] and their performance with luminescence methods [23, 24].

This section focuses on optical transduction principles avoiding labels using direct detection of biomolecular interaction. Furthermore, non-optical methods will be mentioned briefly. Optical transduction techniques will be discussed not in detail, but only to demonstrate feasibility by classifying the many techniques using optical transduction alone. In addition, the section will concentrate on refractometric and reflectometric transduction principles. Nevertheless, a short survey on methods not using labels will be given.

At the beginning out of the many published principles, electrochemical methods first gained interest in biosensing. Potentiometric and amperometric methods are known, and even impedance measurements are used to overcome the disadvantage of single-channel information in comparison to optical methods using spectral information [25]. However, well-developed signal amplification and handling, as well as cheap overall instrumentation, allowed even high parallelization. Although the use of mediators in electrochemical transduction enhances the signals, as in the case of enzyme-linked assays, this approach is outside the scope of this chapter since the interaction process is not detected directly. Nevertheless, electrochemistry is considered one of the promising methods in biosensing, microarraying, and many approaches in the area of -omics [26–30].

Another approach formerly successful in the area of gas sensing is to benefit from mass changes at the sensor by affinity reaction. This change in mass can be detected by use of surface acoustic wave devices, quartz microbalances, or cantilevers. In comparison, piezoelectric mass-sensing devices have reached a better standing. Quartz resonators working at different frequencies allow label-free detection. Many years ago, their state of the art and their perspectives were discussed for analysis of DNA and RNA interactions, observation of changes in cells, and response to pharmacological substances, and their achievements compared to optical sensors [31, 32]. Recent developments are discussed in [33] in combination with some interesting applications [27, 34, 35]. The theory of cantilevers and their future prospects have been reviewed recently [36], and special attention was given to practical design concepts, use of cantilevers in liquid phase and especially in biological applications. Neither the necessity to work in liquids nor the mechanical restraints due to flow systems have been overcome yet.

At present, methods such as atomic force microscopy (AFM) are still a far cry from realization of a biosensor [37]. AFM is a powerful tool for exploring the forces and dynamics of the interaction between individual ligands and receptors, either on isolated molecules or on cellular surfaces. In this review, the current methodology for molecular recognition studies is explained, with an emphasis on strategies available for preparing AFM tips and samples, and on procedures for detecting and localizing single molecular recognition events. An interesting combination between AFM and electrochemistry has been published recently to get high resolution imaging of transports across membranes [38].

An upcoming principle is Terahertz (THz) spectroscopy [39]. Future success in biology will depend on the development of compact, low-cost and flexible systems. The present status of technology of using THz signals for marker-free biomolecular detection on functionalized surfaces in dry and fluid environments is discussed in [40]. The key feature of this method is the possibility of combining quantification of the interaction process with identification. This approach is interesting in the case of amino acid sequencing. However, this method has not yet left the research laboratories.

Regarding mid-infrared technologies, some approaches have been published that combine the ideas of a sensor with the intention to identify the analyte, whereby attenuated total reflectance techniques are used for signal enhancement. This has been done first in the case of chemosensors measuring in the IR, taking advantage of surface-enhanced infrared absorption (SEIRA) and surface-enhanced Raman scattering (SERS) [41] and later in biomedical applications (reviewed in [42]). The requirements of such sensor types are discussed in [43]. Successful real-world applications are known in gas analysis and detection of pollutants in marine water [44].

Meanwhile, applications even in in-vivo measurements have been published [45]. A recently developed infrared attenuated total reflectance (IR-ATR) catheter prototype for measuring atherosclerotic and normal aorta samples was used. Training data were collected ex vivo. In consequence, spectroscopy in combination with multivariate classification strategies allowed identification of normal and atherosclerotic aorta tissue for in vitro and, in the future, in vivo applications.

Recent developments in laser sources, especially laser diodes, Ti-sapphire lasers, or quantum cascade lasers, supply modern light sources in sensing [46, 47]. They provide variable wavelengths, even in near and mid-infrared or UV, high beam quality for in-coupling to waveguides or fibers. Information about properties of such lasers is best obtained by a search on the internet; some aspects are discussed in [48].

Raman spectroscopy has developed into a very reliable method using instrumentation with decreasing complexity. Raman has been successfully applied to many problems in medicine, therefore Raman spectroscopy and especially SERS (surface-enhanced Raman spectroscopy) is now frequently used to examine biomolecular interaction, e.g., for detection of integrins at the nanomolar concentration regime, and to distinguish between two different kinds of integrins [49].

Over the last decade, isothermal titration calorimetry (ITC) has developed from a specialist method that was largely restricted in its use to dedicated experts, to a major, commercially available tool for understanding molecular interactions [50]. The bacterial proteins A and G (SpA and SpG) are immunoglobulin receptors that can be used as probes for monitoring changes in the conformation of heavy chain constant (C(H)) domains. Interaction of anti-(4-hydroxy-3-nitrophenyl)acetyl (NP) antibody (Ab) with SpA and SpG were measured by isothermal titration calorimetry and surface plasmon resonance in order to address the question of whether hapten-binding induces a conformational change in the C(H) domain. Thus, this method proves its feasibility in interaction studies [51]. The correlation between reflectometric and calorimetric determination of binding constants is given in [52].

Without covering all non-optical methods completely and having mentioned optical approaches not yet on their way to commercialization, the following sections will concentrate on refractometric and reflectometric optical principles, approaches to miniaturization, to parallelization, and to overcoming the formerly unfavorable cost–result ratio. Finally, it will be demonstrated that modern optical devices allowing direct (even parallelized), time-resolved monitoring of binding processes are low-priced, offer simple technical setups, are user-friendly and robust, and are thus competitive to electrical sensor arrays.

2.1.1 Optical Fundamentals

Monitoring of electromagnetic radiation is a very successful tool for observing various types of biochemical interaction processes. Radiation provides parameters such as amplitude, frequency (wavelength), phase, and state of polarization [53, 54], which change during interaction processes. Accordingly, the time dependence is a further parameter. This variety of parameters makes superior use of electromagnetic radiation to some other methods and for many applications. Optics can be easily miniaturized, light sources and detection devices have become cheaper and more sophisticated in the last decade, and micromachining and microstructuring offer new interesting devices [53, 55].

This section is restricted to direct optical detection principles, and focuses on using planar transducers to select only a segment of the wide field of optical detection, which, however, is of high interest in bioanalytics at present. Apart from absorbance and emission, reflectance nowadays allows setting up many interesting transducers. For these devices, regular rather than diffuse reflectance is used.

Radiation incident onto an interface between two media with different refractive indices undergoes either regular reflection or even total reflection, besides refraction when passing the interface [54]. Accordingly, an overview of the methods of microrefractometric and microreflectometric principles is given. The benefits and drawbacks of the various approaches are demonstrated using samples from the biosensor field. In all applications reflectance is used. One approach is the measurement of multiple reflections at these layers and their superpositions of the partially reflected radiation at the interfaces. The other approach is to take advantage of total reflectance and the resulting evanescent field [54, 56].

In all realizations of these transduction principles, two requirements have to be achieved. The interaction process either within the sensitive layer or adjacent to it must cause a change in the properties of the electromagnetic radiation either guided in an optical transducer close to the sensitive layer or penetrating this sensitive layer. In either case, the "optical density" of the sensitive layer (containing the biological receptor elements and biopolymers shielding the transducer against non-specific interactions) influences the electromagnetic radiation. Thus, optical thickness is the product of the refractive index and the physical thickness of this layer. Refractometry focuses on measurements of changes in the refractive index and is therefore also temperature-dependent. On the other hand, reflectometry concentrates on monitoring changes of the optical thickness of the sensitive layer, caused either by uptake of molecules into this layer or by affinity reaction with the surface. The principles of reflectometry are shown in Fig. 1a, and those of internal total reflectometry as the basis of refractometry in Fig. 1b.

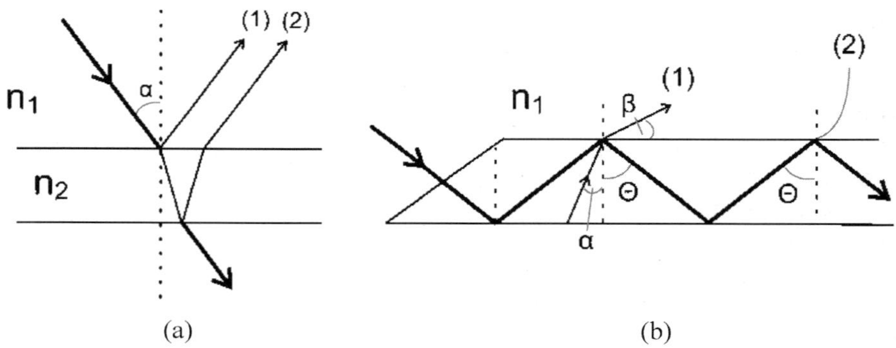

Fig. 1 Reflectometry: **a** Regular reflection; interference at thin film is *1+2* by superposition; incidence from low optical density. **b** Guided radiation; radiation is totally reflected if angle of reflectance θ becomes larger than a critical angle; beam at angle α is refracted according to Snellius law (*1*); *2* demonstrates the evanescent field in case of total reflection. In both cases $n_1 < n_2$

Accordingly, planar transduction is classified into microrefractomectry and microreflectometry. Since the principles and most of the realizations have been reviewed for years [15, 57, 58], the following sections discuss the most common principles only briefly, and concentrate on applications. Irradiation of the sensitive layers or in-coupling of radiation into the transducer as well as the interrogation of the effects are essential. These classify the different principles.

2.1.2 Refractometry

Radiation coupled into a medium with an higher refractive index than the surrounding and propagated at internal angles of reflection θ higher than the critical angle cannot be refracted to the outside of the medium any more. It is totally reflected at the medium interface and is "guided" within the medium (waveguide, fiber). Total reflection within a waveguide generates an evanescent field that penetrates the medium at the interface of the waveguide. It decays exponentially within just half the wavelength of the guided radiation [54] to the layer adjacent to the waveguide. Therefore, "total reflection" does not mean that the electric field vector of the guided radiation is not reduced to zero at the interface. It even exists outside. These devices have the advantage of detecting only effects within this penetration depth (just a few 100 nm) (Fig. 1b).

In all the refractometric methods, the refractive index in the medium in which the evanescent field decays influences the propagation of the guided radiation by means of an effective refractive index. A large number of methods are known. They differ in the interrogation principle of this effective refractive index. The novel design of planar optical waveguide biosensors based on reverse symmetry allows the tuning of the penetration depth of the evanescent electromagnetic field [59].

Integrated optical technology is developed further for biochemical applications. A short review is given on the development steps required to establish an integrated optical technology based on silicon, which includes the steps from the definition of the waveguide structure to the fabrication and characterization procedures. Aspects such as the numerical simulation for the components, the design of the optical element suitable as sensor, the analysis of the light input coupling methods, and the fitting of process parameters according to the required materials are highlighted in [60].

Surface plasmon resonance [61] is the best examined using the evanescent field technique; the theory and application to chemo- and biosensing have been reviewed in many articles [62–65]. The possibility of developing compact SPR probes combining spectral SPR and a planar sensor geometry was investigated. Two miniaturized set-ups were compared to a classical optical-bench device. Performance was assessed by use of the devices as a refractometer. The results were confirmed through two bioassays: (i) a protein multilayer system consisting of a biotin–albumin conjugate and polystreptavidin; and (ii) monitoring the interaction of thrombin and an immobilized thrombin inhibitor. All set-ups showed good long-term stability. Noise values from refractometry in the classical set-up matches those of bioassays, whereas increased noise levels were observed for the miniaturized set-ups, indicating a requirement for improved mechanical device stability [66]. Recently, new developments and applications have been discussed [67–70].

In-coupling of radiation into the waveguide is achieved via a prism that is coated on its base by an approximately 50-nm metal film. The prism takes care of total reflectance of incident radiation, which excites in these metal film plasmons via the evanescent field depending on angle of incidence and wavelength. Resonance of the plasmons takes place at the surface opposite (in case of SPR) to the waveguide interface adjacent to the medium of interest. The resonance condition of these plasmons depends on the refractive index of this medium and reduces the reflected intensity of the *p*-polarized light, resulting in a "dip" in the reflectance diagram. Either a prism or a waveguide with a buffer layer [71] to the metal film is used for achieving total internal reflectance. This direct optical detection method has been commercialized the longest [2]. In most biochemical publications reporting direct optical detection this method has been used.

Another frequently used principle is the grating coupler [72]. A waveguide layer is combined with a layer with a grating structure to couple in radiation (the same or a second grating can be used for the read-out). The grating constant is influenced by the refractive index within the adjacent medium. As for interference filters, this grating condition determines the preferred wavelengths or varies with the angle of incidence. Thus, the guided wave will depend on the gradient in the medium next to the waveguide. Radiation incident on the grating will be reflected or coupled into the waveguide, depending on refractive index, wavelength, and angle of incidence [73]. Either an angle-resolving arrangement or a CCD camera (avoiding mechanical parts) is used to monitor the reflected radiation [74]. This set-up was also commercialized [75] and recently gained new interest in parallelized devices [76]. Bi-diffractive couplers [77] have two gratings with different grating constants superimposed. The out-coupled wave has an angle different from that of the directly reflected wave. Rather tricky are gratings embossed in polycarbonate, which take advantage of non-parallel grooves of the grating or a thickness gradient of the waveguide [78].

Another type of interrogation of the polarization status is applied in prism couplers [79]. The prism is not coated by a metal film but rather by a transparent low refractive index layer. The radiation couples out of the prism via the frustrated total internal reflection of a low refractive index layer (with a thickness of 1000 nm) into the high-refractive-index waveguide. Forty-five degree polarization is chosen, and TM and TE modes travel in the resonant layer (waveguide thickness 100 nm), differently influenced via the evanescent field by changes in the adjacent medium. Thus, the polarization state changes in this "resonant mirror" [80, 81].

Another possibility is demonstrated by devices that combine evanescent field techniques with interferometric principles. Such devices as Mach–Zehnder chips determine the difference between the phases of two waves traveling in two arms of a waveguide [82–85]. Their perspectives are discussed with respect to biochemical applications supplying high-contrast waveguides with very low channel attenuation and modal birefringence [86]. Sensitivity can also be enhanced by phase modulation using magneto-optic material for these interferometric sensors [87].

The Young interferometer is a similar type of interferometer, the waveguide arms not reunifying but rather imaging the interference pattern produced by the two open ends of the waveguide arms on a CCD [74]. Using both TE and TM modes enables internal referencing. This has been improved by Lukosz [88] in its mode beat interferometer, measuring amplitudes and phases of both polarization states. Another mode beat approach is published in [89]. Refractive index rather drastically depends on temperature. Accordingly all devices based on evanescent field techniques require extreme temperature stabilization (in the case of biomolecular interaction analysis as low as 0.02 K) or good referencing.

2.1.3 Reflectometry

Ellipsometry is a long-known method for characterizing thin layers [90–95]. Polarized light is incident to this thin layer and reflected at each interface. They superimpose. The resulting intensity modulation with detection wavelength depends on the thickness of the layer and the refractive index, which influences the phase and/or amplitude of the electromagnetic radiation penetrating this layer or being reflected. Ellipsometry enables separation of the refractive index and the physical thickness when using many wavelengths. Ellipsometry was introduced as far back as in the 1940s and it has regained interest nowadays in modern semiconductor and wafer technology as a simple control technique. In bioanalytics it enables characterization of sensitive layers and is used not only to characterize simple polymer films, but also biopolymers [96, 97].

The use of reflected non-polarized electromagnetic radiation is represented schematically in Fig. 2a [54]. Because one part of the radiation is usually reflected at the interface of a thin layer, whereas the other penetrates the layer and is there reflected at the other interface, these two partial reflected beams can become superimposed and form an interference pattern, resulting in constructive or destructive interference depending on the angle of incidence, wavelength, and optical thickness of the layer. The optical thickness is given by the product of refractive index and physical thickness of the layer. The modulation of this interference pattern, as demonstrated in Fig. 2b, depends on these properties of the layer and changes sensitively in response to changes in or at this layer. The association and dissociation curves can be easily calculated (see Fig. 2c).

This simplified version of ellipsometry, called reflectometric interference spectroscopy (RIfS) [98–101], provides a simple and robust technique in biosensing, as demonstrated later in the Sect. 3. The principle of the arrange-

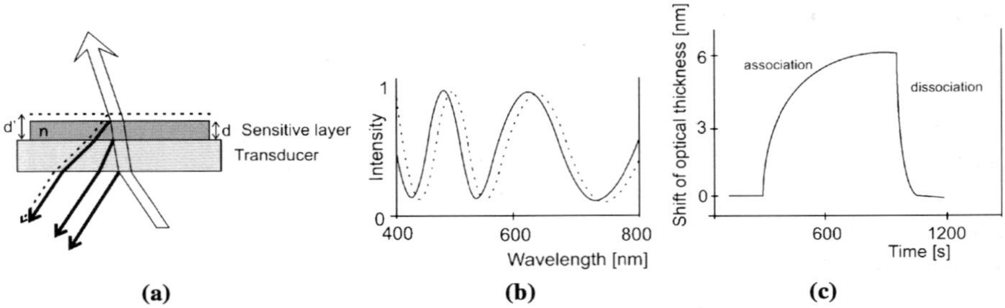

Fig. 2 **a** Sensitive layer with changing physical thickness by binding process. **b** Related shift in the interference pattern. **c** Association curve; dissociation brings the original interference spectrum back

ment is given in Fig. 3 [102]. Any change in the thickness of the layer or in the refractive index influences the superposition of the interfering partial beams and shifts the modulated spectral intensity, the interference spectrum. Principle and applications are reviewed in [58, 103]. A special approach is high-speed optical time domain reflectometry (OTDR) for high spatial resolution and for "smart" structure applications [104]. Similar approaches have been used to observe changes in a sensor layer taking up hydrogen [105]. Meanwhile RIfS has been applied by other groups and some results are published [106–115].

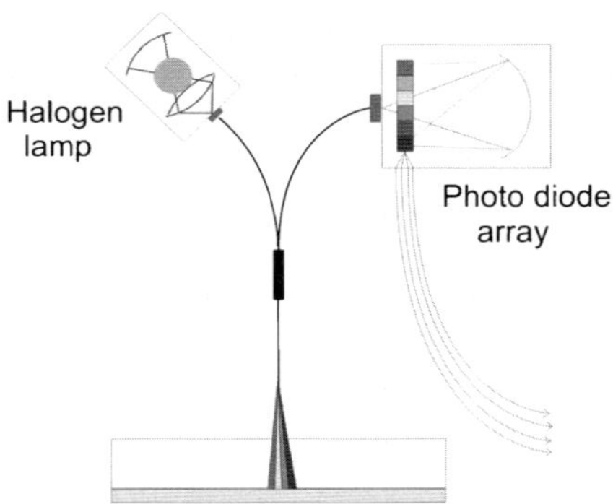

Fig. 3 Reflectometric interference spectroscopy set-up, using a white light source and a diode array spectrometer for monitoring; time resolution is in a range of less than seconds

Fortunately, any increase in layer thickness through temperature is rather well compensated by reduction of refractive index. Thus, the product of physical thickness and refractive index (optical thickness), which is responsible for the shift and is measured by RIfS, is rather independent of temperature. This turns out to be a great advantage compared to all evanescent field techniques [116]. Therefore, RIfS is optically very simple and robust, does not really require temperature stabilization and allows parallelization in an easy way. SPR and RIfS have been compared [117]; and also some evanescent field techniques with RIfS [118, 119]. Furthermore, optical and electrical transducers were compared using a modified cyclodextrin host for enantioselective detection of a halogenated diether as chiral guest [120].

2.2
Trends in Transduction

Developments of direct optical detection techniques are guided by new interesting applications in the field of bioanalytics rather than finding new optical principles, even though sensors in the mid-infrared, using Raman spectroscopy or working in the THz range, are under research. Appropriate citations have been given. Looking at existing methods also successfully in use, two strategies of future development can be identified. One intention is to simplify optics and the sensor system as a total, achieve a simple, cheap, reliable and robust optical equipment to be used for process control or security purposes or even as hand-held home-care applications. The other intention of research and development is directed to parallelized devices, taking advantage of miniaturization and modern CCD-detection equipment. Both developments have many applications in their focus to meet the urgent needs of modern monitoring or fast analysis.

Some interesting new approaches have also been given and many reviews cited above. Thus, analyte-responsive holograms comprise a holographic grating embedded in a smart hydrogel film. The grating acts as a reporter that enables analyte-induced changes in the thickness of the associated polymer films to be accurately detected. These holograms are inexpensive, robust, and have proven suitable for the detection of a wide range of clinically and industrially important analytes [121].

An overview is presented of the main optical biosensors, the operating principle of the different devices, the design of the sensor, the technology of application, the resolution, the dynamic range and detection limit of each device, the most important applications, and the commercial devices on the market. An outlook of the future prospects for this technology is given [122]. A new type of label-free optical biosensor that is inexpensively manufactured from continuous sheets of plastic film and incorporated into standard format microplates to enable highly sensitive, high-throughput detection of small polymer molecules, proteins, and cells. Instrumentation is reviewed in application to two fundamental limiting issues for assays in proteomics research and drug discovery: requirement for quantitative measurement of protein concentration and specific activity, and measurements made with complex systems in highly parallel measurements [123].

2.2.1
Simplified Set-Ups

The first approach to simplify optical sensor systems competes with semiconductor oxides. However, combination of a simplified optical system and a selective sensitive layer can be a successful strategy. Since RIfS is a robust and simple method barely dependent on temperature, stripping down spectral detection to a few wavelengths gains the advantages of a simple device, cheap realization, and selectivity. Accordingly, a simple device has been developed using LEDs at four wavelengths [124, 125] (see Fig. 4). The LEDs have a diameter of 5 mm and a transparent housing (PMMA) with a microlens on top. They achieve the highest possible intensity of radiation. LEDs with a low divergence were chosen: super-bright (Toshiba, Nichia), with wavelengths between 470 and 644 nm and bandwidth between 15 and 30 nm. The spectral detectivity of the photodiode was selected to match the emission wavelength of the LEDs. The set-up was used in comparison to the spectral RIfS approach to determine the feasibility of such simple approach applied to the measurement of ozone-depleting refrigerants and their substitutes [126].

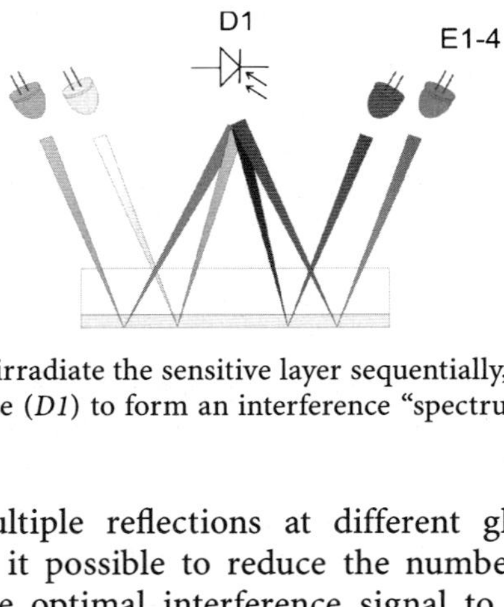

Fig. 4 Four LEDs (*E1-4*) irradiate the sensitive layer sequentially, the intensities are measured with the photodiode (*D1*) to form an interference "spectrum"

Simulation of multiple reflections at different glass-type transducers and polymers made it possible to reduce the number of wavelengths necessary to obtain the optimal interference signal to just one wavelength. Transducers were selected according to the system monitored (biosensors, chemosensors [127]). Theoretical considerations and first applications are given in [128–130].

The instrumental approaches for different set-ups are given in Fig. 5.

Another interesting application of reflectometric interference spectroscopy at total reflection conditions is the use of a Kretschmann configuration and a semicircular prism where the angle of incidence is configurable between 10° and 90°, where the amplitude of interference can be increased, and where there is no background reflection [131–133].

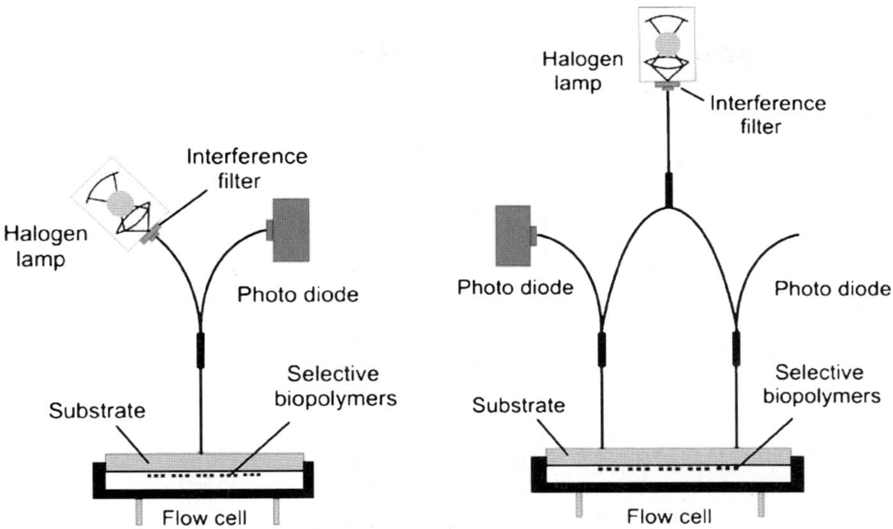

Fig. 5 Two single wavelength arrangements, the one on the *right* is a two-channel approach for internal referencing [127]

Reflectometric interference spectroscopy can also be combined with measurements of zeta potentials and interface conductivity, which yields interesting information on the reactions of physical, chemical, biochemical, and biological reactions at an interface [134].

2.2.2
Parallelized Set-Ups

The other strategy of development of direct optical sensing systems is using recent developments of array detector elements at lower price, better resolution, and advanced low noise to high signal ratio. Since surface plasmon resonance was the first direct optical method to be commercialized, researchers started to try setting up multichannel systems. The first was a multichannel surface plasmon resonance sensor based on the combination of parallel sensing, channel architecture, and the method of spectral discrimination of sensing channels [135]. The potential of this mixed architecture sensor was demonstrated by the detection of monoclonal anti-dinitro-phenyl antibody compensating non-specific adsorption and background refractive index interferences.

A rather new approach is a novel multichannel SPR biosensor with spectro-interrogation and wavelength division, multiplexing of sensing channels. An eight-channel SPR sensor combines the wavelength division multiplexing of zero-sensing channels with a conventional channel architecture.

Its application is described in [136]. Another approach is the technique of SPR imaging, which allows visualization of 2-D areas and provides an alternative method to the exiting detection schemes of biochips [137]. These techniques such as imaging SPR and imaging ellipsometry are reviewed in [138]. Microarray-based assays using these techniques are given to demonstrate both their usefulness and their limits of detection. A discussion of the remaining challenges as well as trends and future application of microarrays are presented in the context of optical sensing.

In parallel, reflectometric interference spectroscopy was developed into a multichannel system [139], and a prototype set-up was developed in co-operation with Carl Zeiss within the project LIBRARIAN [140–142] which sequentially selects eight wavelengths from a white-light source using a filter wheel. These irradiate a microtiter plate system, which can be filled by an automatic dispenser device. For the eight wavelengths in turn, a CCD camera detects intensities of all wells in parallel. During one filter wheel cycle, an "interference spectrum" is obtained for each well within just a few seconds. This set-up allows sequential measurements, at many time points, of the binding curves in each of the wells in parallel. This set-up allows parallel measurement of all wells (up to 384) at any time during the binding process. The feasibility of this set-up will be demonstrated in the next section, "Applications". The prototype of this set-up is shown in Fig. 6. The set-up can be miniaturized and provides a perspective for measuring in microarrays at up to thousands of spots.

Recently, the EPIC system [76] was introduced by Corning and provides a HTS-compatible microplate reader, capable of reading 384 microplate wells. Another rather interesting approach in parallelization is the use of waveguide

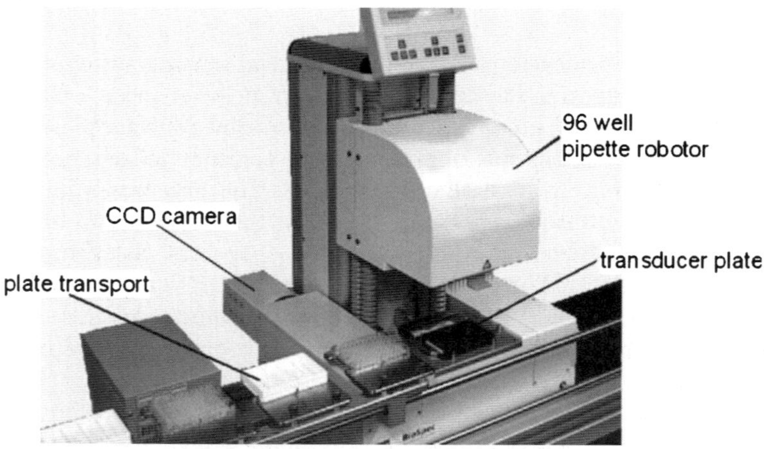

Fig. 6 Prototype of a parallelized RIfS system including dispensing, suitable for measuring 96 or 384 wells

cantilever channels, which allow high integration. For an array of 20 cantilevers, experimental data can be obtained for optical cantilevers coated with an adsorbent material [143].

2.2.3
Hyphenated Techniques

In chromatography, modern instrumentation uses photodiode arrays as a detection device with only poor selectivity. However, many years ago, people already started to combine direct optical detection methods and chromatography to improve selectivity. A first example was the online coupling of a label-free optical biosensor to a HPLC system to detect pesticides in the femtomole range without preconcentration, and comparison with a study using fluorescence-based biochemical detection [144]. In this system, the separation capacity of HPLC was combined with the specifity of a biosensor. Another approach was an immunosensing system based on surface plasmon resonance in the online detection and characterization of carbohydrates during high-performance liquid chromatography measurements. By using weak and readily reversible monoclonal antibodies, the SPR system allows specific online monitoring of the substances [145]. A third interesting approach is the monitoring of antibody affinity chromatography using a label-free optical biosensor technique RIfS [146]. This approach can substitute classical ELISA methods for the characterization of antibodies and their polyclonal fractions with regard to binding capacity, and thus improve antibody purification and separation procedures as well as process control in antibody production.

Another interesting combination of direct optical detection with classical analytical methods is the combination with MALDI-TOF. The development and techniques of biomolecular interaction analytical mass spectrometry (BIA/MS) and application to analyzed proteins from natural systems are discussed. The combination of MALDI-TOF and SPR allows monitoring of the interaction process and identification of the relevant molecules [147]. This unique combination of SPR and matrix-assisted laser desorption time-of-flight mass spectrometry is reviewed with many references [148]. Furthermore, the BIA/MS limit of detection at very low SPR responses have been investigated. For the first time, the detection of in-vivo assembled protein complexes was achieved using BIA/MS [149].

Although MALDI systems require a conductive layer and therefore surface plasmon resonance chips can be used easily, the screening capability of RIfS can also be combined with the identification capability of mass spectrometry. In this case, ITO-coated (indium-tin oxide) transducers instead of simple glass-type surfaces have to be used. This has been verified in the determination of quantitative and qualitative binding of mixtures of vancomycin derivatives [150].

This biomolecular interaction analysis can also be combined with electrophoretic flow conditions. The coupling of BIA and electrophoresis provides many prospects for proteomics and genomics. The advantage of direct optical detection methods is that they work label-free, allow time-resolved process monitoring, and allow miniaturization and parallel measurements. Therefore, methods such as evanescent field techniques, surface plasmon resonance, Mach–Zehnder interferometers, grating couplers, and quartz microbalances could be considered to be used in combination with electrophoresis. However, all these evanescent field techniques are very much temperature-dependent or use gold-coated transducers, which is in direct opposition to the intention of performing electrophoretic measurements.

Since ellipsometry is too complex, reflectrometric interference spectroscopy is the method of choice. Recently, published approaches using enzyme-linked immunosorbent assays [151], surface plasmon resonance [152, 153], or affinity capillary electrophoresis [154, 155] are either restricted to capillary electrophoresis or they are systems that cannot be used under electrophoretic flow conditions due to their sensitivity to temperature changes or their conducting sensor surfaces. However, the combination of reflectometric interference spectroscopy and biomolecular interaction allows combination of both methods and even detection of the separated analytes selectively [156]. In the future, modern analytics will certainly use a strategy to combine these various separation methods with the potential of direct optical detection.

A combination of spectroscopic methods (direct optical with fluorescence) has been published for merging reflectometric interference spectroscopy (RIfS) with total internal reflectance fluorescence (TIRF) [157–159].

2.3 Surface Modification

The key to the functionality of all discussed sensor principles in this article is the introduction of a surface modification on the transducer. In the area of biosensors, a very large number of methods deal with monitoring of interactions occurring at these surfaces, or at least use surface-bound ligands or receptors to monitor changes in analyte concentration, even within a homogeneous phase. For all these detection methods there is a need for very special and smart surface properties. Selectivity, sensitivity, stability, and reversibility are examples of such requirements for these sensor systems, which must be provided in part by the surface properties. In the case of biosensors, the user expects a rather high signal-to-noise ratio, short response times, low limits of detection, high sensitivity, and the possibility to use sensors also in real samples, not just in laboratory applications. An overview on biomaterials and processes already in use as well as an outlook for this important field is given in [160]. A review on immunosensor principles is given in [161].

2.3.1
Biopolymer

For all detection methods based on heterogeneous phase interactions non-specific binding effects are a very important point. To overcome this problem, very sophisticated surface chemistry and modifications have been developed and are a basic requirement to set up an efficient biosensor. Therefore, a large variety of biopolymers [162] have been investigated to achieve surface characteristics and properties designed for a specific application. Especially in the field of DNA and protein arrays 2-D and 3-D surface modifications based on a variety of biopolymers have been reported [163]. The most important biopolymers will be discussed in more detail here.

Dextran hydrogels [164] supply a large number of functional sites within the volume [165]. In addition, these 3-D polymers offer very good properties regarding the suppression of non-specific binding and can be functionalized with different groups such as amino and carboxy groups (AMD and CMD, respectively). An overview on currently available biopolymers based on dextrans is given in [166].

Beside the dextran biopolymers, different approaches based on polyethylene glycol (PEG) with different chain lengths have been reported to be superior in supplying shielded surfaces with reduced non-specific binding properties [167] and allowing functionalization with ligands or receptors [168]. In contrast to the dextran hydrogel, the PEG surfaces form a kind of a 2-D polymer brush, and interaction only occurs at the surface, not in the volume. Therefore, these layers have a reduced number of interaction sites. Also, oligoethylene glycol has been reported to produce functionalized silicon or diamond surfaces with a very good suppression of protein adsorption [169]. The interaction of PEG and phospholipid surfaces with biological systems is reviewed in [170].

The requirements for non-specific binding are enormous, especially in the field of molecular diagnostics via label-free techniques with whole blood, plasma, or serum as matrices. Masson et al. [171] have shown that a simple surface coating with 16-mercaptohexadecanoic acid can also act as effective shielding. Apart from these biopolymers, there are many more polymers (e.g., polydimethylsiloxane, PDMS) or microporous systems or other coatings (e.g., polylysine, dendritic linkers) in use. These polymers can also be used to introduce a pattern of special surface properties on a transducer. PDMS for example can be used to create surfaces that allow cell adhesion in a spatially controlled way [172].

Immobilization at Transducer

The choice of an appropriate immobilization technique for biopolymers on a transducer is predominantly dependent on the material of the solid support [173]. In general, there are the possibilities of adsorption, covalent attachment, physical entrapment, and polyelectrolyte interaction. This section will concentrate on the most commonly used support materials.

Immobilization on gold surfaces is very easy via thiols. In many cases, the thiol containing molecules organize as an orientated and stable self-assembled monolayer (SAM). The phenomenon of SAMs is widely in use, especially in the field of biosensors, and has been reviewed by Schlereth [174]. In the case of glass or quartz transducers one important approach is silanization, with the subsequent covalent binding of various biopolymers supplying stable surfaces and reproducible immobilization with reduced non-specific binding properties.

The first step in this procedure is to activate the surface to form hydroxyl groups on the substrate. Several protocols can be applied to do this. The three most successful and widely used ones are Piranha solution (mixture of conc. H_2SO_4 and H_2O_2), nitric acid, and plasma activation (method of choice for many unusual materials like TiO_2 [175]). The silanization step can be characterized by NMR spectroscopy, ellipsometry [176], and by determining the contact angle. An interesting approach is a protocol of a surface modification based on a mixed silane layer to immobilize proteins for the biosensor with imaging ellipsometry [177]. Recently the group of Kamisetty [178] has shown that glass surfaces can be effectively amine-functionalized by an additional silanization with alkylsilanes.

In the case of silicon nitride as the solid support (e.g., Mach–Zehnder chips) methods for the silanization of semiconductive materials can be applied [179]. Silicon surfaces as used, e.g., in ellipsometry are superior in terms of very low surface roughness and low background fluorescence. Native silicon surfaces react with oxygen under ambient conditions to form a thin layer of silicon oxide. These surfaces are then modified according to the same protocols described for glass and quartz. In general, it can be said that there is a multitude of silanization protocols and follow-up chemical reactions that allow the immobilization of a huge variety of functional molecules in a well-controlled manner [180].

Another approach is to modify glass surfaces via non-covalent immobilization approaches such as hydrophobic interaction or with polycations (e.g., polylysine) and then to attach charged ligands. These approaches are of course very easy to apply but for most applications they do not provide the required stability. In the case of polylysine, the combination with a silanization step seems to be superior [181]. Another very much advanced adsorption-based modification method was proposed for niobium pentoxide-coated silicon wafers with poly(L-lysine)-*g*-poly(ethylene glycol) (PLL-*g*-PEG) copolymers. Surfaces with a high PEG chain density appeared to be superior to non-specific adsorption of proteins [182].

Although originally developed for the modification of quartz surfaces [183], polyelectrolytes also offer a very simple possibility for modifying plastic transducers [184]. Currently, the number of reports and conference contributions on this field is increasing. For example, polyelectrolytes can also be used to build a sensitive layer for the detection of metal ions by SPR [185]. Furthermore, the formation of polyelectrolyte multilayers can lead to very stable modifications with designed properties including an improved biocompatibility, as reported by [186]. Other methods to improve the physicochemical surface properties of materials used in medicine are described in [187]. Out of the variety mentioned there, plasma polymerization should be mentioned here as a very interesting method for creating surfaces with ideal properties for studying cell and protein interactions [188].

In recent years, there has been a trend to use plastic support materials, especially because of mass production aspects. Although plastic transducers are generally used in fluorescence-based assays, they can also be used for label-free methods, as has been shown for RIfS [127]. In general, the problem with plastic surfaces is that they do not offer reactive groups for easy covalent modification. Activation or modification via plasma methods fails in most cases because of the high process temperatures.

Apart from the already mentioned adsorptive or polyelectrolyte-based modification strategies, the current method of choice is photolinking. This method allows the formation of covalent bonds between the biopolymer or other molecules and the surface. To do so, classical techniques derived from photochemistry have been adapted in such a way that so-called photolinkers (photoactivatable heterobifunctional crosslinkers) are used to covalently attach to the surface after irradiation with light (e.g., UV). The selection of appropriate photolinkers and reaction conditions, and the influence of the support material are discussed in [189–192].

A comprehensive overview on surface-modification chemistries, including photolinking, for utilization in microarrays and other applications can be found in [193–195]. The technique of photoimmobilization can also be used to produce arrays with cells. The group of Ito [196] immobilized blood cells for the detection of antibodies. Another interesting approach is micropatterning of glass surfaces via various functional groups using photoactivatable biotin [197]. This allows the immobilization of other biomolecules making use of the biotin/streptavidin affinity system.

2.3.2 Recognition Sites

Recognition sites are introduced to further functionalize a transducer or to make use of them as tools in affinity immobilization. In the latter case, a large number of ideas has been realized, e.g., the use of avidin to immobilize biotinylated biochemical molecules to the transducer surface [198], the use of His-tags [199] or Strep-tags, or the immobilization of membrane structures of lipid double layers [200] to the transducer to model cell walls.

All these various approaches have the intention to reduce non-specific binding, allow a large number of specific binding sites, increase the stability of the layer (which is essential for regeneration strategies), and increase selectivity and sensitivity. The disadvantage of such approaches is that due to the high binging constant the system is not reversible at all. Therefore regeneration strategies have to be introduced to regain the reusability of these sensing surfaces.

The variety of recognition sites to introduce specificity to a transducer seems be unlimited. All kinds of imaginable natural biomolecules are in use: antibodies, receptors, DNA, membranes, proteins, peptides, and even cells. One major drawback of these biomolecules is that although they increase the stability of the biolayer, their stability is not comparable with that of, e.g., polysiloxane films or microporous systems [201]. In order to try to combine the advantages of stability and reversibility with sensitivity and selectivity, a wide field of research for many years has been supramolecular structures [202] and biomimetic layers. To increase selectivity of simple chemosensors supramolecular structures such as calixarene [203] were tried first. Another approach was the use of cyclodextrins [204] or cycloheptapeptide [205] structures.

Hybridization studies were one of the first approaches to introduction of selectivity by immobilizing polynucleotide or peptide sequences. Meanwhile, peptide nucleic acids (PNA, with peptides as a backbone) [206] and locked nucleic acids (LNA, ribose moiety is modified with an extra bridge connecting $2'$ and $4'$ carbons) [207, 208] have proven to be better complementary binding systems than DNA. In the case of PNA this is also because of less repulsion by charges and better backbone stability, thus being stable against DNases and nucleases. The unique properties of both DNA analogs is reviewed in [209].

Another approach is the synthesis of layers of molecular imprinted polymers (MIPs) [210]. During a co-polymerization process, a template is used to form a cavity with certain spacings and some selective binding sites. After removing the template the generated recognition and binding site is highly selective for the template (or molecules similar to the template) as a considered analyte. Although these layers are very stable and selective they often show very slow response times upon analyte exposure. To increase response times in the so-called spreader bar technique [211] the imprinting process is reduced only to the surface instead of the volume. Current developments in this area are more directed towards the recognition of proteins [212] and other molecular receptors [213].

Aptamers represent another promising class of artificial receptors (they are originally derived from natural aptamers). These special nucleic acid species (RNA or DNA) are produced during a process called SELEX (systematic evolution of ligands by exponential enrichment) to specifically bind to a molecular target of choice such as small molecules, proteins, or even cells. Beside therapeutic applications they are used within affinity-based assays comparable to antibodies in immunoassays. Current research on combining the advantages of aptamers with sensor approaches has created the wording "aptasensors". Additional information on this topic is provided in [214–216].

3 Applications

Developments using sensors have been reviewed recently by Niessner [217] as well as in a biannual review covering fiber-optic chemical sensors and biosensors [21]. Every year an increasing number of papers is published dealing with improved or new applications to various fields. Any review can cover only some of these fields, and will be subjective. In the following, typical publications in the areas of environmental analytics, biology, microbiology, medical diagnostics, biotechnology, and in the –omics areas will be cited in brief, with no claim to completeness.

3.1 Sensitive Layer and Assays

Functionality of biosensor arrays based on SPR imaging is mainly defined by the quality of patterned SAM, which is used for sensing of biological substances of interest. It can be shown that changes of the molecular structure in the thin film also induce changes in the SPR signal [218]. Essential is the thermostability of oligonucleotides used as sequences for hybridization studies. Label-free and time-resolved sensing technologies, such as RIfS in an assay format similar to a titration called the binding inhibition assay, are used to discriminate locked nucleic acids (LNA) in comparison to other systems to demonstrate an enhanced thermostability and robustness against nuclease-mediated cleavage [219].

A general review of protein sensing assay formats and devices is given in [220]. Proteins are used as biocatalysts, therapeutic or diagnostic agents, and as such they are biotechnological products as well as biomarkers for health states, diseases, or toxic or other adverse effects, and the intercellular protein for the protein network is essential for the adaptation of an organism to its environment. These proteins have to be examined carefully with their interaction with other systems. Selecting only affinity reactions between an immobilized capture agent and the target protein, the paper reviews the applicability of various methods, the choice of the immobilization substrate, the assay formats, and compares advantages and disadvantages for the different approaches. Examples were chosen that illustrate the potential of the different systems.

Non-specific binding is an urgent problem in biomolecular interaction analysis (BIA). To reduce this problem, binding of a protein to carboxymethylated-dextran (CMD) surfaces was investigated using RIfS. Subsequently, these surfaces were optimized for use as optical biosensors. The influence of immobilization time, concentration of the polymer solution, the amount of carboxylic groups, and mixing the solution during preparation of the surfaces were examined. Three different concentrations of the polymer solution were compared using two different immobilization times for two CMD polymers. For the first time, surface acoustic waves (SAW) were applied for the preparation of CMD surfaces to accelerate immobilization of the polymer without loss of quality. With mixing the solution during silanization and immobilization of the polymer, respectively, the time for preparation of the sensors could be successfully reduced without increase in non-specific adsorption. The relative amount of reactive binding sites was examined via RIfS using the biotin/streptavidin system [221].

3.2
Immunoassays, Environmental and Food Analysis

The characterization of low molecular weight ligand interaction with receptor molecules is of importance for the investigation of biological processes and for drug research. The investigation of the binding of such type of ligands to immobilized receptors by label-free detection is reported. Reflectometric interference spectroscopy allowing the monitoring of a few picograms per square millimeter changes in surface coverage was used to study two model systems. In both cases detection of the binding event was successful. High affinity binding of biotin to immobilized streptavidin was clearly detectable at receptor surface concentrations as low as $1–2 \times 10^{10}$ bindingsites/mm^2. Linear correlation between the receptor surface concentration and the response to biotin binding was observed. Using immobilized DNA, the binding of common intercalators with respect to kinetics and thermodynamics by evaluation of the association and the dissociation part of the binding curve was investigated. Bi-exponential increase and decrease of intercalator loading was observed, indicating complex interaction kinetics. The four structurally different intercalators showed significant distinction in binding kinetics and equilibrium signals. Improvement of experimental parameters is required to obtain more reliable kinetic data [222].

In the last decade, there has been significant progress in the development of known techniques for evaluation of receptor/ligand interactions. With regard to this, SPR-based optical biosensors are extensively used to define the kinetics of a wide variety of micromolecular interactions, and high- and low-affinity small molecular interactions. The brief review describes the SPR technology and a few of its applications in relation to receptor/ligand interactions that have brought about a significant change in the interpretation of SPR signals [223].

Nanotechnology is an interesting approach for immunoassays. Recent applications are discussed, using immunoassays on a chip, nanoparticle-based immunoassays, and nanobiosensors and label-free devices [224].

Another interesting field of increasing research and development activity is the introduction of SPR biosensing technology in environmental monitoring. The development of environmental sensors has been in the focus for the past two decades. Applications and interesting substances are discussed. For the future it is expected that advances and the development of mobile SPR sensor platforms will enable this technique to contribute to the protection of local ecosystems and public health [225]. The applicability of such devices is impeded by harsh work conditions and the complexity of the sample matrixes. To overcome these environmental problems, a microsystem based on glass and polymeric substrates has been developed [226]. RIfS can also be used to determine binding constants and characterize the interaction process, as in the validation of the pollution by benzo[a]pyrene. Kinetic and thermodynamic data can be obtained [227].

Concerted research and development efforts to bring SPR biosensor technology to the field and meet the need for the rapid detection and identification of chemical and biological substances has been realized for human safety and security. Important areas such as medical diagnostics, environmental monitoring, food safety, and security are in the focus of research. These key functions are reviewed in [225]. Of interest are affinity-based label-free biosensors that yield real-time information as well as analyte concentrations in food, agricultural, nutraceutical, and environmental samples. These can realized by way of several transduction techniques used in biosensor devices. They include fiber optic, surface plasmon resonance, optical grating coupler, optical waveguide light mode spectroscopy, and evanescent wave techniques [228].

A review exploits optical surface resonance biosensors for the direct and indirect detection of pathogenic microorganisms in the food chain and environment. Infections with Herpes simplex and human immunodeficiency viruses, Salmonella and Treponema pallidum bacteria were revealed using human sera and avian eggs [229].

Another paper reviews optical sensors and biosensors for environmental monitoring applications. Topics include: optical sensors; new developments in optoelectronic technologies (plastic optical fiber, other special fibers, new developments in light sources, miniature gratings and spectrometers, sol-gels, optical fiber sensors, waveguide and integrated optics sensors); types of optical biosensors (optical immunosensors [evanescent wave, grating coupler, surface plasmon resonance, and reflectance interference spectroscopy immunosensors], Fourier transform IR spectroscopy, microbial sensors, toxicity measurement with Eclox, Aquanox, and Microtox); and optical toxicity and fluorescence sensors (toxicity assay, sensor design) [230].

The sensitive, rapid, and specific detection of microorganisms and toxins that damage the food supply has become increasingly important. Large-scale manufacturing with wide distribution can threaten large populations when contamination occurs. Biosensors can supply a higher level of surveillance. Numerous detection techniques are available, and the authors hope to present a representative cross-section of the useful techniques [231]. An introduction of demanding mycotoxin regulations and respective methods of analysis are given [232]. The biosensors can be integrated in hazard analytics and critical point programs [233]. SPR biosensor techniques are discussed in comparison to ELISA approaches with respect to the development of veterinary drug tests [234]. Even screening assays have been developed [235].

A further application is the information on implants. Reflectometry can be used to characterize the initial pellicle formation on modified titanium dioxide surfaces as a model system for dental implants [236].

3.3 Biotechnology

In biotechnology an application of biosensors is interesting for monitoring processes in bioreactors. Unfortunately, the literature only covers classical sensors measuring parameters such as pH, oxygen concentration, or CO_2 as indirect fermentation parameters. Biorectors are closed systems in which microorganisms can be cultivated under defined, controllable conditions. These have to be optimized with regard to viability, reproducibility, and product-oriented productivity. Optical sensors are an efficient tool to obtain this information [237]. The negligible dependence of RIfS on temperature allows the method to be used to directly monitor the fermentation process. Comparison to HPLC or GC/MS measurements and validation is possible for the emergency antibiotic vancomycin [238]. A review of process analysis technology in biotechnology is given in [239], discussing the advancement from enzyme screening to bioanalytical systems. However, literature on monitoring biotechnological processes lacks modern biosensor techniques.

3.4 Medical Diagnostics

The use of immunosensors in environmental science and in the pharmaceutical and food industries is most common. Similarly, clinical problems may be solved by continuous monitoring of certain analytes. There are many recent developments in the immunosensor field with potential impact. The future role of this technique in intralaboratory and bedside testing will become even more important as the clinical laboratory is faced with increasing pressure to contain costs [240].

SPR biosensors for medical diagnostics are emerging systems. It can be demonstrated that they can be used for a variety of disease biomarkers, hormones, and drugs at clinically relevant levels. It is expected that advances in SPR sensor instrumentation (reduced size, improved sensitivity, increased throughput) in the area of biorecognition elements and methods for their immobilization (increasing sensitivity and specificity) are of interest. They will lead to new systems for rapid detection and identification of disease biomarkers [33, 68, 241].

Surface plasmon resonance biosensor technology can be optimized for a rapid, direct, and low-consumption label-free multianalyte screening of synthetic oligonucleotides with modified internucleotide linkages potentially applicable in antisense therapy [242]. First attempts in antisense diagnostics using direct optical detection were successful on microspots [243].

Conventional SPR is applied in specialized biosensing instruments. These instruments use expensive sensor chips of limited reuse capacity and require complex chemistry for ligand or protein immobilization. However, SPR with colloidal gold particles in buffered solution has successfully been applied. This application offers many advantages over conventional SPR. The substrate is cheap, easily synthesized, and can be coated with various proteins or protein–ligand complexes by charge adsorption. With colloidal gold, the SPR phenomenon can be monitored in any UV-vis spectrophotometer. For high-throughput applications, the technology has been adopted to an automated clinical chemical analyzer. This simple technology finds application in label-free quantitative immunoassay techniques for proteins and small analytes, in conformational studies with proteins, in the real-time association–dissociation measurements of receptor–ligand interactions, and for high-throughput screening and lead optimization [244].

All in all, the area of POCT (point-of-care testing) and bedside as well as home care will require simple instrumentation in the future. At present, only fluorescence-based systems are thought to meet the requirements. However, the first attempts to compete by using direct optical detection look promising.

3.5
–Omics and Parallelization

The field of biosensors and biochips for nucleic acid diagnostics has developed significantly over the last decade. High-throughput techniques offer advantages in the areas of medical diagnostics, forensics, environmental monitoring, and bioterrorism. The necessary considerations for the preparation of immobilized nucleic acid films on a solid sensor substrate and the development of techniques utilized for the detection of selective hybridization of target binding materials is described. Special consideration is given to current methods used for the detection of interfacial DNA hybridization [245].

A new multichannel biosensor combining an optical platform, based on surface plasmon resonance imaging on special multilayers, and polarization contrast with a spatially resolved functionalization is reported. The optical platform offers a considerably higher sensitivity and resolution than conventional SPR imaging. The sensor can perform 64 independent measurements simultaneously, and its limit of detection for 23-mer oligonucleotides is demonstrated to be as low as 100 pM [246]. Recent advances in nucleic acid-based biosensors have led to the development of DNA biosensors for DNA hybridization detection and for nucleic acid–ligand binding studies. Such optical devices include optical fibers, planar waveguide, surface plasmon resonance, resonant mirror, and surface-enhanced Raman scattering. The specificity and response of each optical DNA biosensor are discussed. All in all, the establishment of optical DNA biosensor technology turned out to be a major tool of analytical biochemistry [247]. Another review focuses on improvements in optical detection methods, especially those using direct optical biosensing principles developed for screening of protein–protein interactions in high-throughput screening applications. It summarizes the principles of optical sensors and discusses the advantages of the various techniques. Examples of applications in biomolecular interaction analysis are given for receptor–ligand, protein–protein, DNA–protein, or DNA–DNA affinity binding using real-time measurements [248].

Reflectometric interference spectroscopy allows online monitoring of solid phase peptide syntheses and subsequently the detection of antibody binding to these peptides without cleavage from the support. The step-by-step coupling of different amino acids to the transducer surface was investigated, and complete monitoring of the synthesis of a viral epitope was performed. The success of the synthesis was proven via binding of a specific monoclonal antibody to the transducer-bound product. Since both synthesis and interaction with the biological receptor can be monitored using the same technique, the approach is attractive especially in the field of high-throughput screening [249].

The gastrointestinal disorder coeliac disease (CD) is caused by the ingestion of wheat gluten, and is characterized by damage of the typical structure of the intestinal mucosa. The tissue enzyme transglutaminase (tTGase) was identified as the major target of disease-specific antibodies in patients. An epitope fine-mapping was performed with a series of pentadecapeptides synthesized using parallel multiple peptide synthesis. For the detection of biomolecular interactions, a label-free parallel method, RIfS, was used. This is the first optical label-free method adapted to a high-throughput screening (HTS) format, and the experimental results demonstrate its applicability as a biological screening device. A high titer of anti-tTGase antibodies is found in the serum of coeliac patients. By these means, the first step has been taken towards a fast non-surgical test for the detection of these antibodies. In order to identify and characterize a continuous epitope with high affinity against the anti-tTGase antibody, a screening of 21 pentadecapeptides has been achieved using the parallel RIfS system. A single-channel RIfS system with high resolution was used to detect binding constants of identified peptides with high affinity [250].

Combinatorial triazine libraries were synthesized in the wells of microtiter plates. By this means, it was possible to determine the quality of antibodies against these triazine derivatives with one single experiment in a parallel set-up using RIfS [251]. In the same set-up, an assay of label-free HTS of thrombin inhibitors was successfully achieved [252]. RIfS was also used as a label-free optical immunoprobe for pesticide detection [253].

Using conventional methods such as ELISA to examine even the most obvious interaction of, for example, a therapeutic antibody, can add weeks or months to the development process and clog up R&D programs with thousands of man-hours. With the ever-increasing understanding of the complexity of human protein interactions that have a direct impact on safety and efficacy of therapeutic interventions, the responsibility of researchers and drug companies to fully characterize drug candidates becomes a more and more exhausting task. Failure to fully examine the nature of important interactions may result in costly drug failures further down the pipeline. This is an urgent argument to rely on new techniques in drug screening [254].

Recently, surface plasmon resonance (SPR) biosensor techniques have been examined with respect to their capability in high-throughput formats. Monitoring of biomolecular interactions in real time, measuring affinity and kinetic data, and unique features in applications ranging from protein–peptide interaction analysis to cellular ligation experiments have been demonstrated. Although SPR has historically been limited by its throughput, new methods are emerging that allow the simultaneous analysis of many thousands of interactions. When coupled with new protein array technologies, high-throughput SPR methods provide users with new and improved methods to analyze pathways, screen drug candidates, and monitor protein–protein interactions [255].

Progress in proteomic research is largely determined by the development and implementation of new methods for the revelation and identification of proteins in biological material in a wide concentration range (from 10^{-3} M to single molecules). The most promising approaches involve: (a) nanotechnological physicochemical procedures for the separation of multicomponent protein mixtures; among these, biospecific nanotechnological procedures for selection of proteins from multicomponent protein mixtures with their subsequent concentration on solid support are of particular interest; (b) identification and counting of single molecules by use of molecular detectors. The prototypes of biospecific nanotechnological procedures based on the capture of ligands by immobilized biomolecules and the concentration of the captured ligands on appropriate surfaces are well known; in addition, there are different biosensor methods using magnetic biobead technology. Some aspects are discussed in [256].

THz-wave-based approaches for the label-free characterization of genetic material are interesting, especially time-resolved THz spectroscopy. The analysis of genetic sequences (polynucleotides) demonstrates a distinct complex refractive index in the THz frequency range as a function of the binding state of the analyzed DNA sequences. Integrated THz sensing array developments show high sensitivity and single-base mutation detection capabilities. Recent achievements are illustrated [257].

3.6 Cells and High Content Screening

Optical biosensors are gaining widespread use in drug discovery because of recent advances in instrumentation and experimental design. These advances have expanded the capabilities of optical biosensors to meet the needs at many points in the drug discovery process. Concurrent shifts in drug discovery paradigms have seen the growing use of whole-cell systems for drug screens. Fang reviews important advances in optical biosensor instrumentation and highlights the potential of optical biosensors for drug discovery with an emphasis on whole cell sensing in both high-throughput and high-content fashions [258].

A new reflectometric interference spectroscopy method monitors the adhesion of tissue culture cells to a functionalized surface in a flow system. Interaction of such T cells with other leukocytes or epithelial cells of blood vessels are crucial steps in regulating immune response and inflammatory reactions. Jurkat T cell leukaemia cells are considered in combination with the T cell receptor (TCR)/CD3 complex, monitoring activation-dependent cell spreading. The testing of chemical inhibitors, cell surface molecules and gene products relevant to a key event in T cell immunity illustrates the potential of label-free techniques for the analysis of activation-dependent cell-surface contacts. The results could be compared with confocal fluorescence measurements [259].

The ability to fabricate microfluidic systems with complex structures and with compatible dimensions between the microfluidics and biological cells have attracted significant attention in the development of microchips for analyzing the biophysical and biochemical functions of cells. The cited review focuses on detection methods commonly used in cell-based microfluidic systems, and provides a general survey and an in-depth look at recent developments in optical and electrochemical detection methods for microfluidic applications for biological systems, particularly cell analysis. Selected examples are used to illustrate applications of these detection systems and their merits and drawbacks [260].

4 Outlook

Direct optical detection has proven its feasibility in many applications. To compete with fluorescence-based methods, the limit of detection has to be lowered and non-specific binding as an interfering process has to be reduced. Therefore, the strategy of future development will be the further improvement of technology and, as the most urgent requirement, an improved surface chemistry to increase the loading in order to increase the number of recognition sites and avoid non-specific interaction. Another strategy for the future of direct optical detection methods is the search for applications where labeled systems cannot be used (due to their poor biocompatibility and because of the high costs of the labeling step). Besides, hyphenated techniques in particular open new fields of application and widen the area of bioanalytics. Recently, high-throughput screening approaches have been added to high-content screening methods [261]. Direct optical detection methods can offer new aspects in the examination of multiplexed, functional cell-based screening technologies [262]. These are based on automated digital microscopy [263] and cytometry [264]. An outlook of the strategies and possibilities has been given in the sections on hyphenated techniques (Sect. 2.2.3) and cells/ high content screening (Sect. 3.6). The prospects for direct optical detection methods are promising.

References

1. Udenfriend S (1996) Fluorescence assay in biology and medicine, vol. 2. Molecular biology: an international series of monographs and textbooks. Academic, New York
2. BiaCore (2007) Label-free interaction analysis. http://www.biacore.com/lifesciences/index.html. last visited: 7 Sept 2007
3. Wolfbeis OS, Boisdé GE (1992) Application of optochemical sensors for measuring chemical quantities. In: Göpel W, Hesse J, Zemel JN (eds) Sensors, vol 3. VCH, Weinheim, p 867
4. Trettnak W, Hofer M, Wolfbeis OS (1992) Application of optochemical sensors for measuring environmental and biochemical quantities. In: Göpel W, Hesse J, Zemel JN (eds) Sensors, vol 3. VCH, Weinheim, p 931
5. Lin ZY, Garyantes T (2006) Drug Discov Ser 5:25
6. Templin MF, Stoll D, Schrenkck M, Traub PC, Vohringer CF, Joos TO (2002) Trends Biotechnol 20:160
7. Hertzberg RP, Pope AJ (2000) Curr Opin Chem Biol 4:445
8. Simpson PB, Wafford KA (2006) Drug Discov Today 11(5–6):237
9. De Jon LAA, Uges DRA, Franke JP, Bischoff R (2005) J Chromatogr B: Anal Technol Biomed Life Sci 829(1–2):1
10. Park HG, Song JY, Park KH, Kim MH (2005) Chem Eng Sci 61(3):954
11. Eccleston JF, Hutchinson JP, James DM (2005) Prog Med Chem 43:19
12. Seidel M, Gauglitz G (2003) TRAC 22(6):385
13. Bonetta L (2006) Nat Method 3(7):571
14. Affymetrix (2007) GeneChip system. http://www.affymetrix.com/products/system.affx. last visited: 7 Sept 2007
15. Nagl S, Schaeferling M, Wolfbeis OS (2005) Microchim Acta 151(1–2):1
16. Klimant I, Huber C, Liebsch G, Neurauter G, Stangelmayer A, Wolfbeis OS (2001) Springer Ser Fluoresc 1:257

17. Gauglitz G (1996) Sensors Update 1:1
18. Gauglitz G (2000) Curr Opin Chem Biol 4:351
19. Eddowes MJ (1987) Biosensors 3:1
20. Piehler J, Brecht A, Giersch T, Hock B, Gauglitz G (1997) J Immunol Method 201:189
21. Wolfbeis OS (2006) Anal Chem 78(12):3859
22. Cooper MA (2003) Anal Bioanal Chem 377(5):834
23. Ince R, Narayanaswamy R (2006) Anal Chim Acta 569(1–2):1
24. Dacres H, Narayanaswamy R (2005) Recent Res Dev Anal Biochem 4:35
25. Stoecklein WFM (2006) Chemie Unserer Zeit 40(1):32
26. Ju HX, Zhao HT (2005) Frontiers Biosci 10(1):37
27. Bistolas N, Wollenberger U, Jung C, Scheller F (2005) Biosens Bioelectron 20(12):2408
28. Scheller FW (ed) (2002) Biosensors and biochips. Springer, Heidelberg
29. Dryhurst G, Kadish KM, Scheller FW, Renneberg R (1982) Biological electrochemistry, vol 1. Academic, New York
30. Scheller FW, Bistolas N, Liu SQ, Jaenchen M, Katterle M, Wollenberger U (2005) Adv Colloid Interface Sci 116(1–3):111
31. Janshoff A, Steinem C (2001) Sensors Update 9:313
32. Janshoff A, Galla HJ, Steinem C (2000) Angew Chem Int Ed 39(22):4004
33. Orellana G, Moreno-Bondi MC (eds) (2005) Springer Ser Chem Sens Biosens 3:219
34. Teller C, Halamek J, Makower A, Fournier D, Schulze H, Scheller FW (2006) Sens Actuators B 113(1):214
35. Halamek J, Pribyl J, Makower A, Skladal P, Scheller FW (2005) Anal Bioanal Chem 382(8):1904
36. Ziegler C (2004) Anal Bioanal Chem 379(7–8):946
37. Hinterdorferl P, Dufrêne YF (2006) Nat Method 3:347
38. Kueng A, Kranz C, Lugstein A, Bertagnolli E, Mizaikoff B (2005) Angew Chem Int Ed 44:3419
39. Choi MK, Bettermann AD, Van Der Weide DW (2004) Proc SPIE 5268:27
40. Nagel M, Foerst M, Kurz H (2006) J Phys Condens Matt 18(18):S601
41. Kellner R, Mizaikoff B, Jakusch M, Wanzenbock HD, Weissenbacher N (1997) Appl Spectrosc 51(4):495
42. Ilev IK, Waynant RW (2006) Springer Ser Opt Sci 118:615
43. Dobbs GT, Mizaikoff B (2006) Appl Spectrosc 60(6):573
44. Kraft M, Jakusch M, Karlowatz M, Katzir A, Mizaikoff B (2003) Appl Spectrosc 57(6):591
45. Wang LQ, Chapman J, Palmer RA, Alter TM, Hooper PA, van Ramm O, Mizaikoff B (2006) Appl Spectrosc 60(10):1121
46. Charlton C, Katzir A, Mizaikoff B (2005) Anal Chem 77:4398
47. Charlton C, Temelkuran B, Dellemann G, Mizaikoff B (2005) Appl Phys Lett 86:194102
48. Benson TM, Boriskina SV, Sewell P, Vukovic A, Greddy SC, Nosich AI (2006) Preprint Archive pp 39–70
49. Chowdhury MH, Gant VA, Trache A, Baldwin A, Meiniger G, Coté GL (2006) J Biomed Opt 11(2):024004
50. Cliff MJ, Gutierrez A, Ladbury JE (2004) J Mol Recognit 17(6):513
51. Sagawa T, Oda M, Morii H, Takizwawa H, Kozoni H, Azuma T (2005) Mol Immunol 42(1):9
52. Roth G, Freund S, Möhrle B, Wöllner K, Brünjes J, Gauglitz G, Wiesmüller KH, Jung G (2007) ChemBioChem 8(3):323

53. Bergmann L, Schäfer C, Niedrig H (2004) Experimentalphysik: vol 3, Optik, 10th edn. Walter de Gruyter, Amsterdam
54. Hecht E, Zajek A (2003) Optics. Addision-Wesely, Reading
55. Born M, Wolf E (1980) Principles of optics. Pergamon, New York
56. Harrick NJ (1967) Internal reflection spectroscopy. Interscience, New York
57. Marazuela M, Morno-Bondi M (2002) Anal Bioanal Chem 372(5–6):664
58. Gauglitz G (2005) Anal Bioanal Chem 381(1):141
59. Horvath R, Skivesen N, Larsen NB, Pedersen HC (2005) Springer Ser Chem Sens Biosens 3:279
60. Dominguez C, Llobera A, Zinoviev K (2005) Proc SPIE Int Soc Opt Eng 59560T/1-59560T/15
61. Liedberg B, Nylander C, Lundström I (1983) Sens Actuators B 4:229
62. Karlsson R (2004) J Mol Recognit 17:151
63. Loefas S (2004) Biochemistry 26:27 (Moscow, Russ Fed)
64. Homola J, Yee S, Myszka D (2002) Surface plasmon resonance biosensors. In: Ligler FS, Rowe T, Chris A (eds) Optical biosensors present and future. Elsevier, Amsterdam, p 207
65. Homola J, Yee SS, Gauglitz G (1999) Sens Actuators B 54:3
66. Stemmler I, Brecht A, Gauglitz G (1999) Sens Actuators B 54:98
67. Zhao J, Zhang XY, Yonzon CR, Haes AJ, Van Dyne RP (2006) Nanomedicine 1(2):219
68. Homola J (ed) (2006) Springer Ser Chem Sens Biosens 4:251
69. Navratilova I, Myshka DG (2006) Springer Ser Chem Sens Biosens 4:155
70. Lechuga LM (2000) Quim Anal 19(1):54
71. Piraud C, Mwarania E, Wylangoswski G, Wilkinson J, O'Dwyer K, Schiffrin DJ (1992) Anal Chem 64:651
72. Clerc D, Lukosz W (1994) Sens Actuators B 19:581
73. Nellen PM, Lukosz W (1993) Biosens Bioelectron 8:129
74. Brandenburg A, Hinkov V, Konz W (1992) In: Göpel W, Hesse J, Zemen JN (eds) Sensors, vol 6. VCH, Weinheim, p 399
75. NeoSensors (2007) IAsys biosensors. http://www.affinity-sensors.co.uk/. last visited: 7 Sept 2007
76. Corning (2007) Epic system overview. http://www.corning.com/lifesciences/US-Canada/en/epicsystem/. last visited: 7 Sept 2007
77. Fattinger C, Mangold C, Gale MT, Schuetz H (1995) Opt Eng 34:2744
78. Kunz RE (1991) Proc SPIE Int Soc Opt Eng 1587:98
79. Cush R, Cronin JM, Stewart WJ, Maule CH, Molloy J, Goddard NJ (1993) Biosens Bioelectron 8:347
80. Davies RJ, Edwards PR (2002) IAsys: the resonant mirror biosensor. In: Gizeli E, Lowe CR (eds) Biomolecular sensors. Taylor & Francis, London, p 269
81. Kinning T, Edwards P (2002) The resonant mirror optical biosensor. In: Ligler F, Rowe T, Chris A (eds) Optical biosensors. Elsevier, Amsterdam, p 253
82. Fabricius N, Gauglitz G, Ingenhoff J (1992) Sens Actuators B 7:672
83. Drapp B, Piehler J, Brecht A, Gauglitz G, Luff BJ, Wilkinson JS, Ingenhoff J (1997) Sens Actuators B 39:277
84. Brandenburg A, Henniger R (1994) Appl Opt 33:5941
85. Lechuga LM (2000) Quim Anal 19(1):61
86. Heideman RG, Walker JA, Lionix BV (2006) Proc SPIE Int Soc Opt Eng 61250S/1-61250S/9
87. Sepulveda B, Armelles G, Lechuga LM (2007) Sens Actuators A 134(2):339
88. Lukosz W, Stamm C (1991) Sens Actuators A 25:185

89. Klotz A, Brecht A, Gauglitz G (1997) Sens Actuators B 39:310
90. Drude P (1888) Ann Phys 34:489
91. Vedam K (1998) Thin Solid Films 313–314:1
92. Kretschmann E (1971) Z Phys 241:313
93. Raether H (1977) Phys Thin Films 9:145
94. Azzam RMA, Bashara NM (1989) Ellipsometry and polarized light. North-Holland Physics, Amsterdam
95. Arwin H, Aspnes DE (1986) Thin Solid Films 138:195
96. Mutschler T, Kieser B, Frank R, Gauglitz G (2002) Anal Bioanal Chem 374:658
97. Heideman RG, Kooyman RPH, Greve J (1993) Sens Actuators B 10:209
98. Gauglitz G, Krause-Bonte J, Schlemmer H, Matthes A (1988) Anal Chem 60:2609
99. Gauglitz G, Nahm W (1991) Fresenius Z Anal Chem 341:279
100. Brecht A, Gauglitz G, Nahm W (1992) Analysis 20:135
101. Brecht A, Gauglitz G, Kraus G, Nahm W (1993) Sens Actuators B 11:21
102. Nahm W, Gauglitz G (1990) GIT Fachz Lab 34:889
103. Gauglitz G (2005) Rev Sci Instrum 76(6):062224/1
104. Garside B (1996) Advances in high-speed OTDR (optical time domain reflectometry) detection techniques. In: Culshaw B, Dakin J (eds) Optical fiber sensors. Artech House, Norwood, Mass, p 145
105. Butler MA (1994) Sens Actuators B 22:155
106. Duschl C, Hall EAH (1991) J Colloid Interface Sci 144:368
107. Lin VSY, Moesharei K, Dancil KPS, Sailor MJ, Ghadiri MR (1997) Science 278:840
108. Yu F, Yao DF, Qian WP (2000) Clin Chem 46:1489
109. Jenison R, Yang S, Haeberli A, Polisky B (2001) Nature 19:62
110. Lü X, Huang Y, Ma C (2001) Sensors 1:148
111. VoDinh T, Ascari M (2001) Curr Genomics 2:399
112. Piehler J, Schreiber G (2001) Anal Biochem 289:173
113. Lue XY, Huang Y, Qian WP, Tang ZM, Lu ZH (2003) J Biomed Mater Res 66A:722
114. Lamken P, Lata S, Gavutis M, Piehler J (2004) J Mol Biol 341:303
115. Xia B, Xiao SJ, Wang J, Guo DJ (2005) Thin Solid Films 474:306
116. Pröll F, Möhrle B, Kumpf M, Gauglitz G (2005) Anal Bioanal Chem 382:1889
117. Hänel C, Gauglitz G (2002) Anal Bioanal Chem 372:91
118. Piehler J, Brandenburg A, Brecht A, Wagner E, Gauglitz G (1997) Appl Opt 36:6554
119. Brecht A, Gauglitz G (1995) Biosens Bioelectron 10:923
120. Kieser B, Fietzek C, Schmidt R, Belge G, Weimar U, Schurig V, Gauglitz G (2002) Anal Chem 74:3005
121. Marshall AJ, Kabilan S, Blyth J, Lowe CR (2006) J Phys Condens Matter 18(18):S619
122. Lechuga LM (2005) Optical biosensors. In: Lechuga LM (ed) Comprehensive analytical chemistry, vol 44. Elsevier, Amsterdam, p 209
123. Cunningham BT, Laing L (2006) Exp Rev Proteomics 3(3):271
124. Reichl D, Krage R, Krummel C, Gauglitz G (2000) Appl Spectrosc 54:583
125. Gauglitz G, Reichl D, Seemann JW, Brecht A (2000) DE Patent 19 830 727.6 – 52
126. Dieterle F, Belge G, Betsch C, Gauglitz G (2002) Anal Bioanal Chem 374:858
127. Frank R (2005) PhD thesis, University of Tuebingen
128. Frank R, Möhrle B, Fröhlich D, Gauglitz G (2005) Proc SPIE-Int Soc Opt Eng 5993:49
129. Frank R, Möhrle B, Fröhlich D, Gauglitz G (2005) Proc SPIE-Int Soc Opt Eng 5826:551
130. Frank R, Gauglitz G (2005) Poster 3 presented at the 5th Biosensor symposium, Regensburg, Germany, 13–16 March 2005
131. Reichl D (2000) PhD thesis, University of Tuebingen

132. Jäger S, Gauglitz G (2005) In: 7th Dresdner sensor symposium, Dresdner Beiträge zur Sensorik, vol 24. TUD, Dresden, p 275
133. Jäger S, Gauglitz G (2006) In: Book of abstracts for Europt(r)ode VIII, Tübingen, 2–5 April 2005, p 114
134. Zimmermann R, Werner C, Gauglitz G, Eichhorn KJ (2003) DE Patent 10 205 775
135. Homola JV, Lu HB, Nenninger GG, Dostalek J, Yee SS (2001) Sens Actuators B 76(1–3):403
136. Dostalek J, Vaisocherova H, Homola JV (2005) Sens Actuators B 108(1–2):758
137. Gorodkiewicz E, Fernandez-Gonzalez A, Akkoyun A, Steiner G, Salzer R (2005) Chem Anal 50(1):103 (Warsaw, Pol)
138. Bally M, Halter M, Voros J, Grandin HM (2006) Surf Interface Anal 38(11):1442
139. Rothmund M, Brecht A, Berthel G, Gräfe D, Schütz A, Gauglitz G (1997) Fresenius J Anal Chem 359:15
140. BMBF (2000) Leitstruktur-Identifikation in komplexen Bibliotheken über Rezeptor-Affinitäts-Reaktionen in parallelen Analysesystemen (LIBRARIAN), Teilproject: Interferometrische und Evaneszentfeld-Transducer, BMBF project no 0310838
141. Brecht A, Gauglitz G, Gräfe D, Fuchs W (2006) DE Patent 19 615 366
142. Brecht A, Burckhardt R, Rickert J, Stemmler I, Schütz A, Fischer S, Friedrich T, Gauglitz G, Göpel W (1996) J Biomol Screen 1:191
143. Zinoviev K, Plaza JA, Cadorso V, Dominguez C, Lechuga LM (2007) Proc SPIE-Int Soc Opt Eng 6477:64771A
144. Haake HM, De Best L, Irth H, Abuknesha R, Brecht A (2000) Anal Chem 72(15):3635
145. Jungar C, Strandh M, Ohlson S, Mandenius CF (2000) Anal Biochem 281(2):151
146. Proll G, Kumpf M, Mehlmann M, Tschmelak J, Griffith H, Abuknesha R, Gauglitz G (2004) J Immunol Method 292:35
147. Nelson RW, Nedelkov D, Tubbs KA (2000) Anal Chem 72(11):404A
148. Nedelkov D, Nelson RW (2000) J Mol Recognit 13(3):140
149. Nedelkov D, Nelson RW (2001) Biosens Bioelectron 16(9–12):1071
150. Mehlmann M, Garvin AM, Steinwand M, Gauglitz G (2005) Anal Bioanal Chem 382:1942
151. Zhang R, Barker L, Pinchev D, Marhall J, Rasamoelisolo M, Smith C, Kupchak P, Kireeva I, Ingratta L, Jackowski G (2004) Proteomics 4:244
152. Waseda S, Shimosaka T, Uchiyama K, Hobo T (1999) Chem Lett 11:1195
153. Whelan RJ, Zare RN (2003) Anal Chem 75:1542
154. Guijt-van Dujin RM, Frank J, van Dedem GWK, Baltussen E (2000) Electrophoresis 21:3905
155. Huang CC, Cao Z, Chang HT, Tan W (2004) Anal Chem 76:6973
156. Kumpf M, Gauglitz G (2006) Anal Bioanal Chem 384:1129
157. Kröger K, Jung A, Reder S, Gauglitz G (2002) Anal Chim Acta 469:37
158. Reder S, Dieterle F, Jansen H, Alcock S, Gauglitz G (2003) Biosens Bioelectron 19:447
159. Gavutis M, Lata S, Lamken P, Müller P, Piehler J (2005) Biophys J 88:4289
160. Ratner BD, Bryant SJ (2004) Annu Rev Biomed Eng 6:41
161. Knopp D, Niessner R (2004) Waste Management Ser 4:505
162. Dell A, Famulok M, Imperiali B (2000) Curr Opin Chem Biol 4(6):599
163. Oh SJ, Hong BJ, Choi KY, Park JW (2006) OMICS 10(3):327
164. Loefas L, Johnsson B (1990) J Chem Soc Chem Commun, p 1526
165. Piehler J, Brecht A, Geckeler KE, Gauglitz G (1996) Biosens Bioelectron 11:579
166. Heinze T, Liebert T, Heublein B, Hornig S (2006) Adv Polym Sci 205:199
167. Piehler J, Brecht A, Valiokas R, Liedberg B, Gauglitz G (2000) Biosens Bioelectron 15:473

168. Feldmann K, Hähner G, Spencer ND, Harder P, Grunze M (1999) J Am Chem Soc 121:10134
169. Clare TL, Clare BH, Nichols BM, Abbott NL, Hamers RJ (2005) Langmuir 21(14):6344
170. Vermette P, Meagher L (2003) Colloids Surf B 28(2–3):153
171. Masson JF, Battaglia TM, Khairallah P, Beaudoin S, Booksh KS (2007) Anal Chem 79(2):612
172. Patrito N, McCague C, Norton PR, Petersen NO (2007) Langmuir 23(2):715
173. Kusnezow W, Pulli T, Witt O, Hoheisel JD (2005) Solid supports for protein microarrays and related devices. In: Schena M (ed) Functional genome analysis. Jones and Bartlett, Sudbury, Mass, p 247
174. Schlereth DD (2005) Biosensors based on self-assembled monolayers. In: Gorton L (ed) Comprehensive analytical chemistry, vol 44. Elsevier, Amsterdam, p 1
175. Länge K, Herold M, Scheideler L, Geis-Gerstorfer J, Wendel HP, Gauglitz G (2004) Dental Mater 20(9):814
176. Raitza M, Herold M, Ellwanger A, Gauglitz G, Albert K (2000) Macromol Chem Phys 201:825
177. Wang ZH, Jin G (2004) Colloids Surf B 34(3):173
178. Kamisetty NK, Pack SP, Nonogawa M, Devarayapalli KC, Dokaki T, Makino K (2006) Anal Bioanal Chem 386(6):1649
179. Diao JP, Ren DC, Engstrom JR, Lee KH (2005) Anal Biochem 343(2):322
180. Xu QC, Lam KS (2003) J Biomed Biotechnol 5:257
181. Wu Q, Ma W, Shi R, Zhang B, Mao X, Zheng W (2005) Eng Life Sci 5(5):466
182. Michel R, Pasche S, Textor M, Castner DG (2005) Langmuir 21(26):12327
183. Musabekov KB, Omarova KI, Izimov AI (1983) Acta Phys Chem 29(1–2):89
184. Dauginet L, Duwez AS, Legras R, Demoustier-Champagne S (2001) Langmuir 17(13):3952
185. Palumbo M, Nagel J, Petty MC (2005) IEEE Sens J 5(6):1159
186. Yu DG, Jou CH, Lin WC, Yang MC (2007) Colloids Surf B 54(2):222
187. Ratner BD, Hoffman AS (2004) Physicochemical surface modification of materials used in medicine. In: Ratner BD (ed) Biomaterials science. Elsevier, San Diego, p 201
188. Ratner BD, Cheng XH, Wang YB, Hanein Y, Bohringer KF (2003) Polymer Preprints (Am Chem Soc, Div Polym Chem) 44(1):198
189. Sigrist H, Collioud A, Clemence JF, Gao H, Luginbuehl R, Saenger M, Sundarababu G (1995) Opt Eng 34(8):2339
190. Dorman G, Prestwich GD (2000) Trends Biotechnol 18(2):64
191. Dankbar D, Gauglitz G (2006) Anal Bioanal Chem 386(7–8):1967
192. Ito Y (2006) Biotechnol Prog 22(4):924
193. Sigrist H (2005) Polym Mater Sci Eng 1:389
194. Gao H, Guinchard S, Crevoisier F, Angeloni S, Sigrist H (2003) Chimia 57(10):651
195. Lodhi MA, Opperman GW, Wall JV, Anderson AB (2005) Surface modification chemistries and their utilization in microarrays, diagnostics and drug delivery. In: Laudon M, Romanowicz B (eds) NSTI Nanotech 2005, NSTI nanotechnology conference and trade show, Anaheim CA, Nano Science and Technology Institute, Cambridge, Mass, p 389
196. Ito Y, Yamauchi T, Uchikawa M, Ishikawa Y (2006) Biomaterials 27(11):2502
197. Choi HJ, Kim NH, Chung BH, Seong GH (2005) Anal Biochem 347(1):60
198. Birkert O, Haake HM, Schütz A, Mack J, Brecht A, Jung G, Gauglitz G (2000) Anal Biochem 282:200
199. Gershon PD, Khilko S (1995) J Immunol Method 183:65

200. Tien HT (1985) Prog Surf Sci 19:169
201. Park J, Groves WA, Zellers ET (1999) Anal Chem 71:3877
202. Lehn JM, Ball P (2000) Supramolecular chemistry. In: Hall N (ed) New chemistry. Cambridge University Press, Cambridge, p 300
203. Dickert FL, Schuster O (1995) Microchim Acta 119:55
204. Schurig V, Grosenick H (1994) J Chromatogr A 666:617
205. Jung G, Hofstetter H, Feiertag S, Stoll D, Hofstetter O, Wiesmüller KH, Schurig V (1996) Angew Chem Int Ed Engl 35:2148
206. Wang J (1999) Curr Issue Molec Biol 1(2):117
207. Singh SK, Wengel J (1998) Chem Commun 12:1247 (Cambridge)
208. Koshkin AA, Nielsen P, Meldgaard M, Rajwanshi VK, Singh SK, Wengel J (1998) J Am Chem Soc 120(50):13252
209. Karkare S, Bhatnagar D (2006) Appl Microbiol Biotechnol 71(5):575
210. Haupt K, Mosbach (2000) Chem Rev 100:2495
211. Mirsky VM, Hirsch T, Piletsky S, Wolfbeis OS (1999) Angew Chem Int Ed Engl 38:1108
212. Bossi A, Bonini F, Turner APF, Piletsky SA (2007) Biosens Bioelectron 22(6):1131
213. Fernandez-Gonzalez A, Guardia L, Badia-Laino R, Diaz-Garcia ME (2006) TrAC, Trends Anal Chem 25(10):949
214. Kato T (2006) Chem Sensors 22(1):15
215. Gross M (2006) Nachr Chem 54(12):1213
216. Deisingh AK (2006) Handbook Experiment Pharmacol 173:341
217. Niessner R, Broekaert JAC, Einax JW, Emons H, Engewald W, Haisch C, Jakuboswski N, Salzer R, Schuhmann W, Weller MG (2006) Nachr Chem 54(4):382
218. Sablinskas V, Steiner G, Salzer R (2006) Proc SPIE-Int Soc Opt Eng 6099:609901/1
219. Möhrle B, Kumpf M, Gauglitz G (2005) Analyt 130:1634
220. Bilitewski U (2006) Anal Chim Acta 568(1–2):232
221. Markovic G, Mutschler T, Wöllner K, Gauglitz G (2006) Surf Coat Technol 201(3–4): 1282
222. Piehler J, Brecht A, Gauglitz G, Maul C, Grabley S, Zerlin M (1997) Biosens Bioelectron 12:531
223. Pattnaik P (2005) Appl Biochem Biotechnol 126(2):79
224. Magliulo M, Michelini E, Simoni P, Guardigli M, Roda A (2006) Anal Bioanal Chem 384(1):27
225. Dostalek J, Ladd J, Jiang SY, Homola J (2006) Springer Ser Chem Sens Biosens 4:177
226. Saint-Pierre G, Malecha MM, Saini S, Setford SJ (2005) Proc SPIE-Int Soc Opt Eng 5836:273
227. Länge K, Griffin G, Vo-Dinh T, Gauglitz G (2002) Talanta 56:1153
228. Indyk HE (2006) Chem NZ 70(2):42
229. Bergwerff AA, van Knapen F (2006) J AOAC Int 89(3):826
230. Scully P, Chandy R, Edwards R, Merchant D, Morgan R (2001) Environ Sci Res 56:175
231. Anderson GP, Taitt CR (2005) Biosensor-based detection of foodborne pathogens. In: Fratamico PM, Bhunia AK, Smith JL (eds) Foodborne pathogens. Caister Academic, Wymondham, UK, p 33
232. Pittet A (2005) Mitt Lebensmittel Hygiene 96(6):424
233. Rasooly A, Herold KE (2006) J AOAC Int 89(3):873
234. Haughey SA, Baxter GA (2006) J AOAC Int 89(3):862
235. Cacciatore G, Petz M (2005) Dtsch Lebensm-Rundsch 101(11):501
236. Länge K, Herold M, Scheideler L, Geis-Gerstorfer J, Wandel HP, Gauglitz G (2004) Dental Mater 20(9):814

237. Ulber R, Frerichs JG, Beutel S (2003) Anal Bioanal Chem 376(3):342
238. Tünnemann R, Mehlmann M, Süssmuth RD, Bühler B, Pelzer S, Wohlleben W, Fiedler HP, Wiesmüller KH, Gauglitz G, Jung G (2001) Anal Chem 73:4313
239. Muffler K, Ulber R (2006) Bioforum 29(3):23
240. Luppa PB, Sokoll LJ, Chan DW (2001) Clin Chim Acta 314(1–2):1
241. Vaisocherova H, Homola J (2006) Springer Ser Chem Sens Biosens 4:229
242. Vaisocherova H, Zitova A, Lachmanova M, Stepanek J, Kralikova S, Liboska R, Rejman D, Rosenberg I, Homola J (2006) Biopolymers 82(4):394
243. Sauer M, Brecht A, Charisse K, Maier M, Gerster M, Stemmler I, Gauglitz G, Bayer E (1999) Anal Chem 72(14):2850
244. Englebienne P, Van Hoonacker A, Verhas M (2003) Spectroscopy 17(2,3):255 (Amsterdam)
245. Massey M, Piunno PAE, Krull UJ (2005) Springer Ser Chem Sens Biosens 3:227
246. Piliarik M, Vaisocherova H, Homola J (2007) Sens Actuators B 121(1):187
247. Kuswandi B, Tombelli S, Marazza G, Mascini M (2005) Chimia 59(5):236
248. Kröger K, Seidel M, Gauglitz G (2003) Optical sensing methods in high-throughput screening. In: Potyrailo RA, Amis EJ (eds) High-throughput analysis. Kluwer Academic/Plenum, New York, p 261
249. Haake HM, Tünnemann R, Brecht A, Austel V, Jung G, Gauglitz G (2002) Anal Biochem 300:107
250. Kröger K, Bauer J, Fleckenstein B, Rademann J, Jung G, Gauglitz G (2002) Biosens Bioelectron 17:937
251. Birkert O, Tünnemann R, Jung G, Gauglitz G (2002) Anal Chem 74:834
252. Birkert O, Gauglitz G (2002) Anal Bioanal Chem 372:141
253. Brecht A, Gauglitz G (1997) Anal Chim Acta 347(1–2):219
254. Franklin G (2005) Biochemistry 27(3):28
255. Boozer C, Kim G, Cong SX, Guan HW, Londergan T (2006) Curr Opin Biotechnol 17(4):400
256. Ivanov YD, Govorun VM, Victor A, Archakov AI (2006) Proteomics 6(5):1399
257. Bolivar PH, Nagel M, Richter F, Brucherseifer M, Kurz H, Bosserhoff A, Buettner R (2004) Philos Trans R Soc Lond A 362(1815):323
258. Fang Y (2006) Assay Drug Dev Technol 4(5):583
259. Möhrle B, Köhler K, Jaehrling J, Brock R, Gauglitz G (2006) Anal Bioanal Chem 384(2):407
260. Yi CQ, Zhang Q, Li CW, Yang J, Zhao JL, Yang MS (2996) Anal Bioanal Chem 384(6):1259
261. Giuliano KA, Debiasio RL, Dunlay RT, Gough A, Volosky JM, Zock J, Pavlakis GN, Taylor DL (1997) J Biomolec Screen 2(4):249
262. Nichols A (2007) Meth Molec Biol 356:379
263. Rausch O (2006) Curr Opin Chem Biol 10(4):316
264. Bowen WP, Payne SL (2006) Nat Meth 3(9):iii

Adv Biochem Engin/Biotechnol (2008) 109: 433–453
DOI 10.1007/10_2007_087

Published online: 6 November 2007

DNA Microarrays

Frank F. Bier (✉) · Markus von Nickisch-Rosenegk ·
Eva Ehrentreich-Förster · Edda Reiß · Jörg Henkel · Rothin Strehlow ·
Dennie Andresen

Fraunhofer Institute for Biomedical Engineering, Branch Potsdam, Am Mühlenberg 13,
14476 Potsdam, Germany
frank.bier@ibmt.fraunhofer.de

Abstract Microarray technology provides new analytical devices that allow the parallel and simultaneous detection of several thousands of probes within one sample. Microarrays, sometimes called DNA chips, are widely used in gene-expression analysis, genotyping of individuals, analysis of point mutations and single nucleotide polymorphisms (SNP) as well as other genomic or transcriptomic variations. In this chapter we give a survey of common microarray manufacturing, the selection of support material, immobilisation and hybridisation and the detection with labelled complementary strands. However, DNA arrays may also serve as the basis for more complex analysis based on the action of enzymes on the immobilized templates. This property gives DNA microarrays the potential for being the template for whole PCR and transcription experiments with high parallelism, as will be discussed in the last section of this chapter.

Keywords Active arrays · DNA microarray · DNA chips · On-chip PCR · On-chip transcription

1 Introduction

Desoxyribonucleic acid (DNA) is the most important biomolecule in nature because of its role as carrier of the genetic information. Many kinds of genetic analysis may be done on the gene level. The analysis may vary from single-nucleotide deviations to whole-genome sequencing. Many methods and inventions have been made in recent years to create analytical tools for the analysis of a huge variety of applications. All the methods invented for DNA analysis are in principle applicable to RNA as well. This is of interest for RNA especially in its role as information carrier, the messenger RNA. Using reverse transcription it is possible to get a DNA copy and therefore all that is said for DNA analysis in this chapter will also be valid for most RNA analysis.

Microarray technology provides new analytical devices that allow the parallel and simultaneous detection of several thousands of probes within one sample. Microarrays, sometimes called DNA chips, are widely used in gene-expression analysis, genotyping of individuals, analysis of point mutations and single-nucleotide polymorphisms (SNP), as well as other genomic or transcriptomic variations.

The high specific base-pair interaction within the DNA or DNA–RNA hybrids with labelled complementary strands makes the microarray technology a powerful analytical tool for monitoring whole genomes. In this chapter we give a survey of common microarray manufacturing, the selection of support material, immobilisation and hybridization, the detection with labelled complementary strands. DNA arrays may also serve as the basis for more complex analysis based on the action of enzymes on the immobilized templates. This property gives DNA microarrays the potential of being the template for whole-PCR and transcription experiments with high parallelism, as will be discussed in the last section of this chapter.

1.1 The Physico-Chemical Properties of Nucleic Acids

A very important feature of DNA is its chemical homogeneous appearance with repeating units, the sugar-phosphate backbone. At neutral pH, DNA is negatively charged via the phosphate groups and attracts positive counter-ions, i.e. DNA is a strong polyelectrolyte [1]. Depending on the counter-ion concentration, double-stranded DNA has locally a more or less stiff rod-like structure. This feature is relevant for all properties regarding surface contact as it occurs in microarray experiments. The physical parameters of the surface become relevant for the attraction or repulsion of DNA. This is true for the immobilisation process and its efficiency as well as for the hybridisation.

On the other hand, the structure of DNA is not identically repeating, but due to the four different bases, adenine, guanine, cytosine and thymine (resp. uracile), the sequence is unique depending on composition (see Fig. 1). This feature makes DNA the ideal storage molecule because the blueprint of any organism may be stored within the "bits of the genome". The high similarity and the ease of transformation from a four-letter code to the dual code of information theory and modern computers has inspired much reasoning about "DNA computing" and nucleic-acid-based computers. However, despite the success of Adelman's first demonstration of a molecular biological computing process [2], up to now there is no real computing machine that makes use of DNA as the coding string.

The most prominent interaction of DNA is the base recognition, based on the hydrogen bonds between adenin and thymin, that form two hydrogen bonds, and guanin and cytosin, that form three hydrogen bonds. The binding of two complementary strands is called hybridisation. The stability of the DNA hybrid depends on its GC content. The more GC pairs involved, the higher the stability of the DNA hybrid.

1.2 Microarrays

Since most biological phenomena are within the context of a multitude of parameters and processes, the correlations and interactions of these processes are at the centre of quantitative biological investigations. The interrogation of a broad variety of genes or their transcripts and their activity at one moment is one of the typical questions; and parallel analysis is needed since it can be performed by the use of microarrays.

DNA microarrays are characterised by a structured immobilisation of DNA targets on planar solid supports allowing the profiling of thousands of genes or interactions in one single experiment. An ordered array of these elements on planar substrates is called a "microarray". Usually, for practical reasons, one distinguishes between microarrays and macroarrays, the difference being the size of the deposited spots. Typical spot sizes of macroarrays are featured by a diameter of more than 300 microns, whereas microarray spots are represented in less than 200 microns.

Hybridisation is the underlining principle of DNA chips. According to the nomenclature recommended by Phimister [3] a "probe" is the immobilised or fixed nucleic acid with known sequence, whereas the "target" is the free nucleic acid sample, usually labelled during the preparation process, which interacts with the probe by hybridisation. Commonly used labels for target nucleic acids are fluorescence dyes, radioactive or enzymatic detection labels.

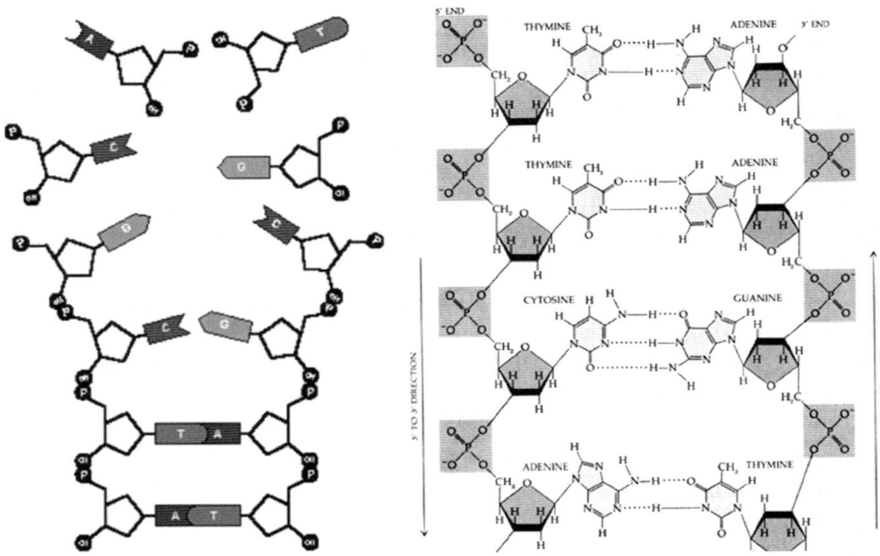

Fig. 1 The building blocks of DNA, base-pairing, the double-helix, and hybridisation

Low expenditure of time, high information content and a minimum of probe volume are attractive features of this technology.

To provide free and accessible functionalities of the DNA strand that is to be immobilised, i.e. which will be the probe, is the essential precondition for a proper immobilisation. The DNA contains three different biochemical components, a base that is substituted on the first carbon of desoxyribose forming a nucleoside, and a negatively charged phosphodiester that connects the sugars to a chain as shown in Fig. 1. In principle, the amines in the bases, the negatively charged backbone, the phosphordiesters within the backbone, and the phosphates on the 5′-end as well as the hydroxyl group on the 3′-end are potential candidates for coupling. However, since hybridisation is the aim of the experiment, it should be noted that in double strands the bases are engaged in hydrogen bonds and thus, are not accessible for the chemical coupling reactions on the surface.

Microarrays can be fabricated with DNA from various sources; the probes may be derived from cDNA, PCR products or synthetic oligomers.

1.3 Enzymes Acting on Immobilised Templates

Biosensors are defined by the close contact of a biomolecular recognition element with a transducer, which converts the biochemically generated signal into an electronic signal. This type of device has been investigated intensively, with a great variety of recognition elements, like enzymes, antibodies, DNA, aptamers, etc. [4]. Enzymes play a major role in this field, since the activity of an enzyme may be used for signal transduction and amplification in one.

More complex biochemical reactions have been introduced into biosensors by use of coupled enzyme reactions or coupled binding and enzymatic reactions. While this is a theme of its own in biosensor research for metabolic biosensors, there are only a few reports on DNA-modifying enzymes (see Sect. 3 in this chapter). When DNA is immobilised properly on a surface, it still keeps most of its physico-chemical properties. Moreover, it turned out that the capability of correct base pairing, the hybridisation process, is not affected by the immobilisation. This enables immobilised DNA to serve as a possible template for even complex enzymatic reactions. Enzymatic digest by restriction endonucleases [5] as well as synthesis on arbitrary templates by polymerases have been shown in real-time and label-free experiments [6, 7]. A second feature of immobilised DNA on a microarray is to use the spots as addresses to couple something else, for example, some proteins that are tagged with the complementary strand [8, 9]. This concept paved the way for the development of "active arrays".

In microarrays this approach has not been used very often. The mere binding event, the hybridisation, is determined using fluorescent labels, but also microarrays bear the potential activity measurement, and the immobilised DNA may serve as a template for DNA–protein interaction analysis as well as a template for chip-coupled PCR. Moreover, whole genes may be immobilised and used for in vitro transcription and translation.

2 Microarray Technology

Microarray technology allows massive parallel determination and multiple measurements of a variety of binding events that are to be carried out simultaneously. In addition, it has the advantage of requiring a small amount of material and a modest investment of labour; moreover, it might save a lot of time and may easily be automated. Microarrays in general consist of many microscopic spots, each containing identical molecules, i.e. receptors, probes or targets. The numbers of spots may vary from less than 100 to many 100 000. The molecules are attached to a solid support that can be made from glass, silicon or a polymer. In the case of nucleic acids, the receptors are usually oligonucleotides or cDNA, and the binding event is simply the hybridisation of complementary strands. Biochip technology (especially DNA chips) is a rapidly developing field with high commercial potential. Introductions to the principles of the technology as well as to various applications are available in many reviews [10, 11], and in the well-written practical approach by Schena [12].

Thus, the combination of molecular biology, microfabrication and bioinformatics has generated novel tools and has the strong potential to bring more products on to the market for genetic analysis purposes; applications of which can be used in all branches of the life sciences.

2.1 The Typical Microarray Experiment

The central process in all biochip or microarray experiments is a binding event, the hybridisation. One binding partner (either receptor or ligand, probe or target) is immobilised within a small but well-defined area on a flat solid support of glass or polymer; referred to as the "spot" or "feature" of the microarray. There are also examples of prestructured slides, structured for example by microcavities (nanotiter plates) or by chemical structuring as well as by electronic features, such as microelectrodes. The features con-

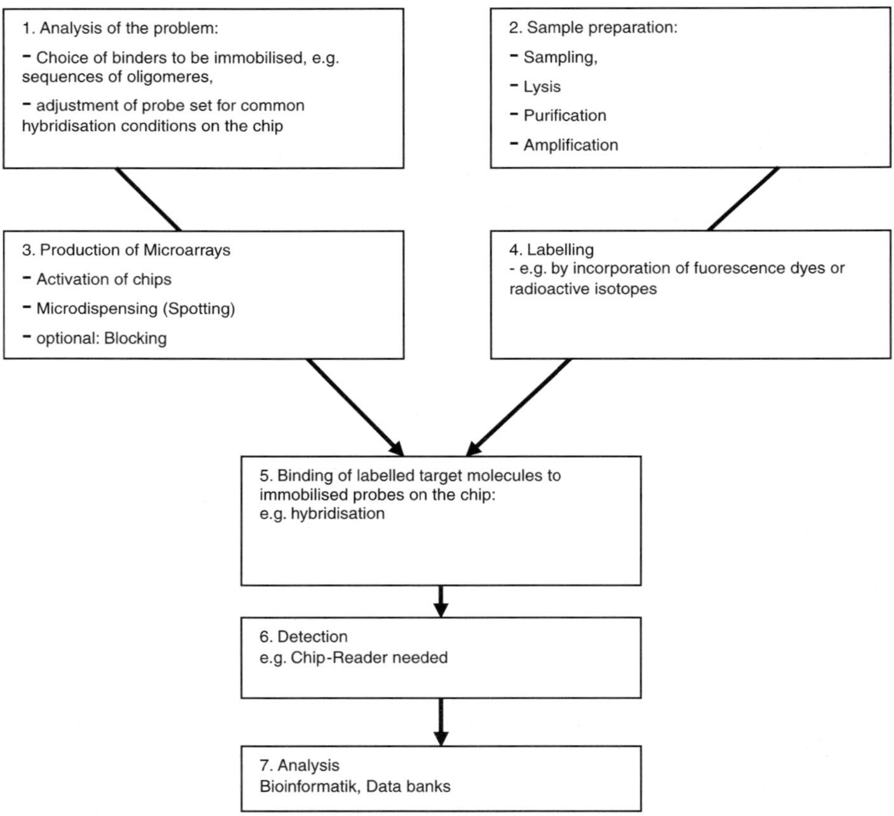

Fig. 2 Workflow of a microarray experiment

tain identical molecules, while the whole array contains arbitrarily chosen or systematically varied compounds.

Once the microarray has been produced, the sample that contains the ligands or targets to be investigated is added and binding occurs at several features on the chip, which results in a characteristic pattern representing the sample. The procedure of a typical biochip experiment is schematically represented in the workflow below.

Fluorescence labels are by far the most often used labels to detect the binding event. The sample itself is labelled in a step prior to incubation, i.e. the hybridisation. After incubation, the chip is washed and then read out by means of a scanning or imaging device, usually in a dry state, giving a snapshot picture of fluorescence intensities. Internal standards for comparison purposes are often incorporated into the procedure.

2.2 Manufacturing of Microarrays

For microarray production, two different approaches are used:

1. Synthesis on the chip; and
2. bulk synthesis with subsequent deposition on the chip.

The first method is applicable to generate chemical libraries, for example, of short oligonucleotides or peptides; the second method may also be adapted to long polynucleotides or proteins, or any of the many receptors. Detailed description of these methods may be found in [13–15].

2.2.1 Synthesis on the Chip

Fodor et al. reported as early as 1991 for the first time the notion of synthesising an array of oligomers on a chip surface by use of step-wise local photo-deprotection [16]. In their first paper, peptide synthesis was in the focus, however, it was soon recognized that oligonucleotide arrays would be much more needed and due to only four bases instead of 20 amino acids would be easier to implement. The process of production is similar to wafer production in microelectronics: a chemically activated silicon wafer surface is covered with photo-labile protecting groups. After UV irradiation through a mask with high spatial resolution, the activated groups are reactive at certain localities and a nucleotide, that again bears the same photo-labile protecting group at its 3′-end, may couple with its 5′-end. Repetition of this procedure with all four bases will end up with defined sequences in each feature of the array, $4n$ steps (and masks) are needed for n nucleotides. Standard microarrays have features of 0.1 mm × 0.1 mm, but the process is optimised down to a feature size of 10 μm × 10 μm, enabling several thousand up to

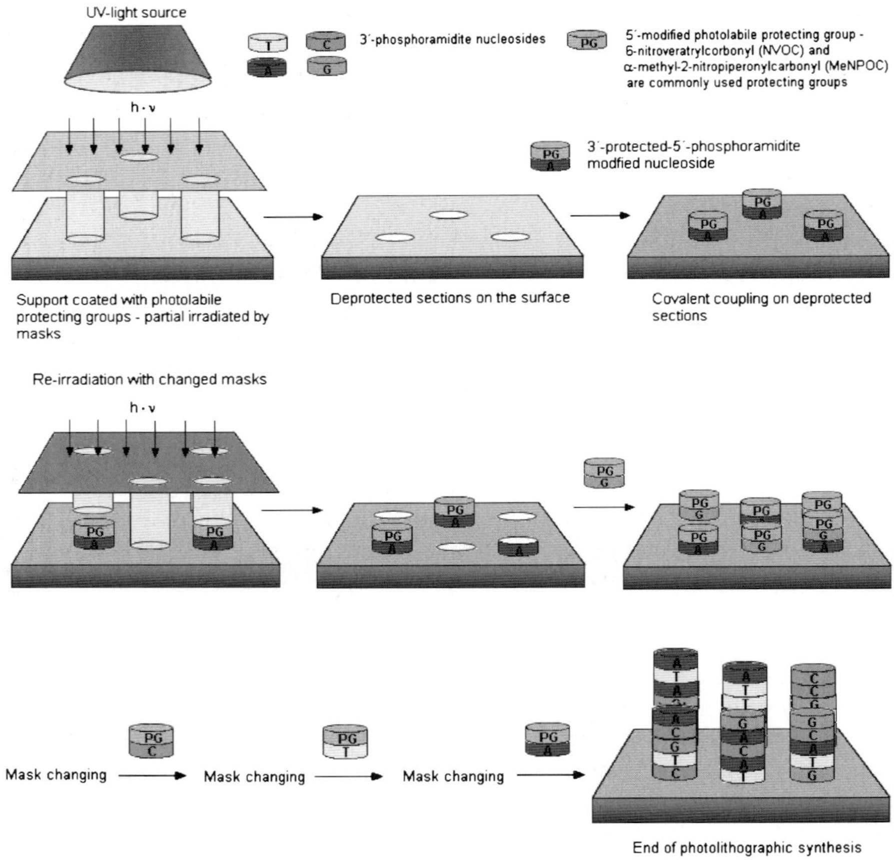

Fig. 3 Photolithographic surface structuring and oligonucleotide synthesis [17]

hundreds of thousands of features on one chip. This method is therefore optimised for high parallel synthesis. The only limitation in absolute number of features comes from feature size and mask resolution. With this type of array, whole-genome screening is possible. This method is very much linked to the name of the company Affymetrix who industrialised the method and currently produces a new series of high-density microarrays for several applications.

The method, however, is limited to oligonucleotides. The length of the oligonucleotides on the chip surface is limited to about 25 nt, since there is no "100%"-chemical reaction and thus the addition of incomplete reactions results in a loss of sequence precision within one feature. Assuming a 99% chemical reaction, a feature with a 25 nt oligomer contains 33% of molecules with an incomplete, i.e. wrong, but nevertheless similar sequence.

2.2.2
Spotting Techniques

An alternative way of creating a microarray is to deposit small amounts of pre-activated oligonucleotides by microdispensing. An advantage compared to on-chip synthesis is the possibility of cleaning up the material that will be deposited. While on-chip synthesis is limited in sequence length by the efficiency of each coupling step during the synthesis, pre-synthesised material may be cleaned up even from a mixed environment like fermentation broth. Moreover, the dispensing methods are not limited to any chemical species, but may be used for any kind of array, including PCR products or complete genomes. On the other hand, pre-synthesised probes have to be processed sequentially and thus the whole procedure (and in consequence the number of features) may be limited by the processing time.

Two different techniques are currently used: contact and non-contact printing.

The first technique uses pin tools, or needles, that are dipped into the probe solution and dispense a certain amount of material by contacting the support material. An advantage of pin tools is that parallelisation is easily

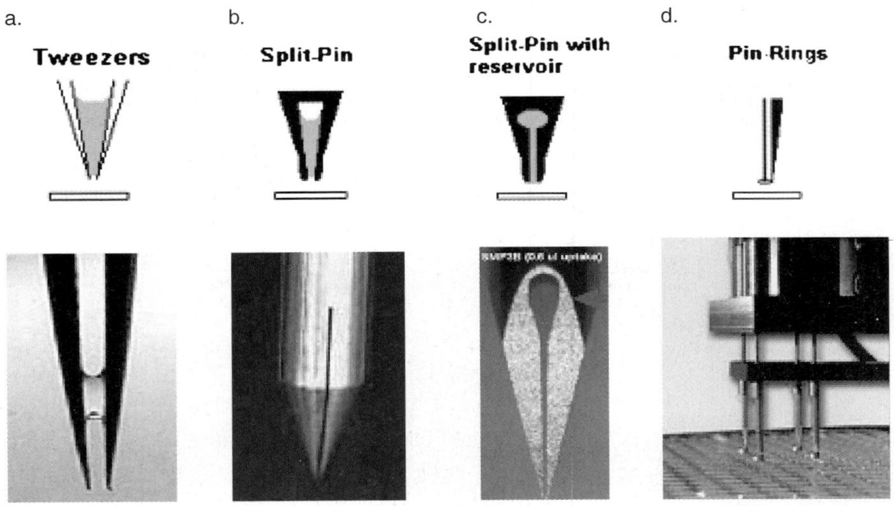

Fig. 4 Pin tools for contact printing: **a** *Tweezers*: Micro-tweezers are loading the sample by capillary action and expels defined spot volumes on the surface by tapping forces. **b** and **c** *Split-Pins*: Through capillary action, a defined target volume is loaded into the split or other cavities (per surface-tapping small spots were deposited). Depending on the amount of spots to be set in one spotting run, the cavaties are variously shaped **d** *Pin rings*: Pin rings load and hold the sample in a ring like a soap-bubble. For spot-deposition, a needle is propelled through the ring while the sample is carried by the needle to contact the surface [12]

done by adding more needles to the printhead. However, reproducibility of this method is limited, since the tiny tips of the pin tools change their shape during repeated contacts to the support. Therefore for high and middle dense arrays for research use, pin tools is the method of choice.

Figure 4 shows a pin tool printer (4d) and some examples of pin tools (4a–c) that have been developed for optimal and repeated deposition of probe material. The slits and holes in the pin tools are reservoirs for repeated printing of the same substance. The print step may be repeated 20–40 times. Typically, pin tools are used for in-house production of a limited number of high specific genomic slides. Reproducibility of this method is limited.

Non-contact techniques rely on the piezo effect. Like in an ink-jet printer, a microfluidic nozzle is activated to release droplets of a size less than 1 nl. The distance of the features can be minimised down to 100 μm. This type of dispensing allows for high accurate manufacturing of low- and medium-density microarrays (up to a few thousand features). Figure 5 shows a one-nozzle non-contact pipette releasing a drop on demand. This kind of pipette may also be used for a variety of different pipetting procedures where small

Fig. 5 Non-contact spotting via piezo technique. **a** Printhead with eight piezo nozzles. **b** Close-up of a piezo nozzle releasing a sample drop. **c** SciFlex Piezo-spotter (Scienion, Berlin)

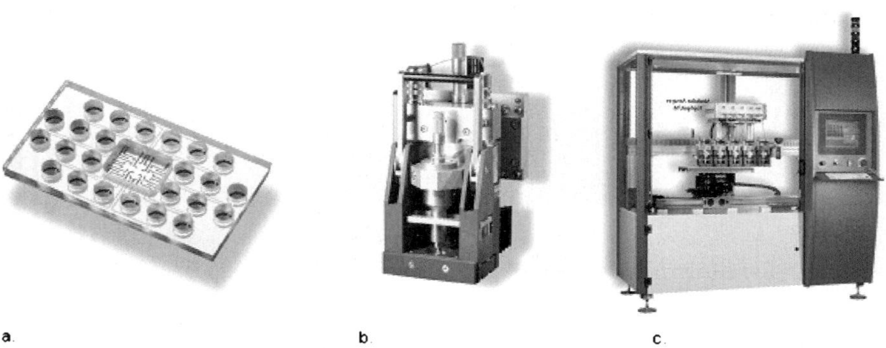

Fig. 6 TopSpot technology. **a** TopSpot Printhead with 24 nozzles. **b** Printing module with actuator that holds the printing head. **c** TopSpot Arrayer with five printing modules for series production of microarrays

volumes as low as 100 pL (pico litre = 10^{-12} L) have to be applied. While machines of this type are capable of producing a lot of spots (features) of the same kind with one loading, reloading and change to the next probe needs intensive washing. Currently, parallelisation of nozzles is limited, suppliers of such machines offer up to 16 parallel nozzles.

A microsystem approach has been made by Zengerle and coworkers [18, 19] which they named "TopSpot". An array of nozzles is produced in a monolithic manner using Si-wafer technology. This array is combined with a glass microfluidic system that incorporates reservoirs and microchannels connecting these reservoirs with the nozzle and thus form a printhead. By external actuation, the nozzles release simultaneously one droplet of each probe loaded in the reservoirs. Printheads with 24 and 96 nozzles are available with a fixed array format, but in principle, any number up to several hundred may be produced. Figure 6 shows printhead and machine for small series production (up to several thousand arrays per hour). The reservoirs can collect up to 5 μL, and thus depending on feature size and support material, more than 20 000 print processes may be done with a single loading.

2.2.3
Performance and Quality Control

Quality control is of great importance for microarray fabrication, especially if diagnostic applications are in the focus of interest. The quality control has to follow each production step, but finally the overall assay performance is the sole criterion for a well-manufactured microarray.

After each spotting step a control is needed, whether all features are properly generated and within defined accuracy limits. The definition of such limits depends on the application. Usually, the diameter of the spot gives a first indication about the quality of the printing process. Figure 7 gives an example from a production series with the TopSpot technology. At this stage, misaligned arrays or arrays with missing or misprinted features may be ex-

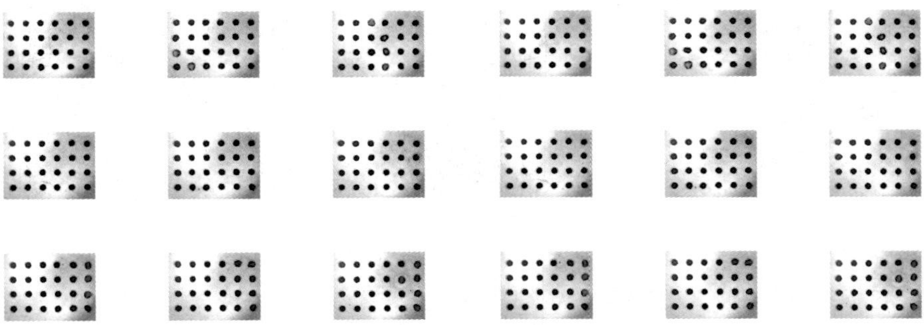

Fig. 7 Production-scale control of sequential printing results

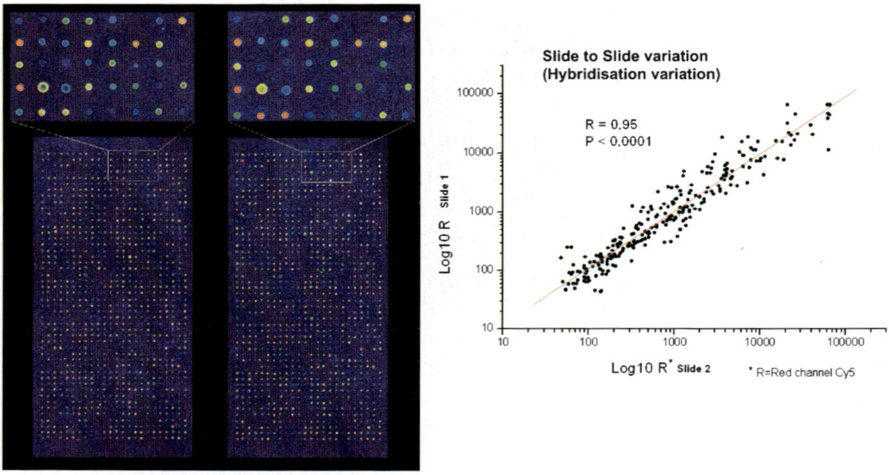

Fig. 8 Comparison between two microarrays for determination of slide-to-slide variation

cluded from further processing. As one measure of quality at this stage the diameter and deviation from the ideal circle may be employed. Since the surface of the slide should be homogeneous and the deposited droplet should have the same amount in each feature, an ideal circle with a well-defined size should occur after each spotting process. Up to now, using this measure the reproducibility of the spots was found to be less than 8% in the ideal case using the TopSpot technology. However, functionality of the features, as measured by hybridisation efficiency, might be even better, since the absolute number of immobilised probe molecules and their accessibility to the sample is the decisive entity, and not the geometric appearance of the feature.

Since a non-invasive test of functionality of all features is not possible, batch control must serve as a quality measure. For this it is necessary to know about the deviations that occur typically within one series. Figure 8 shows a comparison of two microarrays from one batch and the deviation from the mean value after hybridisation. The state of the art is a standard deviation of about 12%.

2.3 Hybridisation

As mentioned before, hybridisation conditions have to be optimised, i.e. buffer conditions, like ionic strength and pH, and temperature have to be chosen in such a way that only complete matching strands form stable hybrids. Since these conditions depend on strand length and on composition, usually different sequences do not bind simultaneously in an equal stringent way. This situation has to be regarded during the microarray design and some bioinfor-

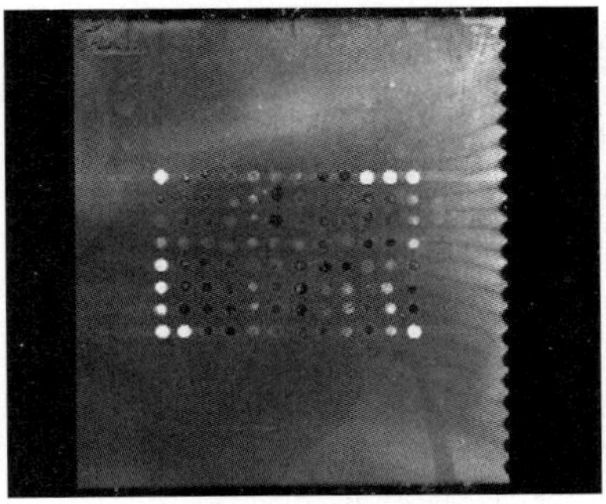

Fig. 9 Microfluidic hybridisation chamber: the special design with multiple inlets allows for a homogeneous sample stream to guarantee equal hybridisation conditions in all places of the microarray

matics tools have been generated to help to find the desired set of sequences for one microarray (e.g. ArrayDesigner 3; Premier Biosoft International).

Hybridisation has to be done under controlled conditions; often incubation over a long period of several hours at elevated temperature is required. Typical are temperatures between 45 and 65 °C. To guarantee the correct incubation, often hybridisation stations are used, some of those speed up the protocol by actuating the reagents during incubation. An example of such actuation is to facilitate gentle and homogeneous sample stream by special design of a flat fluidic chamber. The microfluidic challenge is to cover a plane of relative big size homogeneously with a small volume of fluid. For this purpose, chambers have been developed that solve this problem by a proper design; Fig. 9 gives an example from our own lab (not published).

2.4 Detection and Analysis

Detection fluorescence is by far the most often used method. Usually, fluorescent dyes are incorporated into the sample during preparation, e.g. during amplification. The use of intercalators or similar DNA-specific dyes is also common. Since most instruments for detection are applicable only to dry microscopic slides, fluorescence dyes that work in the dry state are preferred, e.g. dyes from the cyanin family like Cy3 or Cy5. For transcription analysis, two different stages are compared on one slide and therefore two different dyes are used.

Two types of instruments are used: imaging and scanning devices. The principle of the optical arrangement is very similar to a fluorescence microscope in which the fluorescence illumination (excitation) runs the same optical path as does the emitted light. The imaging device takes a picture from the whole or part of the array, while the scanning device works as a laser scanning microscope with a focused excitation beam to enhance performance by reducing the background signal. In both cases, a resolution of about 5 μm is achieved that allows for a high accurate analysis of each single feature. Software tools are usually employed to further analyse the acquired images.

3 The Microrray Experiment

3.1 Transcription Analysis

By far the most often used application of DNA microarrays is transcription analysis. Due to the tenets of molecular biology, the information flow runs from the storage medium, the genomic DNA, via transcription to the messenger RNA (mRNA) and by translation in the ribosome machinery to the protein. Many details of the regulation of gene activity have been discovered in recent years. It is of high scientific (and also medical) interest to learn more about this mechanism and the analysis of transcribed genes as the first step of gene activity is currently in the focus of all life sciences. The "transcriptome", i.e. the manifold of all genes transcribed in a specific cell (cell type) at a given time under defined conditions, can effectively be analysed using DNA microarrays. Usually, a comparison is made between a cell type in status A and an altered state, status B, which might be linked for instance to a disease. Both samples, material extracted from cells of status A and B, are labelled with two different fluorochromes. The difference in gene activity in both cell types can easily be recognised wherever one colour exceeds the other. The comparison method compensates for many shortcomings of the array method, since each feature of the array is referenced in itself. The only information drawn is with regard to the binding of species A or B to the same probe on the surface. Different hybridisation conditions for different features on the array result in various absolute signal values (fluorescence intensities), however, the relative amount of the two compared samples are still valid. Therefore most applications of microarray technology reported today are in the realm of transcription analysis.

3.2 Genotyping with Oligonucleotide Arrays

The first driving force for microarray development was the Human Genome Project (HUGO) conducted as an international exercise to decode the whole human genome for a first example [20].

The microarray approach of sequencing was the so-called method of sequencing by hybridisation (SBH) [21]. It is based on the notion to combine the complete sequence of a sample by presenting all possible sequences as a complement on the chip, these are for 12 bases 4^{12} (=16 777 216) oligomers have to be presented on the chip. This goal was not only difficult to achieve technically in the early state of microarray fabrication technology, but also some basic features of natural sequences have been overlooked. For instance, it would be impossible to get homogeneous hybridisation conditions for all sequences on one single chip since the melting temperature depends strongly on the ratio of GC to AT pairs within this sequence. But even worse, there was no way of overcoming the problem of biologically meaningful redundancies within the genome originating from similar proteins, so it turned out to be impossible to get all the short probed sequences unambiguously linked together.

Today SBH can be used in all those contexts were the sequence of interest is already known to a certain extent and deviations are sought.

4 The Concept of "Active Arrays"

Recently the concept of biosensors to measure binding events and other biomolecular activities using surface-bound molecules has been adapted to microarray technology [22]. By use of time and spatial resolved measurement, the kinetics of biomolecular interactions may be disclosed for a set of immobilised receptors in just one single experiment. A few examples are given in the next two paragraphs.

4.1 Enzymes Acting on Immobilised DNA

In the early 1990s it was shown by several authors that DNA-modifying enzymes are capable of acting on immobilised DNA templates or primers [7, 23, 24]. Especially primers, oligonucleotides of less than 30 bases, are of interest in the context of microarrays.

In our laboratory we achieved the parallel measurement of enzyme activities on several templates by virtue of a microfluidic chamber mounted on top of the microarray slide. As an example, the restriction endonuclease (EcoRI) was chosen acting on different templates simultaneously. In this latter case, a usual microarray covered with immobilised oligonucleotides is exposed to a variety of complementary sequences, each of which is labelled with a fluorochrome (FITC in the actual case). By hybridisation in various spots the endonuclease's cleaving site is formed, forcing the applied enzyme to bind. By addition of the cofactor Mg^{2+} the enzymes starts to cleave and releases the short oligonucleotide with the fluorochrome.

Already in 1996 Buck and Buckle et al. published the first results on the observation of polymerase acting on an immobilised template by use of an SPR device (Biacore) in a single channel. This approach could be extended to the enzyme telomerase that is responsible for the elongation of chromosomal ends during cell proliferation. Since its activity is regulated down in differentiated cells, the amount of activity has been established as a significant tumour marker in recent years [25]. Schmidt et al. succeeded to demonstrate that telomerase acts also on artificial telomeres immobilised on a sensor surface [26, 27].

The use of immobilised templates for polymerase processing was also successful with long DNA including whole genes.

Also, for the analysis of single nucleotide polymorphisms (SNPs), enzymes acting on immobilised templates have been employed. This was first demonstrated by Erdogan et al. who used DNA polymerases to elongate surface-bound primers [28]. These authors also showed the reaction on an oligonucleotide microarray. They were using immobilised allele-specific oligonucleotide primers on a glass slide. Single-stranded PCR products serve as a template that hybridises to the corresponding oligonucleotide probes on the microarray. The match and mismatch primer differ at their 3′-end by a variable base, which is discriminated by the DNA polymerase during elongation process and the incorporation of Cy3-labelled dUTP due to the corresponding signal intensity [28]. In this work, a limit for the template length was stated at 5.7 kb. Recent work by von Nickisch-Rosenegk et al. could not find such a limit and showed on-chip elongation up to 16.5 kb, i.e. the whole mitochondrial genome [29].

4.2 PCR on the Chip

While all reactions described up to now were one-step reactions, a complete PCR on the chip needs reproduction of the polymerisation step several times, and heating up the whole chip to 95 °C for denaturation between every polymerisation step.

A complete microarray-based amplification (on-chip PCR) was first described by Adessi et al. [30]. For the amplification the authors proposed the specific covalent attachment of oligonucleotide primers via their 5′-end to the glass slide, allowing the free 3′-end to prime the DNA synthesis. Template DNA can hybridise to these free 3′-ends and will be available for elongation by a DNA polymerase. After elongation, the resulting amplification product is covalently attached to the surface via the primer. Detection of the immobilised amplificates can be achieved by direct labelling with fluorescently labelled primers or dUTPs. Another possibility may be the use of intercalators like SYBR Green, but due to the resulting background problems, this is not yet successful. The same is true for fluorescent dendrimer labels, e.g. 3DNA Labels from Genisphere, which are available for microarray hybridisations. For further sequence verification, the amplificates can be denatured with alkali and subsequently hybridised with a sequence-specific fluorescent probe.

Primers used for on-chip PCR have to fulfil three criteria besides the usual properties like specificity, G/C-content and melting temperature. They have to be immobilised in a density that allows the detection of the resulting amplificates. Adessi et al. states that for the immobilised primer, a concentration of 50 μM is best suited for on-chip PCR experiments [30], however, they fail to determine the actual primer density on the glass slide they could achieve by their immobilisation procedure. The linkage of the primer to the surface must be stable enough to withstand the extreme temperature conditions during the cycling process. Therefore covalent attachment (e.g. EDC coupling) is the best option. The last important criterion to be considered is the attachment of the primer at the 5′-end and the introduction of a spacer sequence for superior hybridisation efficiency. The length and composition of the spacer sequence can be variable. An optimal length is described to be between 10 and 18 nucleotides, preferentially as polyT-spacer [30, 31].

The amplification process takes place in two different phases: In the liquid and on the solid phase. Figure 10 represents a schematic overview of the general principle of on-chip PCR at the solid-phase state.

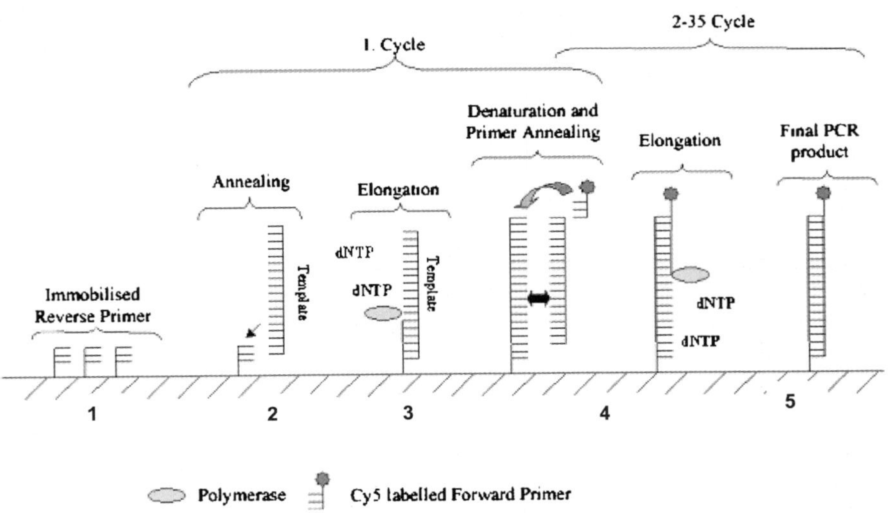

Fig. 10 Principle of the on-chip PCR: solid-phase amplification. In the solid-phase PCR the template DNA, which can be either from the liquid phase or the original template, hybridises to the specific 5′-bound primer (*1*) during the annealing step (*2*). After the annealing, the elongation step by the polymerase follows (*3*). In the next cycle, after denaturation and annealing of the Cy5-labelled primer (*4*), we will find the final Cy5-labelled PCR amplificate 5′-bound to the surface (*5*) and ready for subsequent analysis. This process is repeated during every cycle of the on-chip PCR.

Because the cycling takes place directly on the slide surface, a special containment for the PCR mix is needed. To prevent vaporisation of the PCR mix, this containment has to be absolutely leak-proof even at elevated temperature. For this purpose, there are two different options currently available. The first option is the usage of adhesive reaction chambers, like SealFrame (MJ Research, USA) or HybriWells (Sigma-Aldrich, USA). Another option is the use of the Self-Seal Reagent (MJ Research, USA), which is added directly to the PCR mix and builds a tight seal at the edges of a coverslip by polymerisation during the first denaturation cycle. Both options have been successfully applied for the on-chip PCR [28, 30, 32–34].

The thermocycling of the on-chip PCR is carried out in in situ PCR blocks. Usually, the 16 × 16 Twin Tower in situ block for the PTC 200 thermocycler from MJ Research is used [28, 30, 32–34] for this purpose. Sensitivities for the on-chip PCR are reported to be in the range of pico to nanomolar DNA template concentrations [30] or between 30–100 ng genomic DNA [28, 32, 33].

Due to the fact that on-chip PCR has been developed quite recently, only a few reports are given on applications. Huber et al. were focussing on the detection of single base alterations in genomic DNA (SNP analysis). In advancing this method, the authors focus on sequence-specific genotyping by solid-phase amplification [32, 33]. Because there is no need of template preparation, advantages like time and cost savings are obvious when using the on-chip PCR. Huber et al. could show that their experimental setup is suitable for the detection of SNPs in the human tumour suppressor gene *p*53 [32]. Further integration could be achieved if the single on-chip PCR is extended for parallel detection of several SNP's in one reaction: a multiplex on-chip PCR. Huber et al. were able to show the multiplex on-chip PCR for accessing SNP's in genomic DNA [33]. In this paper the genotyping of ten different polymorphic sites within seven human genes by direct on-chip multiplex PCR has been shown.

Mitterer et al. reports on-chip PCR in combination with universal primer pairs targeting for the Helix 43 and 69 region of the 23s rDNA and species-specific primers immobilised on the chip surface were used to detect 22 common bacteria causing infertility and abortion in mares [34]. Although false-positive signals were obtained in rare occasions, this work shows possible future perspectives for the on-chip PCR in clinical diagnostics.

Recently we demonstrated the multiplex on-chip PCR with a variety of markers for food ingredients [35, 36].

The area of diagnostic applications is still under development in the field of chip-based assays. A technique that could give fast and reliable answers about, e.g. the current state of an individual's viral or pathogen load, would be of great benefit. With further optimisation, the sensitivity of the PCR and the miniaturisation and parallelism of microarrays could join in form of the on-chip PCR and build a core part in diagnostic BioMEMS [37]for future point-of-care devices.

4.3
Transcription on Chip

The immobilised DNA might be amplified, as described above. Recently we demonstrated that even more functionality may be gained on the chip when the sequence immobilised or produced on the surface is a complete gene. When associated with a proper binding site and a promoter, i.e. a complete gene, the action of a transcriptase can be triggered and a messenger RNA was generated [38]. Moreover, it could be shown in these experiments that the mRNA was functional and served as a template for translation. The encoded protein could be produced with an in vitro assay.

Also it was possible to use microarrays the other way around, namely to gain amplified DNA signals from low amounts of mRNA. We demonstrated recently that the reverse transcription PCR (RT-PCR) could be combined with microarray technology for the detection of low abundant mRNA. The experiment runs similar to the above described on-chip PCR, now starting with the reverse transcription in the same incubation chamber.

4.4
Future Prospects

The examples given here show that microarrays are useful tools for molecular biology, especially for transcription analysis. But there are many more possibilities to make use of immobilised arrayed DNA in small dimensions: using microarrays as a template for enzymes helps to detect and investigate a lot of interactions with DNA, to gain insight into novel features and to develop more sophisticated and time-saving tools of increasing complexity. In this way, we gain insight into the processing steps that usually are imbedded into the complex environment of the whole cell.

Acknowledgements Excellent technical assistance by Michaela Schellhase and Dirk Michel is gracefully acknowledged. This work was supported by the German Federal Ministry of Education and Research (BMBF).

References

1. Cantor CR, Schimmel P (1980) Biophysical Chemistry Part 1. Freeman & Co, San Francisco
2. Adleman LM (1998) Rechnen mit DNA. Spektrum der Wissenschaft, pp 70–77
3. Phimister B (1999) Going Global. Nat Genet 21(Supplement):1–6
4. Scheller F, Schubert F (1989) Biosensoren. Akademie Verlag, Berlin
5. Bier FF, Scheller FW (1996) Label-free observation of DNA-hybridisation and endonuclease activity on a wave guide surface using a grating coupler. Biosens Bioelectron 11:669–674

6. Bier FF et al. (1996) An enzymatic amplification cycle for high sensitive immunoassay. Anal Chim Acta 328:27–32
7. Buckle M et al. (1996) Real-time measurements of elongation by a reverse transcriptase using surface plasmon resonance. Proc Natl Acad Sci USA 93:889–894
8. Bier FF et al. (1999) Changing functionality of surfaces by directed self-assembly using oligonucleotides—The Oligo-Tag. Biotechniques 27:752–760
9. Niemeyer CM et al. (1999) Self-assembly of DNA-streptavidin nanostructures and their use as reagents in immuno-PCR. Nucleic Acids Res 27(23):4553–4561
10. The Chipping Forecast 1. Nature Supplement, 1999
11. The Chipping Forecast 2. Nature Supplement, 2002
12. Schena M (2003) Microarray Analysis, 1st ed. Wiley, Hoboken, New Jersey, USA
13. Eisen MB, Brown PO (1999) DNA arrays for analysis of gene expression. Methods Enzymol 303:179–205
14. Sambrook J, Bowtell DDL (2002) DNA Microarrays: A Molecular Cloning Manual. Cold Spring Harbor Laboratory Press, New York, USA, p 375
15. Schena M (1999) DNA Microarrays. A Practical Approach. Oxford University Press, New York, USA, p 232
16. Fodor SPA et al. (1991) Light-directed, spatially addressable parallel chemical synthesis. Science 251:767–773
17. Heise C, Bier FF (2005) Immobilization of DNA on Microarrays. In: Wittmann C (ed) Immobilisation of DNA on Chips II. Springer, Berlin Heidelberg New York
18. de Heij B et al. (2004) Highly parallel dispensing of chemical and biological reagents. Anal Bioanal Chem 378(1):119–122
19. Gutmann O et al. (2004) A highly parallel nanoliter dispenser for microarray fabrication. Biomed Microdevices 6(2):131–137
20. The Human Genome Consortium (2001) The human genome. Nature 409(6822):745–964
21. Drmanac R et al. (2002) Sequencing by hybridization (SBH): Advantages, Achievements, and Opportunities. Adv Biochem Eng Biotechnol 77:75–101
22. Bier FF, Ehrentreich-Forster E, Schmidt PM (2004) Real-time analysis on microarrays. Anal Bioanal Chem 378(1):52–53
23. Bier FF et al. (1996) High sensitive competitive immunodetection of 2,4-dichlorophenoxyacetic acid using enzymatic amplification with electrochemical detection. Anal Bioanal Chem 354(7–8):861–865
24. Bier FF, Ehrentreich-Forster E, Scheller FW (1996) Amplifying bienzyme cycle-linked immunoassays for determination of 2,4-dichlorophenoxyacetic acid. Ann NY Acad Sci 799:519–524
25. Granger MP, Wright WE, Shay JW (2002) Telomerase in cancer and aging. Crit Rev Oncol Hematol 41(1):29–40
26. Schmidt PM et al. (2002) Detection of activity of telomerase in tumor cells using fiber optical biosensors. Biosens Bioelectron 17:1081–1087
27. Schmidt PM et al. (2002) Real-time determination of telomerase activity in cell extracts using an optical biosensor. Biol Chem 383:1659–1666
28. Erdogan F et al. (2001) Detection of mitochondrial single nucleotide polymorphisms using a primer elongation reaction on oligonucleotide microarrays. Nucleic Acids Res 29:e36
29. von Nickisch-Rosenegk M et al. (2005) On-chip PCR amplification of very long templates using immobilized primers on glassy surfaces. Biosens Bioelectron 20(8):1491–1498

30. Adessi C et al. (2000) Solid-phase DNA amplification: characterisation of primer attachment and amplification mechanisms. Nucleic Acids Res 28(20):e87
31. Southern E, Mir K, Shchepinov M (1999) Molecular interactions on microarrays. Nature genetics supplement 21:5–9
32. Huber M et al. (2001) Detection of single-base alterations in genomic DNA by solid-phase polymerase chain reaction on oligonucleotide microarrays. Anal Biochem 299:24–30
33. Huber M et al. (2002) Accessing single-nucleotide polymorphisms in genomic DNA by direct multiplex polymerase chain reaction amplification on oligonucleotide microarrays. Anal Biochem 303:25–33
34. Mitterer G et al. (2004) Microarray-based identification of bacteria in clinical samples by solid-phase PCR amplification of 234S ribosomal DNA sequences. J Clin Microbiol 42(3):1048–1057
35. Andresen D et al. (2005) Microarray-Diagnostik von Allergenen. Laborwelt 6:34–36
36. Andresen D et al. (2005) Multiplex on chip PCR for the analysis of food ingredients. Poster at German Biosensor Symposium Regensburg (Germany)
37. Bier FF, Andresen D, Walter A (2006) DNA-based Bio-Micro-Electronic Mechanical Systems. In: Urban G (ed) BioMEMS. Springer, Berlin Heidelberg New York, p 372
38. Steffen J, von Nickisch-Rosenegk M, Bier FF (2005) In vitro transcription of a whole gene on a surface-coupled template. Lab Chip 5(6):665–668

Adv Biochem Engin/Biotechnol (2008) 109: 455–492
DOI 10.1007/10_2007_082

Published online: 6 October 2007

Scanning Electrochemical Microscopy (SECM) as a Tool in Biosensor Research

Leonard Stoica · Sebastian Neugebauer · Wolfgang Schuhmann (✉)

Analytische Chemie – Elektroanalytik & Sensorik, Ruhr-University Bochum,
Universitätsstr. 150, D-44780 Bochum, Germany
wolfgang.schuhmann@rub.de

Abstract Scanning electrochemical microscopy (SECM) is discussed as a versatile tool to provide localized (electro)chemical information in the context of biosensor research. Advantages of localized electrochemical measurements will be discussed and a brief introduction to SECM and its operation modes will be given. Experimental challenges of the different detection modes of SECM and its applicability for different fields in biosensor research are discussed. Among these are the evaluation of immobilization techniques by probing the local distribution of biological activity, the visualization of diffusion profiles of reactants, cofactors, mediators, and products, and the elucidation of (local) kinetic parameters. The combination of SECM with other scanning-probe techniques allows to maximize the information on a given biosensing system. The potential of SECM as a tool in micro-fabrication aiming for the fabrication of microstructured biosensors will be shortly discussed.

Keywords Biofunctionalized surfaces · Biosensors · Enzyme electrodes · Localized measurements · Scanning electrochemical microscopy · SECM

Abbreviations

a	Radius of the active electrode area
AD/DA	Analog-digital/digital-analog
ADH	Alcohol dehydrogenase
AFM	Atomic force microscopy
AFM-SECM	Combined AFM and SECM instrument

AP	Alkaline phosphatase
c_0	Initial bulk concentration of the redox species
CE	Counter electrode
d	Distance between tip and sample
d/a	Normalized tip to sample distance
D_R	Diffusion coefficient of R
dsDNA	Double-stranded DNA
EC	Electrochemistry
ELISA	Enzyme-linked immunosorbent assay
F	Faraday's constant
FcCOOH	Ferrocenecarboxylic acid
FMA/FMA^+	Ferrocene methanol (reduced and oxidized forms, respectively)
Γ_{enz}	Enzyme coverage per unit of area
GOD	Glucose oxidase
HRP	Horseradish peroxidase
i_{ss}	Steady-state current
i_T–d curve	Approach curve; dependence of tip current on the tip-to-sample distance
k_{cat}	Rate of enzymatic reaction
M-DNA	metalated DNA
$MV^{2+/+}$	Methyl viologen$^{2+/+}$
n	Number of electrons transferred in a redox reaction
NADH	Nicotinamide adenine dinucleotide (reduced form)
O/R	Oxidized/reduced state of redox species
OCP	Open circuit potential
pAP	*p*-aminophenol
pAPP	*p*-aminophenyl phosphate
pQI	p-quinone imine
PQQ	Pyrroloquinoline quinone
QH-ADH	Quinohemoprotein alcohol dehydrogenase
RE	Reference electrode
RG	Ratio of the diameter of the insulating sheath and the diameter of the active electrode area
SECM	Scanning electrochemical microscopy, scanning electrochemical microscope
SEM	Scanning electron microscopy
SPM	Scanning probe microscopy
ssDNA	Single-stranded DNA
TMPD	N,N,N',N'-tetramethyl-*p*-phenylenediamine
UME	Ultramicroelectrode
WE	Working electrode
XPS	X-ray photoelectron spectroscopy

1 Introduction

Scanning electrochemical microscopy (SECM, [1,2]) is a versatile analytical tool used to visualize and quantify heterogeneously distributed (electro)chemical activity on a sample surface. SECM belongs to the scanning probe microscopies (SPM, [3,4]), microscopic techniques which are sequentially collecting information about a specific interphase property by scanning a sharp tip (named probe) at close distance over the surface of the investigated sample (Fig. 1). The specific interactions occurring between the tip and the sample surface are recorded with respect to the *x*-,*y*- and often *z*-positions of the tip. Depending on the nature of the tip, the distance and the specific distance-dependent physical or chemical interactions between tip and sample, a variety of surface properties can be visualized, such as electronic states [5], optical [6], and force interactions [7]. In SECM, a disk-shaped ultramicroelectrode (UME) is used as probe, and the Faradaic steady-state current established at the ultramicroelectrode can be modulated by the (electro)chemical properties of the sample surface.

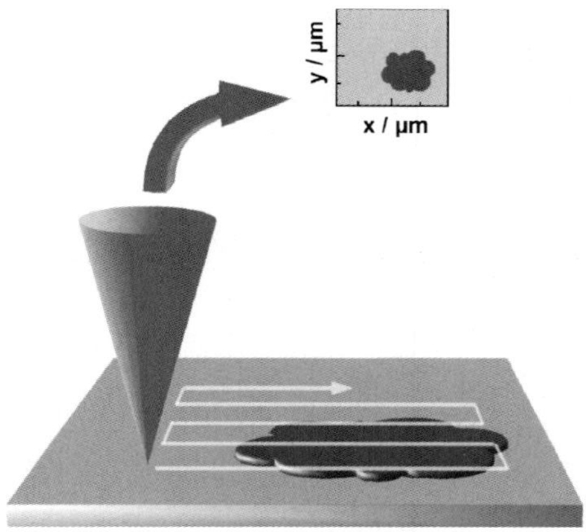

Fig. 1 Principle of scanning probe microscopy

As a consequence of the increasing miniaturization of sensing devices and the increasing use of nano-sized objects in all areas of chemistry, the need for high-resolution visualization of local properties of surfaces is obvious. As a matter of fact, conventional global techniques are averaging the impact of potentially very different surface domains to a single overall value that does not provide all necessary information for rational optimization steps. Thus, localized techniques are the only option for achieving local surface modification procedures or to proof new strategies of selectively modifying micro- and nano-scale objects. Biosensor research is often directed to the elucidation of suitable schemes for the immobilization of biological recognition elements on transducer surfaces or, even more importantly, to the design of pathways for transducing biological recognition events into physical and subsequently electronic signals. Information such as the topography of a surface is not providing insight into complex signal transduction cascades since it does not provide any information on the activity of the surface. Even high-resolution imaging of the topography of single protein molecules on surfaces does not provide insight into its activity. SECM, as one of the few chemical microscopy techniques, can provide information on the local electrochemical activity of a surface and, hence, can especially address problems in the area of electrochemical biosensors namely amperometric, coulometric, impedimetric, or potentiometric sensors. Additionally, SECM can address problems occurring with non-electrochemical sensors such as elucidation of immobilization protocols and homogeneity of the immobilized surface activity, etc. As a matter of fact, mainly amperometric biosensors or biosensor principles involving electrochemical read-out and/or the production and/or consumption of electrochemically active species were investigated by means of SECM. Obviously, the visualization of (bio)catalytic surface activity is providing chemical information that is complementary to the information obtained from techniques displaying mainly the topography and morphology of the sample surface such as high-resolution scanning electron microscopy (SEM), atomic force microscopy (AFM), and high-vacuum surface analysis techniques (e.g., XPS). Most of these techniques require non-physiological conditions such as high vacuum or surface conductivity and are hence generally visualizing denatured biological objects. In contrast, local electrochemical activity is measured in SECM with minimal perturbation of the biologically active system due to the possibility to investigate the biologically modified surfaces under physiological conditions. Moreover, in contrast to well-established microscopic techniques, which are based on fluorescence detection and hence need labeling of the biomolecules with suitable fluorescence dyes, SECM measurements often can be performed without labeling.

In typical biosensor architectures, the complexity is mainly caused by the heterogeneity of the biocatalytic layer comprising the biological recognition element, a suitable immobilization matrix, redox mediators, and diffusion layers, etc., in combination with the transducer surface. The applied immobilization method may lead to active centers (hot spots) in the sensor structure while other areas are not biocatalytically active or show reduced biocatalytic activity. Thus, in presence of the analyte, the biological recognition followed by the biocatalytic conversion of the substrate under formation of the related products occurs with a different rate at microscopically different areas of the sensing layer. This will cause an inhomogeneous diffusion profile with areas where the substrate is already depleted while at other domains the substrate is still available. Thus, the integrated sensor signal shows time-dependent contributions from the different sensing sites that are hidden in the overall signal. Obviously, the concentrations of all components that are involved in the generation of the sensor signal are fluctuating both in time and space. It becomes even more complicated if the product of the biocatalytic reaction is influencing the reaction rates of the biocatalytic process in a feedback loop (e.g., modulation of the local pH value due to liberation or consumption of protons in the biocatalytic reaction).

From this brief discussion it becomes even more obvious that it is indispensable to evaluate a sensor architecture taking into account the varying concentrations of all chemical species in space and time. SECM has the potential to assist these investigations by the visualization of complex concentration gradients, to display local biocatalytic activity, to locally influence the concentrations of selected compounds, etc., with high spatial and temporal resolution. In addition, SECM can be used to visualize the remaining biological activity after immobilization and, hence, is able to study the impact of the immobilization process on the function of the biological recognition elements. Moreover, it can be applied for investigating the local biocatalytic activity over time, information which allows for drawing conclusions on the operational stability of a specific sensor design.

With the availability of ultramicroelectrodes with diameters in the sub-μm to even nanometer range, SECM can be operated with a resolution that is sufficient to obtain an in-depth insight into the complex reaction sequences within the sensing layer. Moreover, quantitative mathematical treatment of SECM data was developed over the past two decades and is now readily available. Thus, even the complicated kinetics of biological recognition processes can be modeled. This does not only allow to study the spatial distribution of surface activity but provides the necessary basis to develop models concerning the function of macro-sized biosensors and/or to provide new insights into the overall mechanisms finally leading to the sensor signal. Even if quantitative data treatment is often too complicated, visualization of activity distribution with high lateral resolution by color coded three-dimensional images provides valuable information about the functions and the complex interplay of all sensor components.

In this review, we will try to look at biosensor research from a SECM perspective (Fig. 2). After a brief introduction into the theory of SECM, the reader will be provided with a basic understanding of the principles of SECM and of the two main working modes typically used in SECM measurements. By discussing concepts and important results obtained in these two working modes, the capability of SECM for biosensor research will become obvious. It will be shown how SECM was employed to study properties of surfaces that were modified with a variety of different biological recognition elements and how it was used to visualize localized activity of (immobilized) biomolecules. As chemically modifying surfaces is seen as a fundamental prerequisite for the design of complex biosensor architectures, the capability of SECM to provide chemical images of modified surfaces is evidently important for understanding fundamental problems in biosensor research. In addition, SECM was successfully employed to study applications of biosensors. The read-out of biosensors by SECM in different configurations was demonstrated and micro biosensors were used as probes in SECM measurements. SECM as a tool in micro- and nanofabrication will be subject of a subchapter demonstrating the potential to modify a surface and to probe the surface modification in-

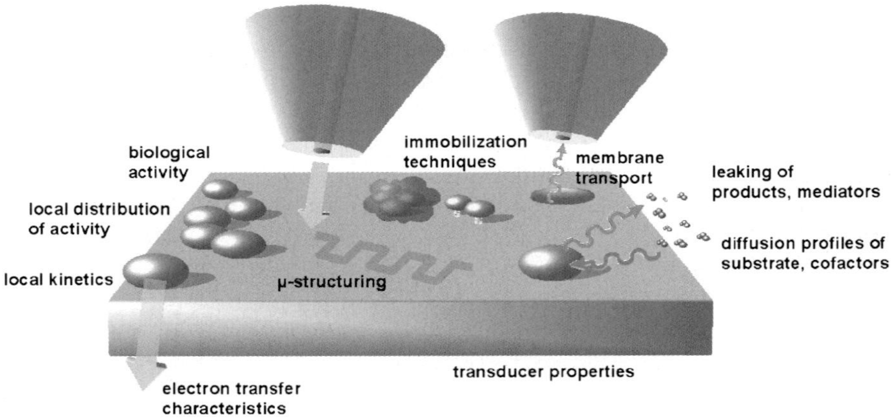

Fig. 2 Fields of interest in biosensor research that can be addressed by means of SECM. Measurements are carried out in electrolyte solution

situ. Due to the significant amount of published results over the past 15 years in this area, it is beyond the scope of this review to cover all SECM studies related to biosensor research. The aim of this chapter is to highlight representative contributions of employing SECM in the study of biosensors and to exemplify concepts for gaining a different perspective for the evaluation of biosensors.

2
SECM Theory

The experimental set-up of scanning electrochemical microscopy is rather simple and it is similar to those of other scanning probe microscopies (Fig. 3). Either sample or probe (tip) are mounted on a x, y, z-positioning system. Mainly, stepmotor-driven translation stages, inchworm motor-driven ones, or piezoceramic systems are employed for an accurate control of the tip position. In SECM, a potentiostat is used to control the tip potential. The set-up is computer controlled using AD/DA cards or similar devices for communication. In order to reduce electronic noise and disturbances by air ventilation or vibration, most SECM systems are built in a Faraday cage mounted on a vibration damping system. Commercial instruments are available from a number of companies, such as CH Instruments [8], Heka [9], Sensolytics [10], UniScan [11], and Windsor Scientific [12].

A core element in any kind of scanning probe microscopy is the probe (tip) and especially its dimensions, which mainly determine the spatial resolution. The probe in SECM is an ultramicroelectrode (UME). Conventionally, disk electrodes with active electrode diameters between 25 µm and 10 µm

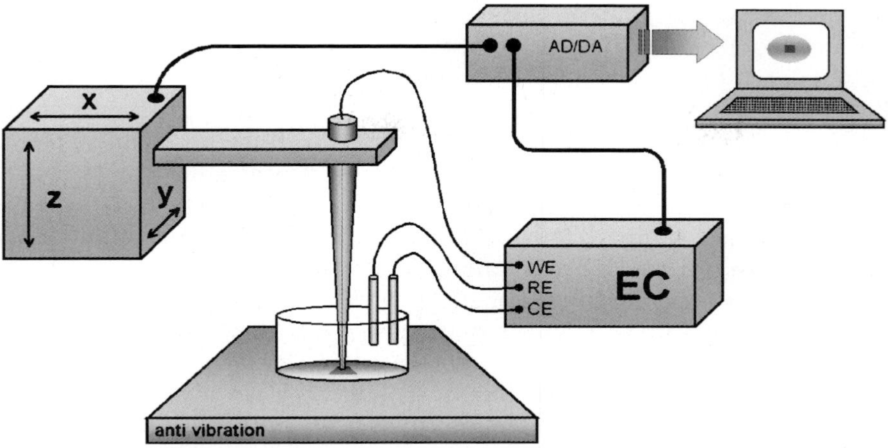

Fig. 3 Typical setup of a SECM: the sample is mounted on a vibration damping system, the electrochemical cell is formed by working, reference, and counter electrodes (WE, RE, CE). Electrode movement is controlled by a x,y,z-translation stage, electrochemistry (EC) is controlled by a potentiostat. All analog signals have to be converted using an AD/DA card or a similar device in order to be displayed on a computer system

are used. Different electrode sizes, geometries, and materials have been described [13]. High-resolution SECM using electrodes with sub-μm dimensions has been demonstrated [14–18]. SECM is generally carried out as a potentiostatic experiment and less often using a galvanostatic or potentiometric approach. Modulations in probe current as induced by the (electro)chemical properties of the sample surface are measured as a function of the x- and y-coordinates of the scanned UME. UMEs are chosen as SECM tips not only because of their small dimensions but also due to their unique electrochemical properties [19]. If an ultramicroelectrode is polarized at a potential sufficient to drive an electrochemical reaction of a species in solution, i.e.,

$$\mathrm{R} \rightarrow \mathrm{O} + \mathrm{e}^- \quad (1)$$

in a diffusion-controlled way, a steady-state current, i_{ss}, is established at the ultramicroelectrode given by equation (2) ([20] and references therein):

$$i_{\mathrm{ss}} = 4nFD_{\mathrm{R}}c_0a\,, \quad (2)$$

where n, is the number of electrons transferred in the electrochemical reaction, F is Faraday's constant, D_{R} is the diffusion coefficient of the reduced form of the redox species, c_0 is the initial bulk concentration of the species, and a is the radius of the active electrode area. Due to a quasi-hemispherical diffusion profile at the ultramicroelectrode, mass transport is very efficient and the steady-state current is established very fast. The basic principle of SECM measurements is based on the modulation of the steady-state current with respect to the properties of the sample surface. To be able to perturb the steady-state current measured at the tip, the tip and surface have to be close to each other. In terms of scanning probe microscopy, the tip has to be within the nearfield region of the surface. Only when the tip is located within the nearfield, the surface able to interact with the tip by modulating the reactions occurring at the tip. Typically, the nearfield distance is about 3–5 times the diameter of the active electrode area. Two basic cases have to be distinguished: approach towards an electrochemically inactive surface and approach towards an electrochemically active surface. When approaching an electrochemically inactive sample, the diffusion of redox species towards the ultramicroelectrode is hindered and fewer species can reach the active electrode area in time. Hence, the current at the SECM tip drops to values below the steady-state current. Ideally, the current decreases down to zero in the case of an optimal coplanar approach between the plane of the UME tip and the sample surface. This effect, which is caused by blocking of the diffusional mass transport of the redox species, is called *negative feedback* (Fig. 5b) [1, 13]. In contrast, the current at the tip increases if the tip approaches an electrochemically active surface. Species that have been oxidized at the tip can be re-reduced (or vice versa) at the sample surface, which is possible either if the sample is at a Nernstian potential sufficiently negative (or positive) to drive the re-reaction or the sample is actively poised to a sufficiently negative (or positive) potential in a bipotentiostatic experiment. If the kinetics of the reaction at the sample surface are faster than the diffusional mass transport of the redox species from the bulk of the electrolyte, an increase in current at the ultramicroelectrode is observed, which is called *positive feedback* (Fig. 5c) [1, 13]. The related implementation of these two effects into SECM experiments is called *feedback mode* (FB-mode; Fig. 4). A sample exhibiting both electrochemically active and inactive domains will

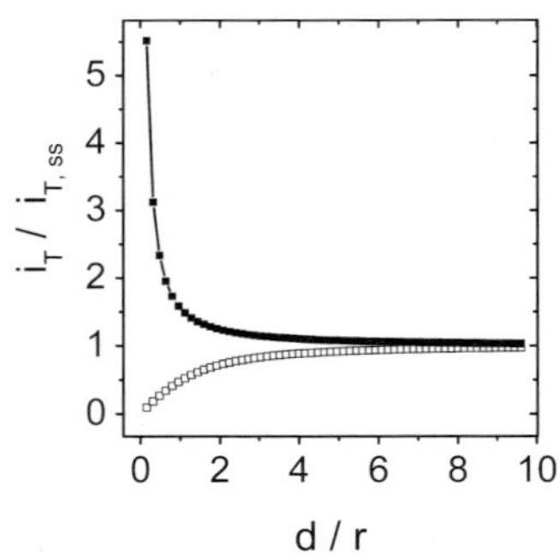

Fig. 4 Feedback approach curves of an electrode towards an electrochemically inactive sample (*open squares*) and towards an electrochemically active surface (*filled squares*). Approach curves were calculated using equations given in [21] assuming a RG value of 10 (for definition of the RG value see below)

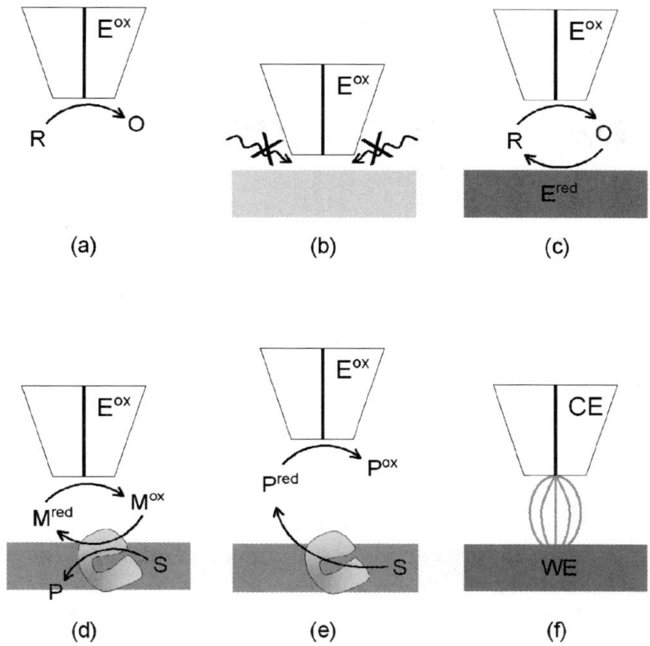

Fig. 5 Schematic representations of the different working modes in SECM. **a** diffusion-limited conversion of redox species in bulk volume with the tip far away from any surface, **b** negative feedback mode, **c** positive feedback mode, **d** enzyme mediated (positive) feedback mode, **e** (enzyme-based) generation-collection mode, **f** direct mode of SECM

exhibit both positive and negative feedback effects and consequently show differences in the tip current at known tip-to-sample distance with respect to the *x*,*y*-positions of the tip. Visualization of the change in tip current due to feedback effects plotted over the *x*,*y*-positions of the tip maps the local electrochemical activity of the sample surface.

It is obvious that the rate at which the mediator (O) is regenerated at the substrate is ultimately decisive for the type of SECM feedback response. Thus, even though the substrate might be able to regenerate the reduced mediator (but at a much slower kinetic rate than the oxidation reaction at the tip) the regeneration at the substrate would not be able to fully compensate the modulation of the diffusion profile from a hemispherical diffusion to a radial diffusion profile. So, even a conductive substrate might lead to a negative feedback mode. In contrast, even an insulating substrate might lead to a positive feedback mode for the case when a catalytically active component is immobilized on the surface of an insulator and that is able to support the regeneration reaction of mediator in the absence of an electron communication with the substrate. For the particular case when the catalytically active component is an enzyme, the related SECM mode is called *enzyme-mediated feedback mode* (Fig. 5d) [22, 23]. In this mode, the enzyme regenerates the mediator species, R, consumed at the tip and the rate of the enzymatic reaction is directly responsible for the resulting type of feedback mode. Positioning the tip over an immobilized enzyme could then switch on the enzymatic reaction by supplying the electron mediator in the appropriate redox state, resulting in a feedback current at the tip. The increase of the tip current in this case would then be related rather to localized enzymatic activity than to electrochemical activity. The improved lateral resolution of the feedback mode over the generator collector mode (see below) of SECM is only valid for the enzyme-mediated feedback mode for a very active enzyme (k_{cat}) and/or a high enzyme loading (Γ_{enz}). In addition, a significant current increase in the enzyme-mediated feedback mode can only be expected at rather close tip-to-sample distance. This may cause problems concerning tip crash or superimposed limitation of substrate and product diffusion.

Equally important as the enzyme-mediated feedback *mode* in biosensor research (but somewhat more straightforward) is the generation-collection mode of SECM (GC-mode, Fig. 5e, [1, 13]). In the sample generation-tip collection version of this mode, an electrochemically active species is produced at the sample and collected at the tip polarized to an appropriate potential. In contrast, in the tip-generation sample collection mode of SECM, a redox species is produced at the tip and collected at the sample surface. The latter approach is not used extensively since currents are often very small and lack sensitivity due to high background currents at the polarized sample. In both cases, however, an increase in current can be observed when the electrode is positioned over an electrochemically active domain and no current is observed when the electrode is positioned over an inactive area. The GC-mode as sample generation-tip collection mode is superior to the feedback mode in sensitivity but inferior in resolution [24 and references therein].

A configuration significantly different from the two main cases described above is the *direct mode* of SECM (Fig. 5f). In this mode, the SECM tip functions as (microscopic) counter electrode for the (macroscopic) sample which is used as working electrode. Due to a limitation through the counter electrode reaction and a confinement of the electric field between the tip and the perpendicular surface area of the sample, electrochemical processes at the sample surface are restricted to an area approximately the size of the tip and located opposite to the SECM tip position.

A number of additional practical considerations have to be discussed in order to understand limitations and possibilities of SECM experiments as discussed in this chapter. The resolution of SECM is mainly limited by the active area of the ultramicroelectrode which is used as SECM tip. However, the size of the insulating sheath, which is found in almost any ultramicroelectrode design applicable in SECM experiments, needs to be added to the active electrode area to determine the whole size of the tip. The ratio of the diameter of the insulating sheath and the diameter of the active electrode area is called the

RG value:

$$RG = \frac{r_{\text{sheath}}}{r_{\text{active area}}} . \qquad (3)$$

Differences in the RG value of the SECM tip result in differences in the feedback approach curves since the ability to block diffusion towards the active electrode area is depending on the RG value [21]. In addition, the tip size and its geometry can become an important issue in experiments aiming at the mapping of locally varying concentrations in diffusion layers. Depending on the size of the tip and the speed of movement, the diffusion layer might be disturbed.

A basic issue in understanding a specific scanning probe microscopy is knowledge of the signal-distance relationship. Interactions of tip and sample become stronger with decreasing tip-to-sample separation. It is not only important to understand how tip and sample interact (feedback, generation-collection) but also to what extent. Evidently, an increase in feedback current can be attributed to either a higher electrochemical activity or a smaller tip-to-sample separation. When using relatively large active electrode areas (10 or 25 μm disk diameter), the change of signal attributed to a change in tip-to-sample separation of a "flat" surface is often negligible. Moreover, a tilt in the mounting of the sample with respect to the scanning plane of the tip can be easily recognized. However, if the tilt or the surface topography becomes similar or equal to the overall working distance, changes in signal caused by changes in topography are no longer negligible. In this case, the local electrochemical activity can no longer be clearly separated from the influence of topographic variations. As a matter of fact, this problem, which is inherent for all scanning probe techniques, is even more pronounced when using very small tip areas with their concomitantly very short working distances. Hence, the tip might crash into protruding features or "loose" the feedback signal over recessed features when the *z*-position is not changed in accordance with changes in topography.

To overcome these limitations, different attempts to integrate height-control systems in SECM have been described. Systems relying on shearforce interactions between a vibrating tip and the surface were described ([14, 15, 25, 26] and references therein) with the tip being either vibrated at a fixed frequency by means of a tuning fork [25] or at one of its resonance frequencies using piezoceramic agitators [14, 15, 26]. Changes in the vibration amplitude or phase shifts caused by shearforce interactions of the tip with the sample surface can be detected optically [26] or by piezo–piezo read-out [14, 15] using phase-sensitive amplification by means of a lock-in amplifier. Shear-force interactions can be detected within distances of 50 to 500 nm from the sample surface and do not significantly interfere with the electrochemical processes occurring at the active electrode area. Other attempts for providing a *constant-distance mode* for SECM make use of an ac voltage overlaying the dc potential to which the tip is polarized [27]. SECM modes using constant current signals [28] and combined atomic force microscopy (AFM, [7, 29]) and scanning electrochemical microscopy (AFM-SECM) [18, 30] also provide possibilities to control the tip-to-sample separation.

In summary, SECM applications are confined to systems that include at least one electrochemically active component. It might be either an intrinsic component (reactant or product of reactions) or an electrochemically active component added on purpose to provide information by interacting, not necessarily reacting, with the system of interest. SECM provides information on localized (electro)chemical activities on surfaces. Electrochemically active domains of a sample surface will induce an increase in current measured at the probe UME due to an additional flux of redox species to the perpendicular positioned SECM tip. Electrochemically inactive domains will provoke a decrease in current due to the restriction of the diffusional flux of redox species towards the active electrode area.

Even though applications of SECM seem to be limited, the prospective of different operating modes and various arrangements of the setup (e.g., different modes of coupling the working and the counter electrodes, different working potentials or potential pulse sequences, the initial redox state of the redox components in bulk solution, etc.) makes SECM a flexible and powerful tool for characterization of surfaces and interfaces, but yet, a sophisticated technique with respect to instrumentation, sample preparation, and operational skills.

3
The Generation-Collection Mode of SECM in Biosensor Research

Despite the fact that first applications of SECM for the investigation of biological systems that are related to biosensor research were performed in the feedback mode of SECM [22] (see below), the generation-collection mode has later been employed intensively in biosensor research or for the visualization of local immobilized enzymatic activity. The first description of the generation-collection mode by Bard and Heller and coworkers [31] can already be seen as being related to biosensor research. In this work, a micro biosensor for H_2O_2 detection was constructed by electrochemically "wiring" horseradish peroxidase (HRP) to the surface of a 8 μm diameter carbon electrodes via a cross-linked Os-complex containing redox polymer. In addition to elaborated theoretical considerations, the micro biosensors were used as SECM tips for visualization of H_2O_2 concentrations in the diffusion zone in front of noble metal microelectrodes embedded within an insulating surface (Fig. 6). The formation of H_2O_2 could be correlated with the local reduction of molecular O_2. Different kinetic regimes and mechanisms were elucidated at different electrode materials. Moreover, the H_2O_2 selective tips were ap-

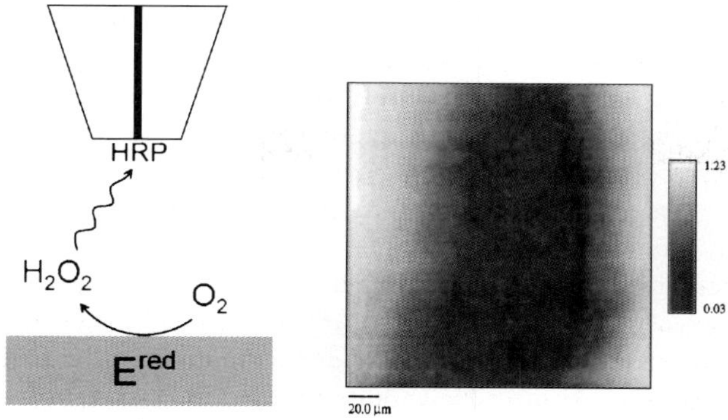

Fig. 6 *Left*: Application of a HRP-modified microbiosensor as SECM tip in the generation-collection mode. *Right*: SECM image of the H_2O_2 concentration profile at a 25 µm Pt disk electrode acquired with a HRP-modified carbon fiber electrode (from [31] with permission)

plied for mapping the enzymatic activity of glucose oxidase immobilized within an Os-complex based redox hydrogel.

Already in this early work advantages and disadvantages of the generation-collection mode became evident. Especially for extracting kinetic parameters, a precise knowledge about the distance dependence of the tip current and the exact tip-to-sample separation is mandatory. In generation-collection experiments, only a negligible background current is measured with either the tip positioned far away from the sample surface or above an electrochemically inactive region of the sample. Thus, tip positioning cannot be easily done by simply approaching the electrode towards the sample surface (z-scan) and recording an i_T–d-curve. Obviously, either the current remains unchanged or it is unpredictably increasing over an active area. Hence, alternatives for positioning of the tip within the nearfield working distance have to be found. For example, in [31], the increasing solution resistance in the gap between tip and sample surface was used as a method for tip positioning. In contrast to the feedback mode of SECM, which is relying on the regeneration of a tip-produced redox species by a biocomponent and hence limited to the oxidoreductases, in the generation-collection mode it is sufficient if one species (substrate, cofactor, product) is electroactive or can be selectively oxidized or reduced at the SECM tip. An example for the straightforward use of the generation-collection mode can be found in [32]. The fundamental problem of mediator leaking from amperometric biosensors [32, 33] was monitored for a number of different electrodes modified with various electrocatalysts for the oxidation of NADH. Mapping of the redox mediator concentration profiles in the diffusion layers of the catalyst-modified electrodes by slow z-approach curves strongly supported assumptions made concerning the mechanism of NADH oxidation at the particular electrodes. Findings from both cyclic voltammetry and SECM resulted in the construction of optimized ethanol biosensors.

The optimization of immobilization strategies for biological recognition elements is in focus of many efforts employing SECM in biosensor research. The visualization of the local distribution of active binding sites in immobilized antibody layers by means of the generation-collection mode of SECM has been described by Wittstock and coworkers (Fig. 7) [34]. Antibodies immobilized onto glass substrates were allowed to complementarily recognize their antigen that was conjugated to alkaline phosphatase (AP) as enzyme label. AP is converting electrochemically inactive p-aminophenyl phosphate (pAPP) into electroactive p-aminophenol (pAP) which can be then electrochemically oxidized at the SECM tip under formation of the related quinone imine (pQI) [35, 36]. The approach could be successfully applied for the visualization of the distribution of active binding sites for the antigen. Here, a general problem in the optimization of immobilization strategies can be addressed. If, for example, an antibody is directly immobilized on a transducer surface, the antibody loading may have an impact on the sensor response due to blocking of substrate access, electron transfer pathways, or passivation of the transducer surface. SECM is able to quantify the local immobilized binding capacity and hence is able to distinguish between biomolecule loading and other influences that may alter the sensor response [34].

A similar effect could be studied for a biotin/avidin-patterned surface [37]. Affinity interaction employing biotin/avidin technology was used as immobilization method for different dehydrogenase enzymes, e.g., glutamate dehydrogenase or alcohol dehydrogenase. However, the high loading of the electrode surface with the biotin/avidin conjugate was proven to dramatically limit the diffusion of the redox mediator (NADH) towards the electrode, and

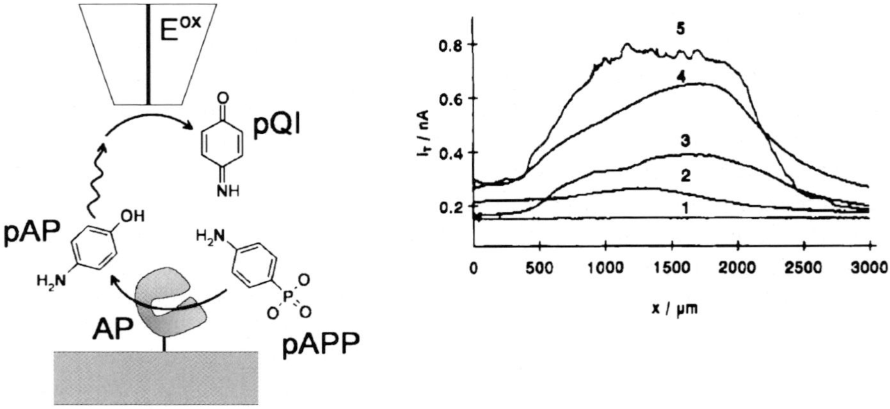

Fig. 7 *Left*: Localized visualization of immobilized AP activity in the generation-collection mode of SECM. *Right*: Line scans over spots of immobilized antibodies. Increasing numbers represent an increasing concentration of the AP-labeled antigen (from [34] with permission)

consequently, decreasing the sensor response. As a solution, the authors suggested a spatial separation of the sites for attachment of the biotinylated enzyme and a non-modified active electrode area for the oxidation of NADH using laser photolithography (Fig. 8). Microchannels were created that permit a facile diffusion of the redox mediator between the active site of the immobilized enzyme and the surface of a carbon electrode. Visualization of the local enzyme activity at the photopatterned glassy carbon electrode surface shows the potential of SECM to distinguish the influence of protein loading and defects.

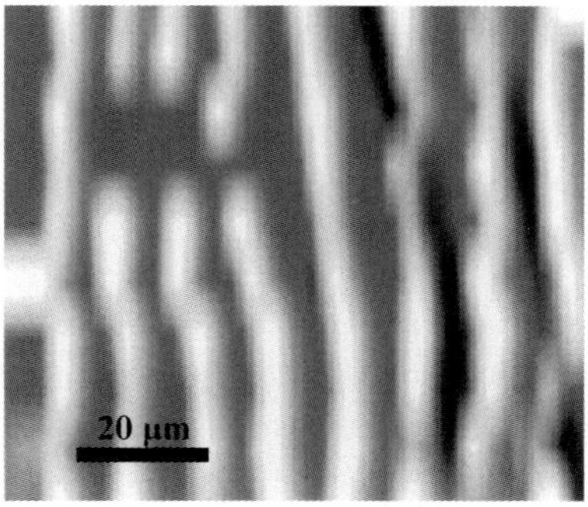

Fig. 8 Local visualization of kinetic activity and defects in patterning at a biotin/avidin modified glassy carbon electrode using the generation-collection mode of SECM (from [37] with permission)

Similar concepts to examine specific features of immunoassays by means of SECM can be found in [38–41]. Enzyme-labeled antibodies were used to investigate their immobilization onto magnetic beads [38]. Breast-cancer related antigen CA15-3 was probed by means of a sandwich immunoassay using monoclonal antibodies against the antigen that were immobilized via streptavidin-biotin chemistry and a secondary HRP-labeled antibody. HRP is able to convert hydroquinone into electrochemically reducible benzoquinone in the presence of H_2O_2. Binding of the antigen only occurs at sites of the immobilized primary antibody preparing these sites for the later binding of the HRP-labeled secondary antibody. Upon addition of H_2O_2, benzoquinone is locally generated, which can be detected in the generation-collection mode of SECM [39]. The use of SECM in ELISA-type immunoassays was also described for a cytokine assay on a cellular chip [41]. In this study, a cell culture embedded in a collagen gel matrix was spotted on an antibody-modified surface. Cytokines produced from active leukocytes in the cell culture were captured by the immobilized antibodies. Again, HRP was used as the enzyme label for a secondary antibody in a sandwich-type assay. In SECM measurements, the cofactor for the enzymatic reaction (ferrocenylmethanol) was produced at the tip to switch on the enzymatic reaction when the tip was scanned over an area where the biological recognition reaction could take place.

In addition to SECM investigations where the detection scheme is based on the redox conversion of free-diffusing products consumed or generated in a sensing layer, SECM is capable of analyzing the concentration profiles *inside* operating biocatalytic layers. Oxidation of H_2O_2 or reduction of molecular O_2 at a bare SECM tip penetrating an enzyme layer was demonstrated [42]. Digital simulation in combination with an in-depth evaluation of the obtained results from generation-collection experiments allows for establishing guidelines for the design of optimized biosensors. The combination of two scanning probe techniques also leads to improved knowledge or at least facilitated data interpretation. As mentioned before, a general problem in SECM is the dependence of the tip current on the tip-to-sample distance. Topographic features may overlap with electrochemical features and may lead to artifacts in the visualization of electrochemical activity [14, 28, 43]. The combination of atomic force microscopy (AFM, [7, 29]) with SECM allows for a straightforward separation of electrochemistry and topography [30]. AFM detects the surface topography while SECM simultaneously maps electrochemical activity at the same site of the sample surface. AFM-SECM tips bearing a sub-micrometer working electrode in direct proximity to the topography sensor as described by Kranz and coworkers [18] were used to simultaneously image the topography and activ-

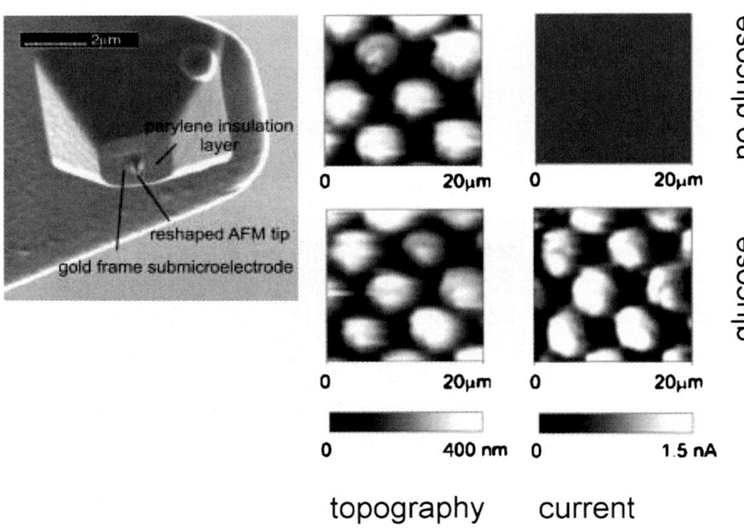

Fig. 9 *Left*: SEM image of an AFM-SECM tip. *Right*: Series of AFM-SECM images of glucose oxidase immobilized in a porous structure. Images on the left show the topography of the sample as recorded by AFM, images on the right show the current response of the integrated micro biosensor. Images were obtained in the absence (*top*) and presence (*bottom*) of glucose. (from [18] with permission)

ity of glucose oxidase immobilized within a polymer in 1 μm diameter pores of a micromachined substrate (Fig. 9). H_2O_2 produced in the enzymatic oxidation of glucose was collected at the working electrode while the AFM was operated in tapping mode [44]. Similar AFM-SECM tips were employed to image glucose oxidase activity at enzyme-modified graphite electrodes [45], and HRP activity on microstructured gold samples [46].

Shearforce positioning of needle-type tips enables the use of *non-electrode* tips in SECM measurements such as enzyme-filled capillaries producing a species detectable at the sample electrode [47]. Active sites at Pt micro-electrodes and poly-(methylene blue)-modified electrodes could be visualized using enzyme-filled capillaries in tip generation-sample collection mode experiments.

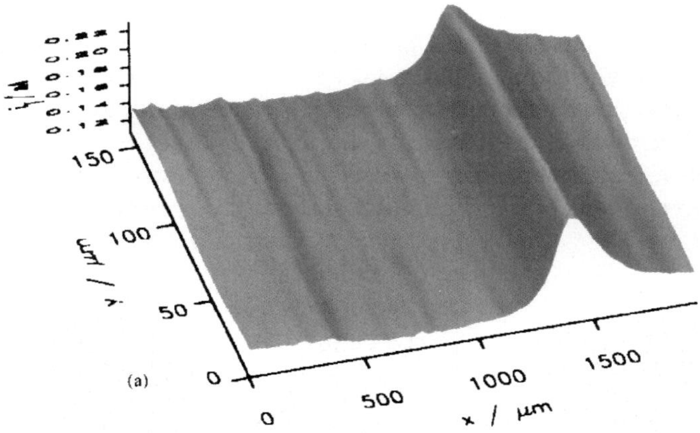

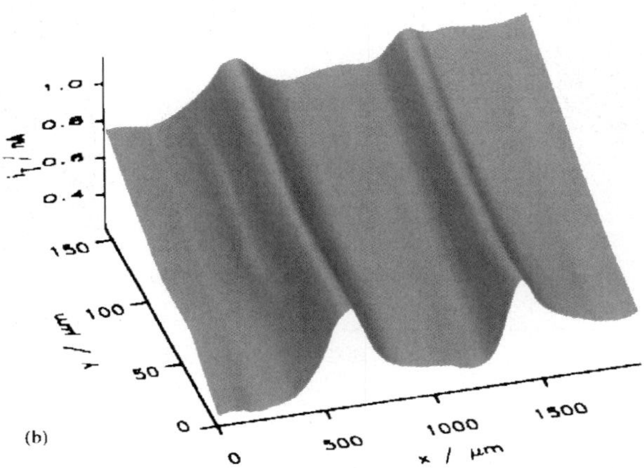

Fig. 10 SECM images visualizing the enzymatic activity of glucose oxidase (*left line*) and lactate oxidase (*right line*) immobilized within a piezo-spotted polymer microstructure. Images were obtained in generation-collection mode in the presence of only lactate (**a**) and both lactate and glucose (**b**). (From [48] with permission)

So far, only results from relatively simple sensor concepts have been described. However, the same easy principle of the generation-collection mode of SECM can be employed to examine more complex sensor architectures. Microstructures made by piezo-dispensing of different enzyme-containing solutions onto self-assembled monolayers (SAMs) have been studied [48]. Two different immobilization strategies using amide bonds between carboxylic acid side chains or lysine residues of the enzymes and functional headgroups of the SAMs were used to achieve microstructured enzyme patterns. Lines made of different enzymes could be visualized in the presence of the enzymes' substrates (Fig. 10).

Even more complex enzyme microstructures were investigated. Multiple enzymes were immobilized in a grid structure while expanding the range of used immobilization methods from covalent binding of biomolecules to SAM-modified surfaces to crosslinking and immobilization within a polymer matrix [49] (Fig. 11). Enzyme amplification and enzyme competition in coupled enzyme reactions were visualized in bienzyme or multienzyme structures using the generation-collection mode in the presence of the substrates of the immobilized enzymes.

The use of multi-enzyme structures opens up possibilities for multi-analyte detection and interference elimination by choosing suitable multi-

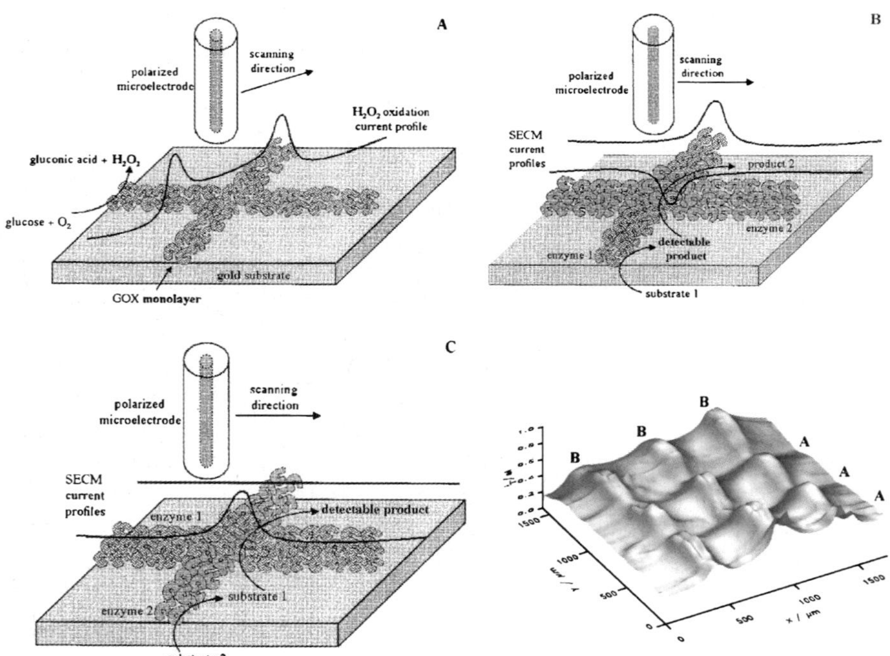

Fig. 11 SECM visualization of multienzymes structures and SECM image of an α-glucosidase-mutarotase-GOD containing grid structure. (From [49] with permission)

enzyme systems. The application of SECM detects the interaction of different enzymes and enables for a rapid screening of a variety of multi-enzyme systems in order to elucidate the best system for a given analytical problem. Optimization of sensor architectures often requires a detailed understanding of the impact of the different components of a given sensor architecture on the sensor response and hence of the ratio of the sensor components. Gradients of enzyme loading in polymer grid structures could be obtained by means of microdispensing, and the local enzymatic activity could be visualized using the generation collection mode of SECM [50]. The selectivity of the electrochemically induced modification of miniaturized sensing elements by means of electrodeposition paints could be proved using SECM [51]. Moreover, optimization of "wired" enzyme electrodes for biofuel cells catalysts that might be useful for biosensor applications has been described [52].

4
The Feedback Mode of SECM in Biosensor Research

Already in 1992, at the early stages of SECM development, Pierce and Bard [22] published the first study that proved the potential of SECM to examine the electron-transfer kinetics of a non-conductive surface modified with a redox enzyme using the feedback mode. The influences of various immobilization methods which were supposed to induce major differences in terms of immobilized enzymatic activity, substrate diffusion, etc., were investigated. In order to avoid inaccuracy in the determination of the distance between tip and sample (d), two redox mediators exhibiting different formal potentials were used. Ferrocenecarboxylic acid (FcCOOH) served as electron acceptor for the immobilized glucose oxidase while methylviologen$^{2+/+}$ (MV^{2+}) was used as calibrant for the determination of the exact tip-to-sample separation. Three different methods for immobilization of glucose oxidase were compared: (1) covalent binding to a Nylon 66 membrane; (2) entrapment into a polymer hydrogel on an aminated glass slide using spin coating; and (3) immobilization within Langmuir-Blodgett films. The modification of Nylon 66 membrane showed a poor reproducibility with respect to the amount of immobilized glucose oxidase leading to a local feedback response varying between positive and negative feedback. The variation in the feedback mode approach curves (Fig. 12) is caused by the concentration of the substrate glucose leading to a more efficient recycling of the FcCOOH at high glucose concentrations overcoming in part the diffusional blocking and to negative feedback at low glucose concentrations which is, however, less pronounced than in absence of glucose.

Similar variability of feedback responses were observed for other parameters that alter the enzymatic activity within the film such as thickness of the hydrogel layer or the enzyme-to-hydrogel ratio. Moreover, after crosslinking

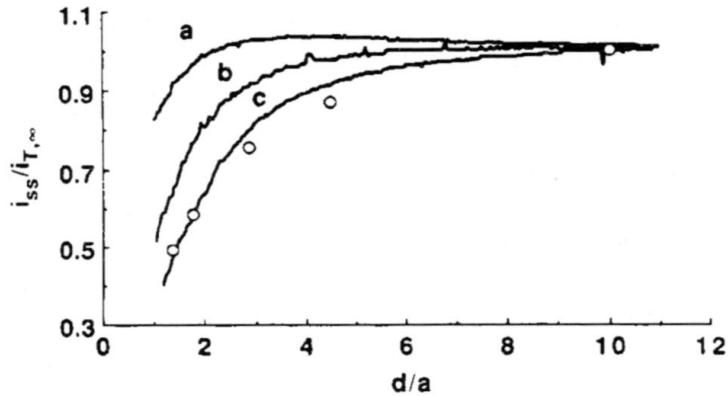

Fig. 12 Variability of feedback mode *z*-approach curves obtained with glucose oxidase-modified Nylon 66 membranes at high glucose concentrations **a**, low glucose concentrations **b**, and in the absence of glucose **c**. (From [22] with permission)

the hydrogel by means of a bifunctional reagent, only pure negative feedback was observed as a consequence of either a chemical modification of the active site of the immobilized enzyme or the significantly slowed down diffusional mass transport of the substrate and the redox mediator.

Besides imaging localized (bio)catalytic activity, SECM provides possibilities for discriminating between zero- and first-order reaction kinetics for different redox mediators such as hydroquinone, FcCOOH or ferrocyanide at various normalized distances (d/a). This is achieved by evaluating the percentage deviation from the average heterogeneous rate constants of the two processes [22]. As expected, at concentrations of the mediator below 50 μM, the two processes—catalytic oxidation of glucose and mediator recycling by glucose oxidase in the hydrogel membrane—are limited by zero-order processes.

As already mentioned above, the immobilization procedure is a key step for constructing and optimizing a biosensor, thus numerous studies are aiming to identify the role and importance of each parameter that affects the immobilization process. The immobilization technique itself, the components/matrix, determination of the surface coverage of all involved catalytic components and, even more important, the spatial distribution of the immobilized catalytic activity, evaluation of the changes of the catalytic activity that occurs during each step of the immobilization procedure, testing of the operational and storage stability of the resulting biosensor, etc., were investigated. Many of these parameters are difficult to characterize as long as the sensor is yielding a single value by averaging overall heterogeneous characteristics of all catalytic sites on the transducer surface. In this respect, SECM can provide complementary information by visualizing localized (bio)catalytic activity for elucidating the importance of the immobilization procedure on various aspects of the sensor performance.

Glucose oxidase was immobilized within cylindrical pores (8–10 μm diameter) of polycarbonate membranes [23], and its activity was detected in positive feedback mode using the recycling of hydroquinone as free diffusing redox mediator. The tip current representing the oxidation of hydroquinone in the presence of glucose is enhanced as the SECM tip is positioned in close proximity of the immobilized glucose oxidase that converts the quinone back to the corresponding hydroquinone. The localization of the enzyme-filled membrane pores was visualized in the negative feedback mode in absence of glucose using reduction of methylviologen (MV^{2+}) at the tip at an applied potential of –0.95 V. In a similar experiment, the activity of NADH-cytochrome *c* reductase, which is localized at the outer membrane of individual rat liver mitochondria [23], could be visualized using NADH as substrate and *N*,*N*,*N*′,*N*′-tetramethyl-*p*-phenylenediamine (TMPD) as mediator (Fig. 13). A potential of +200 mV vs. SCE was applied to the tip to probe the oxidation of TMPD formed during the enzymatic reaction. The average concentration of NADH-cytochrome *c* reductase at the outer membrane of a single mitochondrion was estimated to be 1.1×10^{-12} mol cm^{-2}.

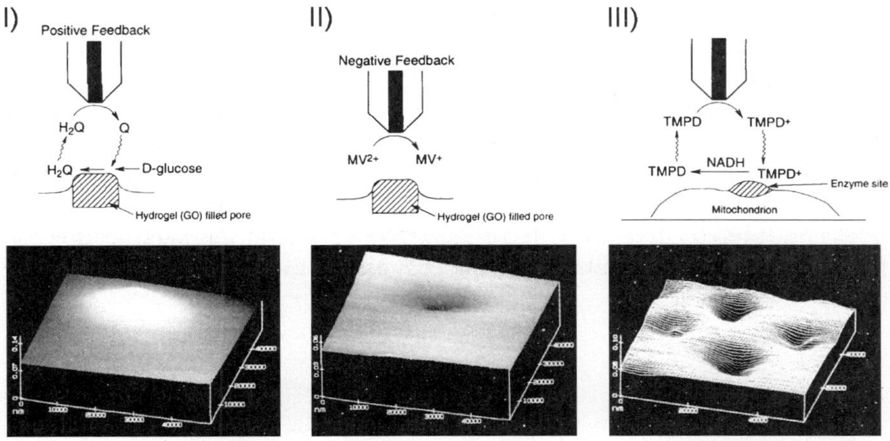

Fig. 13 Schematic representation of the SECM experiments (*upper row*) for visualization of local biocatalytic activity (*lower row*). Visualization of local activity of glucose oxidase immobilized in membrane pores using positive feedback (I) and negative feedback mode (II). Visualization of the activity of NADH-cytochrome *c* reductase localized at mitochondria membranes (III) using positive feedback mode. (From [23] with permission)

Pulsed electrodeposition in the direct mode of SECM was used to deposit microstructured *N*-substituted polypyrrole lines on a gold surface. The polymer lines could be imaged in the feedback mode of SECM using $[Ru(NH_3)_6]^{3+}$ as mediator [53]. The deposition of the polymer lines was confirmed by their low permeability for highly charged mediator, reflected by the negative feedback current obtained at tip positions above the polymer lines. Following this, the anchor groups at the functionalized polypyrrole lines were further employed for covalent attachment of glucose oxidase. Feedback images were recorded both in the absence and presence of glucose, using $[(Os(bpy)_2fpy)Cl]^{2+}$ as redox mediator. As expected, in the absence of glucose, the regeneration of mediator occurs exclusively at the underlying gold surface. The diffusion of the mediator is decreased by the polymer/enzyme layer, and hence a decrease of the positive feedback response in comparison to the uncovered gold area is seen over the microstructure. In the presence of glucose, however, the regeneration of the mediator occurs both at the gold surface and within the reduced active site of glucose oxidase. The contribution of the enzyme-mediated positive feedback can be obtained by subtraction of the response obtained in the absence of substrate from the one recorded in the presence of substrate.

One has to keep in mind that it is of crucial importance to choose the correct oxidation state of the redox mediator since, otherwise, different SECM modes may be selected. For example, Shiku and Matsue [54] (Fig. 14) demonstrated that depending on the redox state of the mediator ferrocene methanol (FMA), the recognition of microspotted carcinoembryonic antigen and HRP-labeled antibody was performed in either the generation-collection mode or the feedback mode of SECM. If oxidized FMA^+ is used as redox mediator (Fig. 14A), the enzymatic reaction continuously proceeds in the presence of H_2O_2 irrespective of the tip position. Thus, the reduced form FMA is accumulated in bulk solution and the background current is slowly increasing with time, which is typical for the generation-collection mode. On the other hand, if FMA was initially added to the bulk solution (Fig. 14B), a constant diffusion-limited current was recorded at the tip, which was significantly enhanced only at tip positions above the enzyme spot.

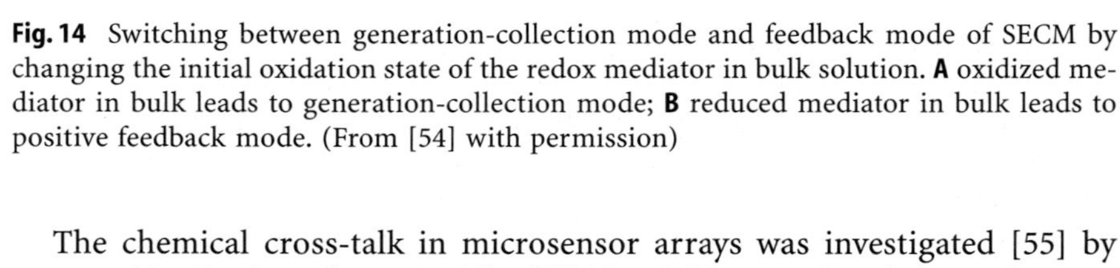

Fig. 14 Switching between generation-collection mode and feedback mode of SECM by changing the initial oxidation state of the redox mediator in bulk solution. **A** oxidized mediator in bulk leads to generation-collection mode; **B** reduced mediator in bulk leads to positive feedback mode. (From [54] with permission)

The chemical cross-talk in microsensor arrays was investigated [55] by means of feedback mode SECM. The diffusion fields over a band electrode array were imaged. It could be demonstrated that the diffusion zones were overlapping for the investigated system, leading to a number of effects such as cross-talk, recycling and shielding that were already observed previously [56, 57].

Feedback mode SECM was used to monitor the activity of the molybdoenzyme nitrate reductase (NaR) (E.C. 1.7.99.4) from *Pseudomonas stutzeri*, using methylviologen (MV^{2+}) as mediator [58], which is known not to adsorb on Pt surfaces. In the presence of the enzyme substrate NO_3^-, an increase of the feedback current was observed due to reduction of MV^{2+} at the Pt-tip (Fig. 15).

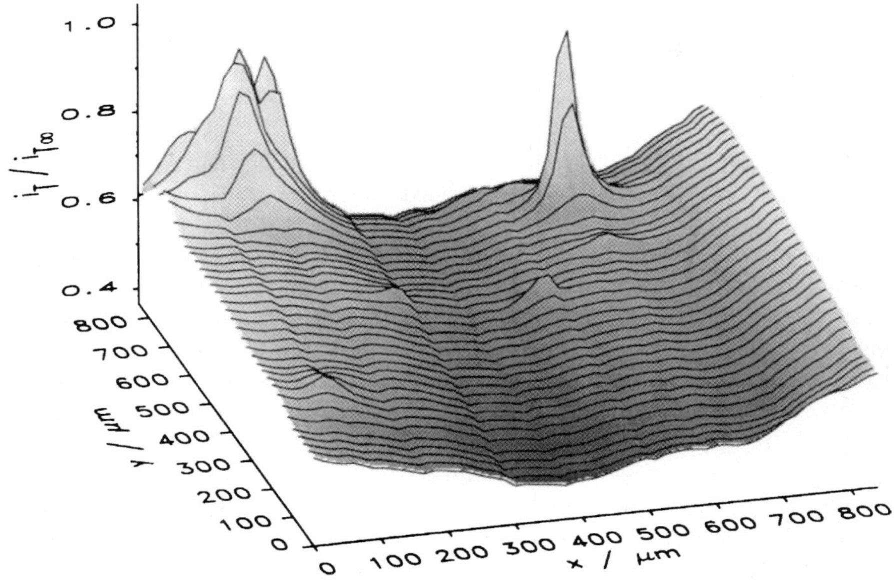

Fig. 15 SECM feedback image showing localized enzymatic activity of membrane-bound nitrate-reductase, recorded in presence of $NaNO_3$ with MV^{2+} as redox mediator

The mediator concentration plays a decisive role in the imaging capabilities of SECM in feedback mode, as exemplified by Wittstock et al. in a study using nitrate reductase immobilized on magnetic beads [59]. In this respect, the possibility to use the SECM tip as a microscopic spatula has to be mentioned. Depending on the tip-to-sample distance, the SECM tip might be used to arrange lines of enzyme-modified microbeads, to move the beads to specific locations at the surface, or to generate patterns.

A further progress of the SECM experiments in feedback mode became possible after introducing the above-mentioned simultaneous topography imaging in the constant-distance mode of SECM [43]. Due to the fact that the shearforce interaction between tip and sample occurs only at very close distances (50–500 nm), a much closer distance becomes feasible allowing a significantly increased amplification by means of redox cycling in the feedback mode and simultaneously avoiding any accidental contact between tip and sample. Improved sensitivity and contrast for the local enzymatic activity of diaphorase [60] immobilized by physical adsorption on a SAM-modified gold surface could be demonstrated using constant-distance mode SECM. Ferrocenylmethanol (FMA) was oxidized at the tip under formation of FMA^+ which was reduced back to FMA in the presence of NADH by means of diaphorase.

For the optimization of reagentless amperometric biosensors, the design of efficient electron-transfer pathways between the immobilized enzyme and the electrode surface via polymer-bound redox relays is a vital issue. In order to address the complex influence of all parameters that affect the overall sensor characteristics such as the enzyme-to-polymer ratio, the loading of the polymer with the redox relays, the degree of cross-linking, and the film thickness, a large number of different sensors would need to be prepared. As a matter of fact, one would try to vary only one parameter while keeping the other parameters constant, an attempt which would demand a high reproducibility in the sensor preparation. As an alternative, the preparation of microscopic patterns of an enzyme containing immobilization layer with a gradient of one parameter along the pattern would offer the advantage of characterizing the influence of the investigated parameter on the biosensor characteristics in a single experiment. Here, SECM can serve as a tool for locally visualizing biosensor properties with the required lateral resolution. Most work in this respect is based on either generation-collection mode SECM [48–50] or on acquiring complementary information using both generation-collection mode and feedback mode SECM [24, 61, 62]. Niculescu et al. [63] (Fig. 16) used feedback mode SECM to investigate immobilized enzymatic microstructures and variations in their localized biochemical activity with respect to the composition of an Os-complex modified hydrogel matrix. PQQ-dependent glucose dehydrogenase (PQQ-GDH) or quinohemoprotein alcohol dehydrogenase (QH-ADH) were crosslinked with an Os-complex containing redox polymer and dispensed in lines of 100 μm width by means of

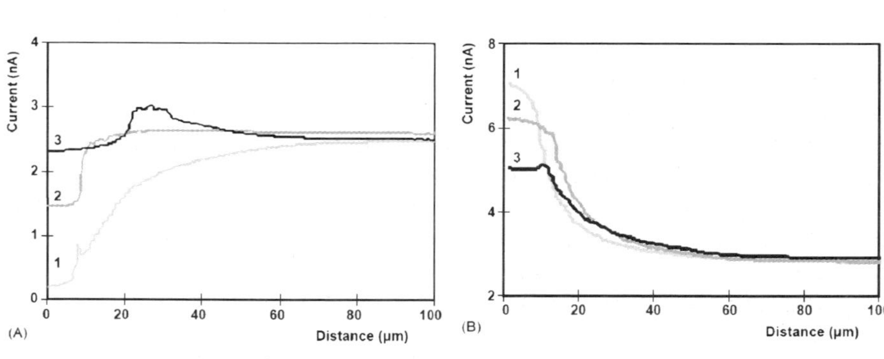

Fig. 16 SECM approach curves obtained with a 25 μm Pt tip moving towards an insulating glass (**A**), or conductive graphite (**B**). Surfaces were modified with (1) BSA, polymer, crosslinker, (2) QH-ADH, polymer, crosslinker, and (3) QH-ADH, Os-complex modified redox polymer, crosslinker. (From [63] with permission)

a piezo microdispenser. In feedback mode, SECM experiments tip generated $[Fe(CN)_6]^{3-}$ can be reduced back either at the polymer-bound Os-complexes, in the active site of the enzyme, or directly at the conductive support. If the sample surface does not participate in the redox reaction of the mediator (Fig. 16A), the feedback approach curves are changing from pure negative feedback for polymer-integrated BSA to a mixed-feedback signal in the presence of redox polymer integrated QH-ADH. If the surface contributes to the recycling of the mediator (Fig. 16B), positive feedback approach curves were obtained in all cases with the fastest redox cycling rate at the redox polymer-integrated QH-ADH line.

The ability to detect the presence of low abundant proteins using SECM in feedback mode was shown by Carano et al. [64]. The detection is based on the oxidation of silver nanoparticles aggregated with the protein molecules. A detection limit of 0.1 ng of BSA was reported. The same principle using Cu deposition instead of Ag was later used for detection of proteins in membranes [65].

The potential of feedback mode SECM was not only exploited for the visualization of local enzymatic activity but also for the visualization of DNA hybridization events. First investigations reporting the visualization of DNA microarrays by means of SECM were published by Yamashita et al. [66] in 2001 although DNA fragments were imaged in constant-current mode SECM (topography) on mica in humid air already in 1999 [67]. Spots of double-stranded DNA (dsDNA) were visualized by SECM due to an enhanced response of the intercalating ligand ferrocenylnaphthalene diimide with the tip positioned above the dsDNA spots [66]. The observed positive feedback was assumed to be caused by the cationic atmosphere containing some extra-amount of ligand other than those bound directly through DNA intercalation. However, in 2006, the same group compared the SECM response of intercalating ferrocenylnaphthalene diimide with ferrocenecarboxylic acid and (ferrocenylmethyl)trimethylammonium iodide, that do not intercalate into dsDNA [68]. It was found that in the presence of those ferrocene derivatives that do not intercalate into dsDNA, negative feedback approach curves were obtained, while in presence of the intercalator ferrocenylnaphtalene diimide the approach curve exhibits an initial decrease of the tip current with an atypical sharp amplification at shorter distances. The authors excluded their previous assumption about the extension of the influence of the ionic atmosphere around the dsDNA, which would obviously be limited to the Debye length. The effect was explained by a direct contact of the dsDNA with the SECM tip excluding the direct physical contact between tip and surface.

The influence of the polyelectrolyte properties of DNA were exploited by Turcu et al. [69] (Fig. 17), demonstrating electrostatic repelling of highly negatively charged anions such as $[Fe(CN)_6]^{3-/4-}$ by surface immobilized ssDNA and dsDNA. Label-free electrochemical recognition of DNA hybridization be-

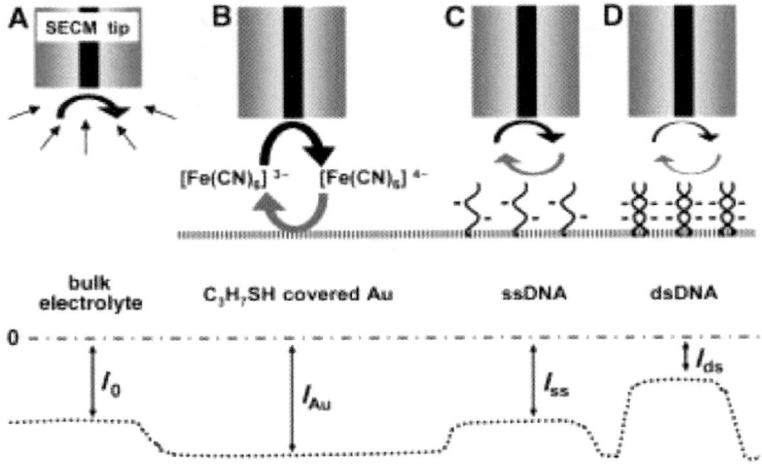

Fig. 17 Schematic representation of repelling-mode SECM. (From [69] with permission)

came possible by modulating the diffusional mass transport and hence the recycling rate of negatively charged anions at a gold surface in the presence of a DNA monolayer. The proposed repelling mode of SECM was used to recognize the presence and to characterize the status (single stranded vs. double stranded) of immobilized DNA. Reduction of $[Fe(CN)_6]^{3-}$ at the polarized tip in bulk solution leads to a diffusion-controlled steady-state current value, i_0 (Fig. 17A). After approaching the tip within feedback distance to a propane thiol-modified Au surface positive feedback amplification of the current is observed, i_{Au} (Fig. 17B). Above the ssDNA spot, the anionic phosphate groups repel the tip-generated $[Fe(CN)_6]^{4-}$ and reduce the rate of redox cycling causing a decrease of the reduction current at the tip, $i_{ss} < i_{Au}$ (Fig. 17C). After hybridization, the formation of dsDNA increases the number of negative charges at the surface, which leads to a further drop in the tip current, $i_{ds} < i_{ss} < i_{Au}$ (Fig. 17D). Influences of the mediator concentration, pH and ionic strength of the buffer, the concentration of DNA and the effect of the tip-to-sample distance on SECM in repelling-mode were presented.

Later it was shown that the formation of metalated DNA (M-DNA) monolayers facilitates the penetration of the mediator towards the surface thus leading to an enhanced positive feedback [70]. Using the above-mentioned Ag staining reaction it was demonstrated that feedback mode SECM is able to detect hybridization of 30 amol of a 17mer target DNA per spot [71] (Fig. 18), which is comparable to fluorescence or coulometric detection methods. The target DNA strand was labeled with biotin, which allowed for a further conjugation of streptavidin-modified colloidal gold after formation of the DNA duplex. The Au particles were stained with Ag, leading to a locally increased

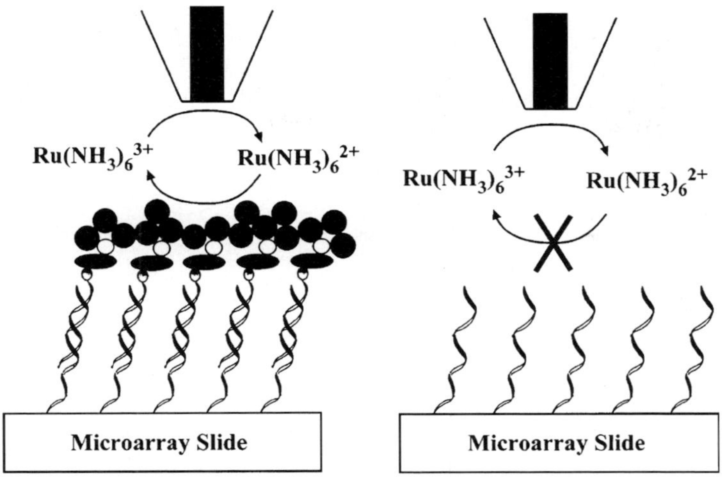

Fig. 18 Detection of DNA hybridization by means of feedback mode SECM using silver staining. (From [71] with permission)

surface conductivity at spots where successful DNA hybridization has occurred, which can be detected using feedback mode SECM.

5 Combined Generation-Collection and Feedback-Mode SECM in Biosensor Research

In general, sensitivity of a SECM measurement and lateral resolution depend also on the SECM mode. As stated earlier, the feedback mode is superior in resolution and also less dependent on time since no accumulation of species in bulk volume takes place. On the other hand, feedback experiments suffer from typically large background currents due to the continuous conversion of redox species at the tip [24, 61]. Additionally, the (enzyme-mediated) feedback mode can only be applied to image the activity of oxidoreductases. A detailed discussion of the advantages and disadvantages of the two basic modes of SECM applied in the visualization of enzymatic activity can be found in a series of papers dealing with the visualization of the localized activity of immobilized enzymes [24, 72–76].

The same system can often be imaged in both the generation-collection and the feedback mode. Immobilized HRP, for example, will reduce H_2O_2 and is capable of oxidizing redox mediators such as ferrocene methanol [24] (Fig. 19). Thus, if ferrocene methanol in its reduced form (Fc) is present in solution, the enzyme will continuously convert H_2O_2 under generation of Fc^+ that can be re-reduced at the SECM tip in a typical generation-collection

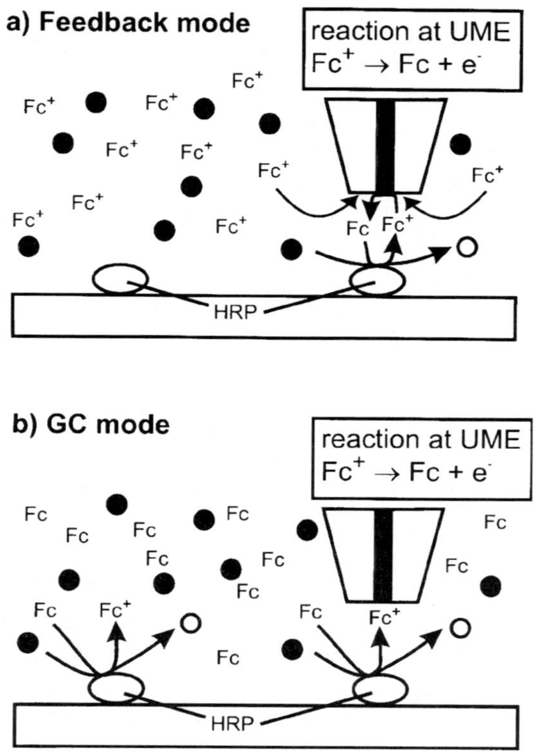

Fig. 19 Image of immobilized enzymatic activity in different SECM modes. Depending on the potential at the working electrode and the oxidation state of the used mediator (original notations Fc/Fc^+ equivalent to herein used FMA/FMA^+, respectively), the system can be imaged in either feedback mode (**a**) or generation-collection mode (**b**). (From [24] with permission)

mode experiment. If, on the other hand, the oxidized form Fc^+ is initially present in solution, no reaction happens at locations of immobilized HRP activity. However, if Fc^+ is reduced at the tip, continuous regeneration of Fc^+ will occur when scanning over a feature-bearing localized enzymatic activity leading to a positive feedback current ([54] and Fig. 14).

The influence of the concentrations of solution components [76], the geometry of the electrode, the tip potential [73], the oxidation state of the mediator and the electron transfer mechanism [72] on SECM measurement were investigated. Depending on the concentration of substrate, cofactor, or redox mediator, the enzymatic reaction might not reach saturation. In this case, the kinetics of the enzymatic reaction determines the rate of mediator regeneration. This effect would limit the capability of the feedback mode since very small currents are expected that might not be measurable on top of high background currents. The insulating sheath of the electrode will not only block diffusion of the redox mediator and of the substrate of the enzyme towards the active tip but also towards the sample surface. This latter effect will limit the capability of the generation-collection mode, since the substrate concentration at the site of the immobilized enzyme is diminished, which will cause a decreased amount of electroactive product to be formed, leading to a small current detected at the tip. Kinetic parameters extracted from measurement in feedback mode and generation-collection mode can differ significantly for the same system [72]. For example, the enzymatic oxidation of glucose by means of glucose dehydrogenase immobilized on magnetic microbeads followed pseudo-first-order kinetics when studied in feedback mode and zero-order kinetics when studied in generation-collection mode. The combination of generation-collection and feedback modes was applied in order to improve both lateral resolution and sensitivity [75] (Fig. 20). In a multi-enzyme system, galactosidase (GAL) and PQQ-dependent glucose dehydrogenase (PQQ-GDH) were bound to paramagnetic microbeads that were mixed together and immobilized to form domains containing both GAL and PQQ-GDH activity. GAL is often used as a labeling enzyme in immunoassays, however, due to the fact that its substrate is intrinsically electrochemically inactive, it cannot be imaged in feedback mode. The product of the enzymatic reaction, p-aminophenol, can be converted to p-iminoquinone (PQI) at the SECM tip. In addition, pQI can serve as redox mediator for PQQ-GDH to induce a feedback current at the tip. It could be shown that the current response of the combined feedback and generation-collection modes was superior to the conventional generation-collection mode and lateral resolution could be significantly improved.

In addition to examples that discuss advantages and disadvantages of the feedback and generation-collection mode and how a combination of the two modes may help to overcome shortcomings of one particular mode, both modes can be employed to access different characteristics of the same sample or to enable the analysis of multi-enzyme patterns and interactions therein. In a pattern containing GOD and HRP, the activities of both enzymes could be visualized independently or interacting [62]. If only glucose and O_2 are

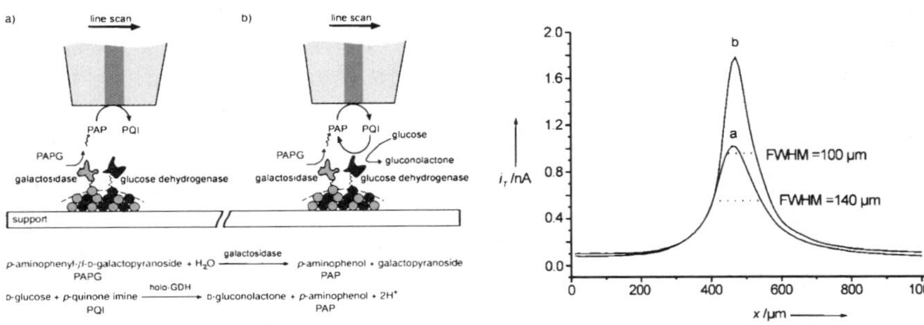

Fig. 20 Detection schemes for conventional GC-mode and combined FB and GC-modes and linescans over GAL/PQQ-GDH domains. (From [75] with permission)

present in solution, only GOD is active and its activity can be probed by means of generation-collection mode. Concomitantly, GOD is producing the substrate of HRP, H_2O_2. In absence of a suitable redox mediator HRP stays, however, inactive. On the other hand, HRP activity can be observed exclusively if a redox mediator such as ferrocene methanol and H_2O_2 are simultaneously present in solution. The activity of HRP can be observed either in the feedback mode or generation-collection mode. In the presence of glucose, O_2, and ferrocene methanol, both enzymes are active and their interaction can be studied in both modes.

The use of multi-enzyme systems represents one trend in the development of biosensors while the application of multi-electrode arrays represents another one. With the parallel use of a high number of electrodes, the effect of (chemical) crosstalk becomes an important concern. The potential of SECM to elucidate crosstalk in microelectrode arrays was shown in [55]. The feedback mode of SECM was used to locate individual electrodes in an array of four Pt micro-band electrodes. Crosstalk between enzyme-modified bands could be investigated in the generation-collection mode. Effects of the bias of one electrode band on the neighboring electrodes could be visualized. Also, considerations were made regarding the limitations of SECM in generation-collection mode. The size of the tip and its disturbance of the diffusion layers of the investigated electrodes cannot always be neglected. A solution to the problem was presented in suggesting the use of electrodes with smaller overall sizes or potentiometric microelectrodes that do not consume any species produced at the sample surface.

Another series of papers employing both feedback and generation-collection modes was aiming at the elucidation of the distribution of sensor components with high spatial variability [77–79]. It is assumed that only in-depth knowledge about the distribution of the individual sensor components within the immobilization layer and about their interaction enables a stringent biosensor optimization.

Depending on the predefined grid dimensions, a SECM image of a biosensor surface consists of a high number of individual current measurements representing the local activity. Each individual data point can be treated as an independent micro biosensor [78]. Glucose oxidase was entrapped in polymer spots of about 300 μm diameter and the local generation of H_2O_2 in the presence of different concentrations of glucose was monitored using the generation collection mode. The data points taken at different grid points above the enzyme spot exhibited large variations, suggesting a significant variability of the properties of the individual micro biosensors. Due to the dependence of SECM measurements from the tip-to-sample distance, however, the convolution of topographic and biocatalytic information has to be broken. Thus, the topography of the polymer spots was determined from a negative feedback mode image obtained in absence of glucose. The topography information could then be used to normalize the data from the generation-collection mode image [79] (Fig. 21). With the knowledge of the local differences in the biosensor characteristics from a single enzyme/polymer spot, an artificial neural network could be trained to find locations of highest selectivity within the spots [77]. Further use of principle component analysis in combination with artificial feed-forward neural networks allowed for elimination of interferences [77,79].

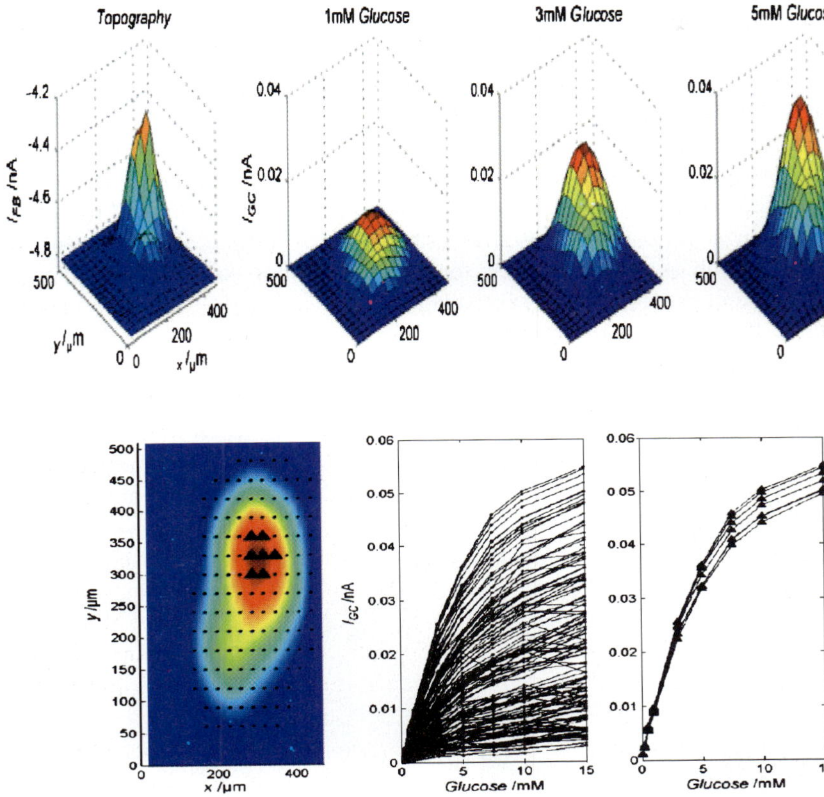

Fig. 21 Imaging the topography of a GOD-containing polymer spot in feedback mode and enzyme activity at different concentrations of glucose in generation-collection mode (*top*). Representation of the scanned raster as discrete micro biosensor (*bottom, left*) and most sensitive biosensors as found by an artificial neural network. (From [79] with permission)

Combined AFM-SECM was applied using both the feedback and the generation collection modes of SECM making use of the inherent advantages especially when using small working electrodes and topographically demanding samples. Modified AFM-SECM tips were described integrating a biofunctionalized ring electrode around the topography sensor [80]. For example, the ring electrode surrounding the AFM tip was modified with GOD entrapped within a polymer layer, or modified with HRP covalently bound via the headgroups of a thiol monolayer, respectively. The obtained glucose sensor was exclusively operated in generation-collection mode, while the H_2O_2 sensor was used in feedback mode. Both types of sensors were used to study membrane transport through a biomimetic membrane.

Besides AFM-SECM and shearforce-based constant distance positioning, the control of the tip-to-sample distance could be achieved by sequential feedback mode approach curves and generation-collection mode imaging using double-barrel electrodes. In double-barrel electrodes two individually addressable microdisk electrodes are arranged in close proximity [81] (Fig. 22). Glucose oxidase and hexokinase were co-entrapped within an electrodeposited polymer layer [51] on the surface of a 25 μm Pt disk electrode for the detection of ATP by a competitive enzymatic reaction. A neighboring 10 μm Pt disk electrode served as distance sensor using reduction of molecular oxygen in a negative feedback measurement. A simple distance control is achieved when the reduction current for O_2 is kept constant by varying the z-position of the double-barrel electrode tip. The double-barrel electrodes were used to study the membrane transport of ATP through the pores of a polycarbonate membrane [82].

In summary, a variety of studies made use of the particular advantages of the two basic modes of SECM. A combination of feedback mode and

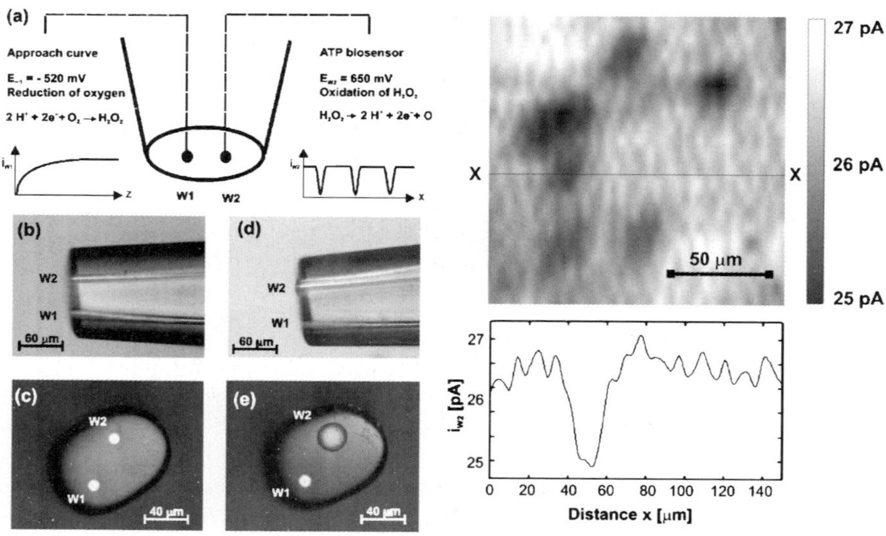

Fig. 22 *Left*: Double-barrel microelectrodes before and after electrodeposition of a polymer layer on top of the 25 μm Pt electrode (W2). The 10 μm Pt electrode (W1) served as distance sensor. *Right*: SECM image of ATP transport through a polycarbonate membrane. (From [81] with permission)

generation collection mode experiments is especially useful for the study of immobilized multi-enzyme systems or the investigation of crosstalk in microelectrode arrays.

6
The Direct Mode of SECM in Biosensor Research

The direct mode of SECM (see Fig. 5f) implies that the tip and sample electrodes are working and counter electrodes of a three-electrode arrangement. Usually, the tip electrode serves as the counter electrode, thus restricting the electric field by limiting the counter electrode reaction to an area at the sample surface opposite to the positioned counter electrode tip. In direct mode, a patterning of the sample surface can be obtained, thus offering possibilities for localized immobilization of biocomponents on the patterned sample surface.

Direct mode SECM was applied making use of a number of different spatially confined modulations of chemical properties leading to the envisaged patterning of the sample surface. The reversibility of deactivation/activation of surface-bound alcohol dehydrogenase (ADH) was induced by the local electrochemically induced modulation of the pH value at the tip by Engstrom et al. [83]. It could be shown that the enzymatic activity was locally turned on as long as the tip was inducing a spatially confined shift of the pH from the value of the bulk (pH 6), at which the enzyme is inactive, to pH 9, at which the enzyme becomes fully active. After activation, the immobilized alcohol dehydrogenase converted ethanol in the presence of NAD^+ under simultaneous formation of NADH, which was detected using fluorescence microscopy. The diameter of the fluorescent sphere was shown to be dependent on the time of the potential pulse at the UME (−1.35 V for 30 s), and even more important, to be limited by the buffer capacity of the electrolyte.

Electrochemical generation of highly reactive substances was applied for the local deactivation of surface-immobilized enzyme molecules [84, 85]. Highly reactive species (e.g., HOBr) were used by Oyamatsu et al. [84] to locally denature diaphorase molecules immobilized on a glass surface underneath the SECM tip. Diaphorase patterns were generated by applying a potential of +1.5 V vs. Ag/AgCl to the tip leading to oxidation of Br to Br_2, which in turn is instantaneously reacting with H_2O to generate HOBr. Diffusion of tip-generated HOBr within the gap between tip and sample to the diaphorase-modified surface leads to inactivation of the enzyme. The deactivated regions were visualized in feedback mode of SECM and scanning confocal microscopy. The duration of the potential pulse drastically affects the area of the inactivated enzyme region. The patterning of enzyme-modified surfaces by generation of highly reactive species, in this case chlorine, was already earlier used by Shiku et al. [85].

Direct mode SECM was used by Turyan et al. [86] for the formation of small Au microstructures by electrochemically induced dissolution of an Au-UME at a potential of +1 V in 10 mM HCl solution. The gold at the interface of the UME is oxidized (fast process) under formation of $AuCl_4^-$, which can be reduced at the Si surface (slow process) under formation of Au nanoparticles. The initially formed nanoparticles are growing to a continuous Au structure as long as the tip electrode is supplying a locally increased concentration of $AuCl_4^-$. The obtained Au microstructures were further modified with a thiol monolayer for covalent binding of glucose oxidase. Subsequently, the enzyme activity was visualized by means of generation-collection mode SECM.

Procedures already established for the use of conducting polymers in biosensor architectures [87] were localized using the direct mode of SECM. Local deposition of polypyrrole lines on the surface of a gold substrate was successfully demonstrated already in 1995 [88] using a 10 μm Pt-tip as counter electrode. A pulse deposition protocol was established, which was shown to be indispensable for the local formation of the polypyrrole film. A short pulse is generating a locally high concentration of pyrrole radical cations in the gap between the Au surface and the tip. During a resting phase, bulk concentration of pyrrole is re-established in the gap avoiding any diffusional depletion of the monomers. Thus, the critical chain length for polymer precipitation is reached and a polypyrrole microstructure can be obtained. An example of a locally deposited polypyrrole line of 1 mm length and about 50–60 μm width is shown in Fig. 23.

Localized deposition of conducting polymers by means of direct-mode SECM was used to interconnect gold microelectrodes leading to an organic transistors [89]. The local electrodeposition of conducting polymers, particularly polypyrrole, under simultaneous entrapment of enzymes [53, 88–90] or as a support for covalent binding of enzyme molecules or DNA strands [91–93] was demonstrated.

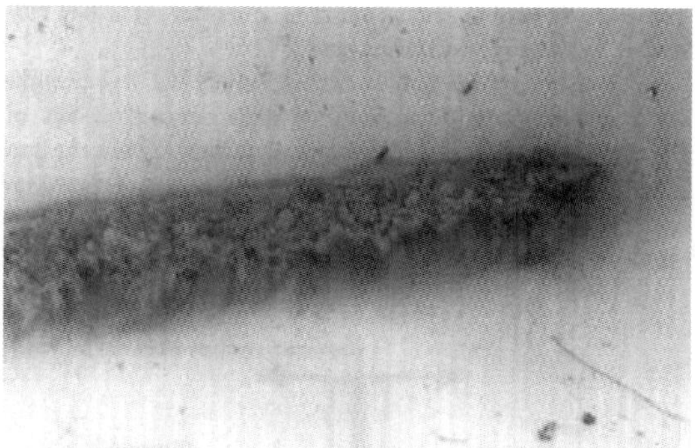

Fig. 23 Polypyrrole line deposited locally in direct mode of SECM. (From [88] with permission)

Direct-mode SECM was employed by Wittstock et al. [76, 94, 95] for formation of microscopic enzymatic spots on a alkanethiolate-covered gold electrode. Combination of electrochemically induced localized desorption of the thiol monolayer and subsequent derivatization of the renewed uncovered gold areas with an amino ω-functionalized alkanethiolate provides functionalized surface areas for the covalent binding of enzyme molecules [76]. The influence of the type of the monolayer (hydrocarbon chain length, functional groups) and the method used for the electrochemically induced desorption of the thiol layer were evaluated [94]. The principle of successive adsorption-desorption-adsorption of the thiol layer, followed by covalent coupling of the enzyme, was successfully demonstrated for the preparation of HRP microstructures [95] (Fig. 24).

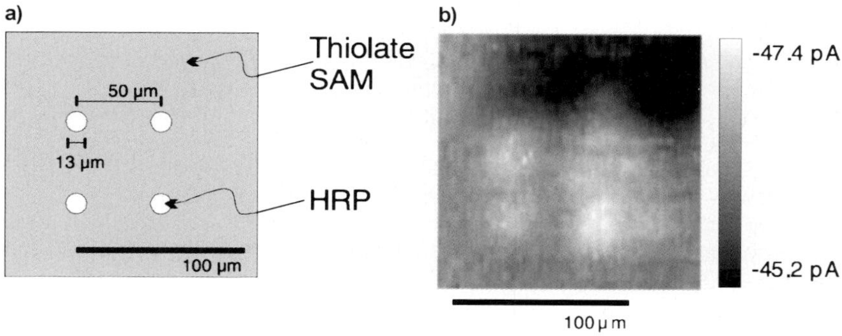

Fig. 24 **a** Schematic representation of the enzyme microstructure formed by localized desorption of a thiol monolayer. **b** SECM visualization of localized HRP activity. (From [95] with permission)

7
Conclusions and Outlook

Scanning electrochemical microscopy has been successfully employed for studying the immobilized biological activity caused by immobilized enzymes, DNA strands, antibodies, or whole cells. Especially, SECM was used as a unique tool to visualize local variations in the immobilized biological activity providing information about the reproducibility of the immobilization process. Extraction of kinetic data provided a better understanding of changes of the immobilized biomolecules imposed by the immobilization process which form a basis for a straightforward optimization of a given immobilization technique. Products and by-products of the biological recognition reaction(s) which are leaking from a biosensor surface could be detected and the recycling of free-diffusing redox mediators could be visualized using

SECM. In addition to the evaluation of biosensors and optimization of biosensor architectures, SECM was used as a tool for microstructuring. Further improvements in the technique such as the implementation of an accurate control of the tip-to-sample separation and the combination with other scanning probe techniques will open the route to an in-depth understanding of mechanisms in biosensors and the complex interplay between all influencing parameters. In conclusion, SECM has already proven to be a powerful and versatile tool in biosensor research which in the near future will develop from a tool in research laboratories to a routine tool for answering a large variety of questions in biosensor research.

References

1. Bard AJ, Fan FRF, Kwak J, Lev O (1989) Anal Chem 61:132
2. Engstrom RC, Pharr CM (1989) Anal Chem 61:A1099
3. Friedbacher G, Fuchs H (1999) Pure Appl Chem 71:1337
4. Poggi MA, Gadsby ED, Bottomley LA, King WP, Oroudjev E, Hansma H (2004) Anal Chem 76:3429
5. Bining G, Rohrer H, Gerber C, Weibel E (1982) Phys Rev Lett 49:57
6. Pohl DW, Novotny L (1994) J Vac Sci Technol B 12:1441
7. Binning G, Quate CF, Gerber C (1986) Phys Rev Lett 56:930
8. http://www.chinstruments.com/chi900.html, last visited: 09-27-07
9. http://www.heka.com/chem/equipment/elproscan.html, last visited: 09-27-07
10. http://www.sensolytics.com/secm.html, last visited: 09-27-07
11. http://www.uniscan.co.uk/scanning_electrochemical_microscope.html, last visited: 09-27-07
12. http://www.windsorscientific.co.uk, last visited: 09-27-07
13. Bard AJ, Mirkin MV (eds) (2001) Scanning Electrochemical Microscopy. Marcel Dekker, New York
14. Katemann BB, Schulte A, Schuhmann W (2003) Chem Eur J 9:2025
15. Katemann BB, Schulte A, Schuhmann W (2004) Electroanalysis 16:60
16. Katemann BB, Schuhmann W (2002) Electroanalysis 14:22(1)
17. Mirkin MV, Fan FRF, Bard AJ (1992) J Electroanal Chem 328:47
18. Kueng A, Kranz C, Lugstein A, Bertagnolli E, Mizaikoff B (2003) Angew Chem Int Ed 42:3238
19. Stulik K, Amatore C, Holub K, Marecek V, Kutner W (2000) Pure Appl Chem 72:1483
20. Arca M, Bard AJ, Horrocks BR, Richards TC, Treichel DA (1994) Analyst 119:719
21. Amphlett JL, Denuault G (1998) J Phys Chem B 102:9946
22. Pierce DT, Unwin PR, Bard AJ (1992) Anal Chem 64:1795
23. Pierce DT, Bard AJ (1993) Anal Chem 65:3598
24. Wittstock G (2001) Fresenius Z Anal Chem 370:303
25. Garay MF, Ufheil J, Borgwarth K, Heinze J (2004) Phys Chem Chem Phys 6:4028
26. Ludwig M, Kranz C, Schuhmann W, Gaub HE (1995) Rev Sci Instr 66:2857
27. Alpuche-Aviles MA, Wipf DO (2001) Anal Chem 73:4873
28. Wipf DO, Bard AJ, Tallman DE (1993) Anal Chem 65:1373
29. Giessibl FJ (2003) Rev Mod Phys 75:949
30. Macpherson JV, Unwin PR (2000) Anal Chem 72:276

31. Horrocks BR, Schmidtke D, Heller A, Bard AJ (1993) Anal Chem 65:3605
32. Grundig B, Wittstock G, Rudel U, Strehlitz B (1995) J Electroanal Chem 395:143
33. Munteanu FD, Mosbach M, Schulte A, Schuhmann W, Gorton L (2002) Electroanalysis 14:1479
34. Wittstock G, Yu K, Halsall HB, Ridgeway TH, Heineman WR (1995) Anal Chem 67:3578
35. Thompson RQ, Barone GC, Halsall HB, Heineman WR (1991) Anal Biochem 192:90
36. Thompson RQ, Porter M, Stuver C, Halsall HB, Heineman WR, Buckley E, Smyth MR (1993) Anal Chim Acta 271:223
37. Nowall WB, Dontha N, Kuhr WG (1998) Biosens Bioelectron 13:1237
38. Wijayawardhana CA, Wittstock G, Halsall HB, Heineman WR (2000) Electroanalysis 12:640
39. Zhang X, Peng X, Jin W (2006) Anal Chim Acta 558:110
40. Zhou HF, Kasai S, Matsue T (2001) Anal Biochem 290:83
41. Kasai S, Shiku H, Torisawa Y, Nagamine K, Yasukawa T, Watanabe T, Matsue T (2006) Anal Chim Acta 566:55
42. Csoka B, Kovacs B, Nagy G (2003) Biosens Bioelectron 18:141
43. James PI, Garfias-Mesias LF, Moyer PJ, Smyrl WH (1998) J Electrochem Soc 145:L64–L66
44. Schmitz I, Schreiner M, Friedbacher G, Grasserbauer M (1997) Anal Chem 69:1012
45. Hirata Y, Yabuki S, Mizutani F (2004) Bioelectrochem 63:217
46. Kranz C, Kueng A, Lugstein A, Bertagnolli E, Mizaikoff B (2004) Ultramicroscopy 100:127
47. Hengstenberg A, Kranz C, Schuhmann W (2000) Chem Eur J 6:1547
48. Mosbach M, Zimmermann H, Laurell T, Nilsson J, Csöregi E, Schuhmann W (2001) Biosens Bioelectron 16:827
49. Gaspar S, Mosbach M, Wallman L, Laurell T, Csöregi E, Schuhmann W (2001) Anal Chem 73:4254
50. Turcu F, Hartwich G, Schäfer D, Schuhmann W (2005) Macromol Rapid Commun 26:325
51. Kurzawa C, Hengstenberg A, Schuhmann W (2002) Anal Chem 74:355
52. Fernandez JL, Mano N, Heller A, Bard AJ (2004) Angew Chem Int Ed 43:6355
53. Kranz C, Wittstock G, Wohlschlager H, Schuhmann W (1997) Electrochim Acta 42:3105
54. Shiku H, Matsue T, Uchida I (1996) Anal Chem 68:1276
55. Strike DJ, Hengstenberg A, Quinto M, Kurzawa C, Koudelka-Hep M, Schuhmann W (1999) Mikrochim Acta 131:47
56. Bard AJ, Crayston JA, Kittlesen GP, Shea TV, Wrighton MS (1986) Anal Chem 58:2321
57. Shea TV, Bard AJ (1987) Anal Chem 59:2101
58. Zaumseil J, Wittstock G, Bahrs S, Steinrucke P (2000) Fresenius Z Anal Chem 367:352
59. Wittstock G, Wilhelm T, Bahrs S, Steinrucke P (2001) Electroanalysis 13:669
60. Oyamatsu D, Hirano Y, Kanaya N, Mase Y, Nishizawa M, Matsue T (2003) Bioelectrochem 60:115
61. Gaspar S, Schuhmann W, Laurell T, Csöregi E (2002) Rev Anal Chem 21:245
62. Wilhelm T, Wittstock G (2003) Angew Chem Int Ed 42:2247
63. Niculescu M, Gaspar S, Schulte A, Csöregi E, Schuhmann W (2004) Biosens Bioelectron 19:1175
64. Carano M, Lion N, Abid JP, Girault HH (2004) Electrochem Commun 6:1217
65. Carano M, Lion N, Girault HH (2007) J Electroanal Chem 599:349
66. Yamashita K, Takagi M, Uchida K, Kondo H, Takenaka S (2001) Analyst 126:1210

67. Fan FRF, Bard AJ (1999) PNAS 96:14222
68. Komatsu M, Yamashita K, Uchida K, Kondo H, Takenaka S (2006) Electrochim Acta 51:2023
69. Turcu F, Schulte A, Hartwich G, Schuhmann W (2004) Biosens Bioelectron 20:925
70. Liu B, Bard AJ, Li CZ, Kraatz HB (2005) J Phys Chem B 109:5193
71. Wang J, Fayi S, Zhou F (2002) Langmuir 18:6653
72. Zhao CA, Wittstock G (2004) Anal Chem 76:3145
73. Wilhelm T, Wittstock G, Szargan R (1999) Fresenius Z Anal Chem 365:163
74. Zhao C, Wittstock G (2005) Biosens Bioelectron 20:1277
75. Zhao C, Wittstock G (2004) Angew Chem Int Ed 43:4170
76. Wittstock G, Schuhmann W (1997) Anal Chem 69:5059
77. Maciejewska M, Schäfer D, Schuhmann W (2006) Electroanalysis 18:1916
78. Maciejewska M, Schäfer D, Schuhmann W (2006) Electrochem Commun 8:1119
79. Schäfer D, Maciejewska M, Schuhmann W (2007) Biosens Bioelectron 22:1887
80. Kueng A, Kranz C, Lugstein A, Bertagnolli E, Mizaikoff B (2005) Angew Chem Int Ed 44:3419
81. Kueng A, Kranz C, Mizaikoff B (2005) Biosens Bioelectron 21:346
82. Compagnone D, Guilbault GG (1997) Anal Chim Acta 340:109
83. O'Brien JC, Shumaker-Parry J, Engstrom RC (1998) Anal Chem 70:1307
84. Oyamatsu D, Kanaya N, Shiku H, Nishizawa M, Matsue T (2003) Sens Act B 91:199
85. Shiku H, Takeda T, Yamada H, Matsue T, Uchida I (1995) Anal Chem 67:312
86. Turyan I, Matsue T, Mandler D (2000) Anal Chem 72:3431
87. Schuhmann W, Kranz C, Huber J, Wohlschlager H (1993) Synth Met 61:31
88. Kranz C, Ludwig M, Gaub HE, Schuhmann W (1995) Adv Mater 7:38
89. Kranz C, Ludwig M, Gaub HE, Schuhmann W (1995) Adv Mater 7:568
90. Kranz C, Gaub HE, Schuhmann W (1996) Adv Mater 8:634
91. Evans SAG, Brakha K, Billon M, Mailley P, Denuault G (2005) Electrochem Commun 7:135
92. Szunerits S, Bouffier L, Calemczuk R, Corso B, Demeunynck M, Descamps E, Defontaine Y, Fiche JB, Fortin E, Livache T, Mailley P, Roget A, Vieil E (2005) Electroanalysis 17:2001
93. Fortin E, Mailley P, Lacroix L, Szunerits S (2006) Analyst 131:186
94. Wilhelm T, Wittstock G (2000) Mikrochim Acta 133:1
95. Wilhelm T, Wittstock G (2001) Electrochim Acta 47:275

Author Index Volumes 101–109

Subject Index